IV		V		VI		VII		VIII		
	a	b	a	b	a	b	a	b	a		b
										^{2}He	
	^{6}C		^{7}N		^{8}O		^{9}F			^{10}Ne	
	^{14}Si		^{15}P		^{16}S		^{17}Cl			^{18}Ar	
	^{2}Ti		^{23}V		^{24}Cr		^{25}Mn		^{26}Fe ^{27}Co ^{28}Ni		
	^{32}Ge		^{33}As		^{34}Se		^{35}Br			^{36}Kr	
	^{0}Zr		^{41}Nb		^{42}Mo		^{43}Tc		^{44}Ru ^{45}Rh ^{46}Pd		
	^{50}Sn		^{51}Sb		^{52}Te		^{53}I			^{54}Xe	
	^{2}Hf		^{73}Ta		^{74}W		^{75}Re		^{76}Os ^{77}Ir ^{78}Pt		
	^{82}Pb		^{83}Bi		^{84}Po		^{85}At			^{86}Rn	
	^{04}Ku		105								

^{66}Dy	^{67}Ho	^{68}Er	^{69}Tm	^{70}Yb	^{71}Lu
^{98}Cf	^{99}Es	^{100}Fm	^{101}Md	^{102}No	^{103}Lw

Rösler / Lange   GEOCHEMICAL TABLES

GEOCHEMICAL TABLES

H. J. Rösler / H. Lange

ELSEVIER PUBLISHING COMPANY

Amsterdam — London — New York

1972

Distribution of this book is being handled by the following publishers

for the U.S.A. and Canada
American Elsevier Publishing Company, Inc.
52 Vanderbilt Avenue
New York, New York 10017

for the East European Countries, China, North
Korea, Cuba, Vietnam and Mongolia
Edition Leipzig, 701 Leipzig, Karlstraße 20

for all remaining areas
Elsevier Publishing Company
335 Jan van Galenstraat
P.O.Box 211, Amsterdam, The Netherlands

Library of Congress Card Number 79-132143
ISBN 0-444-40894-0
With 136 illustrations, 216 tables, and 1 foldout table

Copyright in original German edition by Deutscher Verlag für Grundstoffindustrie, Leipzig

Copyright in this English edition 1972 by Edition Leipzig
Printed in the German Democratic Republic

Preface to the English Edition

Geochemistry affects on an increasing scale all branches of the geosciences and exerts an influence on many neighbouring fields of study and practice, such as mining, metallurgy, agriculture, medicine, and space research. Its importance is due to the accumulation of a wealth of factual information about the composition of the earth and the cosmos, and to the utilisation of this knowledge for the interpretation of geological processes, whereas its dynamic aspects manifest themselves especially in the concepts of migration and geological-geochemical cycles. Today, after 50 years of geochemical activities, it is no longer possible to include the whole subject matter of geochemistry in the syllabuses of university courses. For this reason, fundamental data were compiled, tabulated and presented in this book in the form of Geochemical Tables in order to promote the education of geoscientists at the Freiberg Mining Academy. The book is designed in such a way that it forms a combination of a textbook and a reference book. To comply with the requirements of modern education and research work, both the student and the scientist concerned with geochemistry need a source of information about the essentials of geochemistry and a guide to the literature in this field, in order to familiarise themselves quickly with the special subjects in which they are particularly interested. That is the purpose of this book. From this it follows that it was not written for experts specialised in particular branches of study such as chemical analysis, isotope geochemistry or thermodynamics.

In consideration of the fact that this work is also intended for use as a textbook for students, it should be noted that the historical development of geochemistry and its subdivisions had to be taken into account; this naturally has a particular bearing on the selection of bibliographic data.

In contrast with the German edition, Chapters 1, 8, 9 and 10 in particular were throroughly revised. A great part of the data were adjusted to the latest findings and a great number of publications added to the bibliography.

The authors wish to express their gratitude to a number of their Freiberg colleagues for contributing to the revision of the German edition. These contributors are Dr. D. Harzer (physical geochemistry), Dr. K. Koch (salts), Dr. G. Mathé (bibliography), Dr. J. Pilot (isotope geochemistry and age determination), Dr. W. Schrön (spectroscopy and standards), and Dr. H. Thiergärtner (mathematical methods). Further suggestions, information and help were given by Dr. M. Guntau, Dr. G. Tischendorf and, in a book review, Dr. H. Gundlach (Hannover).

To Professor U. Petersen of Harvard University, who had the kindness to read the manuscript and who made many criticisms and amendments, we owe respect and thanks.

Due to the rapid development of geochemistry, and the continuous expansion of knowledge, the authors are not always in a position to survey all branches and regional units. Therefore, they invite comments and suggestions from their colleagues in this field.

Prof. Dr. habil. H. J. Rösler

Dr. H. Lange

Table of Contents

1. History, Tasks, Position and Divisions of Geochemistry

Geochemistry is one of the geological sciences or geosciences. It has developed especially from chemistry, geology and mineralogy and can today be considered a branch in study of the mineralogical-geochemical sciences.

1.1. History

Geochemistry is a young science. Its development can be divided into three stages:

1. the *preparatory period (up to about 1900)*, in which the first fundamental relations between chemism and geological substance became obvious and were turned to practical use;
2. the *classical period (from 1900 to 1950)* in which the accessible terrestrial and extra-terrestrial matter was chemically investigated and classified systematically, employing the methods of induction (analytic and descriptive geochemistry);
3. the *modern period (from 1950)*, in which increased efforts have been made to study and corroborate the geochemical laws and regularities, both experimentally and employing the method of deduction.

The most important pioneers of modern geochemistry include:

1. C. F. SCHÖNBEIN (1799—1868), chemist (born in S. W. Germany, teacher in Basle), discoverer of ozone and nitrocellulose, promoted the boundary fields of chemistry (e.g. physical chemistry, biochemistry), coined the term "geochemistry" and advocated its use.
2. K. G. BISCHOF (1792—1870), German geologist, author of the "Textbook of Chemical and Physical Geology" (4 volumes from 1863—1871).
3. J. L. A. ROTH (1818—1892), German geologist, author of the book "General and Chemical Geology".
4. J. H. VOGT (1858—1931), Norwegian geologist and expert in the science of ore deposits, noted for researches in the field of the physico-chemical behaviour of magmas and the distribution of the elements.
5. F. W. CLARKE (1847—1931), American chemist, founder of classical geochemistry, compiled the average contents of the chemical elements in the earth's crust in the book "Data of Geochemistry" (first publication in 1889); this work has been republished many times and continued up to this day.
6. H. S. WASHINGTON (1867—1934), American chemist, one of the initiators of petro-chemistry, collaborated with CLARKE on the distribution of the chemical elements.
7. G. BERG (1876—1946), German geologist, author of a great number of books which give a chemical and geochemical interpretation of ore deposits and mineral raw materials.

8. V. I. VERNADSKIĬ (1863—1945), Soviet mineralogist and geochemist, founder of the Russian and Soviet geochemical school, initiator of biogeochemistry and radiogeology (isotope geochemistry), gave the clearest definition of the position and framework of geochemistry, underlined especially the dynamics of geochemical processes (migration, cycles), author of many papers and books.

8a. A. E. FERSMAN (1883—1945), Soviet mineralogist and geochemist, one of the founders of the Russian and Soviet geochemical school, gave the first lecture on geochemistry in 1912, initiator of regional geochemistry, encouraged the popularising of geochemical knowledge, worked especially on problems of lattice energy and pegmatites, activated geochemical prospecting, author of a very great number of publications.

9. N. L. BOWEN (1887—1956), American physico-chemist, initiator of physical geochemistry.

10. V. M. GOLDSCHMIDT (1888—1947), mineralogist and geochemist (born in Zurich, delivered lectures in Oslo and Göttingen), initiator of modern analytic geochemistry, encouraged the use of thermodynamics and crystal chemistry in the solution of mineralogical-geochemical problems.

11. P. NIGGLI (1888—1953), Swiss mineralogist, initiator of modern petrochemistry, promoted mineral chemistry.

12. W. NODDACK (1898—1966) and I. NODDACK, German chemists and geochemists, discoverers of rhenium, promoters of the geochemistry of trace elements.

13. A. A. SAUKOV (1902—1964), Soviet mineralogist and geochemist. Initiator of historical geochemistry, encouraged geochemical prospecting, worked on the regularities of migration and the geochemistry of individual elements.

Mention should also be made in this connection of: F. J. LEWINSON-LESSING (1861—1939), A. N. ZAVARITZKY (1884—1952), F. PANETH (1887—1958), I. I. GINZBURG (1881—1965).

1.2. Tasks of Geochemistry

Several classic authors in the field of geochemistry defined the tasks of geochemistry as follows:

CLARKE

Any rock can be considered a chemical system where chemical changes can be induced by various means. Any such change causes a disturbance of the equilibrium with the ultimate formation of a new system which again is stable under the new conditions. The study of these changes is the purpose of geochemistry. The tasks of the geochemist are to determine the changes which are possible, how and when they may occur, the phenomena by which they are accompanied and the ultimate result to which they may lead.

GOLDSCHMIDT

1. Determination of the abundance of the elements and atoms in the earth.
2. Exploration of the distribution of the elements in the geochemical spheres of the earth, i.e. in the minerals and rocks of the lithosphere and in natural products of various kinds.

3. Discovery of laws which govern the abundance and distribution of the elements.

VERNADSKIĬ

Geochemistry investigates the chemical elements, i.e. the atoms of the earth's crust and, as far as possible, of the whole globe. It studies their history, their distribution and migration in space and time, and their genesis

FERSMAN

Geochemistry studies the history of the chemical elements — the atoms in the earth's crust and their behaviour under different thermodynamic and physicochemical natural conditions.

Today we summarise the tasks of geochemistry as follows:

"Geochemistry studies the distribution laws of the chemical elements and isotopes of the earth, in the past and at present."

Geochemistry studies

1. the quantitative distribution of the chemical elements and isotopes in the individual geologic formations,
2. the laws according to which the distribution occurs, and
3. the changes of these laws of distribution in the course of the earth's development.

The knowledge of these laws enables us to make predictions regarding the occurrence of elements and isotopes in certain geological bodies and to draw conclusions regarding the formation of minerals, rocks and ore deposits (specific scope of "applied geochemistry").

1.3. Position and Division

Geochemistry uses the knowledge and the working methods of many neighbouring branches of study (Fig. 1, upper part). On the other hand, it stimulates many of these branches by its own results.

Geochemistry can be subdivided according to two points of view:

  a) according to its (geological) object of investigation (see Fig. 1, upper and lower parts)

  Lithogeochemistry = geochemistry of the earth's solid crust

   a) main components of the rocks = *petrochemistry*
   b) chemistry of soils = *pedogeochemistry*
   c) main and trace elements of the minerals = *mineral chemistry*

  Hydrogeochemistry = geochemistry of waters (surface and ground waters)

  Biogeochemistry = geochemistry of the living matter including their fossil products *(organic geochemistry)*

  Atmogeochemistry = geochemistry of gases, especially of the atmosphere

  Cosmochemistry = chemistry of the extraterrestrial matter. Geochemistry may also be considered a branch of cosmochemistry.

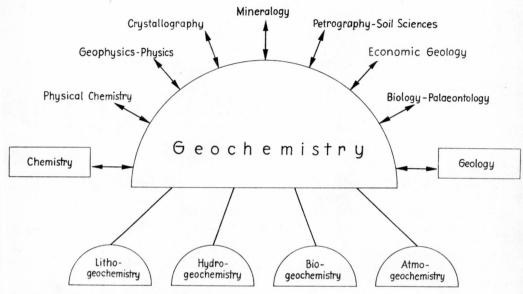

Fig. 1   Position and divisions of geochemistry

b) according to other specific points of view (e.g., method of investigation, practical applicability)

Isotope geochemistry = extension of the "normal" geochemistry to isotopes. Detection and interpretation of the natural variations in isotope distribution.

Isotope geology (including *radiogeology*) may be understood as the direct application of isotope-geochemical knowledge and results to geological problems. Special fields: radiogeochemistry (geochemistry of radioactive elements and isotopes and the physical determination of age by means of stable and unstable isotopes).

Physical geochemistry = use of physicochemical laws for the clarification of geochemical distribution laws, especially by experimental determination *(experimental geochemistry)* of equilibrium between mineral phases and solutions.

Geochemistry of trace elements = determination of the laws of the distribution of trace elements in minerals and rocks for the clarification of crystal-chemical and genetic problems, and for the practical utilisation of these elements.

Regional geochemistry = determination of geochemical particularities of regional geologic units, especially of regional Clarke-values, of geochemical provinces, petrochemical and deposit provinces.

Exploration geochemistry = use of the general and regional geochemical laws in geological exploration, especially in the exploration of ore deposits *(geochemical prospecting)*.

Historical geochemistry = determination of the dependence of geochemical laws and factors on the geological history of the earth.

Bibliography

History of Geochemistry

FISCHER, W.: Gesteins- und Lagerstättenbildung im Wandel der wissenschaftlichen Anschauung (Abschnitt Geochemie), Stuttgart 1961, 255—314

HELINGER, E.: Über die neuere Entwicklung der Geochemie. Fortschr. Miner. Kristallogr. und Petrogr. *12* (1927), 253—336

LARSEN, E. S.: Geology 1888 to 1938 (Chapter Geochemistry). J. Geol. Soc. Amer., 1941, 391—413

MANTEN, A. A.: Historical Foundations of Chemical Geology and Geochemistry. Chem. Geol. *1* (1966) 1, 5—31

SAUKOV, A. A.: Fundamentals of Historical Geochemistry. Internat. Geol. Congr. 22nd Sess. India 1964, Vol. Abstracts, 243—244

VINOGRADOV, A. P.: Entwicklungswege der Geochemie. Sowjetwissenschaft, Naturwiss. Beiträge 2 124; Berlin 1956

VINOGRADOV, A. P.: Half a Century of Geochemistry. Geokhimiya 1967 (11), 1259—1262 (Russian)

Biographies of Geochemists

BISCHOF, K. G.: Z. dtsch. Geol. Ges. *94* (1942), 55—63

BOWEN, N. L.: Amer. Mineralogist *42* (1957), 242—248

CLARKE, F. W.: 1. Amer. Mineralogist *16* (1931); 2. Biograph. Mem. National Acad. Sciences XV, 4. (1932), 146—165

FERSMAN, A. E.: 1. Lyudi russkoĭ nauki, Moscow 2 (1962), 222—233; 2. Aleksander Evgenevich Fersman, Izd. "Nauka", Moscow 1965

GOLDSCHMIDT, V. M.: 1. Izvestiya Akad. Nauk SSSR. *6* (1924), 490—493; 2. Naturwissenschaften *34* (1947), 129—131; 3. Geol. Rdsch. *35* (1948), 179f.

NIGGLI, P.: 1. Geologie 2 (1953), 124—130; 2. Schweiz. miner. petrogr. Mitt. *33* (1953), 9—20; 3. American Mineralogist *39* (1954), 280—283; 4. Z. angew. Math. und Physik *4* (1953), 415—506, Gedenkheft

NODDACK, W.: Physikalische Blätter *14* (1958), 370f.

ROTH, J. L. A.: 1. N. Jb. Miner. 1893, II, 14—18; 2. Leopoldina XXVIII (1892), 115

SAUKOV, A. A.: Geokhimiya (1965) 2, 248—249

SCHÖNBEIN, CH. F.: 1. Hagenbachs Rectoratsprogramm 4°, Basle 1868 (with complete bibliography); 2. Mem. Acad. Sciences Metz XLIX, 2. 1869, 3—32

VERNADSKIĬ, V. I.: 1. Amer. Mineralogist *32* (1947), 181—188; 2. Lyudi russkoĭ nauki, Moscow, *2* (1962), 135—157; 3. Ocherki po istorii geologicheskiĭ znaniĭ, Moscow *11* (1963), 153 pages

WASHINGTON, H. S.: 1. Z. f. Vulkanologie XVI, 1934/35, 1—6; 2. Amer. Mineralogist *20* (1935), 3. Miner.-Petrogr. Mitt. *47* (1936); 4. Miner. Magazine *24* (1936)

Subject, Position, Problems and Aims of Geochemistry

Besides discussions of these problems in standard works and textbooks, there are the following publications:

GOLDSCHMIDT, V. M.: Probleme und Methoden der Geochemie. Gerlands Beitr. Geophys. *15* (1926), 38—50

GUNTAU, M.: Zum Problem der Klassifizierung der geologischen Wissenschaften. Ber. Geol. Ges. DDR, Sonderheft 1 (1963), 5—30

KREJCI-GRAF, K.: Geochemie — Anwendung und Abgrenzung. Naturwiss. Rdsch. *10* (1957), 294

KRÜGER, P.: Bemerkungen zum Gegenstand und zur Stellung der Radiogeologie im System der Wissenschaften. Freiberger Forschungs-H., D, *53* (1967), 109—113

LANGE, H.: Gegenstandsbestimmung der Geochemie. Freiberger Forsch. H., D, 51, 103—117, Leipzig 1965

LEUTWEIN, F.: Geochemie und Lagerstättenkunde. Freiberger Forsch. H., C, 17, Berlin 1955

MOLINA, B. R.: Qué es la geoquimica? Rev. Soc. quim. Mexico *9* (1965), 3, 103—108

NIEUWENKAMP, W.: Géochimie Classique et Transformiste. Bull. soc. geol. France *6* (1956), 407—429

NODDACK, W., and J. NODDACK: Aufgaben und Ziele der Geochemie. Freiburger Wiss. Ges., Heft 26, Speyer-Verlag, Freiburg 1937

RANKAMA, K.: What is Geochemistry? Amer. J. Sci. *245*, (1947), 7, 458—461

RÖSLER, H. J.: 1. Entwicklung, Stand und Perspektive der mineralogisch-geochemischen Wissenschaft. Ber. dtsch. Ges. geol. Wiss. *13* (1968), 33—41; 2. Die Bedeutung der Geochemie in der Volkswirtschaft. Bergakademie *10* (1965), 585—588

SAUKOV, A. A.: Historicism in Geochemistry. From: The Combined Action of Sciences in the Exploration of the Earth. Izd. "Nauka", Moscow 1963, 285—308 (Russian)

TRUSOV, U. P.: Subject and Method of Geochemistry and a Few Problems of the Combined Action of Sciences in the Present Stage of Development of Natural Sciences. From: The Combined Action of Sciences in the Exploration of the Earth. Izd. "Nauka", Moscow 1963, 234—284 (Russian)

WEIBEL, M.: Geochemie. Naturwiss. Rdsch. *15* (1962), 140—145

Classical Works and Papers

BEHREND, F., and G. BERG: Chemische Geologie. Verlag Enke, Stuttgart 1927, 595 pp.

BERG, G.: Vorkommen und Geochemie der mineralischen Rohstoffe. Akad. Verlagsbuchhandlung, Leipzig 1929

BISCHOF, G.: Lehrbuch der chemischen und physikalischen Geologie, Vol. 1—4, 1863, 1864, 1866, 1871

BOWEN, N. L.: The Evolution of the Igneous Rocks. Princeton 1928

BRAUNS, R.: Chemische Mineralogie. Leipzig 1896

CLARKE, F. W.: 1. The Relative Abundance of the Chemical Elements. Bull. Phil. Soc., Washington *11* (1889), 131; 2. The Data of Geochemistry. U.S. Geol. Surv. Bull. 770 (5th edition, 1924), 841 pp.

CLARKE, F., and H. WASHINGTON: The Composition of Earth's Crust. U.S. Geol. Surv. Prof. papers 127, Washington 1927

DOELTER, C.: Handbuch der Mineralchemie. Th. Steinkopff, Leipzig, Vol. I (1912)—Vol. IV/3 (1931)

FERSMAN, A. E.: 1. Geokhimiya, Vol. I—IV. Akad. Nauk USSR, 1933, 1934, 1937, 1939; 2. Geochemische Migration der Elemente. Abh. prakt. Geologie Bergwirtsch. *18* (1929), *19* (1930), Halle

GOLDSCHMIDT, V. M.: 1. Der Stoffwechsel der Erde. Vid. Selsk. Skr., Oslo 1922; 2. Geochemie. Handwörterbuch der Naturwissenschaften, 2nd edition, Jena 1932; 3. Grundlagen der quantitativen Geochemie. Fortschr. Min., Krist. u. Petrogr. *17* (1933), 112

GOLDSCHMIDT, V. M. et al: Geochemische Verteilungsgesetze I—IX. Skrift. Norske Vid., Oslo 1923 to 1938

NIGGLI, P.: 1. Gesteine und Minerallagerstätten. 2 Vols. 1948—1952; 2. Das Magma und seine Produkte. Zurich 1937

NODDACK, J., and W. NODDACK: 1. Die Häufigkeit der chemischen Elemente. Naturwissenschaften *18* (1930), 757; 2. Die geochemischen Verteilungskoeffizienten der Elemente. Svensk Kemisk Tidskrift *46* (1934), 173

TAMMANN, G.: Zur Analyse des Erdinnern. Z. anorg. allg. Chem. *131* (1923), 96

VOGT, J. H. L.: 1. The Physical Chemistry of the Magmatic Differentiation of Igneous Rocks. Oslo 1926—1931; 2. The Average Composition of the Earth's Crust etc. Norske Vid. Selsk. Skr., Oslo 1931; 3. Über die relative Verbreitung der Elemente, besonders der Schwermetalle usw. Z. prakt. Geol. (1898), 225, 314, 377, 413, (1899), 10

WASHINGTON, H. S.: 1. The Chemical Analysis of Rocks. J. Wiley a. Sons, New York 1930; 2. The Chemistry of the Earth Crust. J. Franklin Inst. *190* (1920), 757 pp.

VERNADSKIĬ, V. I.: Geochemie. Akad. Verlagsbuchhandlg., Leipzig 1930

Modern Introductions, Textbooks, and Reference Books

AHRENS, L. H. (Edit.): Origin and Distribution of the Elements. Pergamon Press, Oxford-New York 1968

ENERGLYN, Lord, and L. BREALEY: Analytical Geochemistry. Elsevier, Amsterdam 1971

EUGSTER, H. P.: Researches in Geochemistry. John Wiley, New York 1959

FERSMAN, A. E.: Entertaining Geochemistry. (a) Soviet edition; (b) German edition: Verlag Neues Leben, Berlin 1953

FYFE, W. S.: Geochemistry of Solids. Mc Graw Hill Book Co., New York-San Francisco-Toronto-London 1964

GOLDSCHMIDT, V. M.: Geochemistry. Clarendon Press, Oxford 1954

KRAUSKOPF: Introduction to Geochemistry. Mc Graw Hill Book Comp., New York 1967

LUKASHEV, K. I.: Lithology and Geochemistry of the Weathering Crust. I.P.S.T., Jerusalem 1970

MASON, B.: Principles of Geochemistry. John Wiley, New York 1966 (3rd edition)

MIYAKE, Y.: Elements of Geochemistry. Maruzen Co. Ltd., Tokyo 1965

POLANSKI, A.: Geochemia Isotopov. Wydawnistwa Geologiczne, Warsaw 1961 (Polish)

POLANSKI, A., K. SMULIKOWSKI: Geochemia. Wydawnistwa Geologiczne, Warszawa 1969 (Polish)

RANKAMA, K.: Progress in Isotope Geology. John Wiley, New York-London 1963

RANKAMA, K., and TH. SAHAMA: Geochemistry. Univers. Chicago Press, Chicago/London 1950

RILEY, J. P., and R. CHESTER: Introduction to Marine Chemistry. Academic Press, London 1970

SAUKOV, A. A.: Geochemistry. (a) Soviet edition: Gosgeolizdat, Moscow 1952; (b) German edition: Verlag Technik, Berlin 1953

SMALES, A. A., and L. R. WAGER, (Edit.): Methods in Geochemistry. Interscience Publisher, New York 1960

SMITH, F. G.: Physical Geochemistry. Addison-Wesley Publ. Comp., Palo Alto-London 1963

SWAIN, F. M.: Non-Marine Organic Chemistry. Cambridge Univ. Press, London 1970

SZÁDECZKY-KARDOSS, E.: Geochemistry. Akademia Kiado, Budapest 1955 (Hungarian)

WEDEPOHL, K. H.: Geochemie. Sammlung Göschen, Verlag de Gruyter u. Co., Berlin 1967

WEDEPOHL, K. H.: Handbook of Geochemistry. Vol. Ia. II/1, Springer Verlag, Berlin-Heidelberg-New York 1969

2*

2. Fundamental Chemical and Physical Concepts

2.1. The Chemical Elements

A chemical element is defined as a substance which is composed exclusively of atoms having the same nuclear charge.

This section gives the most important information about the chemical elements. Table 1 shows the atomic masses of all elements up to element 104, whereas Table 2 shows further details of the transuranic elements. Table 3 shows the names of the chemical elements in the most important languages and Fig. 2 gives a graphic representation of the history of the discovery of the chemical elements.

Table 1. Atomic Masses of the Elements
(1959: 0 = 16.000,0; 1961: ^{12}C = 12.000,00)[2]); Explanations see p. 22.

Element	Symbol	Atomic Number[1] Z	Atomic Mass A 1959	A 1961[2])
Actinium	Ac	89	[227]	
Aluminium	Al	13	26.98	26.9815
*Americium	Am	95	[243]	
Antimony	Sb	51	121.76	121.75
Argon	A	18	39.944	39.948
Arsenic	As	33	74.92	74.921,6
*Astatine	At	85	[210]	
Barium	Ba	56	137.36	137.34
*Berkelium	Bk	97	[247]	
Beryllium	Be	4	9.013	9.012,2
Bismuth	Bi	83	208.99	208.980
Boron	B	5	10.82	10.811[a])
Bromine	Br	35	79.916	79.909[a])
Cadmium	Cd	48	112.41	112.40
Calcium	Ca	20	40.08	40.08
*Californium	Cf	98	[251]	
Carbon	C	6	12.011	12.011,15[a])
Cerium	Ce	58	140.13	140.12
Cesium	Cs	55	132.91	132.905
Chlorine	Cl	17	35.457	35.453[b])
Chromium	Cr	24	52.01	51.996[b])
Cobalt	Co	27	58.94	58.933,2
Columbium	Cb (= Nb)	41	92.91	92.906
Copper	Cu	29	63.54	63.54
*Curium	Cm	96	[247]	
Dysprosium	Dy	66	162.51	162.50

Table 1 continued

Element	Symbol	Atomic Number[1] Z	Atomic Mass	
			A 1959	A 1961[2]
*Einsteinium	Es	99	[252]	
Erbium	Er	68	167.27	167.27
Europium	Eu	63	152.0	151.96
*Fermium	Fm	100	[253]	
Fluorine	F	9	19.00	18.998,4
*Francium	Fr	87	[223]	
Gadolinium	Gd	64	157.26	157.25
Gallium	Ga	31	69.72	69.72
Germanium	Ge	32	72.60	72.59
Gold	Au	79	197.0	196.967
Hafnium	Hf	72	178.50	178.49
Helium	He	2	4.003	4.002,6
Holmium	Ho	67	164.94	164.930
Hydrogen	H	1	1.008,0	1.007,97[a]
Indium	In	49	114.82	114.82
Iodine	I	53	126.91	126.904,4
Iridium	Ir	77	192.2	192.2
Iron	Fe	26	55.85	55.847[b]
Krypton	Kr	36	83.80	83.80
Kurchatovium	Ku	104		
Lanthanum	La	57	138.92	138.91
*Lawrentium	Lw	103	[257]	
Lead	Pb	82	207.21	207.19
Lithium	Li	3	6.940	6.939
Lutetium	Lu	71	174.99	174.97
Magnesium	Mg	12	24.32	24.312
Manganese	Mn	25	54.94	54.938,1
*Mendelevium	Md	101	[256]	
Mercury	Hg	80	200.61	200.59
Molybdenum	Mo	42	95.95	95.94
Neodymium	Nd	60	144.27	144.24
Neon	Ne	10	20.183	20.183
*Neptunium	Np	93	[237]	
Nickel	Ni	28	58.71	58.71
Niobium	Nb (= Cb)			
Niton	Nt (= Rn)			
Nitrogen	N	7	14.008	14.006,7
*Nobelium	No	102	[253]	
Osmium	Os	76	190.2	190.2
Oxygen	O	8	16.000	15.999,4[a]
Palladium	Pd	46	106.4	106.4
Phosphorus	P	15	30.975	30.973,8
Platinum	Pt	78	195.09	195.09
*Plutonium	Pu	94	[244]	
Polonium	Po	84	210	210
Potassium	K	19	39.100	39.102
Praseodymium	Pr	59	140.92	140.907
*Promethium	Pm	61	[145]	
Protactinium	Pa	91	231	231
Radium	Ra	88	226	226.05

Table 1 continued

Element	Symbol	Atomic Number[1]	Atomic Mass	
		Z	A 1959	A 1961[2]
Radon	Rn	86	222	222
Rhenium	Re	75	186.22	186.2
Rhodium	Rh	45	102.91	102.905
Rubidium	Rb	37	85.48	85.47
Ruthenium	Ru	44	101.1	101.07
Samarium	Sm	62	150.35	150.35
Scandium	Sc	21	44.96	44.956
Selenium	Se	34	78.96	78.96
Silicon	Si	14	28.09	28.086[a]
Silver	Ag	47	107.873	107.870[b]
Sodium	Na	11	22.991	22.989,8
Strontium	Sr	38	87.63	87.62
Sulfur (Sulphur)	S	16	32.066	32.064[a]
Tantalum	Ta	73	180.95	180.948
⋆Technetium	Tc	43	[97]	
Tellurium	Te	52	127.61	127.60
Terbium	Tb	65	158.93	158.924
Thallium	Tl	81	204.39	204.37
Thorium	Th	90	232	232.038
Thulium	Tm	69	168.94	168.934
Tin	Sn	50	118.70	118.69
Titanium	Ti	22	47.90	47.90
Tungsten	W	74	183.86	183.85
Uranium	U	92	238.07	238.03
Vanadium	V	23	50.95	50.942
Xenon	Xe	54	131.30	131.30
Ytterbium	Yb	70	173.04	173.04
Yttrium	Y	39	88.91	88.905
Zinc	Zn	30	65.38	·65.37
Zirconium	Zr	40	91.22	91.22

⋆ Artificially preserved elements
Atomic masses in square brackets [ ] are the mass numbers of the isotopes of longest known life.

[1] Also known as nuclear charge number

[2] Explanations regarding the atomic masses (1961):

The atomic masses are equal to the values established by the International Union of Pure and Applied Chemistry (I.U.P.A.C.) at their Montreal Meeting in August 1961 (cf. H. REMY: Die neue Atomgewichtstabelle. Angew. Chem. 74 (1962), 69—74). These values are referred to the value 12 assigned to the relative atomic mass of the carbon isotope ^{12}C.

The atomic masses marked by a) are liable to variations due to natural changes in their isotopic composition. The observed ranges of variation are

Boron	±0.003	Sulphur	±0.003
Carbon	±0.000.05	Silicon	±0.001
Oxygen	±0.000,1	Hydrogen	±0.000,01

Atomic masses marked by b) show the following experimental uncertainties:

Bromine	±0.002	Iron	±0.003
Chlorine	±0.001	Silver	±0.003
Chromium	±0.001		

Table 2. The Transuranic Elements (Atomic Numbers from 93 to 104)

Atomic Number	Element	Symbol	Production	First Identified Isotope and its Half-life		
93	Neptunium	Np	irradiation of uranium 238 by neutrons	^{239}Np	2.35	d
94	Plutonium	Pu	bombardment of uranium 238 by deuterons	^{238}Pu	86.4	a
95	Americium	Am	irradiation of plutonium 239 by neutrons	^{241}Am	458	a
96	Curium	Cm	bombardment of plutonium 239 by helium ions	^{242}Cm	162.5	d
97	Berkelium	Bk	bombardment of americium 241 by helium ions	^{243}Bk	4.5	h
98	Californium	Cf	bombardment of curium 242 by helium ions	^{245}Cf	44	m
99	Einsteinium	Es	irradiation of uranium 238 by neutrons from a thermonuclear explosion	^{253}Es	20	d
100	Fermium	Fm	irradiation of uranium 238 by neutrons from a thermonuclear explosion	^{255}Fm	22	h
101	Mendelevium	Md	bombardment of einsteinium 253 by helium ions	^{256}Md	1.5	h
102	Nobelium	No	bombardment of curium 246 by carbon ions	^{254}102	3	s
103	Lawrentium	Lw	bombardment of californium 252 by boron ions	^{257}Lw	8	s
104	Kurchatovium[1])	Ku	bombardment of plutonium 242 by ^{22}neon nuclei	^{264}Ku ^{260}Ku	immediate disintegration 0.3	s

[1]) Is not included in the group of actinides, as are the other transuranic elements (elements 90 to 103), but belongs to the 4th group of the periodic system of elements

Table 3. Symbols and Names of the Chemical Elements in the Most Important Languages, Year of Discovery (cf. Fig. 2) and Discoverer

Symbol	Latin	English	German	Russian	French	Spanish	Remarks	Discovery Date	Discoverer
A	Argon	Argon	Argon		Argon		German, Russian and Spanish see Ar		
Ac	Actinium	Actinium	Actinium	Актиний	Actinium	Actinio		1899	Debierne
Ag	Argentum	Silver	Silber	Серебро	Argent	Plata		ancient	Wöhler
Al	Aluminium	Aluminium	Aluminium	Алюминий	Aluminium	Aluminio		1827	Seaborg, James, Morgan
Am	Americium	Americium	Americium	Америций	Americium	Americio		1944	
Ar	Argon	Argon	Argon	Аргон	Argon	Argón or Argo	English and French A	1894	Ramsay, Rayleigh
As	Arsenicum	Arsenic	Arsen	Мышьяк	Arsenic	Arsénico or Arsenio		Middle Ages	
At	Astatine	Astatine	Astat	Астатин	Astatine	Astato or Ástato		1940	Corson, McKenzie, Segrè
Au	Aurum	Gold	Gold	Золото	Or	Oro		ancient	Gay-Lussac, Thénard, Davy
B	Borum	Boron	Bor	Бор	Bore	Boro		1808	Scheele
Ba	Barium	Barium	Barium	Барий	Baryum	Bario	French and Spanish also Gl	1774	Vauquelin
Be	Beryllium	Beryllium	Beryllium	Бериллий	Béryllium	Berilio		1797	Valentinus
Bi	Bismutum	Bismuth	Wismut	Висмут	Bismuth	Bismuto		15th cent.	Seaborg, Ghiorso, Thompson
Bk	Berkelium	Berkelium	Berkelium	Беркелий	Berkelium	Berkelio		1950	Balard
Br	Bromum	Bromine	Brom	Бром	Brome	Bromo		1826	
C	Carboneum	Carbon	Kohlenstoff	Углерод	Carbone	Carbono		ancient	Davy
Ca	Calcium	Calcium	Calcium	Кальций	Calcium	Calcio	see Nb	1808	
Cb	Columbium	Columbium			Columbium				Stromeyer
Cd	Cadmium	Cadmium	Cadmium	Кадмий	Cadmium	Cadmio		1817	Berzelius
Ce	Cerium	Cerium	Cer	Церий	Cérium	Cerio		1803	Seaborg, Thompson, Street, Ghiorso
Cf	Californium	Californium	Californium	Калифорний	Californium	Californio		1950	
Cl	Chlorum	Chlorine	Chlor	Хлор	Chlore	Cloro		1774	Scheele (1810: Davy)

Table 3 continued

Symbol	Latin	English	German	Russian	French	Spanish	Remarks	Discovery Date	Discoverer
Cm	Curium	Curium	Curium	Кюрий	Curium	Curio		1944	Seaborg, James, Ghiorso
Co	Cobaltum	Cobalt	Kobalt	Кобальт	Cobalt	Cobalto		1735	Brandt
Cp	Cassiopeium		Cassiopeium				see Lu	1907	Auer v. Welsbach
Cr	Chromium	Chromium	Chrom	Хром	Chrome	Cromo		1797	Vauquelin
Cs	Cesium	Cesium	Caesium	Цезий	Césium	Cesio		1860	Bunsen, Kirchhoff
Ct	Celtium				Celtium		German, Russian, English and Spanish see Hf		
Cu	Cuprum	Copper	Kupfer	Медь	Cuivre	Cobre		ancient	
(D)	Deuterium	Deuterium	Deuterium	Дейтерий	Deuterium	Deuterio	Heavy hydrogen isotope, also denoted ^2H		
Dy	Dysprosium	Dysprosium	Dysprosium	Аиспрозий	Dysprosium	Disprosio		1886	Lecoq de Boisbaudran
Es	Einsteinium	Einsteinium	Einsteinium	Эйнштейний	Einsteinium	Einsteinio		1954	discovered together with Fm
Em							see Rn (up to 1923 called radium emanation EM)		
Er	Erbium	Erbium	Erbium	Эрбий	Erbium	Erbio		1843	Mosander
Eu	Europium	Europium	Europium	Европий	Europium	Europio		1896	Demarcay
F	Fluorum	Fluorine	Fluor	Фтор	Fluor	Fluor		1886	Moissan
Fa	Francium						see Fr		
Fe	Ferrum	Iron	Eisen	Железо	Fer	Hierro		ancient	
Fm	Fermium	Fermium	Fermium	Фермий	Fermium	Fermio		1954	discovered together with Es
Fr	Francium	Francium	Francium	Франций	Francium	Francio	also Fa	1939	Perey
Ga	Gallium	Gallium	Gallium	Галлий	Gallium	Galio		1875	Lecoq de Boisbaudran

Table 3 continued

Symbol	Latin	English	German	Russian	French	Spanish	Remarks	Discovery Date	Discoverer
Gd	Gadolinium	Gadolinium	Gadolinium	Гадолиний	Gadolinium	Gadolinio		1880	MARIGNAC
Ge	Germanium	Germanium	Germanium	Германий	Germanium	Germanio		1885	WINKLER
Gl	Glucinium				Glucinium	Glucinio	see Be		
H	Hydrogenium	Hydrogen	Wasserstoff	Водород	Hydrogene	Hidrógeno		1766	CAVENDISH
He	Helium	Helium	Helium	Гелий	Helium	Helio		1864	LORYER (1894), RAMSAY, CLEVE
Hf	Hafnium	Hafnium	Hafnium	Гафний		Hafnio	French Ct	1922	COSTER, HEVESY
Hg	Hydrargyrum	Mercury	Quecksilber	Ртуть	Mercure	Mercurio		ancient	
Ho	Holmium	Holmium	Holmium	Гольмий	Holmium	Holmio		1879	CLEVE
I	Jodum	Iodine			Iode	Yodo	see J		
In	Indium	Indium	Indium	Индий	Indium	Indio		1863	REICH, RICHTER
Ir	Iridium	Iridium	Iridium	Иридий	Iridium	Iridio		1803/1804	TENNANT
J	Jodum		Jod	Иод			English and French I	1811	COURTOIS
K	Kalium	Potassium	Kalium	Калий	Potassium	Potassio		1807	DAVY
Kr	Krypton	Krypton	Krypton	Криптон	Krypton	Kriptón		1898	RAMSAY
Ku	Kurcatovium	Kurchatovium	Kurcatovium	Курчато-виум	Kurcatovium			1964	ZVARA (in Dubna)
La	Lanthanum	Lanthanum	Lanthan	Лантан	Lanthane	Lantano		1839	MOSANDER
Li	Lithium	Lithium	Lithium	Литий	Lithium	Litio		1817	ARFVEDSON
Lu	Lutetium	Lutetium	Lutetium	Лютеций	Lutécium	Lutecio	sometimes also Cp	1907	
Lw	Lawrentium	Lawrentium	Lawrentium	Lawrentium	Lawrentium	Lawrencio		1961	GHIORSO, SIKELAND, LARSH, LATTMER (in Berkeley)
Ma	Masurium						see Tc		
Md	Mendelevium	Mendelevium	Mendelevium	Менделевий	Mendelevium	Mendelevio		1955	GHIORSO
Mg	Magnesium	Magnesium	Magnesium	Магний	Magnésium	Magnesio		ancient	
Mn	Manganum	Manganese	Mangan	Марганец	Manganèse	Manganeso		1774	GAHN
Mo	Molybdaenum	Molybdenum	Molybdän	Молибден	Molybdène	Molibdeno		1778	SCHEELE

Table 3 continued

Symbol	Latin	English	German	Russian	French	Spanish	Remarks	Discovery Date	Discoverer
N	Nitrogenium	Nitrogen	Stickstoff	Азот	Azote Nitrogène	Nitrógeno		1772	RUTHERFORD
Na	Natrium	Sodium	Natrium	Натрий	Sodium	Sodio		1807	DAVY
Nb	Niobium	Niobium	Niob	Ниобий		Neobio	English and French Cb	1801	HATCHETT
Nd	Neodymium	Neodymium	Neodym	Неодим	Néodyme	Neodimio		1885	AUER v. WELS-BACH
Ne	Neon	Neon	Neon	Неон	Néon	Neón		1898	RAMSAY
Ni	Niccolum	Nickel	Nickel	Никель	Nickel	Niquel		1751	CRONSTEDT
No	Nobelium	Nobelium	Nobelium	Нобелий	Nobelium	Nobelio		1957	in the Nobel-Institute, Stockholm
Np	Neptunium	Neptunium	Neptunium	Нептуний	Neptunium	Neptunio		1940	McMILLAN, ABELSON
Nt	Niton	Niton					see Rn		
O	Oxygenium	Oxygen	Sauerstoff	Кислород	Oxygène	Oxígeno		1774	PRIESTLEY, SCHEELE
Os	Osmium	Osmium	Osmium	Осмий	Osmium	Osmio		1803	TENNANT
P	Phosphorus	Phosphorus	Phosphor	Фосфор	Phosphore	Fósforo		1669	BRAND
Pa	Protactinium	Protactinium	Protactinium	Протакти-ний	Protactinium	Protactinio		1918	HAHN, METTNER SODDY, CRANSTON
Pb	Plumbum	Lead	Blei	Свинец	Plomb	Plomo		ancient	
Pd	Palladium	Palladium	Palladium	Палладий	Palladium	Paladio		1803	WOLLASTON
Pm	Promethium	Promethium	Promethium	Прометий	Promethium	Prometeo		1945	CORYELL, MARINSKY, GLENDENIN
Po	Polonium	Polonium	Polonium	Полоний	Polonium	Polonio		1898	CURIE
Pr	Praseo-dymium	Praseo-dymium	Praseodym	Празодим	Praséodyme	Praseodimio		1885	AUER v. WELS-BACH
Pt	Platinum	Platinum	Platin	Платина	Platine	Platino		1750	WATSON (1748 ANTONIO DE ULLOA)

Table 3 continued

Symbol	Latin	English	German	Russian	French	Spanish	Remarks	Discovery Date	Discoverer
Pu	Plutonium	Plutonium	Plutonium	Плутоний	Plutonium	Plutonio		1940	Seaborg, McMillan, Wahl, Kennedy
Ra	Radium	Radium	Radium	Радий	Radium	Radio		1898	M. and P. Curie, Bunsen,
Rb	Rubidium	Rubidium	Rubidium	Рубидий	Rubidium	Rubidio		1861	Kirchhoff
Re	Rhenium	Rhenium	Rhenium	Рений	Rhénium	Renio		1925	Noddack, Tacke
Rh	Rhodium	Rhodium	Rhodium	Родий	Rhodium	Rodio		1803	Wollaston
Rn	Radon	Radon	Radon	Радон	Radon	Radón	up to 1923 denoted Em; English also Nt	1900	Rutherford, Soddy
Ru	Ruthenium	Ruthenium	Ruthenium	Рутений	Ruthénium	Rutenio		1844	Claus
S	Sulfur	Sulfur Sulphur	Schwefel	Сера	Soufre	Azufre		ancient	
Sb	Stibium	Antimony	Antimon	Сурьма	Antimoine	Antimonio		ancient	
Sc	Scandium	Scandium	Scandium	Скандий	Scandium	Escandio		1879	Nilson
Se	Selenium	Selenium	Selen	Селен	Sélénium	Selenio		1817	Berzelius
Si	Silicium	Silicon	Silizium	Кремний	Silicium	Silicio		1822	Berzelius
Sm	Samarium	Samarium	Samarium	Самарий	Samarium	Samario		1879	Lecoq de Boisbaudran
Sn	Stannum	Tin	Zinn	Олово	Étain	Estaño		ancient	
Sr	Strontium	Strontium	Strontium	Стронций	Strontium	Estroncio		1793	Klaproth
T	Tritium	Tritium	Tritium	Тритий	Tritium	Tritio	Heavy hydrogen isotope; also denoted ^{3}H		
Ta	Tantalum	Tantalum	Tantal	Тантал	Tantale	Tantalio		1802	Ekeberg
Tb	Terbium	Terbium	Terbium	Тербий	Terbium	Terbio		1843	Mosander
Tc	Technetium	Technetium	Technetium	Технеций	Technetium	Tecnetio	(cf. Ma)	1937	Perrier, Segrè
Te	Tellurium	Tellurium	Tellur	Теллур	Tellure	Telurio		1782	Müller v. Reichenstein
Th	Thorium	Thorium	Thorium	Торий	Thorium	Torio		1828	Berzelius
Ti	Titanium	Titanium	Titan	Титан	Titane	Titanio		1795	Klaproth
Tl	Thallium	Thallium	Thallium	Таллий	Thallium	Talio		1861	Crookes
Tm	Thulium	Thulium	Thulium		Thulium	Tulio	Russian see Tu	1879	Cleve

Table 3 continued

Symbol	Latin	English	German	Russian	French	Spanish	Remarks	Discovery	Discoverer
Tu	Thulium			Тулий		Tulio	German, English, French Tm		
U	Uranium	Uranium	Uran	Уран	Uranium	Uranio		1789	Klaproth
V	Vanadium	Vanadium	Vanadin	Ванадий	Vanadium	Vanadio		1830	Sefström
W	Wolfram	Tungsten or Wolfram	Wolfram	Вольфрам	Tungstène	Tungsteno or Wolframio		1781	Scheele
X	Xenon				Xénon		German, Russian, English, Spanish Xe		
Xe	Xenon	Xenon	Xenon	Ксенон		Xenón		1898	Ramsay
Y	Yttrium	Yttrium	Yttrium	Иттрий	Yttrium	Itrio		1794	Gadolin
Yb	Ytterbium	Ytterbium	Ytterbium	Иттербий	Ytterbium	Iterbio		1878	Marignac
Zn	Zincum	Zinc	Zink	Цинк	Zinc	Cinc or Zinc		ancient	
Zr	Zirconium	Zirconium	Zirkonium	Цирконий	Zirconium	Circonio		1787	Klaproth

29

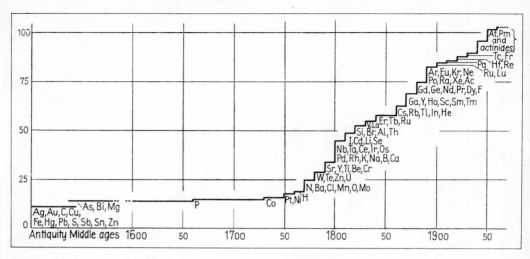

Fig. 2   Discovery of the chemical elements in chronological order

Bibliography

Textbooks of Inorganic Chemistry

COTTON, F. A., and G. WILKINSON: Advances Inorganic Chemistry, 2nd ed. J. Wiley and Sons, New York/London 1966

DAWSON, R. F.: Practical Inorganic Chemistry. Methuan and Co., London 1963

HESLOP, R. B., and P. ROBINSON: Inorganic Chemistry. Elsevier, Amsterdam/London/New York 1967

HOFMANN-RÜDORFF, W.: Lehrbuch der anorganischen Chemie, 19th ed. Vieweg und Sohn, Braunschweig

HOLLEMAN, A. F., and E. WIBERG: Lehrbuch der anorganischen Chemie, 57th to 60th ed. Walter de Gruyter, Berlin 1964

HÜCKEL, W.: Lehrbuch der Chemie. 1st part: Anorganische Chemie, 7th ed. Akademische Verlagsgesellschaft Geest & Portig, Leipzig 1957

KOLDITZ, L.: Anorganikum. Lehr- und Praktikumsbuch der anorg. Chemie mit einer Einführung in die physikal. Chemie. Deutsch. Verlg. Wiss., Berlin 1967

PAULING, L.: General Chemistry, 2nd. ed. Freeman and Co., San Francisco 1954

PAULING, L.: Chemie — Eine Einführung. Verlag Chemie, Weinheim/Bergstraße 1962

REMSEN-REIHLEN, H., and G. RIENÄCKER: Einleitung in das Studium der Chemie, 21st ed. Theodor Steinkopf, Dresden 1963

REMY, H.: Lehrbuch der anorganischen Chemie, Vol. I, 12th ed. 1965, Vol. II, 11th ed. Akademische Verlagsgesellschaft Geest & Portig, Leipzig 1961

REMY, H.: Grundriß der anorganischen Chemie, 12th ed. Akademische Verlagsgesellschaft Geest & Portig, Leipzig 1964

TRZEBIATOWSKI, W.: Lehrbuch der anorganischen Chemie, 4th ed. VEB Deutscher Verlag der Wissenschaften, Berlin 1968

WESTERMANN, K., K.-H. NÄSER, and K.-H. GRUHL: Anorganische Chemie, Vol. I, 6th ed. 1966, Vol. II, 4th ed. 1963. VEB Deutscher Verlag für Grundstoffindustrie, Leipzig

Pocket-books:

Gmelins Handbuch der anorganischen Chemie, 8th ed. Verlag Chemie, Weinheim/Bergstraße 1964

HALLERMEYER-REBEL: Taschenbuch der Chemie, 2nd ed. Lindauer Verlag (Schaefer), Munich 1964

30

PERELMAN, W. J.: Taschenbuch der Chemie, 2nd ed. VEB Deutscher Verlag der Wissenschaften, Berlin 1959

VOGEL, H. U. von, W. BUBAM, and H. NAHME: Chemiker-Kalender, 56th ed. Springer-Verlag, Berlin/Göttingen/Heidelberg 1956

Group of authors: Kolloidchemisches Taschenbuch, 5th ed. Akademische Verlagsgesellschaft Geest & Portig, Leipzig 1960

LAX, E. (Edit.): Taschenbuch für Chemiker und Physiker. Vol. 1, Makroskop. physik.-chem. Eigenschaften, 3rd ed., Springer-Verlag, Berlin/Göttingen/Heidelberg 1967

Books of Tables:

KÜSTER, F. W., K. THIEL, and K. FISCHBECK: Logarithmische Rechentafeln, 93rd ed. Walter de Gruyter, Berlin 1962

LANDOLDT, H., and R. BOERNSTEIN: Zahlenwerte und Funktionen aus Physik, Chemie, Astronomie, Geophysik und Technik, Vol. I, 6th ed. Springer-Verlag, Berlin/Göttingen/Heidelberg 1950—1955

NIKOLSKI, B. P.: Handbuch des Chemikers, Vol. I, 2nd ed. 1959, Vol. II, 1st ed. 1957, Vol. III, VEB Verlag Technik, Berlin, 1st ed. 1959.

RAUSCHKE, K., J. VOIGT, I. WILKE, and K.-TH. WILKE: Chemische Tabellen und Rechentafeln für die analytische Praxis, 4th ed., VEB Deutscher Verlag für Grundstoffindustrie, Leipzig 1968

2.2. Structural Units of Matter

Elementary particles are basic structural units of the atoms and thus basic structural units of matter. At present, about 100 elementary particles are known. Table 4 gives a selection of them. The "elementary" character of these particles is rather conditional because any particle can change into other particles (or radiation) or develop from other particles.

The different types of atoms (elements), of which 104 could be detected up to now (Table 2), are brought into existence by combinations of these "elementary particles".

Table 4. Survey of a Few Important Elementary Particles
(from HERTZ, Vol. II, 1960; RANKAMA, 1963)

Group	Positive Particle Charge e^+	Neutral Particle Charge 0	Negative Particle Charge e^-
Radiation quanta		Photon (γ) $M\ 0$ $T_{1/2}\ \infty$ stable	
Leptons		Neutrino (ν) $M\ 0$ $T_{1/2}\ \infty\ (?)$ stable (?)	
		Antineutrino $(\bar{\nu})$ $M\ 0$ $T_{1/2}\ \infty\ (?)$ stable (?)	

Table 4 continued

Group	Positive Particle Charge e^+	Neutral Particle Charge 0	Negative Particle Charge e^-
	Positron, Antielectron (e^+) M 1 $T_{1/2}$ ∞ stable		Electron (e^-) M 1 $T_{1/2}$ ∞ stable
	positive μ-Meson (Muon) (μ^+) M 206.86 $\pm$ 0.11 $T_{1/2}$ $(2.22 \pm 0.02) \cdot 10^{-6}$ $e^+ + \nu + \bar{\nu}$		negative μ-Meson (μ^-) M } see $T_{1/2}$} positive $e^- + \nu + \bar{\nu}$
L-Mesons (partly π-Mesons)	positive π-Meson (Pion) (π^+) M 273.27 $\pm$ 0.11 $T_{1/2}$ $(2.56 \pm 0.05) \cdot 10^{-8}$ $\mu^+ + \nu$ $e^+ + \nu$ $\pi^0 + e^+ + \nu$	neutral π-Meson (π^0), ($\tilde{\pi}^0$) M 264.27 $\pm$ 0.31 $T_{1/2}$ $<4 \cdot 10^{-16}$ $\gamma + \gamma$ $e^+ + e^- + \gamma$ $2e^+ + 2e^-$ $e^+ + e^-$	negative π-Meson (π^-) M } see $T_{1/2}$} positive $\mu^- + \nu$ $e^+ + \nu$ $\pi^0 + e^+ + \nu$
K-Mesons	positive K-Meson (K^+) M 966.76 $\pm$ 0.4 $T_{1/2}$ $(1.224 \pm 0.013) \cdot 10^{-8}$ $\pi^+ + \pi^+ + \pi^-$ $\pi^+ + \pi^0 + \pi^0$ $\pi^+ + \pi^0$ $\mu^+ + \pi^0 + \nu$ $\mu^+ + \nu$ $e^+ + (\pi^0) + \nu$	neutral K_1-Meson (K_1^0), ($\tilde{K}_1^0$) M 974.4 $\pm$ 1.2 $T_{1/2}$ $(1.00 \pm 0.38) \cdot 10^{-10}$ $\pi^+ + \pi^-$ $\pi^0 + \pi^0$ ——————————— neutral K_2-Meson (K_2^0), ($\tilde{K}_2^0$) M 974.4 $\pm$ 1.2 $T_{1/2}$ $\left(6.1 \begin{smallmatrix}+1.6\\-1.1\end{smallmatrix}\right) \cdot 10^{-8}$ $\pi^+ + \pi^- + \pi^0$ $\pi^0 + \pi^0 + \pi^0$ $\mu^{\pm} + \pi^{\pm} + \nu$ $e^{\pm} + \pi^{\pm} + \nu$	negative K-Meson (K^-) M 965.6 $\pm$ 1.2 $T_{1/2}$ $(1.25 \pm 0.11) \cdot 10^{-8}$ $\pi^+ + \pi^- + \pi^-$ $\pi^- + \pi^0$ $\mu^- + \nu$ $e^- + (\pi^0) + \nu$

Barions

		Positive Particle	Neutral Particle	Negative Particle
	Nucleons	Proton (p^+) M 1,836.118 $\pm$ 0.02 $T_{1/2}$ ∞ stable	Neutron (n) M 1,838.645 $\pm$ 0.02 $T_{1/2}$ $(1.04 \pm 0.13) \cdot 10^3$ $e^- + p + \tilde{\nu}$	Antiproton (p^-) M } see $T_{1/2}$} Proton stable (?)
			Antineutron ($\tilde{n}$) M } see $T_{1/2}$} Neutron $e^+ + p^- + \nu$?	
	Hyperons or Y-particles		Λ-Hyperon (Λ^0), ($\tilde{\Lambda}^0$) M 2,182.9 $\pm$ 0.2 $T_{1/2}$ $(2.505 \pm 0.086) \cdot 10^{-10}$ $p + \pi^-$ $n + \pi^0$	

Table 4 continued

Group	Positive Particle Charge e^+	Neutral Particle Charge 0	Negative Particle Charge e^-
Hyperons or Y-particles	Σ^+-Hyperon (Σ^+), $(\widetilde{\Xi^-})$ M 2,328.0 $\pm$ 0.4 $T_{1/2}$ $(0.83 \pm 0.05) \cdot 10^{-10}$ $p + \pi^0$ $n + \pi^+$ $n + \pi^-$ $\Lambda^0 + \gamma$	Σ^0-Hyperon (Σ^0), $(\widetilde{\Sigma^0})$ M 2,332.6 $\pm$ 1.0 $T_{1/2}$ $<10^{-11}$; $\approx 10^{-19}$ $\Lambda^0 + \gamma$	Σ^--Hyperon (Σ^-), $(\widetilde{\Xi^+})$ M 2,341.3 $\pm$ 0.7 $T_{1/2}\left(1.59 \begin{array}{c} + 0.1 \\ - 0.09 \end{array}\right) \cdot 10^{-10}$ $n + \pi^-$
(Barions)		Ξ^0-Hyperon (Ξ^0), $(\widetilde{\Xi^0})$ M 2,579 $\pm$ 16 $T_{1/2}$ 1.5 $\cdot$ 10^{-10} $\Lambda^0 + \pi^0$	Ξ^--Hyperon (Ξ^-), $(\widetilde{\Xi^+})$ M 2,581 $\pm$ 1 $T_{1/2}\left(1.9 \begin{array}{c} + 0.6 \\ - 0.4 \end{array}\right) \cdot 10^{-10}$ $\Lambda + \pi^-$

Explanation:

e elementary charge, M rest mass related to the rest mass of the electron (mass of the electron $9.107,2 \cdot 10^{-28} = 1$), $T_{1/2}$ half-life in seconds (related to the spontaneous disintegration of particles, but not to their annihilation by reaction with other particles, see also PAULING, 1962); the decay products follow after the half-life. The symbols given in brackets ($\sim$) denote the relevant antiparticles.

2.3. Atomic Structure

Roughly outlined, atoms and ions are spherical, more or less incompressible structures. According to BOHR's classical theory (atom model, 1913), atoms consist of a positively charged nucleus and a negatively charged electron shell which in principle is unlimited. The major part of the density charge concentrates in a finite space which is characterized as the sphere of action.

The electrons of the entire shell of the atom are distributed to seven spatial electron shells or orbits or orbitals (characterizing the different energy levels of the electron); see Fig. 3, Tables 5 and 6. For further information, especially about modern views on the atomic structure (wave mechanics in particular) see textbooks of inorganic chemistry and EVANS (1954), KLEBER (1961, 1963), PAULING (1962, 1964).

Electron shells

These are denoted according to the principal quantum number (n) from 1 to 7 (or K to Q). The maximum number of electrons which can be contained in any given shell is $2n^2$.

Azimuthal quantum number

Is also known as the orbital angular momentum quantum number or secondary quantum number (l) and represented by the symbols s, p, d, f.

The following relations can be derived from the periodic system:

Atomic number

Equals the total number of electrons in the shell or the number of positive nuclear charges.

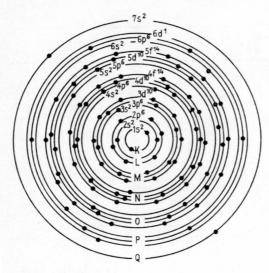

$7s^2$
$6s^2$ — $6p^6$ $6d^1$
$5d^{10}$ $5f^{14}$
$5s^2$ $5p^6$
$4s^2$ $4p^6$ $4d^{10}$ $4f^{14}$
$3s^2$ $3p^6$ $3d^{10}$
$2s^2$ $2p^6$
$1s^2$
K
L
M
N
O
P
Q

Fig. 3 Electron-shell structure of the element 103 (the $5f$ shell of this element has been filled up; in the element 104, one electron is added to the $6d$ shell); after SEABORG and FRITSCH, 1964

Group number

Indicates the number of outer electrons of the atoms of a certain group.

Number of period

This is the number of the outer electron shell where the number of outer electrons characterized by the group number of the atom are to be found.

The properties of the atom depend largely on the electrons contained in the outer electron shell (outer electrons). The atoms belonging to a group of the periodic system or the elements made up of these atoms possess the same number of outer electrons and, consequently, show a very similar chemical behaviour (regarding the elements of the main groups; elements of the subgroups are less related chemically with those of the relevant main groups —filling of the last but one electron shell). GOLDSCHMIDT (1924): The geochemical distribution of a given element is conditioned by the structure of the electron shell (among other things, for an explanation of the geochemical character of the elements due to the atomic structure, see also BETECHTIN, 1952).

The electronic structure of noble gases with $2, 10, 18, 36, 54$ or 86 electrons about one nucleus is very stable (cf. the very high ionisation energy, negative electron affinity, slowness to react). For information about the filling of shells, stability conditions, Pauli principle, electron spin, resonance, etc. see the publications given in 2.1.

Table 5. Grouping and Characterization of Electron Shells (after PAULING, 1964)

Spectroscopic Characterization					Chemical Characterization				
K-shell	$1s^2$				Helium shell	$1s^2$			
L-shell	$2s^2$	$2p^6$			Neon shell	$2s^2$	$2p^6$		
M-shell	$3s^2$	$3p^6$	$3d^{10}$		Argon shell	$3s^2$	$3p^6$		
N-shell	$4s^2$	$4p^6$	$4d^{10}$	$4f^{14}$	Krypton shell	$3d^{10}$	$4s^2$	$4p^6$	
O-shell	$5s^2$	$5p^6$	$5d^{10}$	$5f^{14}$	Xenon shell	$4d^{10}$	$5s^2$	$5p^6$	
P-shell	$6s^2$	$6p^6$	$6d^{10}$		Radon shell	$4f^{14}$	$5d^{10}$	$6s^2$	$6p^6$
Q-shell	$7s^2$	$7p^6$			Eka-Radon shell	$5f^{14}$	$6d^{10}$	$7s^2$	$7p^6$

Table 6. Electronic Configuration of the Atoms in Their Ground State (from PAULING, 1964)

		He	Neon		Argon		Krypton			Xenon			Radon				Eka-Radon			
		1s	2s	2p	3s	3p	3d	4s	4p	4d	5s	5p	4f	5d	6s	6p	5f	6d	7s	7p
1	H	1																		
2	He	2																		
3	Li	2	1																	
4	Be	2	2																	
5	B	2	2	1																
6	C	2	2	2																
7	N	2	2	3																
8	O	2	2	4																
9	F	2	2	5																
10	Ne	2	2	6																
11	Na				1															
12	Mg				2															
13	Al	10			2	1														
14	Si	Neon			2	2														
15	P	core			2	3														
16	S				2	4														
17	Cl				2	5														
18	Ar	2	2	6	2	6														
19	K							1												
20	Ca							2												
21	Sc						1	2												
22	Ti						2	2												
23	V						3	2												
24	Cr						5	1												
25	Mn						5	2												
26	Fe	18					6	2												
27	Co		Argon-core				7	2												
28	Ni						8	2												
29	Cu						10	1												
30	Zn						10	2												
31	Ga						10	2	1											
32	Ge						10	2	2											
33	As						10	2	3											
34	Se						10	2	4											
35	Br						10	2	5											
36	Kr	2	2	6	2	6	10	2	6											
37	Rb										1									
38	Sr										2									
39	Y									1	2									
40	Zr									2	2									
41	Nb									4	1									
42	Mo									5	1									
43	Tc				36					5	2									
44	Ru				Krypton-core					7	1									
45	Rh									8	1									

Table 6 continued

		He Neon 1s 2s 2p	Argon 3s 3p	Krypton 3d 4s 4p	Xenon 4d 5s 5p	Radon 4f 5d 6s 6p	Eka-Radon 5f 6d 7s 7p
46	Pd				10		
47	Ag				10 1		
48	Cd				10 2		
49	In				10 2 1		
50	Sn				10 2 2		
51	Sb		36		10 2 3		
52	Te		Krypton-core		10 2 4		
53	I				10 2 5		
54	Xe	2 2 6	2 6	10 2 6	10 2 6		
55	Cs					1	
56	Ba					2	
57	La					1 2	
58	Ce					1 1 2	
59	Pr					2 1 2	
60	Nd					3 1 2	
61	Pm					4 1 2	
62	Sm					5 1 2	
63	Eu					6 1 2	
64	Gd					7 1 2	
65	Tb					8 1 2	
66	Dy					9 1 2	
67	Ho					10 1 2	
68	Er					11 1 2	
69	Tm			54		12 1 2	
70	Yb			Xenon-core		13 1 2	
71	Lu					14 1 2	
72	Hf					14 2 2	
73	Ta					14 3 2	
74	W					14 4 2	
75	Re					14 5 2	
76	Os					14 6 2	
77	Ir					14 7 2	
78	Pt					14 9 1	
79	Au					14 10 1	
80	Hg					14 10 2	
81	Tl					14 10 2 1	
82	Pb					14 10 2 2	
83	Bi					14 10 2 3	
84	Po					14 10 2 4	
85	At					14 10 2 5	
86	Rn	2 2 6	2 6	10 2 6	10 2 6	14 10 2 6	
87	Fr						1
88	Ra						2
89	Ac			86			1 2
90	Th			Radon-core			2 2
91	Pa						3 2
92	U						4 2
118	Eka-Rn	2 2 6	2 6	10 2 6	10 2 6	14 10 2 6	14 10 2 6

Bibliography

CORRENS, C. W.: Einführung in die Mineralogie und Petrologie. Springer-Verlag, Berlin/Göttingen/ Heidelberg 1949 and 1969

EVANS, R. C.: Einführung in die Kristallchemie. Johann Ambrosius Barth, Leipzig 1954

FRANK-KAMENECKY, V. A.: The Nature of the Structure Elements in Minerals. Leningrad University 1964 (Russian)

HESLOP, R. B., and P. L. ROBINSON: Inorganic Chemistry. 3rd ed. Elsevier. Amsterdam/London/New York 1967

KLEBER, W.: Kristallchemie. B. G. Teubner, Leipzig 1963

MACHATSCHKI, F.: Grundlagen der Allgemeinen Mineralogie und Kristallchemie. Springer, Vienna 1946

MAKAROV, E. G.: Crystal Chemistry. Akad. Nauk, Moscow 1958 (Russian)

PAULING, L.: Die Natur der chemischen Bindung, 2nd ed. Verlag Chemie, Weinheim/Bergstraße 1964

SEABORG, G., and A. R. FRITSCH: Synthetische Elemente. Umschau *64* (1964), 12, 372—376

2.4. Structure of Atomic Nuclei, Isotopes

The atomic nucleus is a combination of protons and neutrons. The size of the nucleus is about 10^{-13} cm.

Nuclear charge number (atomic number): The number of protons in the nucleus; it characterizes the type of atom. The proton number varies between 0 (neutronium) and 104 (Kurchatovium).

The charge of a proton is $+1$, i.e. 1 positive elementary charge $\approx 1.602 \cdot 10^{-19}$ Coulomb ($4.803 \cdot 10^{-10}$ electrostatic units). The positive charge of the atomic nucleus is always a whole multiple of the elementary charge. Whereas the number of protons is constant for any one atomic species the number of neutrons may vary within certain limits. Atoms whose nuclei have the same number of protons (as indicated by their charge number), but a different number of neutrons are called *isotopes* (isotopy discovered by Soddy in 1910). Since the mass of one proton and that of one neutron is equal to about one mass unit each, the isotopes of one element have different masses in accordance with their different number of neutrons.

Isotopes

Atoms having the same nuclear charge (the same number of protons), hence constituting the same element, but differing in mass; the mass number differentiates the various isotopes. The *mass number* (A) is equal to the sum of protons and neutrons in the nucleus; also to the number of nucleons (notation e.g. ^{16}O, ^{17}O, ^{18}O).

Mass number A less nuclear charge number Z (proton number P) results in the number of neutrons N.

$$A - Z \text{ (or } P) = N.$$

Isobars

Nuclei with the same mass number, i.e. having different Z, hence different elements, but having the same number of nucleons (e.g. ^{40}K, ^{40}Ar, ^{40}Ca or ^{87}Rb, ^{87}Sr).

Isotones

Nuclei having the same number of neutrons N (e.g. ^{13}C, ^{14}N or ^{14}C, ^{15}N).

Isodiapheres

Nuclei having the same difference between the number of neutrons and protons, also known as neutron excess, $N - Z = A - 2Z$ (e.g. ^{238}U, ^{236}U).

Isomers

Nuclei having the same N and the same Z, one of them being in a metastable state of excitation. The range of life time which, at present, can be directly measured experimentally extends to about 10^{-14} s (cf. HERTZ 1960). Isomery discovered by Hahn in 1922.

Table 7. Elements up to $Z = 83$ with Natural Radioactivity

Z	Element	Radioactive Isotope	Relative Isotopic Abundance [%]	Half-life [(years)]
1	Hydrogen	^3T	$10^{-19}...10^{-20}$	12.4
6	Carbon	^{14}C	10^{-14}	5,570
19	Potassium	^{40}K	0.012	$1.31 \cdot 10^9$
20	Calcium	^{48}Ca	0.185	$2 \cdot 10^{16}$
23	Vanadium	^{50}V	0.25	$4.8 \cdot 10^{14}$
30	Zinc	^{70}Zn	0.62	$>10^{15}$
34	Selenium	^{79}Se	<0.01	$6.5 \cdot 10^4$
37	Rubidium	^{87}Rb	27.85	$5 \cdot 10^{10}$
40	Zirconium	^{96}Zr	2.80	$6 \cdot 10^{16}$
42	Molybdenum	^{92}Mo	15.86	$>4 \cdot 10^{18}$
		^{100}Mo	9.62	$\geqq 3 \cdot 10^{17}$
48	Cadmium	^{106}Cd	1.22	$\geqq 6 \cdot 10^{16}$
		^{116}Cd	7.58	$\geqq 6 \cdot 10^{16}$
49	Indium	^{115}In	95.77	$6 \cdot 10^{14}$
50	Tin	^{124}Sn	5.98	$2 \cdot 10^{17}$
52	Tellurium	^{130}Te	34.49	$1.4 \cdot 10^{21}$
56	Barium	^{138}Ba	≈ 70	$>10^{15}$
57	Lanthanum	^{138}La	0.089	$7 \cdot 10^{10}$
58	Cerium	^{142}Ce	11.07	$5 \cdot 10^{15}$
60	Neodymium	^{144}Nd	23.8	$5 \cdot 10^{15}$
		^{150}Nd	5.6	$>2 \cdot 10^{15}$
62	Samarium	^{147}Sm	15.07	$1.25 \cdot 10^{11}$
64	Gadolinium	^{152}Gd	0.20	$9.5 \cdot 10^{14}$
71	Lutetium	^{176}Lu	2.60	$2.4 \cdot 10^{10}$
72	Hafnium	^{174}Hf	≈ 0.18	$4.3 \cdot 10^{15}$
73	Tantalum	^{180}Ta	0.012	$2 \cdot 10^{13}$
74	Tungsten	^{186}W	28.4	$6 \cdot 10^{15}$
75	Rhenium	^{187}Re	62.93	$\approx 5 \cdot 10^{10}$
76	Osmium	^{192}Os	41.0	$>10^{14}$
78	Platinum	^{190}Pt	0.012,7	$5.9 \cdot 10^{11}$
		^{192}Pt	0.78	$\approx 10^{15}$
		^{198}Pt	7.19	$>10^{15}$
82	Lead	^{204}Pb	1.55	$\approx 1.4 \cdot 10^{17}$
83	Bismuth	^{209}Bi	100	$2.7 \cdot 10^{17}$

a) Depending on the mass and structure of the nucleus: atomic mass, radioactivity and its forms.

b) Depending on the outer layers of the electron shell (Section 2.3.):
the chemical and, to some extent, the physical properties.

If, for example, one atom enters a chemical combination, the atomic nucleus remains untouched, only the structure of the outermost electron shell changes.

From a) and b) the result for isotopes is:

1. their stability or different radioactivity,
2. their similar chemical behaviour.

The identification of isotopes is carried out by means of mass spectrometers and other means (see Section 4.13.).

Basically, isotopes can be grouped as follows:

1. stable isotopes (267 of them have been found in nature, Table 13); only in elements up to $Z = 82$ (lead) with the exception of technetium (Z 43) and promethium (Z 61) which do not possess a stable isotope. Elements having an odd atomic number (Z) possess not more than two and 19 of them possess no stable isotopes, whereas elements having an even atomic number possess considerably more stable isotopes (exception: natural Be has no isotopic composition). Of the stable nuclei, 78% belong to elements with even Z (Table 8). The light nuclei whose ratio of mass number/atomic number (A/Z ratio) is equal to or approximately 2 or where A is a multiple of four excel in having special stability.

2. unstable radiogenic isotopes

a) Naturally occurring unstable isotopes (natural radioactivity discovered by BECQUEREL 1896).
The natural radioactive elements having atomic numbers over 82 do not possess a stable isotope (mainly due to the increase in the electrical forces of repulsion between the protons); U, Th, Ac, Np see decay series of these elements, Tables 9 to 12 and Figs. 4 to 7. The isotopes of the lighter elements shown in Table 7 are naturally radioactive.

b) Artificial isotopes (artificial radioactivity).
Besides the more than 1170 artificially made isotopes of natural elements, all isotopes of the artificial elements technetium (Z 43), promethium (Z 61) and astatine (Z 85) as well as the majority of the transuranic elements (Table 2) come under this heading.

A	$Z = P$	N	Number of Nuclei
Even	even	even	157
Odd	even	odd	56
Odd	odd	even	48
Even	odd	odd	6

Table 8. Relations between A (Mass Number), Z (Atomic Number, Number of Protons), N (Number of Neutrons), and Number of Stable Nuclei (after HERTZ, 1960)

Table 9. The ^{238}U (uranium) Radioactive Decay Series

Radioactive Element	Nuclides	Type of Decay	Half-life
Uranium I	^{238}U	α	$4.51\cdot10^9$ a
Uranium X1	^{234}Th	β^-	24.10 d
Uranium X2	^{234m}Pa	β^-	1.175 m
Uranium Z	^{234}Pa	β^-	6.66 h
Uranium II	^{234}U	α	$2.48\cdot10^5$ a
Ionium	^{230}Th	α	$8.0\cdot10^4$ a
Radium	^{226}Ra	α	1622 a
Radium emanation, Radon	^{222}Rn	α	3.8229 d
Radium A	^{218}Po	α, β^-	3.05 m

99.98% | 0.02%
α β^-

| Radium B | ^{214}Pb | β^- | 26.8 m |
| Astatine | ^{218}At | α, β^- | 1.5 ··· 2 s |

99.9 % | 0.1 %
β^-

| Radium C | ^{214}Bi | β^-,α | 19.7 m |
| Radon | ^{218}Rn | α | 0.019 s |

0.04 % | 99.96 %
α β^-

Radium C'	^{214}Po	α	$1.64\cdot10^{-4}$ s
Radium C''	^{210}Tl	β^-	1.32 m
Radium D	^{210}Pb	β^-	19.4 a
Radium E	^{210}Bi	β^-, α	5.013 d

$5\cdot10^{-5}$ % | ≈100 %
α β^-

Thallium	^{206}Tl	β^-	4.19 m
Radium F	^{210}Po	α	138.401 d
Radium G	^{206}Pb	stable	–

Explanation: a = annum; d = day; h = hour; m = minute; s = second

Table 10. The ^{232}Th-(Thorium-) Radioactive Decay Series

Radioactive Element	Nuclides	Type of Decay	Half-life
Thorium	^{232}Th	α	$1.39 \cdot 10^{10}$ a
↓			
Mesothorium I	^{228}Ra	β^-	6.7 a
↓			
Mesothorium II	^{228}Ac	β^-	6.13 h
↓			
Radiothorium	^{228}Th	α	1.910 a
↓			
Thorium X	^{224}Ra	α	3.64 d
↓			
Thorium emanation, Thoron	^{220}Rn	α	51.5 s
↓			
Thorium A	^{216}Po	α	0.158 s
↓			
Thorium B	^{212}Pb	β^-	10.64 h
↓			
Thorium C	^{212}Bi	β^-, α	60.5 m
36,2% 63,8%			
α β^-			
Thorium C″	^{208}Tl	β^-	3.10 m
Thorium C′	^{212}Po	α	$3.04 \cdot 10^{-7}$ s
Thorium D	^{208}Pb	stable	—

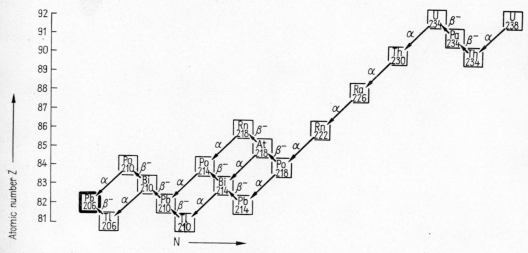

Fig. 4   The ^{238}U Radioactive Decay Series (after RANKAMA, 1963)

Table 11. The ^{235}U (actinium) Radioactive Decay Series

Radioactive Element	Nuclides	Type of Decay	Half-life
Actinouranium	^{235}U	α	$7.1 \cdot 10^8$ a
↓			
Uranium Y	^{231}Th	β^-	25.64 h
↓			
Protactinium	^{231}Pa	α	$3.43 \cdot 10^4$ a
↓			
Actinium	^{227}Ac	β^-,α	21.6 a

1.2 % │ 98.8 %

α ↓

Actinium K β^-	^{223}Fr	β^-,α	22 m
≈6·10⁻³ % │ 99 + % Radioactinium	^{227}Th	α	18.17 d

α ↓

Astatine β^-	^{219}At	α, β^-	0.9 m

≈97 % │ ≈3 %

α β^-

Actinium X	^{223}Ra	α	11.68 d
Bismuth	^{215}Bi	α^-	8 m
Actinon	^{219}Rn	α	3.92 s

Actinium A	^{215}Po	α, β^-	$1.83 \cdot 10^{-3}$ s

» 99 % │ $5 \cdot 10^{-4}$ %

α β^-

Actinium B	^{211}Pb	β^-	36.1 m
Astatine	^{215}At	α	≈10^{-4} s
Actinium C	^{211}Bi	α, β^-	2.16 m

99.7 % │ 0.3 %

α ↓ β^-

Actinium C''	^{207}Tl	β^-	4.79 m
Actinium C'	^{211}Po	α	0.52 s
Actinium D	^{207}Pb	stable	—

Table 12. The ^{237}Np (neptunium) Radioactive Decay Series

Nuclides				Type of Decay	Half-life
^{237}U				β^-	6.75 d
↓					
^{237}Np				α	$2.20 \cdot 10^6$ a
↓					
^{233}Pa				β^-	27.0 d
↓					
^{233}U				α	$1.62 \cdot 10^5$ a
↓					
^{229}Th				α	7340 a
↓					
^{225}Ra				β^-	14.8 d
↓					
^{225}Ac				α	10.0 d
↓					
^{221}Fr				α	4.8 m
↓					
^{217}At				α	0.018 s
↓					
^{213}Bi				β^-, α	47 m

2 %   │   98 %

α ↓
^{209}Tl β^- β^- 2,2 m

                     ^{213}Po α $4,2 \cdot 10^{-6}$ s

^{209}Pb β^- 3,30 h
↓
^{209}Bi $\alpha(?)$ $2 \cdot 10^{17}$ a or
                                                          $> 2 \cdot 10^{18}$ a

^{205}Tl   stable   —

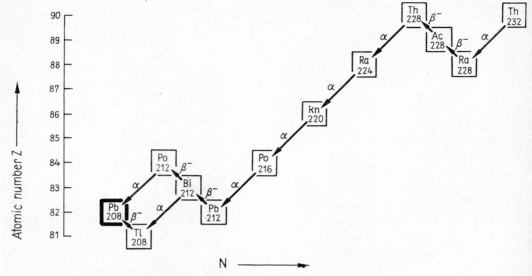

Fig. 5   The ^{232}Th radioactive decay series (after RANKAMA, 1963)

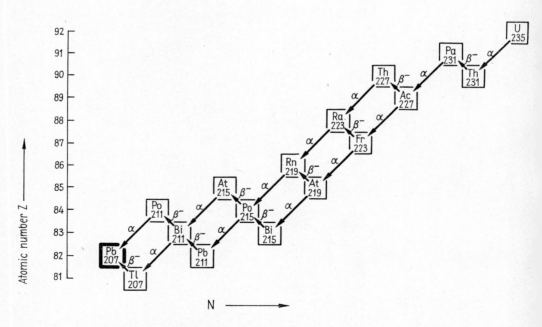

Fig. 6   The ^{235}U (actinium) radioactive decay series (after RANKAMA, 1963)

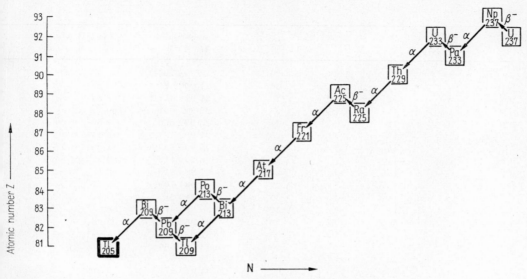

Fig. 7   The ^{237}U (neptunium) radioactive decay series (after RANKAMA, 1963)

Table 13. The Stable Atomic Nuclei and a Few Long-lived Unstable Nuclei. Explanations see p. 53 (from G. Hertz: Lehrbuch der Kernphysik, Ed. Teubner, Leipzig 1960)

Z	Symbol	N	N — Z	A	Mass Value M (AMU)	Natural Abundance [%]	
0	n	1	1	1 *)	1.008,983,0?	—	
					1.008,986,1?		
1	H	0	—1	1	1.008,145,1	99,985,1	ν
1	D	1	0	2	2.014,742,5	0.014,9	ν
1	T	2	1	3 *)	3.016,998?	$<10^{-10}$	ν
					3.017,000?		
2	He	1	—1	3	3.016,979?	0.000,13	ν, r
					3.016,970		
		2	0	4	4.003,873?	99.999,87	ν, r
					4.003,874,2?		
3	Li	3	0	6	6.017,028?	7.18— 7.52	ν
		4	1	7	7.018,222?	92.82—92.48	ν
4	Be	5	1	9	9.015,041?	100	
5	B	5	0	10	10.016,121	18.45—19.58	ν
		6	1	11	11.012,802	81.55—80.42	ν
6	C	6	0	12	12.003,815,6	98.88—98.95	ν
		7	1	13	13.007,490,0	1.12— 1.05	ν
7	N	7	0	14	14.007,525,8	99.635	ν
		8	1	15	15.004,882,0	0.365[1])	ν
8	O	8	0	16	16	99.758,7	ν
		9	1	17	17.004,537,4	0.037,4	ν
		10	2	18	18.004,885,4	0.203,9	ν
9	F	10	1	19	19.004,445,3	100	

Table 13 continued

Z	Symbol	N	$N-Z$	A	Mass Value M (AMU)	Natural Abundance [%]	
10	Ne	10	0	20	19.998,798,2	90.92	
		11	1	21	21.000,524,8	0.257	
		12	2	22	21.998,377,1	8.82	
11	Na	12	1	23	22.997,091	100	
12	Mg	12	0	24	23.992,669	78.60—78.98	ϱ
		13	1	25	24.993,781,9	10.11—10.05	ϱ
		14	2	26	25.990,853,8	11,29—10.97	ϱ
13	Al	14	1	27	26.990,113	100	
14	Si	14	0	28	27.985,819	92.16—92.41	ϱ
		15	1	29	28.985,700	4.71— 4.57	ϱ
		16	2	30	29.983,283	3.13— 3.01	ϱ
15	P	16	1	31	30.983,612,6	100	
16	S	16	0	32	31.982,238,8	95.0[2])	ϱ
		17	1	33	32.981,947	0.75	ϱ
		18	2	34	33.978,664	4.2[2])	ϱ
		20	4	36	35.978,525	0.017	ϱ
17	Cl	18	1	35	34.979,972	75.5	
		20	3	37	36.977,657	24.5	
18	Ar	18	0	36	35.978,983	0.337	ϱ
		20	2	38	37.974,802	0.063	ϱ
		22	4	40	39.975,093	99.6	ϱ, r
19	K	20	1	39	38.976,100	93.08—93.58	ϱ
		21	2	40 *)	39.976,709	0.011,9	ϱ
		22	3	41	40.974,856	6.91— 6.41	ϱ
20	Ca	20	0	40	39.975,293	96.92—96.97	ϱ, r
		22	2	42	41.971,967	0.64	ϱ
		23	3	43	42.972,444	0.132—0.145	ϱ
		24	4	44	43.969,471	2.13— 2.06	ϱ
		26	6	46	45.968,298	0.003,2—0.003,3	ϱ
		28	8	48	47.967,777	0.179—0.185	ϱ
21	Sc	24	3	45	44.970,212	100	
22	Ti	24	2	46	45.967,242	7.99	
		25	3	47	46.966,686	7.32	
		26	4	48	47.963,191	73.99	
		27	5	49	48.963,429	5.46	
		28	6	50	49.960,669	5.25	r
23	V	27	4	50 *)	49.963,045	0.25	
		28	5	51	50.960,175	99.75	
24	Cr	26	2	50	49.961,931	4.31	
		28	4	52	51.957,026	83.76	
		29	5	53	52.957,482	9.55	
		30	6	54	53.956,023	2.38	
25	Mn	30	5	55	54.955,523	100	
26	Fe	28	2	54	53.956,759	5.81[1])	ϱ
		30	4	56	55.952,725	91.6	ϱ
		31	5	57	56.953,513	2.21[1])	ϱ
		32	6	58	57.951,736	0.34	ϱ

[1]) In organic nitrogen, the relative abundance of ^{15}N may come up to 0.375
[2]) $^{32}S/^{34}S = 21.60—23.05$

Table 13 continued

Z	Symbol	N	N — Z	A	Mass Value M (AMU)	Natural Abundance [%]	
27	Co	32	5	59	58.951,920	100	
28	Ni	30	2	58	57.953.772	67.8	
		32	4	60	59.949,825	26.2	
		33	5	61	60.950,463	1.25	
		34	6	62	61.948,029	3.66	
		36	8	64	63.948,338	1.16	
29	Cu	34	5	63	62.949,604	69.1	
		36	7	65	64.948,427	30.9	
30	Zn	34	4	64	63.949,472	38.9 [2]	$\wp$
		36	6	66	65.947,014	27.8	$\wp$
		37	7	67	66.948,419	4.11	$\wp$
		38	8	68	67.946,458	18.6 [2]	$\wp$
		40	10	70	69.947,576	0.62	$\wp$
31	Ga	38	7	69	68.947,63	60.5	
		40	9	71	70.947,37	39.5	
32	Ge	38	6	70	69.946,23	20.38—20.64	$\wp$
		40	8	72	71.944,47	27.37—27.50	$\wp$
		41	9	73	72.946,53	7.78— 7.72	$\wp$
		42	10	74	73.944,50	36.65—36.43	$\wp$
		44	12	76	75.945,42	7.82— 7.71	$\wp$
33	As	42	9	75	74.945,54	100	
34	Se	40	6	74	73,946,04	0.87	
		42	8	76	75.943,41	9.02	
		43	9	77	76.944,54	7.58	
		44	10	78	77.942,16?	23.52	
		45	11	79 *)	78.943,58?	(<0.01)	
		46	12	80	79.941,88	49.82	
		48	14	82	81.942,68	9.19	
35	Br	44	9	79	78.943,49	50.54	
		46	11	81	80.942,15	49.46	
36	Kr	42	6	78	77.944,96	0.354	
		44	8	80	79.942,27	2.27	
		46	10	82	81.939,51	11.56	
		47	11	83	82.940,42	11.55	
		48	12	84	83.938,19	56.90	
		50	14	86	85.938,11	17.37	
37	Rb	48	11	85	84.939,03	72.15	
		50	13	87 *)	86.936,94	27.85	
38	Sr	46	8	84	83.939,93	0.56	$\wp$
		48	10	86	85.936,68	9.86 [3]	$\wp$
		49	11	87	86.936,62	7.02	$\wp, r$
		50	12	88	87.933,96	82.56 [3]	$\wp$
39	Y	50	11	89	88.933,99	100	
40	Zr	50	10	90	89.932,9	51.46	
		51	11	91	90.934,7	11.23	
		52	12	92	91.933,8?	17.11	
		54	14	94	93.937,0	17.40	
		56	16	96	95.939,1	2.80	

[1] $^{54}Fe/^{57}Fe = 2.60$—2.86. [2] $^{64}Zn/^{68}Zn = 2.61$—2.65. [3] $^{86}Sr/^{88}Sr = 0.114,0$—0.122,0.

Table 13 continued

Z	Symbol	N	N — Z	A	Mass Value M (AMU)	Natural Abundance [%]	
41	Nb	52	11	93	92.935,21	100	
42	Mo	50	8	92	91.935,2?	15.86	
		52	10	94	93.935,8	9.12	
		53	11	95	94.935,7?	15.70	
		54	12	96	95.935,4	16.50	
		55	13	97	96.937,2?	9.45	
		56	14	98	97.937,2?	23.75	
		58	16	100	99.938,2	9.62	
43	Tc	56	13	99 *)	98.938,5?	—	
44	Ru	52	8	96	95.938,4	a) 5.50 b) 5.57	
		54	10	98	97.937,1?	a) 1.91 b) 1.86	
		55	11	99	98.938,2?	a) 12.70 b) 12.7	
		56	12	100		a) 12.69 b) 12.6	
		57	13	101		a) 17.01 b) 17.1	
		58	14	102	101.935,6	a) 31.52 b) 31.6	
		60	16	104	103.937,5	a) 18.67 b) 18.5	
45	Rh	58	13	102	102.937,2	100	
46	Pd	56	10	103	101.937,28	0.96	
		58	12	104	103.936,31	10.97	
		60	14	106	105.936,5	27.33	
		62	16	108	107.937,77	26.71	
		64	18	110	109.939,47	11.81	
47	Ag	60	13	107	106.938,9?	a) 51.35 b) 51.92	
		62	15	109	108.939,3?	a) 48.65 b) 48.08	
48	Cd	58	10	106	105.939,66	1.215	
		60	12	108	107.938,38	0.875	
		62	14	110	109.938,40	12.39	
		63	15	111	110.939,55	12.75	
		64	16	112	111.938,62	24.07	
		65	17	113	112.940,38	12.26	
		66	18	114	113.939,77	28.86	
		68	20	116	115.941,97	7.58	
49	In	64	15	113	112.940,22	4.33	
		66	17	115 *)	114.940,16	95.67	
50	Sn	62	12	112	111.940,3?	0.95	
		64	14	114	113.939,5?	0.65	
		65	15	115	114.939,9	0.34	r
		66	16	116	115.939,05	14.24	
		67	17	117	116.940,28	7.57	
		68	18	118	117.939,54	24.01	

Table 13 continued

Z	Symbol	N	N — Z	A	Mass Value M (AMU)	Natural Abundance [%]	
		69	19	119	118.940,97	8.58	
		70	20	120	119.940,33	32.97	
		72	22	122	121.942,24	4.71	
		74	24	124	123.944,64	5.98	
51	Sb	70	19	121	120.942,0?	57.25	
		72	21	123	122.943,1?	42.75	
52	Te	68	16	120	119.942,63	0.089	
		70	18	122	121.941,68	2.46	
		71	19	123	122.943,4	0.87	
		72	20	124	123.942,52	4.61	
		73	21	125	124.944,3	6.99	
		74	22	126	125.943,91	18.71	
		76	24	128	127.946,2	31.79	
		78	26	130	129.948,27	34.49	
53	I	74	21	127	126.945,02	100	
		76	23	129 *)	128.945,8?	<0.002	
54	Xe	70	16	124	123.945,52	0.096	
		74	20	128	127.944,19	1.919	
		75	21	129	128.945,70	26.44	
		76	22	130	129.944,81	4.08	
		77	23	131	130.946,70	21.18	
		78	24	132	131.946,11	26.89	
		80	26	134	133.947,99	10.44	
		82	28	136	135.950,42	8.87	
55	Cs	78	23	133	132.947,38	100	
		80	25	135 *)		<0.002	
56	Ba	74	18	130	129.948,1	a) 0.101 b) 0.13	
		76	20	132	131.947,06	a) 0.097 b) 0.19	
		78	22	134	133.946,83	a) 2.41 b) 2.60	
		79	23	135	134.948,46	a) 6.59 b) 6.73	
		80	24	136	135.947,59	a) 7.81 b) 8.07	
		81	25	137	136.949,07	a) 11.32 b) 11.9	
		82	26	138	137.948,74	a) 71.66 b) 70.4	r
57	La	81	24	138 *)	137.950,6	0.089	
		82	25	139	138.950,21	99.911	
58	Ce	78	20	136	135.950,3	0.193	
		80	22	138	137.949,9	0.250	r
		82	24	140	139.949,77	88.48	r
		84	26	142 *)	141.954,42	11.07	
59	Pr	82	23	141	140.952,28	100	
60	Nd	82	22	142	141.952,61	a) 27.09 b) 27.3	

Table 13 continued

Z	Symbol	N	N — Z	A	Mass Value M (AMU)	Natural Abundance [%]	
		83	23	143	142.955,02	a) 12.14	r
						b) 12.32	
		84	24	144 *)	143.955,56	a) 23.83	
						b) 23.8	
		85	25	145	144.958,1	a) 8.29	
						b) 8.29	
		86	26	146	145.959,09	a) 17.26	
						b) 17.10	
		88	28	148	147.963,51	a) 5.74	
						b) 5.67	
		90	30	150	149.968,49	a) 5.63	
						b) 5.56	
62	Sm	82	20	144	143.957,41	a) 3.16	
						b) 3.15	
		85	23	147 *)	146.961,20	a) 15.07	
						b) 15.09	
		86	24	148	147.961,5	a) 11.27	r
						b) 11.35	
		87	25	149	148.964,2	a) 13.84	
						b) 13.96	
		88	26	150	149.964,56	a) 7.47	
						b) 7.47	
		90	28	152	151.967,66	a) 26.6	
						b) 26.6	
		92	30	154	153.970,87	a) 22.5	
						b) 22.4	
63	Eu	88	25	151	150.967,53	a) 47.86	
						b) 47.77	
		90	27	153	152.969,2	a) 52.14	
						b) 52,23	
64	Gd	88	24	152 *)		a) 0.205	
						b) 0.20	
		90	26	154	153.969,9	a) 2.23	
						b) 2.15	
		91	27	155	154.972,0	a) 15.10	
					154.970,9?	b) 14.73	
		92	28	156	155.971,7	a) 20.6	
						b) 20.47	
		93	29	157	156.972,5	a) 15.7	
					156.971,2?	b) 15.68	
		94	30	158	157.973,5	a) 24.5	
						b) 24.87	
		96	32	160	159.978,1	a) 21.6	
						b) 21.90	
65	Tb	94	29	159		100	
66	Dy	90	24	156		a) 0.052,4	
						b) 0.057	
		92	26	158		a) 0.090,2	
						b) 0.100	
		94	28	160	159.974,8	a) 2.294	

Table 13 continued

Z	Symbol	N	N — Z	A	Mass Value M (AMU)	Natural Abundance [%]	
						b) 2.35	
		95	29	161	160.976,9	a) 18.88	
						b) 19.0	
		96	30	162	161.977,1	a) 25.53	
						b) 25.5	
		97	31	163	162.979,6	a) 24.97	
						b) 24.9	
67	Ho	98	32	164	163.980,6	a) 28.18	
68	Er					b) 28.1	
		98	31	165	164.981,3	100	
		94	26	162		0.136	
		96	28	164	163.981,9	1.56	
		98	30	166	165.981,5	33.4	
		99	31	167	166.983,6	22.9	
		100	32	168	167.984,2	27.1	
		102	34	170	169.989,6	14.9	
69	Tm	100	31	169		100	
70	Yb	98	28	168		a) 0.135	
			99			b) 0.140	
		100	30	170		a) 3.14	
						b) 3.03	
		101	31	171		a) 14.40	
						b) 14.31	
		102	32	172	171.984?	a) 21.9	
						b) 21.82	
		103	33	173		a) 16.2	
						b) 16.13	
		104	34	174	173.981?	a) 31.6	
						b) 31.84	
		106	36	176		a) 12.60	
						b) 12.73	
71	Lu	104	33	175		97.412	
		105	34	176 *)	175.997?	2.588	
72	Hf	102	30	174 *)		a) 0.163	
						b) 0.18	r
		104	32	176	175.996,0	a) 5.21	
						b) 5.15	
		105	33	177	176.998,5	a) 18.56	
						b) 18.39	
		106	34	178	177.999,6	a) 27.10	
						b) 27.1	
		107	35	179	179.002,3	a) 13.75	
						b) 13.78	
		108	36	180	180.003,8	a) 35.22	r
						b) 35.44	
73	Ta	107	34	180 *)	180.003,3?	0.012,3	
		108	35	181	181.004,3?	99.987,7	
74	W	106	32	180	180.002,8?	a) 0.16	r
						b) 0.135	
		108	34	182	182.004,9	a) 26.35	

Table 13 continued

Z	Symbol	N	N — Z	A	Mass Value M (AMU)	Natural Abundance [%]	
						b) 26.4	
		109	35	183	183.006,8	a) 14.32	
						b) 14.4	
		110	36	184	184.008,3	a) 30.68	
						b) 30.6	
		112	38	186	186.012,0	a) 28.49	
						b) 28.4	
75	Re	110	35	185		37.07	
		112	37	187 *)	187.010,2	62.93	
76	Os	108	32	184		0.018	
		110	34	186	186.012,9	1.59	r
		111	35	187	187.015,1	1.64	r
		112	36	188	188.015,4	13.27	r
		113	37	189	189.018,6	16.14	
		114	38	190	190.018,5	26.4	
		116	40	192	192.022,0	41.0	
77	Ir	114	37	191	191.021?	38.5	
		116	39	193	193.025?	61.5	
78	Pt	112	34	190 *)		0.012,7	
		114	36	192	192.023?	0.78	
		116	38	194	194.025,3	32.9	
		117	39	195	195.025,5	33.8	
		118	40	196	196.029,7?	25.2	
					196.026,9		
		120	42	198	198.027,9	7.19	
79	Au	118	39	197	197.028,0?	100	
80	Hg	116	36	196	196.028,136	0.146	
		118	38	198	198.029,692	10.02	
		119	39	199	199.031,487	16.84	
		120	40	200	200.031,911	23.13	r
		121	41	201	201.034,192	13.22	
		122	42	202	202.034,845	29.80	
		124	44	204	204.038,328	6.85	
81	Ti	122	41	203	203.035,0?	a) 29.50	
						b) 29.46	
		124	43	205	205.037,5?	a) 70.50	
						b) 70.54	
82	Pb	122	40	204 *)	204.037,935	1.55	
		124	42	206	206.039,928	22.51	r
		125	43	207	207.041,695	22.60	r
		126	44	208	208.042,779	53.34	r
83	Bi	126	43	209	209.044,9?	100	
90	Th	142	52	232 *)	232.109,8?	100	
91	Pa	140	49	231 *)	231.108,27?	—	
92	U	142	50	234 *)	234.113,1?	a) 0.005,6	
						b) 0.005,8	
		143	51	235 *)	235.116,6?	a) 0.718	
						b) 0.720	
		146	54	238 *)	238.124,3?	a) 99.276	
						b) 99.274	

Explanations of Table 13 (page 45-52)

Z atomic number or nuclear charge number (number of protons in the nucleus), N number of neutrons, $N - Z$ neutron excess (also denoted $A - 2Z$), A mass number (number of nucleons — sum of protons and neutrons); an asterisk after the mass number means that the nucleus is unstable.

Mass value in atomic mass units (AMU) of the physical mass-value scale (related to the oxygen isotope $^{16}O = 16.000$). The data represent the atomic masses of the neutral atoms, especially those measured quantities which were determined by mass spectrometry. Data which were derived from the Q-values of nuclear reactions or from the disintegration energy of radioactive transformations are marked by a question-mark (?).

The *values of the relative natural abundance* in [%] of the relevant nucleus of an element may be very variable in the various naturally occurring materials (e.g. by isotopy effects). Such values are marked by a v following the figure; in the case of a great variability, the lower and the upper abundance limits are stated. An r indicates that the nucleus in question can also be produced by natural radioactive processes. From this there also results a more or less variable isotopic composition of the element under consideration in various naturally occurring materials.

2.5. Ionization Potential

The ionization potential I in [V] is numerically equal to the ionizing energy [eV].
I [eV] $\cdot$ 23.053 $= I$ [kcal/mol].
Directly measurable quantity, characteristic of any given ion; mirrors the influence of the outer-shell electron number; of importance to an assessment of the chemical activity and the inclination to diadochy (see Section 3.2.3.2.); important to the spectroscopic excitability of an element (see Section 4.7.); ionization of atoms by radioactive radiation. Different degrees of ionization are possible in the various elements. The work to be done to cause an atom to pass form the normal state to the electrically charged state by the removal of electrons is called the ionization potential or ionisation energy (I). The first ionization potential (I_1) is the energy required to remove one electron from the neutral atom.

Table 14. Properties of Ions as a Function of the Ionization Potential I

I	Properties
With increasing I	in the case of cations of the same valence, increase in electron affinity, increase in complex stability and electro-negativity
Very high I	form soluble anions or they are inert
Low I	form soluble cations
Roughly equal I	geochemically similar
in the Periodic System:	
I increases	with increasing group number (hence from the left to the right) e.g. alkali metals have a low I → inert gases a high I, inclination to form ions decreases; with increasing I of the constituents, a compound approaches the covalent type
I decreases	within each group with increasing atomic number (i.e. from top to bottom); Fig. 8

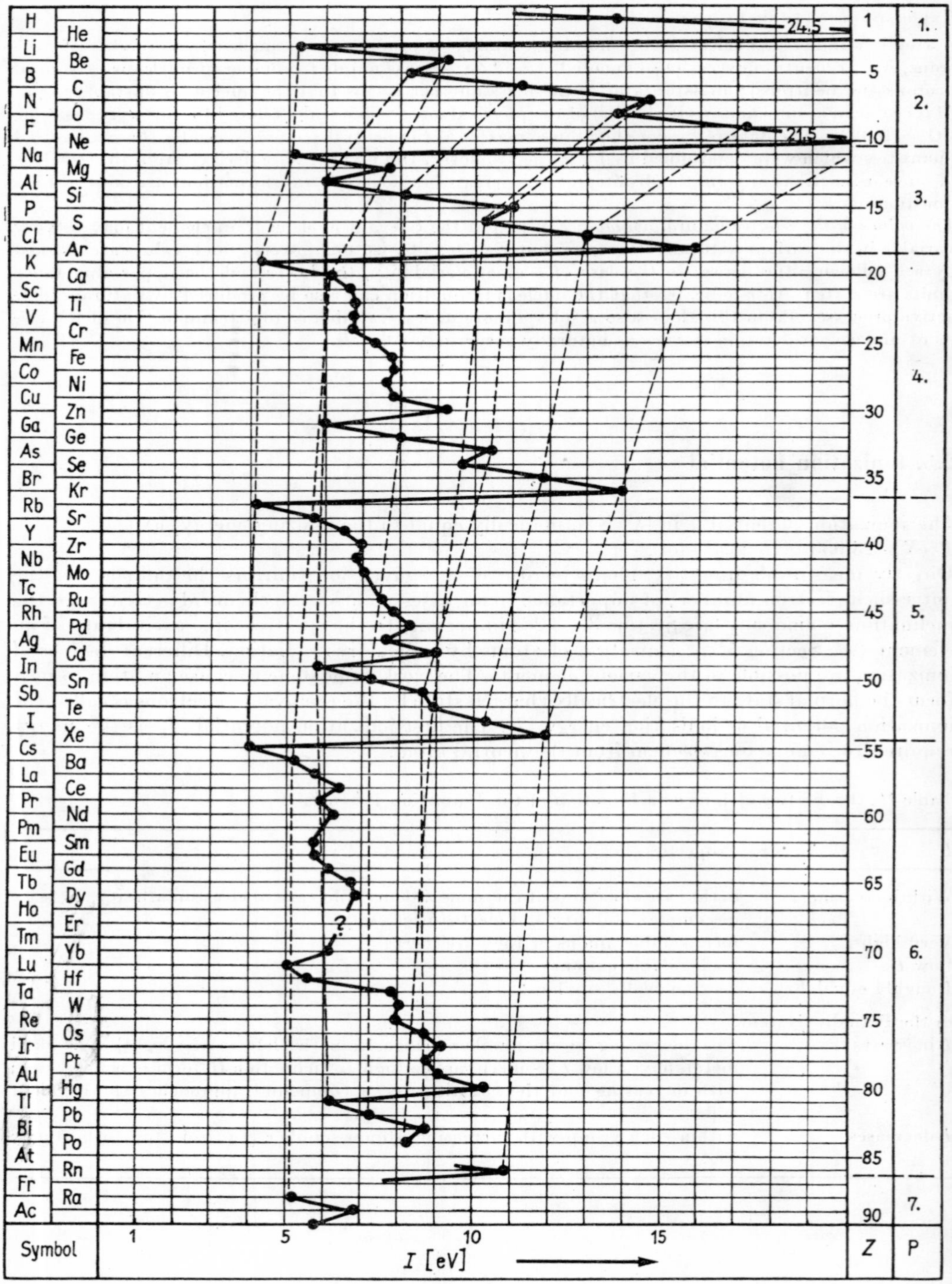

H
He
Li
Be
B
C
N
O
F
Ne
Na
Mg
Al
Si
P
S
Cl
Ar
K
Ca
Sc
Ti
V
Cr
Mn
Fe
Co
Ni
Cu
Zn
Ga
Ge
As
Se
Br
Kr
Rb
Sr
Y
Zr
Nb
Mo
Tc
Ru
Rh
Pd
Ag
Cd
In
Sn
Sb
Te
I
Xe
Cs
Ba
La
Ce
Pr
Nd
Pm
Sm
Eu
Gd
Tb
Dy
Ho
Er
Tm
Yb
Lu
Hf
Ta
W
Re
Os
Ir
Pt
Au
Hg
Tl
Pb
Bi
Po
At
Rn
Fr
Ra
Ac
24.5
21.5
?
Symbol
1
5
10
15
I [eV]
Z
P
1
5
10
15
20
25
30
35
40
45
50
55
60
65
70
75
80
85
90
1.
2.
3.
4.
5.
6.
7.

Table 15. First and Second Ionization Potentials (I_1 and I_2) of the Elements in [eV] and [kcal/mol] (from LANDOLT-BÖRNSTEIN and PAULING, 1964)

Element Symbol	Z	I_1 [eV]	I_1 [kcal/mol]	I_2 [eV]	I_2 [kcal/mol]
Ag	47	7.58	174.6	21.4	495
Al	13	5.97	137.9	18.8	433.9
Ar	18	15.76	363.2	27.5	636.7
As	33	10.5	226	≈20	429
Au	79	9.23	213	20.0	473
B	5	8.28	191.2	25.1	579.8
Ba	56	5.21	120.1	10.0	230.6
Be	4	9.32	214.9	18.2	419.7
Bi	83	8.0	168.0	16.6	384.5
Br	35	11.76	272.9	19.2	497.9
C	6	11.27	259.5	24.8	561.9
Ca	20	6.11	140.9	11.9	273.6
Cd	48	8.99	207.3	16.9	389.7
Ce	58	6.54	150.8	14.8	341.2
Cl	17	12.90	300	23.7	548.7
Co	27	7.84	181	17.4	393
Cr	24	6.74	155.9	16.7	380.1
Cs	55	3.89	89.7	23.4	578.6
Cu	29	7.72	178.1	20.2	467.7
Dy	66	≈6.8	≈157		
Eu	63	5.64	130.0	11.2	252.2
F	9	17.43	401.5	34.9	806
Fe	26	7.83	181	16.5	373
Ga	31	5.97	138	20.5	473
Gd	64	6.7	154		
Ge	32	8.13	182	16.0	367
H	1	13.527	313.4		
He	2	24.56	566.7	54.1	1,254.2
Hf	72	6.94	160	(14.8)	343
Hg	80	10.44	240	18.8	432.3
I	53	10.44	241.0	19.4	440.0
In	49	5.79	133.4	18.9	434.8
Ir	77	9.2	210		
K	19	4.34	100.0	31.7	733.3
Kr	36	14.00	322.6	24.5	566.2
La	57	5.6	129	11.4	263
Li	3	5.40	124.3	75.7	1,743.2
Lu	71	5	115.3		
Mg	12	7.64	176.2	15.0	346.5
Mn	25	7.43	171.3	15.64	360.5
Mo	42	7.06	164	16.13	372
N	7	14.55	335	29.6	682.2
Na	11	5.14	118.4	47.3	1,090

Fig. 8 Ionization energies (I_1) of the elements as a function of the atomic number (Z). Data from AHRENS and S. R. TAYLOR (1961) and PAULING (1964)

Table 15 continued

Element Symbol	Z	I_1 [eV]	[kcal/mol]	I_2 [eV]	[kcal/mol]
Nb	41	6.88	159	(13)	330
Nd	60	≈6.3	≈145		
Ne	10	21.56	497.0	40.9	947
Ni	28	7.63	176.0	18.2	418.4
O	8	13.62	313.8	35.2	809.3
Os	76	≈8.7	200	16.92	390
P	15	10.90	241.7	19.7	455
Pb	82	7.42	170.9	15.0	346.4
Pd	46	8.1	192	≈19.8	448
Po	84	8.42	194		
Pr	59	≈5.8	≈134		
Pt	78	≈8.9	210	10.76	247.9
Ra	88	5.21	121.7	10.1	233.8
Rb	37	4.17	96.3	27.3	634.0
Re	75	≈8	181	16.62	383
Rh	45	7.7	172	18.09	417
Rn	86	10.698	246.7		
Ru	44	≈7.5	169.8	16.89	389.4
S	16	10.36	238.8	23.4	539
Sb	51	8.64	199.2	(18)	380.4
Sc	21	6.7	151	12.8	295
Se	34	9.73	225	(21)	495.6
Si	14	8.15	187.9	16.4	377
Sm	62	≈5.6	≈129	≈11.4	262.8
Sn	50	7.3	169.3	14.6	337.2
Sr	38	5.69	131.2	11.0	254.2
Ta	73	7.7	182	16.18	373
Tb	65	6.7	154		
Tc	43	7.29	168	15.27	352
Te	52	8.96	208	18.61	429
Th	90	≈5.7	≈131		
Ti	22	6.84	157	13.6	313
Tl	81	6.12	140.8	20.3	470.7
U	92	≈4	≈92		
V	23	6.71	155	14.1	338
W	74	7.94	184	17.70	408
Xe	54	12.13	279.6	21.2	488.7
Y	39	6.5	147	12.3	282
Yb	70	6.22	143.4	≈12	≈277
Zn	30	9.39	216.5	18.0	414.0
Zr	40	6.95	158	14	303

Bibliography

AHRENS, L. H.: The Use of Ionization Potentials, I. Ionic Radii of the Elements. II. Anion Affinity and Geochemistry. Geochim. cosmochim. Acta [London] 2 (1952), 155—169; 3 (1953) 1—29

AHRENS, L. H., and S. R. TAYLOR: Spectrochemical Analysis, 2nd ed., Addison-Wesley Publ. Comp. Massachusetts, London 1961

AHRENS, L.: The Evaluation of Ionization Potentials in Geochemistry. Compilation Probl. Geochim. Izd-vo "Nauka", Moscow 1965 (Russian)

BAILER, J. C.: The Chemistry of Coordination Compounds. Reinhold, New York 1956

BARNES, H. L., H. C. HELGESON, and A. J. ELLIS: Ionization Constants in Aqueous Solutions. Geol. Soc. Amer., Mem. 97 (1966), 401—413

BETECHTIN, A. G.: On the Fundamental Law of Geochemistry. Izv. Akad. Nauk SSSR, ser. geol. 3 (1952), 6—26 (Russian)

LEUTWEIN, F., and K. DOERFFEL: Über einige Verfahren zur theoretischen Klärung geochemischer Prozesse, unter besonderer Berücksichtigung der Gitterenergie. Geologie 5 (1956), 65—100

MOELLER, T.: Inorganic Chemistry. An Advanced Textbook. John Wiley & Sons, New York 1952

MOORE, C. B.: Atomic Energy Levels as Derived from the Analyses of Optical Spectra. Circular of the Nat. Bureau of Standards 467, Government Print. Off., Washington, D. C. 1949—1958, Vol. III

PAULING, L.: Die Natur der chemischen Bindung. 2nd ed. Verlag Chemie, Weinheim/Bergstraße 1964

SZÁDECZKY-KARDOSS, E.: Das Verbindungspotential und seine Beziehungen zum Schmelzpunkt und zur Härte. Acta geol. Acad. Sci. Hung. 8 (1955), 115—161

WILLIAMS, R. J. P.: Metal Ions in Biological Systems. Biol. Rev. Cambridge philos. Soc. 28 (1953), 381—415

2.6. Size of Atoms and Ions

Starting from the working hypothesis that atoms and ions (i.e. their spheres of action) are rigid balls which contact each other in crystal lattices, their distance from each other is equal to the sum of their radii. The closest distance of approach between the centres of mass of atoms and ions in the unit cell can be determined by X-ray measurements and the atomic or ionic radii derived from them (the so-called "apparant" radii in contrast with the "true" radii which are determined by physical calculations and which are smaller in every case).

Following work by BRAGG (1920), NIGGLI (1921), and GRIMM (1922), GOLDSCHMIDT (1926) determined the first useful atomic and ionic radii, taking as a basis the data for the ionic radii of F^- of the order of 1.33 Å and O^{2-} of 1.32 Å, found by WASASTJERNA (1922).

PAULING (1927) started from quantum-mechanical considerations for his determinations (radii related to O^2 1.40 Å).

For additional information on and corrections of atomic and ionic radii see: e.g. ZACHARIASEN (1931), WYCKOFF (1948), AHRENS (1952), cf. Table 18.

LEBEDEV (1967) reports on a new system of calculating atomic and ionic radii.

2.6.1. General Laws

The radius of the structural unit of a crystal depends primarily on:

1. the atomic number,
2. the state of the particle (mainly on the degree of ionization, valence),
3. the coordination number,
4. the polarization properties.

In the periodic system (cf. Figs. 10 and 11):

1. Increase in the atomic or ionic radii within the *groups* from top to bottom, i.e. with increasing atomic number at the same charge (because with any new period, another electron shell is added).

2. Decrease in the atomic or ionic radii within the *periods* from the left to the right (increase in the number of electrons without the formation of a new shell and, at the same time, increase in the nuclear charge → causes a higher attraction to the nucleus, i. e. contraction).

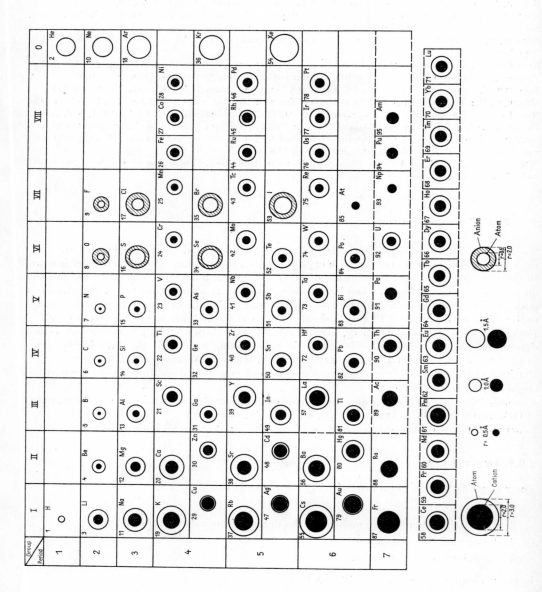

Fig. 9 The sizes of atoms and ions depend on their position in the periodic system

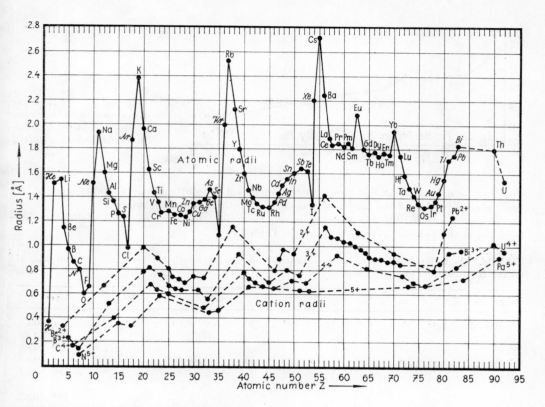

Fig. 10   Periodicity of the atomic radii (for coordination number 12) and cation radii (mainly for KZ 6). Values from PAULING (1964), AHRENS (1952) and GOLDSCHMIDT (1926); the geochemical character of the elements after GOLDSCHMIDT (1924); symbols in boldface — siderophil, in italics — chalcophil, in Roman types — lithophil, in Latin types — atmophil, cf. Section 7.2

Diagonal Rule

From 1 and 2, the fact results that two ions which are diagonally one beneath the other in the periodic system, have approximately the same atomic or ionic radius (of importance to the crystallochemical and geochemical properties of the elements, e.g. isomorphism and diadochism).

The rare earths occupy a special position. Within the lanthanide group, the ionic radii decrease with increasing atomic number (lanthanide contraction; Fig. 11). When considering the tripositively charged ions of these elements we find that the number of outer-shell electrons (3) and thus the valence remain unchanged with increasing atomic number. The added electron is incorporated at the *4f* level (*N*-shell) instead of the *5d* or *6s* level (outermost shells, cf. Table 6). Since, with increasing atomic number, the nuclear charge also increases, the outer

59

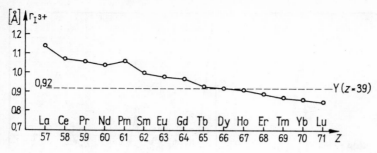

Fig. 11 Decrease in the ionic radii r_j^{3+} [Å] of the lanthanons with increasing atomic number Z (ionic radii after AHRENS, 1952)

shell can be drawn closer and closer to the nucleus due to attraction. The consequence is a successive reduction of the radius of action from lanthanide to lanthanide.

The influence exerted by this contraction on the radii of the elements which follow the lanthanides in the periodic system is shown in Table 16.

Table 16. The Influence of the Lanthanide Contraction (atomic numbers from La 57 through to Lu 71) on the Radii of the Subgroup Elements in the 5th and 6th Periods

Subgroup	I	III	IV	V	VI	VIII
5th Period	Ag^{1+} 1.34	Y^{3+} 0.92	Zr^{4+} 0.79	Nb^{5+} 0.69	Mo^{6+} 0.62	Ru^{4+} 0.67
						Rh^{3+} 0.68
						Pd^{4+} 0.65
6th Period	Au^{1+} 1.49	La^{3+} 1.18 to Lu^{3+} 0.85	Hf^{4+} 0.68	Ta^{5+} 0.68	W^{6+} 0.62	Os^{4+} 0.69
						Ir^{3+} 0.68
						Pt^{4+} 0.65

Table 17. The Atomic Radii of the Elements (from PAULING, 1964)

Symbol	Z	r_A	Symbol	Z	r_A	Symbol	Z	r_A
Ag	47	1.44	Hf	72	1.59	Rb	37	2.48
Al	13	1.43	Hg	80	1.51	Re	75	1.37
Ar	18	(1.92)	Ho	67	1.76	Rh	45	1.34
As	33	1.48	In	49	1.58	Ru	44	1.34
Au	79	1.44	Ir	77	1.36	S	16	1.27
B	5	0.98	J	53	(1.36)	Sb	51	1.66
Ba	56	2.22	K	19	2.35	Sc	21	1.62
Be	4	1.12	Kr	36	(1.98)	Se	34	1.40
Bi	83	1.78	La	57	1.87	Si	14	1.38
Br	35	(1.11)	Li	3	1.55	Sm	62	1.80
C	6	(0.86)	Lu	71	1.74	Sn	50	1.62
Ca	20	1.97	Mg	12	1.60	Sr	38	2.15
Cd	48	1.51	Mn	25	1.27	Ta	73	1.46
Ce	58	1.82	Mo	42	1.39	Tb	65	1.77
Cl	17	(0.99)	N	7	(0.8)	Tc	43	1.36
Co	27	1.25	Na	11	1.90	Te	52	1.60
Cr	24	1.28	Nb	41	1.46	Th	90	1.80
Cs	55	2.67	Nd	60	1.82	Ti	22	1.47

Table 17 continued

Symbol	Z	r_A	Symbol	Z	r_A	Symbol	Z	r_A
Cu	29	1.28	Ne	10	(1.60)	Tl	81	1.60
Dy	66	1.78	Ni	28	1.24	Tm	69	1.76
Er	68	1.76	O	8	(0.6)	U	92	1.52
Eu	63	2.08	Os	76	1.35	V	23	1.34
F	9	(0.64)	P	15	1.28	W	74	1.39
Fe	26	1.26	Pb	82	1.80	Xe	54	(2.18)
Ga	31	1.40	Pd	46	1.37	Y	39	1.80
Gd	64	1.80	Pm	61	1.83	Yb	70	1.93
Ge	32	1.44	Po	84	(1.41)	Zn	30	1.34
H	1	(0.46)	Pr	59	1.82	Zr	40	1.60
He	2	(1.45)	Pt	78	1.39			

Note: Atomic radii apply to KZ [12]
  Corrections to KZ [8]: —3%; KZ [6]: —4%; KZ [4]: —12%

Table 18. The Ionic Radii of the Elements

Symbol	Atomic Number	Charge	GOLDSCHMIDT 1926	PAULING 1927	ZACHARIASEN 1931	WYCKOFF 1948	AHRENS 1952
Ac	89	+3	—	—	—	—	1.18
Ag	47	+1	1.13	1.26	—	0.97	1.26
		+2	—	—	—	—	0.89
Al	13	+3	0.57	0.50	0.55	0.55	0.51
Am	95	+3	—	—	—	—	1.07
		+4	—	—	—	—	0.92
As	33	+3	0.69	—	—	0.69	0.58
		+5	—	0.49	—	—	0.46
		—3	—	—	—	1.91	—
At	85	+7	—	—	—	—	0.62
Au	79	+1	—	1.37	—	—	1.37
		+3	—	—	—	—	0.85
B	5	+3	—	0.20	0.24	—	0.23
Ba	56	+2	1.43	1.35	1.31	1.38	1.34
Be	4	+2	0.34	0.31	0.39	0.30	0.35
Bi	83	+3	—	—	—	1.20	0.96
		+5	—	0.74	—	—	0.74
		—3	—	—	—	2.13	—
Br	35	—1	1.96	1.95	—	1.96	—
		+5	—	—	—	—	0.47
		+7	—	0.39	—	0.39	0.39
C	6	+4	0.2	0.15	0.19	0.15	0.16
		—4	—	—	—	2.60	—
Ca	20	+2	1.06	0.99	0.98	1.05	0.99
Cd	48	+2	1.03	0.97	—	0.99	0.97
Ce	58	+3	1.18	—	—	1.10	1.07
		+4	1.02	1.01	0.89	1.01	0.94
Cl	17	—1	1.81	1.81	1.81	1.80	—

Table 18 continued

Symbol	Atomic Number	Charge	GOLDSCHMIDT 1926	PAULING 1927	ZACHARIASEN 1931	WYCKOFF 1948	AHRENS 1952
		+5	—	—	—	—	0.34
		+7	—	0.26	—	0.26	0.27
Co	27	+2	0.82	0.72	—	0.78	0.72
		+3	0.64	—	—	0.65	0.63
Cr	24	+2	≈0.83	—	—	—	—
		+3	0.64	—	—	0.70	0.63
		+6	≈0.35	0.52	—	0.52	0.52
Cs	55	+1	1.65	1.69	1.67	1.70	1.67
Cu	29	+1	—	0.96	—	0.58	0.96
		+2	—	—	—	—	0.72
Dy	66	+3	1.07	—	—	—	0.92
Er	68	+3	1.04	—	—	—	0.89
Eu	63	+2	1.24	—	—	—	—
		+3	1.13	—	—	—	0.98
F	9	−1	1.33	1.36	1.33	1.33	—
		+7	—	0.07	—	0.07	0.08
Fe	26	+2	0.82	0.80	—	0.80	0.74
		+3	0.67	—	—	0.67	0.64
Fr	87	+1	—	—	—	—	1.80
Ga	31	+3	0.62	0.62	—	0.65	0.62
Gd	64	+3	1.11	—	—	—	0.97
Ge	32	+2	0.9	—	—	0.65	0.73
		+4	0.44	0.53	—	0.55	0.53
H	1	+1	1.54	2.08	1.36	1.27	—
Hf	72	+4	0.84	—	—	—	0.78
		+1	—	—	—	0.72	—
Hg	80	+2	1.12	1.10	—	0.66	1.10
Ho	67	+3	1.05	—	—	—	0.91
I	53	−1	2.20	2.16	2.19	2.20	—
		+5	0.94	—	—	—	0.62
		+7	—	0.50	—	—	0.50
In	49	+3	0.92	0.81	—	0.95	0.81
Ir	77	+4	0.66	0.64	—	0.65	0.68
K	19	+1	1.33	1.33	1.33	1.33	1.33
La	57	+3	1.22	1.15	1.06	1.15	1.14
Li	3	+1	0.78	0.60	0.68	0.70	0.68
Lu	71	+3	0.99	—	—	—	0.85
Mg	12	+2	0.78	0.65	0.71	0.75	0.66
Mn	25	+2	0.91	0.80	—	0.83	0.80
		+3	0.70	—	—	—	0.66
		+4	0.52	0.50	—	0.52	0.60
		+7	—	0.46	—	—	0.46

Table 18 continued

Symbol	Atomic Number	Charge	GOLDSCHMIDT 1926	PAULING 1927	ZACHARIASEN 1931	WYCKOFF 1948	AHRENS 1952
Mo	42	+4	0.68	0.66	—	0.68	0.70
		+6	—	0.62	—	0.65	0.62
N	7	+3	—	—	—	—	0.16
		+5	0.15	0.11	—	—	0.13
		−3	—	—	—	1.48	—
NH$_4$	—	+1	1.43	—	—	—	—
Na	11	+1	0.98	0.95	0.98	1.00	0.97
Nb	41	+4	0.69	0.67	—	0.67	0.74
		+5	0.69	0.70	—	—	0.69
Nd	60	+3	1.15	—	—	1.07	1.04
Ni	28	+2	0.78	0.69	—	0.74	0.69
Np	93	+3	—	—	—	—	1.10
		+4	—	—	—	—	0.95
		+7	—	—	—	—	0.71
O	8	−2	1.32	1.40	1.40	1.35	—
		+6	—	0.09	—	—	0.10
Os	76	+4	0.67	0.65	—	0.65	—
		+6	—	—	—	—	0.69
P	15	+3	—	—	—	—	0.44
		+5	≈0.35	0.34	—	0.34	0.35
		+3	—	—	—	1.86	—
Pa	91	+3	—	—	—	—	1.13
		+4	—	—	—	—	0.98
		+5	—	—	—	—	0.89
Pb	82	+2	1.32	1.21	—	1.18	1.20
		+4	0.84	0.84	—	0.70	0.84
Pd	46	+2	—	—	—	—	0.80
		+4	—	—	—	—	0.65
Pm	61	+3	—	—	—	—	1.06
Po	84	+6	—	—	—	—	0.67
Pr	59	+3	1.16	—	—	1.09	1.06
		+4	1.00	0.92	—	—	0.92
Pt	78	+2	—	—	—	0.52	0.80
		+4	—	—	—	0.55	0.65
Pu	94	+3	—	—	—	—	1.08
		+4	—	—	—	—	0.93
Ra	88	+2	1.52	—	—	1.42	1.43
Rb	37	+1	1.49	1.48	1.48	1.52	1.47
Re	75	+4	—	—	—	—	0.72
		+7	—	—	—	—	0.56
Rh	45	+3	0.68	—	—	0.75	0.68
Ru	44	+4	0.65	0.63	—	0.65	0.76
S	16	−2	1.74	1.84	1.85	1.82	—
		+4	—	—	—	—	0.37
		+6	0.34	0.29	—	—	0.30
Sb	51	+3	0.90	—	—	0.90	0.76

Table 18 continued

Symbol	Atomic Number	Charge	GOLDSCHMIDT 1926	PAULING 1927	ZACHARIASEN 1931	WYCKOFF 1948	AHRENS 1952
		+5	—	0.62	—	—	0.62
		+3	—	—	—	2.08	—
Sc	21	+3	0.83	—	—	0.83	0.81
Se	34	—2	1.91	1.98	1.96	1.93	—
		+3	0.83	0.81	0.78	—	—
		+4	—	—	—	—	0.50
		+6	≈0.35	0.42	—	—	0.42
Si	14	+4	0.39	0.41	0.44	0.40	0.42
Sm	62	+3	1.13	—	—	—	1.00
Sn	50	+2	—	—	—	1.02	0.93
		+4	0.74	0.71	—	0.65	0.71
Sr	38	+2	1.27	1.13	1.15	1.18	1.12
Ta	73	+5	0.68	—	—	—	0.68
Tb	65	+3	1.09	—	—	—	0.93
		+4	0.89	—	—	—	0.81
Tc	43	+7	—	—	—	—	0.56
Te	52	—2	2.11	2.21	2.18	2.12	—
		+4	0.89	0.81	—	0.84	0.70
		+6	—	0.56	—	—	0.56
Th	90	+4	1.10	1.02	—	1.10	1.02
Ti	22	+2	0.80	—	—	0.76	—
		+3	0.69	—	—	0.70	0.76
		+4	0.64	0.68	0.62	0.60	0.68
Tl	81	+1	1.49	1.44	—	1.50	1.47
		+3	1.05	0.95	—	—	0.95
Tm	69	+3	1.04	—	—	—	0.87
U	92	+4	1.05	0.97	—	1.05	0.97
		+6	—	—	—	—	0.80
V	23	+2	0.72	—	—	—	0.88
		+3	0.65	—	—	0.75	0.74
		+4	0.61	0.59	—	0.57	0.63
		+5	0.4	0.59	—	—	0.59
W	74	+4	0.68	0.66	—	0.68	0.70
		+6	—	—	—	0.65	0.62
Y	39	+3	1.06	0.93	0.93	0.95	0.92
Yb	70	+3	1,00	—	—	—	0.86
Zn	30	+2	0.83	0.74	—	0.83	0.74
Zr	40	+4	0.87	0.80	0.79	0.80	0.79

Note:

Ionic radii related to KZ [6]
Corrections for KZ [4]: — 6%
 KZ [8]: + 3%
 KZ [12]: +12%

2.6.2. Valence

If an ion occurs in different valences, its radius increases with increasing negative charge, whereas it decreases with increasing positive charge (cf. Fig. 12).

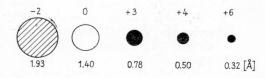

Fig. 12   The atomic radius and the ionic radii of Se as a function of the charge. The values comply with the latest determination

Valence

It determines the number of other atoms with which one atom of a given element can combine;
it depends on the number of outer-shell electrons (valence electrons) of the given original atom.
Table 19 gives information about the possible maximum valences of the various elements. This Table further shows that the chemical properties and especially the valence of a given element exhibit a certain relation to the difference between the atomic number of this element and that of the nearest noble gas. The atoms neighbouring noble gases, when entering a combination, tend to acquire or to give away electrons to reach the electron number of these noble gases and thus also a stable noble gas configuration. This applies especially to the atoms of the main groups. The elements of a few subgroups remoter from the noble gases, show electron numbers varying from those of the noble gases (always caused by giving up electrons), whereas with 28 or 46 electrons, a certain stability seems to be ensured (Fig. 13).

On the definition of valence

The concept of valence is differently defined in various publications. Explanations of theoretical valence concepts were given above all by SEEL (1954); he distinguished between the following terms:

1. *Stoichiometric valence* (=formal valence, REMY, 1955). The stoichiometric valence of a chemical element is the number which indicates the quantity of individual atoms or groups of atoms recognised as monovalent, with which one atom of the element under consideration can combine or which it can replace in other combinations (indication of the different valences by Roman numerals)

$$\text{Valence} = \frac{\text{atomic weight}}{\text{equivalent weight}}$$

or more general (including combinations)

$$= \frac{\text{formular weight}}{\text{equivalent weight}}$$

2. *Ionic valence* (= electrochemical valence, REMY, 1955). Magnitude of the electric charge (measured in elementary charges e) of an ion.

$$\text{Ionic valence} = \frac{\text{ionic weight}}{\text{ionic equivalent weight}} = \text{ionic charge number}$$

It is characterised by a number with positive or negative sign (cations or anions); e.g. the +trivalent iron (III)ion = Fe^{3+}.

3. *Oxidation number.* Stage of oxidation and the oxidation number should be considered equivalent; used especially for oxidation and reduction processes.

As clues to the determination of the stage of oxidation, remember:

Metals, boron and silicon are marked by positive oxidation numbers,
Fluorine always has the oxidation number —1,
Hydrogen +1,
Oxygen —2,
Halogens and chalcogens —1 and —2, respectively.
The sum of the oxidation numbers is equal to the charge of the system.
The following is valid for atom-ions:
The stage of oxidation is equal to the ionic valence,
the oxidation number (sign neglected) is equal to the stoichiometric valence.

4. *Covalence* (number of atomic bonds which are caused by one atom, characterized by valence line formulas). Frequently the covalence is numerically equal to the stoichiometric valence.

If the oxidation number (apart from the charge) or the covalence is numerically equal to the stoichiometric valence, the word "valent" can generally be used for denoting the stage of oxidation or covalence.

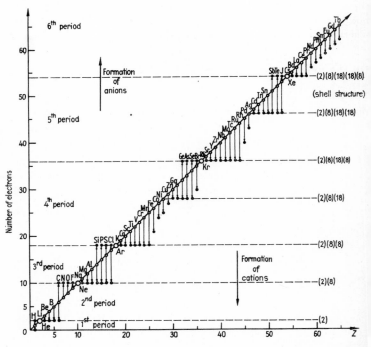

Fig. 13. Maximum valences of the elements and their tendency to assume the electronic configuration of a noble gas (after KOSSEL, 1916, and PASCAL, 1956)

Table 19. Maximum Valences of Atoms and Ions as a Function of Group Number (Exceptions: H only +1 and −1, O only −2, F only −1, Br max. +5, Fe max. +6, Cu and Ag max. +2, Au max. +3) and Some Important Properties (from REMY, 1955)

IV	V	VI	VII	VIII	I	II	III
+4	+5	+6	+7	0	+1	+2	+3
−4	−3	−2	−1				
			(H) 1	He 2	Li 3	Be 4	B 5
C 6	N 7	(O) 8	(F) 9	Ne 10	Na 11	Mg 12	Al 13
Si 14	P 15	S 16	Cl 17	Ar 18	K 19	Ca 20	Sc 21
+4	+5	+6	+7	+8 … …	(+1)	+2	+3
(slightly varying valences)							
Ti 22	V 23	Cr 24	Mn 25	(Fe) 26 · Co 27 · Ni 28	(Cu) 29	Zn 30	Ga 31
Ge 32	As 33	Se 34	(Br) 35	Kr 36	Rb 37	Sr 38	Y 39
Zr 40	Nb 41	Mo 42	Tc 43	Ru 44 · Rh 45 · Pd 46	(Ag) 47	Cd 48	In 49
Sn 50	Sb 51	Te 52	J 53	X 54	Cs 55	Ba 56	La 57…
Hf 72	Ta 73	W 74	Re 75	Os 76 · Ir 77 · Pt 78	(Au) 79	Hg 80	Tl 81
Pb 82	Bi 83	Po 84	At 85	Rn 86	Fr 87	Ra 88	Ac 89
Th 90	Pa 91	U 92	(transuranic elements →)				

Properties:

volatile hydrogen compounds	neither volatile nor saline hydrogen compounds	solid, saline hydrogen compounds
acidifying →	—	basifying ←
non-met. character ↗	metals	
ions are: diamagnetic	part of ions paramagnetic	diamagnetic

(Noble gas column VIII: "—" for hydrogen compounds, diamagnetic)

Bibliography

CLARK, S. P. jr.: Conversion Factors, Numerical Constants, Atomic Constants. Geol. Soc. Amer., Mem. *97* (1966), 579—583

LEBEDEV, V. I.: A New System of Atomic and Ionic Radii. Dokl. Akad. Nauk SSSR *176* (1967) 6, 1407—1410 (Russian)

PASCAL, P. et al: Nouveau Traité de Chimie Minérale. Part 1; Masson et Cie, Paris 1956

SEEL, F.: Valenztheoretische Begriffe, Angew. Chem. *66* (1954), 19, 581—624

POVARENNYCH, A. S.: Problems of Determining the Effective Ionic Radii. Dokl. Akad. Nauk, SSSR *109* (1956), 1167—1170 (Russian)

Group of authors: Tables of Interatomic Distances and Configuration in Molecules and Ions. The Chemical Society, Spec. Publ. No. 18, London 1965

2.6.3. Polarization

Small and highly charged cations exert attractive forces on the electrons of adjacent anions and deform their electron shells displacing them towards the cation; of particular importance to heteropolar bonds.

Quantitatively, measurement of the polarization energy is conditional, therefore, only purely qualitative data are usually given.

The *polarizing effect* increases with increasing positive charge and with decreasing radius of an ion, i.e., with increasing ion potential.

The *polarizability* increases with the ionic radius and, in the case of anions, with the number of the valence electrons gained.

For the periodic system the following applies:

			He	Li^+	Be^{2+}	B^{3+}	C^{4+}
			O^{2-}	F^-	Ne	Na^+	Mg^{2+}	Al^{3+}	Si^{4+}
			S^{2-}	Cl^-	Ar	K^+	Ca^{2+}	Sc^{3+}	Ti^{4+}
			Se^{2-}	Br^-	Kr	Rb^+	Sr^{2+}	Y^{3+}	Zr^{4+}
			Te^{2-}	J^-	X	Cs^+	Ba^{2+}	La^{3+}	Ce^{4+}

decreasing polarizability / increasing polarizing effect (decreasing radii) ↑

increasing polarizing effect (decreasing radii) →

decreasing polarizability ←

Cations with noble gas character show a lower polarizing effect than ions with a radius of the same magnitude which do not possess a noble gas shell.

Results of polarization:

1. Change in the characteristic symmetry of the ion and, hence, of its "spherical" shape.
2. Reduction of the distance between ions so that the ions draw closer together → increase in the lattice energy.
3. Partial discharge of the ions and transition from purely ionic bonds to bonds with portions of homopolar linkages in the lattice. The proportion of heteropolar bonds can only be assessed by means of the difference in electronegativity of the individual kinds of bonds (see Section 2.8., Fig. 16).
4. Reduction of the coordination number by intense polarization effect.
5. Colour intensity increases with increasing polarization.
6. Connection between polarization and refraction of light.

Bibliography

BETECHTIN, A. G.: On the Fundamental Law of Geochemistry. Izv. Akad. Nauk SSSR, ser. geol. *3* (1952), 6—26 (Russian)

DEVORE, G. W.: Crystal Growth and the Distribution of Elements. J. Geology *63* (1955), 471—494

DEVORE, G. W.: The Role of Adsorption in the Fractionation and Distribution of Elements. J. Geology *63* (1955), 159—190

RAMBERG, H.: Chemical Bonds and the Distribution of Cations in Silicates. J. Geology *60* (1952), 331—355

General information can be looked up in textbooks on crystal chemistry.

2.6.4. Ionic Potential Φ

According to Cartledge, this is a measure of the polarizing and related properties of ions.

$$\Phi = \frac{\text{valence}}{\text{ionic radius}} = z/r \; .$$

Φ is the characteristic to which no defined measure (in the physical sense) is assigned; this also applies to the potentials derived from it, e.g. atomic, anionic, cationic and combination potentials (SZÁDECKY-KARDOSS, 1955 and 1958; LEUTWEIN and DOERFFEL, 1956).

For connections between Φ and EK values, respectively the electronegativities etc., see SZÁDECZKY-KARDOSS (1955).

Table 20. Properties of Cations as a Function of Their Ionic Potential [Φ]

Φ	Properties
roughly equal	frequently geochemically associated
$\leqq 4$—5	lose electrons; basifier; soluble at normal pH
between	
4—5 and 10—12	amphoteric; tend to form insoluble oxides and hydroxides
$\geqq 10$—12	gain electrons; acidifier; tend to form anion complexes
high	more intensively polarizing; more hydratised

Bibliography

AHRENS, L. H.: The Use of the Ionization Potentials, I. Ionic Radii of the Elements. Geochim. cosmochim. Acta [London] *2* (1952), 155—169

BAILER, J. C.: The Chemistry of Coordination Compounds. Reinhold, New York 1956

CARTLEDGE, G. H.: Studies on the Periodic System. I. The Ionic Potential as a Periodic Function. II. The Ionic Potential and Related Properties. J. Amer. chem. Soc. *50* (1928), 2855—2863, 2863 to 2872

CARTLEDGE, G. H.: The Correlation of Thermochemical Data by the Ionic Potential. J. Physic. Chem. *55* (1951), 248—256

CARTLEDGE, G. H.: Relation Between Ionization Potential and Ionic Potential. J. Amer. Chem. Soc. *52* (1930), 3076

LEUTWEIN, F., and K. DOERFFEL: Über einige Verfahren zur theoretischen Klärung geochemischer Prozesse, unter besonderer Berücksichtigung der Gitterenergie. Geologie *5* (1956), 65—100

PUFFE, E.: Mineralfolgen auf den Lagerstätten und Atombau. Fortschr. Mineralog. *28* (1951), 64—68
RINGWOOD, A. E.: The Principles Governing Trace Element Distribution during Magmatic Crystalli-
zation, II. The Role of Complex Formation. Geochim. cosmochim. Acta [London] *7* (1955), 242—254
SZÁDECZKY-KARDOSS, E.: Das Verbindungspotential und seine Beziehungen zum Schmelzpunkt und
zur Härte. Acta geol. Acad. Sci. Hung. *3* (1955), 115—161
SZÁDECZKY-KARDOSS, E.: Bemerkungen zu einer Arbeit von F. LEUTWEIN und K. DOERFFEL. Acta geol.
Acad. Sci. Hung. *5* (1958), 359—380

2.6.5. Coordination

Coordination or the coordination number is equal to the number of the nearest, practically
equidistant neighbours (ligands: atoms, ions or molecules) of a central atom or central ion
(commonly a metal); in accordance with the geometric conditions → formation of different
coordination polyhedra (Table 21). Pauling's rule: e.g. "electrostatic valence rule" or parsi-
mony rule; see textbooks on crystal chemistry.
The influence of the coordination on the size of atoms and ions: The higher the coordination
number, the larger the distances between the atoms or ions in the crystal lattices (Fig. 14).
Basic crystallochemical law by GOLDSCHMIDT (1927): "The structure of a crystal is determined
by the ratio of the numbers, the ratio of the sizes, and the properties of polarization of its
structural units. Structural units are atoms (or ions) and groups of atoms".

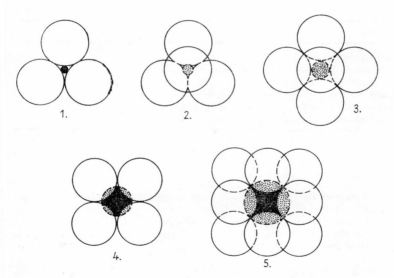

Fig. 14 The most frequent coordination groups. The central atom or ion (usually a cation) in the
arrangement denoted 1. is directly bonded to three neighbours, i.e. coordination of three; in 2., the
group forms a coordination of four (tetrahedral coordination); in 3., a coordination of six (octahedral
coordination); in 4., a coordination of eight (cubic coordination); in 5., a coordination of 12 (close-
packed structure)

Table 21. Relations between Radii Quotient and Coördination

Radii Quotient (r_{cation}/r_{anion})	Coordination Number	Coordination Polyhedra
1.000	[12]	cubo-octahedron (close spherical packing)
0.732—1.000	[8]	cube (hexahedron)
0.645—0.732	[8]	double disphenoid[1])
0.414—0.645	[6]	octahedron
0.225—0.414	[4]	tetrahedron
0.155—0.225	[3]	equilateral triangle

[1]) cf. KLEBER (1963)

2.7. Types of Bonds

The following types of bonds bind the structural units of crystals (atoms, ions, molecules) together:

1. Heteropolar bond (polar, electrovalent or ionic bond, frequently occurring, in over 80% of all minerals, e.g. halogenides, oxides).
2. Homopolar bond (covalent or atomic bond, e.g, diamond).
3. Metallic bond (metal atoms, e.g. Cu).
4. van der Waals' bond (molecular bond, e.g. inert gases or covalent-bonded molecules of non-metals).
5. Hydroxyl bond [together with van der Waals' bond in hydroxides, e.g. gibbsite — $Al(OH)_3$].

Several types of bonds are found almost exclusively, side by side in one crystal lattice, *combined bonds:*
homodesmic lattices, one type of bond is prevalent (e.g. in the diamond);
heterodesmic lattices, several types of bond exist side by side (e.g. in graphite).
Pure types of bond are given in the least number of cases. Mostly, mixed or transitional bonds are given, i.e. combinations and transitional stages of the above types (see Fig. 15).

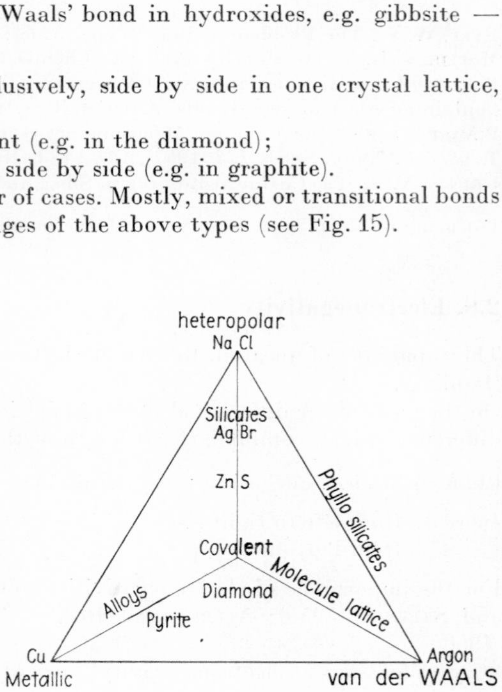

Fig. 15 Diagrammatic representation of the four most important types of bond with examples

As mentioned previously, with increasing polarization effect, the heteropolar bond changes more and more into a covalent bond. However, the reverse is also possible, that is to say, a dipole moment forms in covalent-bonded atoms, one atom attracts a negative charge of another constituent atom. The greater the capability of an atom (in comparison with another one) to attract electrons (electronegativity), the more the heteropolar proportion of the bonds increases. An intermediate state of bonding appears. This phenomenon is called resonance or mesomerism (see textbooks by KLEBER, 1963; PAULING, 1964).

Table 22. Properties of Crystals as a Function of the Type of Bond

Properties	Type of Bond		
	heteropolar (NaCl)	homopolar (diamond)	metallic (Cu)
Electric conductivity	low (good electrolytic conductivity)	none (isolators)	good
Melting and boiling points	high	increase with coordination	vary
Transparency	colourless to coloured-transparent	transparent	opaque (metallic lustre)
Particularly depending on bonding force	compressibility, hardness, capability of being torn		thermal expansion, compressibility, plastic deformation

Bibliography

FYFE, W. S.: The Problem of Bond Type. Amer. Mineralogist. *39* (1954), 991
MULLIKEN, R. S.: Overlap Integrals and Chemical Bonding. J. Amer. Chem. Soc. *72* (1950), 4493
PAULING, L., and N. L. HUGGINS: Covalent Radii of Atoms and Interatomic Distances in Crystals Containing Electron-pair Bonds. Z. Kristallogr. *87* (1944), 205—233
POVARENNYCH, A. S.: Function Relations between the Hardness of Minerals and the Chemical Bond Type. Izv. Nauk. SSSR *112* (1957), 181—183 (Russian)
URUSOV, V. S.: The Chemical Binding in Silica and Silicates. Geokhimiya *4* (1967), 399—412 (Russian)

For general information see textbooks of crystallochemistry.

2.8. Electronegativity

The capability of an atom to attract electrons in covalent bond is called electronegativity $\varkappa$ (Pauling).
On the basis of the different electronegativities of two elements ($\varkappa_A - \varkappa_B$ = electronegativity difference) which combine, the strength of this bond can be assessed.

Changes of electronegativities in the periodic system:

Increase from left to right $\rightarrow$
        from bottom to top $\uparrow$

For the importance of electronegativities to chemistry see: PAULING (1962, 1964); PRITCHARD and SKINNER (1955); ALLRED and ROCHOV (1958); SANDERSON (1958); GORDY and THOMAS (1956).
"Crystallochemical electronegativity" according to KAPUSTINSKIĬ (1949); DOERFFEL (1956).

Table 23. Relations between Electronegativity Difference $(x_A - x_B)$ and the Type and Properties of Given Bonds (also applies to the substitution of elements, cf. VENDEL, 1955)

Electronegativity Difference $x_A - x_B$	Properties of the Bond
increasing (>1.9)	strength increases, a great amount of energy is released in bonding, increase of the partial ionic character (heteropolar proportion of bonds is prevalent)
small (<1.9)	weak and not very stable bond, small amount of heat of formation, primarily covalent bond (Fig. 16)

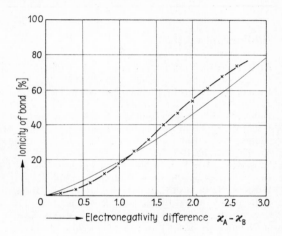

Fig. 16 Fractional ionic character of bonds as a function of the electronegativity difference of combining elements $(x_A - x_B)$. Continuous curve according to PAULING (1962); dotted curve according to HANNAY and SMYTH (1946)

Table 24. Electronegativities (mainly from PAULING, 1964)
(These values are in the main based on the experimental heats of formation of combinations)

Ac	1.1	Er	(1.2)	Mo^{4+}	1.6	Sb^{5+}	2.1
Ag	1.9	Es	1.3	Mo^{6+}	2.1	Sc	1.3
Al	1.5	Eu	(1.1)	N	1.3	Se	2.4
Am	1.3	F	4.0	Na	0.9	Si	1.8
As	2.0	Fe^{2+}	1.8	Nb	1.6	Sm	(1.1)
At	2.2	Fe^{3+}	1.9	Nd	(1.1)	Sn^{2+}	1.8
Au^{1+}	2.3	Fm	1.3	Ni	1.8	Sn^{4+}	1.9
Au^{3+}	2.9	Fr	0.7	No	1.3	Sr	1.0
B	2.0	Ga	1.6	Np	1.3	Ta	1.5
Ba	0.85	Gd	(1.1)	O	3.5	Tb	(1.2)
Be	1.5	Ge	1.8	Os	2.2	Tc	1.9
Bi	1.9	H	2.1	P	2.1	Te	2.1
Bk	1.3	Hf	1.3	Pa	1.5	Th	1.3
Br	2.8	Hg^{1+}	1.8	Pb^{2+}	1.6	Ti^{3+}	1.5
C	2.5	Hg^{2+}	1.9	Pb^{4+}	1.8	Ti^{4+}	1.6
Ca	1.0	Ho	(1.2)	Pd	2.2	Tl^{1+}	1.5
Cd	1.7	I	2.5	Pm	(1.1)	Tl^{3+}	1.9
Ce^{3+}	1.05	In	1.7	Po	2.0	Tm	1.2

73

Table 24 continued

Cf	1.3	Ir	2.2	Pr	1.1	U^{4+}	1.7
Cl	3.0	K	0.8	Pt	2.2	V^{3+}	1.35
Cm	1.3	La	1.1	Pu	1.3	V^{4+}	1.65
Co	1.8	Li	1.0	Ra	0.9	V^{5+}	1.8
Cr^{2+}	1.5	Lu	(1.2)	Rb	0.8	W^{4+}	1.6
Cr^{3+}	1.6	Md	1.3	Re	1.9	W^{6+}	2.1
Cr^{6+}	2.1	Mg	1.2	Rh	2.2	Y	1.3
Cs	0.7	Mn^{2+}	1.4	Ru	2.2	Yb	(1.2)
Cu^{1+}	1.9	Mn^{3+}	1.5	S	2.5	Zn	1.6
Cu^{2+}	2.0	Mn^{7+}	2.3	Sb^{3+}	1.8	Zr	1.4
Dy	(1.2)						

Bibliography

ALLRED, A. L., and E. G. ROCHOV: A Scale of Electronegativity Based on Electrostatic Force. J. Inorg. Nucl. Chem. 5 (1958), 264—268

DRAGO, R. S.: On the Electronegativities of the Group IV Elements. J. Inorg. Nucl. Chem. 15 (1960), 237

DOERFFEL, K.: Das System der Ionenenergien. Freiberger Forschungsheft, C 20, Berlin 1956

FYFE, W. S.: Isomorphism and Bond Type. Amer. Mineralogist 36 (1951), 538—542

GORDY, W.: A Relation between Characteristic Bond Constants and Electronegativities of the Bonded Atoms. Physic. Rev. 69 (1946), 130, 604

HAISSINSKY, M.: Échelle des Electronégativités de Pauling et Chaleurs de Formation des Composés Inorganiques. J. Physique Radium 7 (1946), 7—11

HANNAY, N. B., and C. P. SMYTH: The Dipole Moment of Hydrogen Fluoride and the Ionic Character of Bonds. J. Amer. Chem. Soc. 48 (1946), 171—173

LAKATOS, B., J. BOHUS and G. Y. MEDGYESI: A New Way for the Calculation of the Degree of Polarity of Chemical Bonds: II and IV. Acta chim. Acad. Sci. Hung. 20 (1959), 1, and 21 (1959), 293

LITTLE, E. J., and M. M. JONES: A Complete Table of Electronegativities. J. chem. Educat. 37 (1960), 231

MULLIKEN, R. S.: A New Electronegativity Scale; together with Data on Valence States and on Valence Ionization Potentials and Electron Affinities. J. Chem. Physics 2 (1934), 782

POVARENNYCH, A. S.: Quantitative Determination of the Chemical Kind of Bond in Minerals. Dokl. Akad. Nauk. SSSR 109 (1956), 993—996 (Russian)

PRITCHARD, H. O., and H. A. SKINNER: The Concept of Electronegativity. Chem. Reviews 55 (1955), 745—786

RINGWOOD, R. E.: The Principles Governing Trace Element Distribution during Magmatic Crystallization. I. The Influence of Electronegativity; II. The Role of Complex Formation. Geochim. cosmochim. Acta [London] 7 (1955), 189—202, 242—254

SANDERSON, R. T.: Alternations in Electronegativity and the Pritchard and Skinner Review. J. Inorg. Nucl. Chem. 7 (1958), 157—158

VENDEL, M.: Die Substituierbarkeit der Ionen und Atome vom geochemischen Gesichtspunkt. Acta geol. Sci. Hung. 3 (1955), 245—300

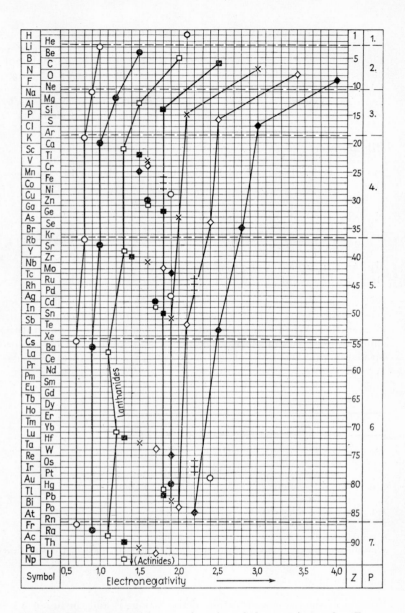

Fig. 17 Electronegativities as a function of the atomic number Z (values from PAULING, 1964)

The elements of the main groups are connected to each other:

○ 1st group   × 5th group
● 2nd group   ◇ 6th group
□ 3rd group   ◆ 7th group
▨ 4th group   + 8th group

2.9. Physical Constants of Atoms and Combinations

The geochemical migration of elements is determined by a number of physical constants. The most important of them are density (d), melting point (Mp), boiling point (Bp), solubility in water, and the solubility product. The element migration in magmas and in hydrothermal and surface waters is governed by these factors. Table 25 shows a selection of these properties for a series of geochemical elements and combinations of elements.

Bibliography

CLARK, S. P., jr.: Handbook of Physical Constants. Memoir 97, The Geol. Soc. America, New York 1966

Explanations of Table 25 (page 77)

d density (at room temperature) or weight per litre of gases in g/l (at 0 °C and 760 mm Hg).

Mp melting point ⎱ in °C at 760 mm Hg
Bp boiling point ⎰ (or at the pressures given in brackets).

dec. = decomposes; subl. = sublimates. When these abbreviations are given before the temperature figure = the element or the combination decomposes or sublimates at the stated temperature. When the abbreviation is given after the temperature — the element or the combination melts or boils at the given temperature and decomposes at the same time. If, for instance, H_2O escapes at one of the given temperatures, $-H_2O$ is written before the temperature figure.

Solubility

a	Meaning	x
∞	miscible in any proportion	
e.s.	easily soluble	<10
s.	soluble	10—30
h.s.	hardly soluble	30—10,000
i.	insoluble	>10,000
Int.	interaction with H_2O	

a symbol adopted for the degree of solubility
x number of parts by mass of the solvent per part by mass of the substance to be solved

Solubility product

Product of the ion concentration of the cation C^+ and the anion A^- (in g-ion/l) of an electrolyte CA of low solubility which is in equilibrium with the solid phase CA.

$$L_{CA} = [C^+] \cdot [A^-],$$

$$\sqrt{L_{CA}} = s = c \text{ (molar concentration of CA)},$$

$$s = \text{solubility of CA in water (in mol/l)}.$$

At a composition of $C_m A_n$, the following applies to an electrolyte: $(C^+)^m \cdot (A^-)^n = L_{C_m A_n}$.

Table 25. Physical Constants of the Elements and of a Few Anorganic Compounds (Explanation See Page 76)

Symbol or Formula	d	Mp	Bp	Solubility [g] in 100 g of Water at 20 °C	at 100 °C	Solubility Product
Ag	10.5	960.8	1,950	i.	i.	
Ag_2CO_3	6.08	218 dec.	—	$3.2 \cdot 10^{-3}$	$5 \cdot 10^{-2}$	$6.2 \cdot 10^{-12}$ (25 °C)
AgCl	5.56	455	1,550	$1.5 \cdot 10^{-4}$	$2.1 \cdot 10^{-3}$	$1.6 \cdot 10^{-10}$ (25 °C)
$AgNO_3$	4.35	210 (212)	444 dec.	218	910 (952)	
Ag_2S	7.2—7.3	825	dec.	$1.4 \cdot 10^{-5}$	—	$1.0 \cdot 10^{-51}$ ($1.6 \cdot 10^{-49}$) (18 °C)
Ag_2SO_4	5.46	660 (652)	1,085 dec.	0.79	1.41	$7.7 \cdot 10^{-5}$ (25 °C)
Al	2.70	659.7	2,057	i.	i.	
$AlCl_3$	2.47	190 (2.5 atm)	180.7 subl.	46	49 Int.	
AlF_3	3.07	1,040	1,260	s.	s.	
Al_2O_3	3.5—4.1	2,050	2,980 (3,500 ± 200)	$1 \cdot 10^{-4}$	i.	
$Al(OH)_3$	2.42	$-2 H_2O$ 150	$-2 H_2O$ 300	$1 \cdot 10^{-4}$	i.	$1.9 \cdot 10^{-33}$ ($3.7 \cdot 10^{-15}$) (25 °C)
$AlPO_4$	2.57	>1,500	—	i.	i.	
$Al_2(SO_4)_3$	2.71	770 dec.	—	36.2	89.1 (98.1)	
Ar	1.78 g (0 °C)	—189.2	—185.7	5.6 cm³ (0 °C)	3.01 cm³ (50 °C)	
As_4, crist.	5.73 (14 °C)	814 (36 atm)	subl. 615	i.	i.	
As_4, amorphous	4.7			i.	i.	
$AsCl_3$	2.16	—18	130.2	Int.	Int.	
As_2O_3	3.3—4.1	—	subl. (193)	2.0_4 (25 °C)	11.46	
As_2S_3	3.43	310 (300)	707	$5 \cdot 10^{-5}$	h.s.	
As_4S_4	α 3.5 / β 3.25	transformation fr. α to β (transformation at 267 β 307	565	i.	i.	
Au	19.3	1,063	2,600	i.	i.	
$AuCl_3$	3.9	dec. >200 (dec. 254)	(subl. 265)	68	e.s.	
B	2.34 / 2.37 amorphous	2,300	2,550	i.	i.	

Table 25 continued

Symbol or Formula	d	Mp	Bp	Solubility [g] in 100 g of Water		Solubility Product
				at 20°C	at 100°C	
B₄C	2.5	2,550 (2,450)	>3,500	h.s. (i.)	h.s. (i.)	
BCl₃	1.434 (0°C)	—108 (107)	18 (12.5)	Int.	Int.	
BN	2.34	≈3,000 (Dr.)	subl.	i.	i.	
	(2.20)	(3,000 subl.)				
BF₃	2.99 g/l	—127	—101	Int.	Int.	
B₂O₃	1.84	≈600	1,860	1.1 (0°C) Int.	15.7 Int.	
(glassy-colloidal)		(about 450)	1,500		s.	
(B₂O₃), crist.	(1.805)	(294 + 1)	—	s.		
Ba	3.5	725	1,140	Int.	Int.	
BaCO₃ α	4.3—4.4	1,740 (90 atm)	dec. >900	2.2 · 10⁻³	6.5 · 10⁻³	8.0 · 10⁻⁹ (25°C)
BaCO₃ β	—	(tranformation at 982 into α)	—	2.2 · 10⁻² (18°C)	65 · 10⁻⁴	
BaCO₃ γ	4.43	tranformation at 811 into β	1,450	2.2 · 10⁻³	65 · 10⁻⁴	
BaCl₂ (cub.)	—	962	1,560	—		
BaCl₂ (monoclinic)	3.86	958	1,560	35.7	58.8	
		(transform. 925)				
BaSO₄	4.5	1,580	(transform. 1,149)	2.3 · 10⁻⁴	3.9 · 10⁻⁴	
Be	1.85	1,278 ±	2,970	i.	s.	
BeCl₂	1.90	≈450	520	e.s.	e.s.	
BeF₂	2.01 (1.99)	800	—	e.s.	e.s.	
BeO	3.03 (3.01)	2,520 (2,530 ± 30)	3,900	2 · 10⁻⁵	—	
Bi	9.80	271.3	1,420—1,560	i.	i.	
BiCl₃	4.75	232	447	Int.	Int.	
Bi₂O₃ (rhomb.)	8.9	820	(1,890i)	i.	i.	
Bi₂O₃ (cub.)	8.20	transform. 704		i.	i.	
Bi₂O₃ (rhomb.)	8.5	860				
Bi(OH)₃	4.36	—H₂O, 100	—4,5 H₂O, 400	1.4 · 10⁻⁴	—	
Bi₂S₃	7.4	685, dec.	—	1.8 · 10⁻⁵	—	1.6 · 10⁻⁷² (25°C)
Br₂	2.928 (59°C)	—7.2	58.78	3.58	3.52 (50°C)	

Table 25 continued

Symbol or Formula	d	Mp	Bp	Solubility [g] in 100 g of Water at 20 °C	at 100 °C	Solubility Product
C,amorphous	1.8—2.1	subl. 3,652—3,697	4,200	i.	i.	
C, Graphite	2.25	subl. 3,652—3,697	4,200	i.	i.	
C, Diamond	3.51	>3,500	4,200	i.	i.	
CO	1.250 g/l	—205 (—207)	—192 (—190)	2.3 ml	1.62 ml (50 °C)	
CO_2	1.977 g/l	—56.6 (5.2 atm)	subl. —78.5	88.0 ml	44 ml (50 °C)	
Ca	1.55	842	1,240	Int.	Int.	
CaC_2	2.22	1,900—2,300 (stable 425—447)	—	Int.	Int.	
$CaCO_3$ (rhomb.)	2.7—2.9	dec. >500	—	h.s.	h.s.	$4.8 \cdot 10^{-9}$ $(0.87 \cdot 10^{-8})$ (25 °C)
$CaCO_3$	(2.93)	(825)		$(1.53 \cdot 10^{-3} at 25 °C)$ $(1.4 \cdot 10^{-3} at 25 °C)$	$(1.9 \cdot 10^{-3} at 75 °C)$ $(1.8 \cdot 10^{-3} at 75 °C)$	
$CaCl_2$	(2.71) 2.15 (2.51)	(1,339 at 1,025 atm) 772	(894.4) >1,600	74.5	159	
CaF_2	3.18	1,403 (1,360)	2,500	$1.6 \cdot 10^{-3}$	—	$4 \cdot 10^{-11}$ (25 °C)
$Ca(NO_3)_2$	2.36	561	—	129 (341 at 25 °C)	364 (376)	
CaO	3.40 (3.35)	2,585	2,850	0.12 Int.	$6 \cdot 10^{-2}$ Int.	
$Ca(OH)_2$	2.1—2.3	$-H_2O$ 580	—	0.165	$7.7 \cdot 10^{-2}$	$\cdot 3.1 \cdot 10^{-5}$ (25 °C)
$Ca_3(PO_4)_2$	3.14	>1,700 (1,670)	—	i. $(2 \cdot 10^{-3}—3 \cdot 10^{-3})$ s. Int.	Int.	
CaS	2.15	—	dec.	$(1.21 \cdot 10^{-2} at 15 °C)$	Int.	
$CaSO_4$	2.5—2.96	1,450	(1,193)	0.200	(0.4,614) 0.162	$6.1 \cdot 10^{-5}$ $(0.95 \cdot 10^{-4})$ (10 °C)
$CaSO_4 \cdot 2H_2O$	2.2—2.3	$-1.5 H_2O$ <170 (128)	$-2 H_2O$ >500 (163)	0.241 (0 °C)	0.222	
Cd	8.642	320.9	767 ± 2	i.	i.	
$CdCO_3$	4.25	dec. <360 (<500)		i.	i.	
$CdCl_2$	4.1	568	>900 (960)	135 (140)	150	$2.5 \cdot 10^{-14}$ (25 °C)
CdS	4.82	1,750 (100 atm)	subl. N_2 (980 °C)	$1.3 \cdot 10^{-4}$	colloidal	$1 \cdot 10^{-29}$ $(3.6 \cdot 10^{-29})$ (18 °C)
Ce (hexagonal)	6.7	804	2,900	i.	i.	
CeO_2	7.3	>2,000 (2,600)	—	i.	i.	

Table 25 continued

Symbol or Formula	d	Mp	Bp	Solubility [g] in 100 g of Water		Solubility Product
				at 20 °C	at 100 °C	
				310 cm³ (10 °C)	177 cm³ (30 °C)	
Cl₂	3.21_4	−103 ± 5	−34.6	310 cm³ (10 °C)	177 cm³ (30 °C)	
Co	8.9	1,495	2,900	i.	i.	
CoCO₃	4.13	dec.	—	i.	i.	$1 \cdot 10^{-12}$ (25 °C)
CoCl₂	3.36	735 (subl.)	subl. (1,049)	53	106.2	
CoS	5.45	>1,100	—	$1 \cdot 10^{-4}$	—	$2 \cdot 10^{-27}$ ($3 \cdot 10^{-28}$) (18 °C)
CoSO₄	3.71	989	—	34.4	76.1	
Cr	7.20 (28 °C)	1,890	2,480	i.	i.	
CrCl₂	2.75	824	—	e.s.	e.s.	
CrF₃	3.8	>1,000	dec.	i.	i.	
Cr₂O₃	5.2	2,275 (1,990)	—	i.	i.	
Cr₂(SO₄)₃·18 H₂O	1.86 (1.7)	−12 H₂O 100	—	e.s. (120)	Int.	
Cs	1.873	28.5	670	Int.	Int.	
Cs₂CO₃	—	dec. 610	—	260.7	e.s.	
CsCl	3.97	638 (646)	1,303 (1,290 subl.)	186.5	270.5	
Cs₂SO₄	4.24	1,019 (1,010)	— (transform. 600)	178.6	220.36	
Cu	8.92	1,083	2,336	i.	i.	
CuCO₃ · Cu(OH)₂	4.0	dec. 200	1,366	i.	Int.	$1.8 \cdot 10^{-7}$ ($1.02 \cdot 10^{-6}$) (18—20 °C)
CuCl	3.5	422	dec. >500 (transform. 993)	h.s.	h.s.	
CuCl₂	3.1	≈500	dec.(−HNO₃170)	73	110	
Cu(NO₃)₂ · 3 H₂O	2.05	114.5	(−O) 1,800	124.8	247.3 (1,270)	
Cu₂O	6.1	1,235	—	i.	i.	
CuO	6.4	dec. >1,000 (1,026)	—	i.	i.	
Cu₂S (rhomb.)	5.6	>1,100	—	i.	i.	$2.5 \cdot 10^{-50}$ ($2 \cdot 10^{-47}$) (16—18 °C)
CuS₂	5.78	1,130	—	i.	—	
CuS	4.6	dec. 220	(dec. 220)	i.	—	$4 \cdot 10^{-38}$ (25 °C)
CuSO₄	3.6	200 (transform. 103)	dec. 650	20.7	75.4	

Table 25 continued

Symbol or Formula	d	Mp	Bp	Solubility [g] in 100 g of Water at 20°C	at 100°C	Solubility Product
Er	4.77?	—	—	—	—	
Eu	—	1,150 ± 50	—	—	—	
F_2	1.69 g/l	—223	—188	Int.	Int.	
Fe	7.86	1,535	3,000	i.	i.	
$FeCl_2$	2.98	672	subl.	64.4 (10 °C)	105.7	
$FeCl_3$	2.8	304 (282)	subl. 303 (315)	91.9	537 (535.7)	
$Fe(CO)_5$	1.46	—20	103	i.	—	
Fe_3O_4	5.1 (5.18)	1,550 (1,538)	dec. FeO	i.	i.	
Fe_2O_3	5.1 (5.24)	1,565	—	i.	i.	
FeO	5.7	1,420	—	i.	i.	
$Fe(OH)_3$	3.4	—1.5 H_2O 500	—	$5 \cdot 10^{-5}$	—	$4 \cdot 10^{-38}$ $(1.1 \cdot 10^{-36})$ (18 °C)
FeS	4.7 (4.84)	1,493	dec.	h.s.	—	$4 \cdot 10^{-19}$ $(3.7 \cdot 10^{-19})$ (18 °C)
FeS_2, Marcasite	4.9 (4.87)	(transform. 450)				
FeS_2, Pyrite	5.00	1,171				
$FeSO_4 \cdot 7 H_2O$	1.9	64	—7 H_2O 300	26.5	50.9 (70 °C)	
Ga (solid)	6.095	29.78	1,983	i.	i.	
(solved)	6.904					
Ge	5.35	958.5	(2,700)	i.	i.	
$GeCl_4$	1.88	—51 (—49.5)	84 (83.1)	Int.	—	
GeO_2	4.70	1,115	1,200	0.4	1	
H_2	0.089,9 g/l	—259.14	—252.8	1.91 cm³ (25 °C)	0.85 cm³ (80 °C)	
H_3BO_3	1.44	dec. >70	—4.5 H_2O 300	5	40.3	
HBr	3.5 g/l	—88	—67.0	210 g (10 °C)	171 g (130 g) (50 °C)	
HCN	0.901 g/l	—14	26	∞	∞	
HCl	1.64 g/l	—112	—84	72.1	56.1 g (60 °C)	
HF	0.921 g/l	—83 (—92.3)	19.5 (19.4)	∞	e.s.	
HNO_3	1.51 g/l	—41	dec. 86	∞	∞	

Table 25 continued

Symbol or Formula	d	Mp	Bp	Solubility [g] in 100 g of Water at 20 °C	at 100 °C	Solubility Product
H_3PO_4	1.87 g/l	42.35	−0.5H_2O 213	into H_3PO_3	e.s.	
H_2S	1.54 g/l	−85.6 (−82.9)	−60.7 (−61.8)	258 ml	139 ml (50 °C)	
H_2SO_4	1.84	10.45	dec. 336.5	∞	∞	
$H_2[SiF_6]$	1.27 (30% aqu. sol.) 1.46 (60% aqu. sol.)	—	dec.		s.	
He	0.178,5 (0 °C)	−272.2 (26 atm)	−268.9	0.94 cm³ (25 °C)	1.21 cm³ (75 °C)	
Hf	13.3	1,700	>3,200	i.	—	
HfO_2	9.68	2,770 (2,810)	—			
Hg	13.546	−38.87	356.6	i.	i.	
Hg_2Cl_2	7.15	525 (subl. 400)	subl. 383.7	2 · 10⁻⁴	1 · 10⁻³ (43 °C)	2.10⁻¹⁸ (25 °C)
$HgCl_2$	5.42 (5.44)	276	301 (302)	6.6	55 (61.3)	
HgS	8.1	subl. 583.5	—	1 · 10⁻⁶	—	4.10⁻⁵⁸ (4.10⁻⁵³) (18 °C)
Hg_2SO_4	7.56	dec.	dec.	6 · 10⁻² (25 °C)	9 · 10⁻²	
$HgSO_4$	6.47	dec.	—	Int.	Int.	
In	7.3	156.4	2,000 ± 10	i.	i.	
Ir	22.421	2,454	>4,800	i.	i.	
J_2	4.93	113.7	184.35	2.9 · 10⁻²	7.8 · 10⁻² (50 °C)	
JCl α	3.82 (3.18)	27 (27.2)	dec. 97 (97.4)	Int.	Int.	
JCl β	(3.24) (34 °C)	(13.92)	(97.4)	Int.		
K	0.86	62.3	760	Int.	Int.	
KBr	2.75	728 (730)	1,376 (1,380; 1,435)	65.8	105 (102)	
KCl	1.99	768 (736)	1,417 (subl. 1,500)	34.0 (34.7)	56.7	
KH_2PO_4	2.34	dec.	—	25 (25 °C)	72.2 (83 °C)	
KJ	3.13	686 (723)	1,330	144	208	
KNO_3	2.11	334	dec. 400	31.7	246	
KOH	2.04	360	1,324	112	178	

Table 25 continued

Symbol or Formula	d	Mp	Bp	Solubility [g] in 100 g of Water		Solubility Product
				at 20°C	at 100°C	
K_2SO_4	2.66	1.069	—	11.7	24.1	
Kr	3.708 g/l	−156.6	−152.9	11.0 (0°C)	4.67 (50°C)	
La	6.15	826	—	Int.	Int.	
La_2O_3	6.50 (6.51)	2,320 (2,315)	—(4,200)	$4 \cdot 10^{-4}$ (29°C)	Int.	
Li	0.534	186	1,336 ± 5	Int.	Int.	
Li_2CO_3	2.11	735 (618)	dec.	1.33	0.72	$1.7 \cdot 10^{-3}$ (25°C)
LiF	2.30 (2.06)	842 (870)	1,676	0.3	h.s.	
$LiNO_3$	2.38	252 (255)	—	74.5	175 (60°C)	
Li_2SO_4	2.22	860	—	34.2	30 (23)	
Lu						
Mg	1.74	651	1,107	i.	Int.	
$MgCO_3$	2.98 (3.04)	dec. >300 (350)	—CO₂ 900	h.s.	—	$1 \cdot 10^{-5}$ ($2.6 \cdot 10^{-5}$) (12°C)
$MgCl_2$	2.33	718 (708)	1,412	54.5	73	
MgO	3.2 (3.58)	>2,800 (2,800)	—	$6.2 \cdot 10^{-4}$	$3 \cdot 10^{-3}$ Int.	
$Mg(OH)_2$	2.4 (2.38)	dec. (—OH 350)	—	$9 \cdot 10^{-4}$	$4 \cdot 10^{-3}$	$5 \cdot 10^{-12}$ ($1.2 \cdot 10^{-11}$) (18°C)
$MgSO_4$	2.66	dec. 1,155 (1,124)	—	36	68 (73.8)	
Mn	7.20	1,260	1,900	Int.	Int.	
$MnCO_3$	3.1–3.7 (3.125)	dec. >100	—	$6.5 \cdot 10^{-3}$ (25°C)	—	$1 \cdot 10^{-10}$ (25°C)
$MnCl_2$	2.98	650	1,190	74	115 (123.8)	
MnO_2	5.03	dec. >535 (—O 535)	—	i.	i.	
$MnSO_4$	3.25	700	dec. 850	64	35	
Mo	10.2	2,620 ± 10	4,800	i.	i.	
$MoCl_5$	2.93	194	268	s. Int.	Int.	
MoO_3	4.5–4.7	795	1,155 (subl.)	0.1	2.1 (80°C)	
MoS_2	4.6–4.8 (4.8)	1,185	—	i.	i.	

Table 25 continued

Symbol or Formula	d	Mp	Bp	Solubility [g] in 100 g of Water at 20 °C	at 100 °C	Solubility Product
N_2	1.25 g/l	−209.86	−195.8	2.33 (0 °C)	1.42 cm³ (40 °C)	
$(NH_4)_2CO_3 \cdot H_2O$	—	dec. 58		100 (15 °C)	Int.	
NH_4Cl	1.54 (1.53)	dec. (subl. 335)	subl.	37.5	77.3 (75.8)	
NH_4HCO_3	1.58	dec. 36—60 (107.5)	subl.	21.6	43 (60 °C)	
NH_4NO_3	1.73	169.6	dec. >190 (210)	178	1,010 (871)	
Na	0.97	97.5	880	Int.	Int.	
$Na_2B_4O_7$	2.37	741	dec. 1,575	2.6	52.5	
Na_2CO_3	2.53	851	dec.	21.5	45.5	
$NaCl$	2.16	800	1,440 (1,413)	36.0	39.1	
NaF	2.79	995	1,700	4.2	5.0	
$NaHCO_3$	2.20	−CO_2 270	—	9.6	Int.	
$NaHS$	1.79	350		e.s.		
$NaNO_3$	2.26	306	dec. 380	87.5	180	
$NaOH$	2.13	328 (318.4)	1,388	109	347	
Na_2S	1.86	1,180	>1,300	19	57.3 (90 °C)	
Na_2SO_4	2.7	885	dec. 1,430	19.4	42.5	
$Na_2S_2O_7$	2.66	401	dec. >460	s.	—	
Nb	8.55	2,500 ± 50	3,700	i.	i.	
Nb_2O_5	4.47	1,520	—	i.	i.	
Nd	6.9	840	—	—	—	
Ne	0.90 g/l	−248.67	−245.9	1.16 cm³ (25 °C)	—	
Ni	8.90	1,455	2,900	i.	i.	
$Ni(CO)_4$	1.31	−25	43	$1.8 \cdot 10^{-2}$ (10 °C)	i.	
$NiCl_2$	3.55	subl. (1,001)	(973)	64.2	87.6	
NiO	6.8 (7.45)	1,990	—	i.	i.	
NiS	5.2—5.7 (5.3—5.65)	797	—	$>1.5 \cdot 10^{-4}$	h.s.	$1.4 \cdot 10^{-24}$ (25 °C)
$NiSO_4$	3.68	−SO_3 840	—	38	77 (83.7)	
O_2	1.429 g/l	−218.4	−182.96	3.16 cm³ (25 °C)	2.3 cm³	

Table 25 continued

Symbol or Formula	d	Mp	Bp	Solubility [g] in 100 g of Water at 20°C	at 100°C	Solubility Product
Os	22.48	2,700	>5,300	i.	i.	
P	1.82—2.70	593	>200	i.	i.	
PCl_3	1.57	—94 (—91)	75	Int.	Int.	
PCl_5	2.11 (4.65 g/l)	163	subl. 159 (subl. 162)	Int.	Int.	
P_2O_5	2.39	569 (563)	subl. 359 (subl. 347)	Int.	Int.	
Pb	11.343,7	327.43	1,620	i.	i.	
$2 PbCO_3 \cdot Pb(OH)_2$	6.14	dec. >200 (400)		i.	i.	
$PbCl_2$	5.85	501	950	0.99	3.34	$1.7 \cdot 10^{-5}$ (25°C)
$PbCrO_4$	6.1 (6.3)	844	dec.	$4.3 \cdot 10^{-6}$	h.s.	$1.8 \cdot 10^{-14}$ (25°C)
PbF_2	8.24	855	1,290	$6.5 \cdot 10^{-2}$	—	$3.7 \cdot 10^{-8}$ (25°C)
$Pb(NO_3)_2$	4.53	dec. ≈200 (470)	—	52.2 (56.5)	127	
PbO	9.4—9.6 (8.0)	888	1,470	$1.7 \cdot 10^{-3}$	—	
Pb_3O_4	9.1	dec. 500	—	i.	i.	
PbO_2	9.4	dec. >280 (290)	—	i.	i.	$1 \cdot 10^{-29}$ ($3.4 \cdot 10^{-28}$)
PbS	7.5	1,114	—	>$3 \cdot 10^{-5}$	i.	(18°C)
						$2 \cdot 10^{-8}$ (25°C)
$PbSO_4$	6.2	dec. 1,000	—	$4.1 \cdot 10^{-3}$	$5.8 \cdot 10^{-3}$ (50°C)	$2 \cdot 10^{-8}$ (25°C)
Pd	11.97 (0°C)	1,549.4	about 2,200	i.	i.	
$PdCl_2$	—	678 (500)	≈800	e.s.	e.s.	
Pr	6.5	940	—	Int.	—	
Pt	21.45	1,773.5	4,300	i.	i.	
$PtCl_4$	—	dec. 370	—	e.s. Int.	e.s. Int.	
Ra	5?	700	1,140	Int.	—	
$RaCl_2$	4.91	—	—	25	s.	
Rb	1.532	38.5	700	Int.	Int.	
Rb_2CO_3	—	837 (CO_2)	dec. >740	450	s.	
RbCl	2.8	717 (715)	1,338 (1,390)	91.2	138.9	

85

Table 25 continued

Symbol or Formula	d	Mp	Bp	Solubility [g] in 100 g of Water at 20 °C	at 100 °C	Solubility Product
RbNO$_3$	3.11	313 (310)	—	53.3	452	
Re	20.53	3,167 ± 60	—	—	—	
Rh	12.4	1,966 ± 3	>2,500	i.	i.	
Rn	9.73 g/l	—71	61.8	51 cm^3 (0 °C)	13 cm^3 (50 °C)	
Ru	12.6	>1,950	—	i.	i.	
Ru (hexagonal)	1.063	2,450	>2,700	i.	i.	
RuO$_4$	3.28 (3.29)	25.5	108 (about 100)	s. (2.033)	Int.	
S α	2.07	112.8	444.6	i.	i.	
S β	1.96	119.25	444.6	i.	i.	
S γ	1.92	about 120	444.6	i.	i.	
SO$_2$	2.927 g/l	—72.7 (—75.46)	—10.08 (—10.0)	3,937 ml	1,877 ml	
Sb	6.68	630.5	1,380	i.	i.	
SbCl$_3$	3.14	73	223	931.5	∞.80 °C Int.	
Sb$_2$O$_3$	5.2	656	subl. ≈1,500	h.s.	h.s.	
	5.7		(1,550)			
Sb$_2$S$_3$	4.1—4.6	548 (550)	—	$1.8 \cdot 10^{-4}$	Int.	
Sb$_2$S$_5$	4.12	dec.	—	i.	i.	
Sc	2.5	1,200	2,400	Int.	Int.	
Se$_8$ (amorphous)	4.82	—220	688	i.	i.	
Se$_8$ (colloidal)	—	—	—	s.	—	
Se$_8$ (solved)	4.28	—	688	i.	i.	
Se$_8$ (monoclinic)	4.46	170—180	688	i.	i.	
Se$_8$ (trigonal)	4.79	217	688	i.	i.	
SeO$_2$	3.95	340 (340—350)	subl. 315	38.4 (14 °C)	82.5 (60 °C)	
Si (cubic)	2.42	1,420	2,355	i.	i.	
Si (amorphous)	2.0	—	2,355	i.	i.	
Si	about 2.4	—	2,355	i.	i.	

Table 25 continued

Symbol or Formula	d	Mp	Bp	Solubility [g] in 100 g of Water at 20 °C	at 100 °C	Solubility Product
SiC	3.2	—(subl. 2,600)	dec. >2,200	i.	i.	
SiCl$_4$	1.483	—70	57 (57.57)	Int.	Int.	
SiF$_4$	4.68 g/l (4.67)	—77 (—97)	subl. —95 (—65)	Int.	Int.	
SiO$_2$	2.65 (2.653—2.660)	1,713 (>1,470)	2,590 (2,230)	i.	i.	
Sm	7.7	>1,300	—	—	—	
Sn α	5.75	231.89	2,260	i.	i.	
Sn β	7.28	231.89	2,270	i.	i.	
Sn γ	6.52—6.56	231.89	2,270	i.	i.	
SnCl$_2$	3.393	247 (246)	652 (623)	83.9 (0 °C)	269.8 (15 °C)	
SnCl$_4$	2.23	—32	114	s.	Int.	
SnO$_2$	6.95	>1,900 (1,127)	subl. 1,800—1,900	i.	i.	
SnS$_2$	4.5	—SO$_2$ 360 (>360)	—	h.s.	18	
SnSO$_4$	—	—	—	19		
Sr	2.6	774	1,150	Int.	Int.	
SrCO$_3$	3.70	1,497	—CO$_2$ 1,340	1 · 10^{-3}	6.5 · 10^{-2}	1·10^{-9} (1.6·10^{-9}) (25 °C)
SrCl$_2$	3.05	868 (873)	—	53	100.8	
Sr(NO$_3$)$_2$	2.99	645 (570)	dec.	70.5	100 (90 °C)	
Sr(OH)$_2$	3.63	375	—	0.41 (0 °C)	21.83	
SrSO$_4$	3.96	≈1,600 (1,580)	dec.	1.14 · 10^{-2}	—	2.8·10^{-7} (3.81·10^{-7}) (17.4 °C)
Ta	16.6 14.49	2,996 ± 50	about 4,100	i.	i.	
Ta$_2$O$_5$	8.74	dec. 1,470	—	i.	i.	
Tb	—	327 ± 5	—	—	—	
Te$_2$ (rhomb.)	6.25	452	1,390	i.	i.	
Te$_2$ (amorphous)	6.0	452	1,390	i.	i.	
TeO$_2$	5.66 (0 °C)	—	subl. 450	7 · 10^{-4}	—	

87

Table 25 continued

Symbol or Formula	d	Mp	Bp	Solubility [g] in 100 g of Water		Solubility Product
				at 20°C	at 100°C	
TeO₂ (rhomb.)	(5.91)	—	—			
Th	11.2	1,845	4,500	i.	i.	
ThCl₄	4.59	770 (820)	921, dec. 1,100	e.s.	e.s.	
ThO₂	9.7 (10.03)	3,050	>4,000 (4,400)	i.	i.	
Ti	4.5	1,800	>3,000	i.	Int.	
TiCl₄	1.73	−23 (−30)	136	s.	Int.	
TiC	4.25 (4.93)	3,140 (3,140 ± 90)	4,300	i.	i.	
TiO₂, Anatase	3.84	—	—	i.	i.	
TiO₂, Rutile	4.26 (4.17)	1,825	—	i.	i.	
Tl	11.85	302	1,457 ± 10	i.	i.	
Tl₂CO₃	7.11	273		4.2 (16°C)	27.2	
TlCl	7.0	430	806 (720)	0.3	1.97	2.10⁻⁴ (25°C)
TlNO₃	5.56 (γ)	206 (α)	430 (α)	9.6 (α)	413 (α)	
Tl₂SO₄	6.77	632	dec.	4.9	18	
Tm	—	—	—	i.	i.	
U	18.7	about 1,133	—	i.	i.	
UCl₄	4.725	subl. 590	—	e.s.	Int.	
UF₄	—	960 (about 1,000)	618	e.s.		
UO₂	10.9	2,176	—	i.	i.	
U₃O₈	7.31 (8.3)	dec.	—	i.	i.	
V	5.96	1,710	3,000	i.	i.	
V₂O₃	4.87	1,970	—	e.s.	s.	
V₂O₅	3.36	675 (690)	dec. 1,750	e.s. 0.8		
W	19.3	3,370	5,900	i.	i.	
W₂C	16.06 (17.15)	2,860	(6,000)	i.	—	
WC	15.7 (15.63)	2,870 (2,870 ± 50)	(6,000)	i.	—	
WO₃	≈7.0 (7.16)	1,473	subl. 1,357	i.	i.	

Table 25 continued

Symbol or Formula	d	Mp	Bp	Solubility [g] in 100 g of Water at 20°C	at 100°C	Solubility Product
Xe	5.85 g/l	—112	—107.1	24.1 (0°C)	7.12 (80°C)	
Y	5.51	1,490	2,500	Int.	Int.	
Yb	—	1,800	—			
Zn	7.14	419.47	907	i.	i.	$6 \cdot 10^{-11}$ (25°C)
ZnCO$_3$	4.4	—CO$_2$ 300	—	$2.06 \cdot 10^{-2}$	—	
ZnCl$_2$	2.9	313 (262)	732	368	614 (615)	
ZnF$_2$	4.84	872	1,500	h.s.	s.	
ZnO	5.5—5.6	2,000	subl. 1,950 dec.	$1.6 \cdot 10^{-4}$ (29°C)	—	
amorphous	(5.47)					
ZnO, Zincite	(5.606)	(>1,800)	(subl. 1,800)			
Zn(OH)$_2$	3.053	dec. 125	—	$5.2 \cdot 10^{-4}$ (18°C)	—	$8 \cdot 10^{-26}$ ($1.2 \cdot 10^{-23}$) (18°C)
ZnS α	4.0—4.1	>1,800	subl. 1,180	h.s.	—	
	(4.087)	(1,850, 150 at)	(subl. 1,185)	($7 \cdot 10^{-4}$)	—	
ZnS β	3.6 (3.74)	transform. 1,020		($7 \cdot 10^{-4}$)		
ZnSO$_4$		dec. 740	—	544	80 (80.8)	
Zr	6.4	1,857	>2,900	i.	i.	
ZrCl$_4$	2.8	—	subl. >300	s.	Int.	
ZrO$_2$	5.8—6.2	≈2,700		i.	i.	
	(5.49)	(2,700)	(4,300)	i.	i.	
	(5.6)	(2,715)	—	i.	i.	

2.10. Thermodynamic Data

Any thermodynamic system can be in different physico-chemical states. When the system comes to thermodynamic equilibrium, these states are completely characterized by variable quantities. There are primary variables of state: pressure, density, temperature, and entropy. Variables of state (or potentials) derived from the primary quantities are, for example: free energy F and enthalpy H.

2.10.1. Entropy S

In a closed system all spontaneously occurring physico-chemical processes take such a course that the initial state passes to a state which shows a lower degree of order. The most probable state of a closed system is a state of complete disorder which is characterized by a perfectly uniform distribution of pressure, density, temperature, energy, etc. The degree of disorder and, thus, the random way of bringing about a certain state is described by the quantity which is called entropy S. The degree of order of a closed system which is left to itself and whose processes are allowed to go on in their own way cannot increase. This is why the entropy always increases with such processes.

2.10.2. Free Energy F

In most cases, we are not dealing with closed systems. Many processes taking place in a laboratory develop because of an exchange of heat with the surrounding at constant temperature. Under this condition, the free energy F is the whole capacity of the system to do work ($\Delta F > 0$ input — endergonic process — or $\Delta F < 0$ output — exergonic process — of external work). Besides work, a specified amount of heat always develops (heat of formation).

In thermochemistry, the whole energy which a process is capable of supplying (free energy + heat of formation) is subsumed under the term of heat tone (negative heat tone in exothermic processes). All spontaneous reactions are exergonic, hence, they take place at a decrease of F ($\Delta F < 0$).

2.10.3. Enthalpy H and Free Enthalpy G

At constant pressure, processes may do external work under certain circumstances (e.g. gas evolved in a chemical reaction does the work of displacement). To describe such processes conveniently, it is advisable to combine internal energy and external work into the term of enthalpy H (Gibbs' heat function). At constant pressure the enthalpy is the total available heat supply of the system.

If, in a system, both pressure and temperature are constant, a free enthalpy G (Gibbs' potential) can be introduced, analogous to the free energy. Spontaneous changes in this system take such a course that the free enthalpy decreases ($\Delta G < 0$).

Explanations of Table 26 (page 91)

$\Delta H°$ and $\Delta G°$ indicate the numerical values of enthalpy and free enthalpy for the formation of the stable modification of a combination at standard conditions (25 °C and 1 atm.).

$\Delta H°$ (also denoted $\Delta I°$) normal enthalpy; the values in brackets apply to 18 °C and are mainly from Perelman (1956).

$\Delta G°$ (in English literature also known as $\Delta F°$) free normal enthalpy (thermodynamic normal potential or Gibbs' potential).

$S°$ normal entropy.

Explanations of Table 26 (continued)
For gases, the data are related to the ideal state of gas (dissociation and association were not taken into consideration). (The majority of data is from ROSSINI: Selected Values of Chemical Thermodynamic Properties, Circul. Nat. Bureau of Standards No. 500, Dept. of Commerce, Washington 1952.) State of aggregation: s solid; li liquid; g gaseous; aq aqueous solution; am amorphous.

Table 26. Thermodynamic Data of the Elements and of Several Combinations (see explanations p. 90)

Symbol Formula	State of Aggregation	ΔH^0 [kcal/g mol]		ΔG^0 [kcal/g mol]	S^0 $\left[\dfrac{\text{kcal}}{\text{kmol deg}}\right]$
Ag	s	0.00		0.00	10.21
	g	69.12		59.84	41.32
Ag_2CO_3	s	—120.97		—104.48	40.0
AgCl	s	— 30.36		— 26.22	22.97
$AgNO_3$	s	— 29.43		— 7.69	33.68
Ag_2S	s	— 7.6		— 9.62	34.8
Ag_2SO_4	s	—170.50		—147.17	47.8
Al	s	0.0		0.00	6.7
	g	75.0	(55.0)	65.3	39.303
$AlCl_3$	s	—166.2		—152.2	40
AlF_3	s	—311	(—329)	—294	23
α-Al_2O_3	s	—399.09		—367.77	12.19
$Al_2O_3 \cdot SiO_2$ (andalusite)	s	—617.0			
$Al_2O_3 \cdot SiO_2$ (disthene)	s	—617.4			
$Al_2O_3 \cdot SiO_2$ (sillimanite)	s	—623.7			
$Al(OH)_3$	am	—304.2			
$Al_2(SO_4)_3$	s	—820.98	(—700)	—738.99	57.2
Ar	g	0.00		0.00	36.983
As grey	s	0.00		0.00	8.4
As_4	g	35.7		25.2	69
$AsCl_3$	g	— 71.5	(—74.7)	— 68.5	78.2
As_2S_2 (realgar)	s	— 31.9			
As_2S_3 (oripigment)	s	— 35	(—20)		
Au	s	0.00		0.00	11.4
	g	82.92	(92)	72.83	43.12
$AuCl_3$	s	— 28.3			
B	s	0.00		0.00	1.56
	g	97.2	(115)	86.7	36.649
B_4C	s				6.47
BCl_3	g	— 94.5		— 90.9	69.29
BN	s	— 32.1	(—28.5)	— 27.2	8
BF_3	g	—265.4		—261.3	60.70
B_2O_3	s	—302.0		—283.0	12.91
Ba	s	0.00		0.00	16
	g	41.96		34.60	40.699
$BaCO_3$ (witherite)	s	—291.3	(—285)	—272.2	26.8
$BaCl_2$	s	—205.56		—193.8	30
$BaSO_4$	s	—350.2	(—304.2)	—323.4	31.6

Table 26 continued

Symbol Formula	State of Aggregation	ΔH^0 [kcal/g mol]		ΔG^0 [kcal/g mol]	S^0 $\left[\dfrac{\text{kcal}}{\text{kmol deg}}\right]$
Be	s	0.00		0.00	2.28
	g	76.63		67.60	32.55
$BeCl_2$	s	—122.3	(—112.6)		21.5
BeF_2	aq	—251.4			
BeO	s	—146	(—138)	—139.0	3.37
Bi	s	0.00		0.00	13.6
	g	49.7		40.4	44.67
$BiCl_3$	s	— 90.61		— 76.23	45.3
Bi_2O_3	s	—137.9		—118.7	36.2
$Bi(OH)_3$	s	—169.6			
Bi_2S_3	s	— 43.8		— 39.4	35.3
Br_2	li	0.00		0.00	36.4
	g	7.34		0.75	58.64
C	g	171.70		160.85	37.76
C (diamond)	s	0.45		0.69	0.58
C (graphite)	s	0.00		0.00	1.36
CO	g	— 26.42		— 32.81	47.30
CO_2	g	— 94.05		— 94.26	51.06
Ca	s	0.00		0.00	9.95
	g	46.04		37.98	36.99
CaC_2	s	— 15.0		— 16.2	16.8
$CaCO_3$ (aragonite)		—288.49		—269.53	21.2
(calcite)		—288.45		—269.78	22.2
$CaCl_2$	s	—190.0		—179.3	27.2
CaF_2	s	—290.3		—277.7	16.46
$Ca(NO_3)_2$	s	—224.0		—177.34	46.2
CaO	s	—151.9		—144.4	9.5
$Ca(OH)_2$	s	—235.80		—214.33	18.2
$Ca_3(PO_4)_2$	s	—986.2	(—983)	—929.7	57.6
CaS	s	—115.3	(—113.4)	—114.1	13.5
$CaSO_4$ (anhydrite)	s	—342.42	(—338.7)	—315.56	25.5
$CaSO_4 \cdot 2\,H_2O$	s	—483.06		—429.19	46.36
Cd	s	0.00		0.00	12.3
	g	26.97		18.69	40.7
$CdCO_3$	s	—178.7	(—182)	—160.2	25.2
$CdCl_2$	s	— 93.00		— 81.88	28.3
CdS	s	— 34.5		— 33.6	17
Ce	s	0.00		0.00	13.8
CeO_2	s	—233			16
Cl_2	g	0.00		0.00	53.29
Co	s	0.00		0.00	6.8
	g	105		94	42.88
$CoCO_3$	s	—172.7		—155.36	
$CoCl_2$	s	— 77.8	(—76.9)	— 67.5	25.4

Table 26 continued

Symbol Formula	State of Aggregation	ΔH^0 [kcal/g mol]		ΔG^0 kcal/g mol]	S^0 $\left[\dfrac{\text{kcal}}{\text{kmol deg}}\right]$
CoS	s	$-$ 20.2		$-$ 19.8	
$CoSO_4$	s	$-$207.5		$-$182.1	27.1
Cr	s	0.00		0.00	5.68
	g	80.5		69.8	41.64
$CrCl_2$	s	$-$ 94.56	($-$103.1)	$-$ 85.15	27.4
CrF_3	s	$-$265.2			
Cr_2O_3	s	$-$269.7	($-$273)	$-$250.2	19.4
Cs	s	0.00		0.00	19.8
Cs_2CO_3	s	$-$267.4			
CsCl	s	$-$103.5	($-$106.3)		24
Cs_2SO_4	s	$-$339.38			
Cu	s	0.00		0.00	7.96
	g	81.52		72.04	39.74
CuCl	s	$-$ 35.5	($-$64.4)	$-$ 28.2	20.2 (43.8)
$CuCl_2$	s	$-$ 49.2	($-$53.4)		27.0
Cu_2O	s	$-$ 39.84		$-$ 34.98	24.1
CuO	s	$-$ 37.1		$-$ 30.4	10.4
Cu_2S	s	$-$ 19.00		$-$ 20.6	28.9
CuS	s	$-$ 11.6		$-$ 11.7	15.9
$CuSO_4$	s	$-$184.00		$-$158.2	27.1
F_2	g	0.00		0.00	48.6
Fe	s	0.00		0.00	6.49
	g	96.68		85.76	43.11
$FeCl_2$	s	($-$ 81.5)			28.6
$FeCl_3$	s	$-$ 96.8	($-$93.6)		32.2
$Fe(CO)_5$	li	$-$187.8			
Fe_3O_4 (magnetite)	s	$-$267.0		$-$242.4	35.0
Fe_2O_3 (hematite)	s	$-$196.5		$-$177.1	21.5
FeO (ferrous oxide)	s	$-$ 63.7		$-$ 58.4	12.9 (14.2)
$Fe(OH)_3$	s	$-$197.0			
FeS	s	$-$ 22.72		$-$ 23.32	16.1
FeS_2 (pyrite)	s	$-$ 42.52		$-$ 39.84	12.7
FeS_2 (marcasite)	s	$-$ 36.88			
$FeSO_4 \cdot 7\,H_2O$	s	$-$718.7			
Ga	s		(0.00)		10.2
	g	66.0	(52)	57.0	40.38
Ge	s	0.00		0.00	10.14
	g	78.44		69.50	40.11
$GeCl_4$	li	$-$130			
GeO_2	s	$-$128.3			
H_2	g	0.00		0.00	31.21
H_3BO_3	aq	$-$255.2		$-$230.24	38.2
	s	($-$251.6)			21.2

Table 26 continued

Symbol / Formula	State of Aggregation	ΔH^0 [kcal/g mol]	ΔG^0 [kcal/g mol]	S^0 $\left[\dfrac{\text{kcal}}{\text{kmol deg}}\right]$
HBr	g	— 8.66	— 12.72	47.44
HCl	g	— 22.06	— 22.77	44.62
	aq	— 40.02	— 31.35	13.17
HF	aq	— 78.66	— 66.08	— 2.3
HNO_3	aq	— 49.37 (—41.7)	— 26.41	35.0
H_3PO_4	s	—306.2		
H_2S	g	— 4.815	— 7.892	49.15
H_2SO_4	aq	—216.90 (—193.8)	—177.34	4.1
$H_2(SiF_6)$	aq	—557.2		
He	g	0.00	0.00	30.13
Hf	s	0.00	0.00	13.1
HfO_2	s	—271.5		14.2
Hg	li	0.00	0.00	18.5
	g	15.54	7.59	41.8
Hg_2Cl_2	s	— 63.32	— 50.35	46.8 (23.5)
$HgCl_2$	s	— 55.0 (—53.4)	— 42.2	30.0
HgS red (cinnabar)	s	— 13.90 (—15.6)	— 11.67	18.6
Hg_2SO_4	s	—177.34	—142.12	47.98
$HgSO_4$	s	—168.3		
I_2	s	0.00	0.00	27.9
	g	14.88	4.63	62.28
ICl	g	— 4.20	— 1.32	59.12
In	g	58.2 (52)	49.6	41.51
Ir	s	0.00	0.00	8.7
	g	165	154	46.25
K	s	0.00	0.00	15.2
	g	21.51	14.62	38.30
$KAl(SO_4)_2 \cdot 12\,H_2O$ (alum)	s	—1,447.74	—1,227.8	164.3
KBr	s	— 93.73	— 90.63	23.05
KCl	aq	—100.06	— 98.82	37.7
	s	—104.18	— 97.592	19.76
KJ	aq	— 73.41	— 79.82	50.6
	s	— 78.31	— 77.03	24.94
KNO_3	s	—117.76	— 93.96	31.77
KOH	aq	—115.00 (—102.0)	—105.06	22.0
K_2SO_4	s	—342.66	—314.62	42.0
Kr	g	0.00	0.00	39.19
La	s	0.00	0.00	13.7
La_2O_3	s	—458		
Li	s	0.00	0.00	6.70
	g	37.07	29.19	33.14

94

Table 26 continued

Symbol Formula	State of Aggregation	ΔH^0 [kcal/g mol]	ΔG^0 [kcal/g mol]	$S^0 \left[\dfrac{kcal}{kmol\ deg}\right]$
Li$_2$CO$_3$	s	-290.54	-270.66	21.60
LiF	s	-146.3	-139.5	8.57
LiNO$_3$	aq	-115.93	$-\ 96.63$	38.4
Li$_2$SO$_4$	aq	$-350.01\ \ (-340.2)$	-317.78	10.9
Lu	s	0.00	0.00	
	g	87		44.14
Mg	s	0.00	0.00	7.77
	g	35.9	27.6	35.5
MgCO$_3$	s	-266	-246	15.7
MgCl$_2$	s	-153.40	-141.57	21.4
MgO	s	-143.84	-136.13	6.4
Mg(OH)$_2$	s	$-221.00\ \ (-218.7)$	-199.27	15.09
MgSO$_4$	s	$-305.5\ \ (-313.1)$	-280.5	21.9
Mn α	s	0.00	0.00	7.59
γ	s	0.37	0.33	7.72
MnCO$_3$	s	$-213.9\ \ (-219.4)$	-195.4	20.5
MnCl$_2$	s	$-115.3\ \ (-111.6)$	-105.5	28.0
MnO$_2$ (pyrolusite)	s	-124.5	-111.4	12.7
MnSO$_4$	s	$-254.24\ \ (-250.3)$	-228.48	26.8
Mo	s	0.00	0.00	6.83
	g	155.5	144.2	43.46
MoCl$_5$	s	$-\ 90.8$		
MoO$_3$	s	-180.33	-161.95	18.68
MoS$_2$	s	$-\ 55.5$	$-\ 53.8$	15.1
N$_2$	g	0.00	0.00	45.77
NH$_4$Cl	s	$-\ 75.38$	$-\ 48.73$	22.6
NH$_4$HCO$_3$	aq	-196.92	-159.31	49.3
NH$_4$NO$_3$	s	$-\ 87.27$		
Na	s	0.00	0.00	12.2
	g	25.98	18.67	36.72
Na$_2$B$_4$O$_7$	s	-777.7		
Na$_2$CO$_3$	s	-270.3	-250.4	32.5
NaCl	s	$-\ 98.23$	$-\ 91.79$	17.30
NaF	s	-136.0	-129.3	14.0
NaHCO$_3$	s	-226.5	-203.6	24.4
NaNO$_3$	s	$-101.54\ \ (-111.54)$	$-\ 87.45$	27.8 (28.8)
NaOH	aq	$-112.24\ \ (-102)$	-100.184	11.9
Na$_2$S	s	$-\ 89.2$		22.5
Na$_2$SO$_4$	s	-330.90	-302.78	35.73
Nb	s	0.00	0.00	8.3
	g	184.5	173.7	44.49
Nb$_2$O$_5$	s	-463.2		32.8
Nd	g	87		
	s	0.00	0.00	17.5

Table 26 continued

Symbol Formula	State of Aggregation	ΔH^0 [kcal/gmol]	ΔG^0 [kcal/gmol]	S^0 $\left[\dfrac{\text{kcal}}{\text{kmol deg}}\right]$
Ne	g	0.00	0.00	34.95
Ni	s	0.00	0.00	7.20
	g	101.61	90.77	43.59
NiCl$_2$	s	— 75.5	— 65.1	25.6
NiO	s	— 58.4	— 51.7	9.22
NiS	s	— 17.5 (—20.7)		
NiSO$_4$	s	—213.0	—184.9	18.6
O$_2$	g	0.00	0.00	49.003
Os	s	0.00	0.00	7.8
	g	174	163	45.97
P (white)	s	0.00	0.00	10.6
P (red)	s	— 4.4		
P	g	75.18	66.71	38.98
PCl$_3$	s	— 73.22 (—75.9)	— 68.42	74.49
Pb	s	0.00	0.00	15.51
	g	46.34	38.47	41.89
PbCl$_2$	s	— 85.85	— 75.04	32.6
PbCrO$_4$	s		—203.6	
PbF$_2$	s	—158.5	—148.1	29
Pb(NO$_3$)$_2$	s	—107.35		
PbO (red)	s	— 52.40	— 45.25	16.2
Pb$_3$O$_4$	s	—175.6 (—170.4)	—147.6	50.5
PbO$_2$	s	— 66.12	— 52.34	18.3
PbS	s	— 22.54	— 22.15	21.8
PbSO$_4$	s	—219.50	—193.89	35.2
Pd	s	0.00	0.00	8.9
	g	93	84	39.91
PdCl$_2$	s	— 45.4		
Pr	s	0.00	0.00	
	g	87		
Pt	s	0.00	0.00	10.0
	g	121.6	110.9	45.96
PtCl$_4$	s	— 62.9		
Ra	s	0.00	0.00	17
	g	31	23	42.15
Rb	s	0.00	0.00	16.6
	g	20.51	13.35	40.63
Rb$_2$CO$_3$	s	—269.6 (—273.7)		
RbCl	s	—102.9 (—105.1)	— 98.48	22
RbNO$_3$	s	—117.04		
Re	s	0.00	0.00	10
	g	189	179	45.13

Symbol Formula	State of Aggregation	ΔH^0 [kcal/g mol]	ΔG^0 [kcal/g mol]	S^0 $\left[\dfrac{\text{kcal}}{\text{kmol deg}}\right]$
Rh	s	0.00	0.00	7.6
	g	138	127	44.39
Rn	g	0.00	0.00	42.10
Ru	s	0.00	0.00	6.9
	g	160	149	44.57
rhombic-	s	0.00	0.00	7.62
monoclinic	s	0.071	0.023	7.78
	g	53.25	43.57	40.085
SO_2	g	— 70.96	— 71.79	59.40
Sb	s	0.00	0.00	10.5
	g	60.8	51.1	43.06
$SbCl_3$	s	— 91.34	— 77.62	44.5
Sb_2O_3	aq	—166.5		
Sb_2S_3 (orange)	am	— 36.0		
Sc	s	0.00		9.0
	g	93		41.76
Se hexagonal	s	0.00	0.00	10.0
	g	48.37	38.77	42.21
SeO_2	s	— 55.0		
Si	s	0.00	0.00	4.47
	g	88.04	77.41	40.12
SiC	s	— 26.7	— 26.1	3.94
$SiCl_4$	g	—145.7 (—150.1)	—136.2	79.2
SiF_4	g	—370 (—360.1)	—360	68.0
SiO_2 (quartz)	s	—205.4 (—208.3)	—192.4	10.00
Sm	g	87		43.74
Sn (white)	s	0.00	0.00	12.3
(grey)	s	0.6	1.1	10.7
	g	72	64	40.25
$SnCl_2$	s	— 83.6		
$SnCl_4$	s	—130.3	—113.3	61.8
SnO_4	s	—138.8	—124.2	12.5
Sr	s	0.00	0.00	13.0
	g	39.2	26.3	39.33
$SrCO_3$ (strontianite)	s	—291.2	—271.9	23.2
$SrCl_2$	s	—198.0	—186.7	28
$Sr(NO_3)_2$	s	—233.25		
$Sr(OH)_2$	s	—229.3		
$SrSO_4$	s	—345.2 (—341.9)	—318.9	29.1

Table 26 continued

Symbol Formula	State of Aggregation	ΔH^0 [kcal/gmol]		ΔG^0 [kcal/gmol]	S^0 $\left[\dfrac{kcal}{kmol\,deg}\right]$
Ta	s	0.00		0.00	9.9
	g	185		175	44.24
Ta$_2$O$_5$	s	—499.9		—470.6	34.2
Tb	s	0.00		0.00	
	g	87			
Te	s	0.00		0.00	11.88
	g	47.6		38.1	43.64
TeO$_2$	s	— 77.69		— 64.60	16.99
Th	s	0.00		0.00	13.6
ThCl$_4$	s	—285			
ThO$_2$	s	—292		—278.4	
Ti	s	0.00		0.00	7.24
	g	112		101	43.07
TiCl$_4$	li	—179.3		—161.2	60.4
TiC	s	— 54		— 53	5.8
TiO$_2$ (rutile)	s	—218.0	(—190.4)	—203.8	12.01
Tl	s	0.00		0.00	15.4
	g	44.5		36.2	43.23
TlCl	s	— 48.99		— 44.19	25.9
TlNO$_3$	s	— 58.01		— 36.07	38.2
Tl$_2$SO$_4$	aq	—214.14	(—221.8)	—192.85	65.1
Tm	s	0.00		0.00	
U	s	0.00		0.00	12.03
	g	125			
UCl$_4$	s	—251.2		—230.0	47.4
UF$_4$	s	—443		—421	36.1
UO$_2$	s	—270		—257	18.6
U$_3$O$_8$	s	—898	(—845.2)		
V	s	0.00		0.00	7.05
	g	120		109	43.55
V$_2$O$_3$	s	—290		—271	23.58
V$_2$O$_5$	s	—373	(—437)	—344	31.3
W	s	0.00		0.00	8.0
	g	201.6		191.8	41.55
WC	s	— 9.09			
WO$_3$	s	—200.84	(—195.7)	—182.47	19.90
Xe	g	0.00		0.00	40.53
Y	s	0.00		0.00	
	g	103			42.87

Table 26   continued

Symbol Formula	State of Aggregation	ΔH^0 [kcal/gmol]	ΔG^0 [kcal/gmol]	S^0 $\left[\dfrac{kcal}{kmol\ deg}\right]$
Yb	s	0.00	0.00	
	g	87		41.30
Zn	s	0.00	0.00	9.95
	g	31.19	22.69	38.45
ZnCO$_3$	s	—194.2	—174.8	19.7
ZnCl$_2$	s	— 99.40	— 88.26	25.9
ZnF$_2$	aq	—187.9	—166	
ZnO	s	— 83.17	— 76.05	10.5
Zn(OH)$_2$	s	—153.5		
ZnS (sphalerite)	s	— 48.5	— 47.4	13.8
ZnS (wurtzite)	s	— 45.3	— 44.2	
ZnSO$_3$	s	—233.88	—208.31	29.8
Zr	s	0.00	0.00	9.18
	g	125	115	43.31
ZrCl$_4$	s	—230	—209	44.5
ZrO$_2$	s	—258.2	—244.4	12.03

Bibliography

Physical Chemistry (Textbooks)

BARROW, G. M.: Physical Chemistry. Mc Graw Hill Book Co., New York/London/Toronto 1961

BRDIČKA, R.: Grundlagen der physikalischen Chemie, 4th ed. VEB Deutscher Verlag der Wissenschaften, Berlin 1963

DANIELS, F., and R. A. ALBERTY: Physical Chemistry, 2nd ed. Wiley and Sons, New York/London 1956

EUCKEN, A., and E. WICKE: Grundriß der physikalischen Chemie, 10th ed. Akademische Verlagsgesellschaft Geest & Portig, Leipzig 1959

FINDLAY, A.: Practical Physical Chemistry (ed. J. A. KITCHENER), 8th ed. Longman, Green and Co., London/New York/Toronto 1955

GARN, P. D.: Thermoanalytical Methods of Investigation. Academy Press, New York, London 1965

KORTÜM, G.: Einführung in die chemische Thermodynamik, 4th ed. Vandenhoeck und Ruprecht, Göttingen 1963

KORTÜM, G.: Lehrbuch der Elektrochemie, 3rd ed. Verlag Chemie, Weinheim/Bergstraße 1962

MOELWYN-HUGHES, D. A.: Physical Chemistry

SCHÄFER, K.: Physikalische Chemie, Springer, Berlin/Göttingen/Heidelberg 1951

ULICH, H., and W. JOST: Kurzes Lehrbuch der physikalischen Chemie, 15th ed. Theodor Steinkopff, Darmstadt 1963

3. Geochemical Migration Factors

3.1. Geochemical Migration

The term "geochemical migration of elements" was coined by VERNADKSIĬ (1924) and practically corresponds to terms such as "geochemical distribution paths of the elements" (GOLD-SCHMIDT, 1924) or "geochemical course of life of the elements" (STRUNZ, 1961).

Geochemical migration is defined as the transfer of elements and ions which leads to their dispersion or concentration, especially in the earth's crust.

VERNADSKIĬ (1924) thought the radioactive processes to be the source of energy causing an element migration. A numerical representation of the migration capability of individual elements is given by RAMBERG (1948).

FERSMAN (1934) groups the factors that govern the migration of elements as follows:

a) Internal factors, associated with the properties of the atoms and their combinations; the so-called inherent atomic factors (STRUNZ, 1961). These migration factors are a function of the atomic structure, especially of the outer electron shell of the atoms. They do not change in the course of the geological development of the earth.

b) External factors which are governed by the circumstances to which the atoms and their combinations are exposed; so-called environmentally conditioned factors (STRUNZ, 1961). The external migration factors distinctly change in the course of the geological development of the earth.

STRUNZ (1961) arranged another group, the "factors of the desired and achieved combination". A strict distribution of all individual factors to the above two groups is impossible.

Below only the most important migration factors are discussed which can be assigned to one of the groups with satisfactory accuracy.

The migration of the chemical elements and isotopes must be considered the most important phenomenon in nature (cf. VERNADSKIĬ, 1924; FERSMAN, 1929, 1953).

The geochemical laws of distribution of elements and isotopes (Section 7), the investigation of which is the main task of geochemistry, result from the interaction of internal and external migration factors.

In accordance with their "internal" properties and according to how the action of the external factors may be, atoms and ions enter combinations (while the temperature should be considered a universal factor; SAUKOV, 1953). Only at very high temperatures (e.g. interior of the earth, pre-geological stage of the earth, sun, stars), are chemical combinations unstable and the elements occur in atomic or ionic state. Of the vast number of possible combinations of elements, only about 2,000 minerals occur in the outer geosphere. This may be attributed to the non-uniform stability of combinations under the different outer conditions.

This means that the migration capacity of the elements on the earth and in its crust is mainly dependent on the properties of the combination.

Bibliography

BERNS, R. et al: Factors which Govern the Distribution Coefficients of Mineral-forming Processes. Compilation. Problem. Geokhim. Izd-vo "Nauka", Moscow 1965 (Russian)

BREWER, L.: The Equilibrium Distribution of the Elements in the Earth's Gravitational Field. J. Geology 95 (1951), 490—497

FERSMAN, A. E.: Geochemische Migration der Elemente. Halle 1929 (I.), 1930 (II)

GOLDSCHMIDT, V. M.: Geochemische Verteilungsgesetze der Elemente. II. Beziehungen zwischen den geochemischen Verteilungsgesetzen und dem Bau der Atome. Vidensk. Skrift. I. Mat.-naturv. Kl. No. 4, 1924, Kristiania

GOLDSCHMIDT, V. M.: The Geochemical Background of Minor-element Distribution. Soil Sci. 60 (1945), 1—7

NEUMANN, H., J. MEAD, and C. VITALIANO: Trace Element Variation during Fractional Crystallisations Calculated from the Distribution Law. Geochim. cosmochim. Acta [London] 6 (1954) 90—99

NIGGLI, P.: Geochemie und Konstitution der Atomkerne. Fennia 50 (dedicated to J. J. Sederholm), No. 6

NODDACK, I., and W. NODDACK: Die geochemischen Verteilungskoeffizienten der Elemene. Svensk. Kem. Tidskrift 46 (1934), 173

OVCHINNIKOV, L. N.: Geochemical Mobility of Elements According to the Experimental Data. Dokl. Akad. Nauk SSSR 109 (1956), 141—143 (Russian)

RAMBERG, H.: Radial Diffusion and Chemical Stability in the Gravitational Field. J. Geology 56 1948), 448—458

SAUKOV, A. A.: Evolution of the Migration Factors of the Elements in Historical Geology. Izvest, AN SSSR, ser. geol. 5 (1951), 3—16 (Russian)

SHAW, D. M.: Element Distribution Laws in Geochemistry. Geochim. cosmochim. Acta [London] 23 (1961), 1/2, 116—134

SHCHERBINA, V. V.: Principles of Combinations of Cations and Anions during their Formation. Geokhimiya 11 (1962), 1361—1369 (Russian)

SZÁDECZKY-KARDOSS, E.: Studien über die geochemische Migration der Elemente. Part I—III. Acta geol. Acad. Sci. Hung. 2 (1954), 135—144, 145—167, 269—283

STRUNZ, H.: Die für den geochemischen Lebenslauf der Elemente verantwortlichen Faktoren. Acta geol. Acad. Sci, Hung. 7 (1961), 1/2, 53—55

URUSOV, V. S.: Direction of Natural Exchange Reactions and the "Affinity" of Elements with One Another. Geokhimiya 5 (1965), 668—673 (Russian)

3.2. Internal Migration Factors

These are factors or properties of atoms and their combinations which result from the structure of the atoms. Included are also the properties which have been mentioned in Section 2 and which have been referred to as fundamental concepts and which are also known as chemical and physical constants.

The quantitative proportion of the elements is conditioned by the structure of the atomic nucleus (ODDO, 1914; HARKINS, 1917).

Oddo-Harkins rule: Of two elements adjacent to each other in the periodic system, the Clarke (see p. 225) of the element having an even atomic number (Z) is higher than that of the element with an odd atomic number.

The geochemical way of distribution of a given quantity of an element is dependent on the structure of the electron shell (GOLDSCHMIDT, 1924).

Bibliography

AHRENS, L. H.: The Significance of the Chemical Bond for Controlling the Geochemical Distribution of the Elements — Part I. Phys. Chem. of the Earth *5* (1964), 1—54

GOLDSCHMIDT, V. M.: Geochemische Verteilungsgesetze der Elemente II (Beziehungen zwischen den geochemischen Verteilungsgesetzen und dem Bau der Atome). Vidensk. Skrifter, I. Mat. naturv. Kl., No. 4, Kristiana 1924

GOLDSCHMIDT, V. M.: The Principles of Distribution of Chemical Elements in Minerals and Rocks. J. Chem. Soc. [London] (1937), 655

HARKINS, W.: J. Amer. Chem. Soc. *39* (1917), 859

HARKINS, W.: Philos. Mag. J. Sci. *42* (1921), 386

ODDO, G.: Z. anorg. Chem. *87* (1914), 266

3.2.1. "Specific Gravity" of Ions and Atoms

The concept of the ionic and atomic specific gravity (in German "Ionenwichte" and "Atomwichte") was coined by SZÁDECZKY-KARDOSS (1954) and is considered by him to be the most influential factor in geochemical migration. He explains the migration of the elements as a differentiation by diffusibility in the gravitational field of the earth, i.e. by gravitative sorting.

Ionic or atomic specific gravity $\varrho = \dfrac{3\,A}{4\,\pi\,r^3} = \dfrac{\text{atomic mass}}{\text{atomic or ionic volume}}$

where A is the atomic mass and r the atomic or ionic radius. Numerical values of ϱ and of the electron density are given in the above-mentioned publication. When ϱ is plotted as a function of the atomic number of the elements, the curve obtained resembles that of a similar representation of the atomic or ionic radii. With the help of ϱ, the sorting of the elements in Goldschmidt's various earth spheres (cf. Section 7), and Korzhinskii's series of the migration capability of the elements in the case of metamorphic processes (Section 3.2.2.3), can be explained rather well.

Restricting the general use of the terms ionic and atomic specific gravities, LEUTWEIN and DOERFFEL (1956) point out that the accuracy of the numerical values of ϱ is conditional, because the ionic radii are not exactly known, and that in addition, it is impossible to extend this concept to-complex ions.

Bibliography

LEUTWEIN, F., and K. DOERFFEL: Über einige Verfahren zur theoretischen Klärung geochemischer Prozesse, unter besonderer Berücksichtigung der Gitterenergien. Geologie *5* (1956), 65—100

SZÁDECZKY-KARDOSS, E.: Studien über die geochemische Migration der Elemente, Part I: Die Ionenwichte und ihre geochemisch-geologische Rolle. Acta geol. Acad. Sci. Hung. *2* (1954), 135—144

3.2.2. Energetics of Crystals

3.2.2.1. Lattice Energy

Lattice energy U is the minimum value of Φ_p at $T = 0°$ abs.; $U = -\Phi_p$ [kcal/mole].
Φ_p means the potential energy between the structural units of an undefined lattice (simplest physical constant of this lattice).

Lattice Energy U (general)

U is defined as the work required to decompose the crystal lattice of a mole of substance into its elementary structural units and to displace them so that there is a sufficiently large distance between them; or the decrease in enthalpy ΔH which occurs when a gram-atom each of the gaseous structural units enters a solid lattice combination:

$$AB]_{\text{solid}} \rightleftharpoons A + B \qquad \Delta H = \pm U .$$

Lattice Energy (in a broader sense)

Energy content of a crystal lattice at $T = 0°$ abs. Total energy of a lattice at temperature T:

$$U_{\mathrm{T}} = U + \Sigma W + \int\limits_0^T \int\limits_0^p C_{\mathrm{p}}\,(T, p)\,dT\,dp .$$

Where:

U is the lattice energy at $T = 0°$ abs., U_{T} the lattice energy at $T°$ abs., W any given heat zone, e.g. heat of transformation, T the temperature, p the pressure and C_{p} the molar heat at $p = $ constant.
The difference between U and U_{T} is usually small and can be neglected.

Table 27. Forces between Particles of Different Chemical Bond Types and Their Magnitude, and Methods of Calculating the Lattice Energies (S heat of sublimation, D heat of dissociation, Q heat of formation, I ionization energy), compiled from Doerffel (1956); Leutwein and Doerffel (1956)

Type of Bond	Kind and Mode of Linkage Forces	Magnitude of Linkage Forces kcal/mole	Calculation of Lattice Energy According to
heteropolar	electrostatic fields, which, in the ideal case, spherically surround any ion	10^2—10^3	Born's equation, Kapustinskiĭ's equation Haber-Born cycle Fersman's energy coefficients
homopolar	caused by pairs of electrons which belong to two atoms in common	roughly of the order of the heteropolar forces; quantitative calculation (on a wave-mechanical base) possible only in very simple cases	for combinations of one kind of atoms: U_{h} heat of sublimation (e.g. diamond); for combinations of different kinds of atoms (e.g. SiO_2): $U_{\mathrm{h}} = \Sigma S + \Sigma D + Q$ (cf. cycle according to Haber-Born)
metallic	positive metal residues and free electrons, the latter do not belong to a definite atom ("electron gas")	$\approx 10^1$—10^2	$U_{\mathrm{m}} = S + I$ (cf. cycle according to Haber-Born)
van der Waals	electrically neutral particles bonded by loose, so-called non-classical dispersive powers. Dipole moments cause attractive forces (resonance forces)	1—12	$U_{\mathrm{w}} = S + 9/8\,R\vartheta$ R gas constant ϑ characteristic temperature or $U_{\mathrm{w}} \sim S$

There are interrelations between the lattice energy and the following properties (and others) (see also DOERFFEL, 1956):

a) Melting point;
b) Solubility;
c) Hardness;
d) Sequence of mineral crystallization (mineral paragenesis, age sequences) see Table 28; see also SAUKOV (1953), DOERFFEL (1956), LEUTWEIN and DOERFFEL (1956); whereas this is rejected by SZÁDECZKY-KARDOSS (1958): only being applicable to certain simple sulphides, oxides and a few halogens, and absolutely not applicable to combinations whit anion complexes and others;
e) Isomorphism (energy gain determines the direction of the exchange process);
f) Lattice geometry;
g) Probability of the existence of minerals (according to the Thomson and Berthelot principle).

SAUKOV (1953), for example, considers the energy of the crystal lattice the most important geochemical constant.

Table 28. Crystallization Sequence as a Function of Lattice Energy

Beginning ——————————— *Crystallization Sequence* ——————————— ► End		
high ——————————— *Lattice Energy U* ——————————————— ► small		
high negative values (minerals with a high melting point and poor solubility)	$U < 0^1)$	low negative values (minerals with a low melting point and good solubility)

Examples:

a) Crystallization sequence from melts: BOWENS' series

	Olivine	Pyroxene	Amphibole	Biotite	Feldspar	Quartz
$U_I{}^2)$	$-4.9 \cdot 10^3$	$-4.0 \cdot 10^3$	$-3.9 \cdot 10^3$	$-3.8 \cdot 10^3$	$-2.9 \cdot 10^3$	$-3.0 \cdot 10^3$

b) Crystallization sequence from solutions: Freiberg "kiesig-blendige (kb) Bleierzformation" (according to DOERFFEL, 1956), further see KRAFT (1959), etc.

	SiO_2	SnO_2	FeS	FeS_2	(Zn, Fe)S	$CuFeS_2$	Cu_3SbS_3	PbS	$FeCO_3$	$CaCO_3$
U_I	-1501	-1409	-838	-838	-855	-836	-791	-718	-743	-648

[1]) These are always negative values considered from the viewpoint of the system!
[2]) U_I: Lattice energy related to the "structural unit" (Fersman and Lemmlein), that is to say, U is divided by the number of the atoms forming the complex.

The relations between the crystallization sequence and the lattice energy shown in Table 28 were modified by FERSMAN, 1935; SZÁDECZKY-KARDOSS, 1955a, 1958; KLIBURSZKY, 1958. For example: The lattice energy is not a coefficient of the measure of the free energy or the free enthalpy and thus does not indicate the direction of geochemical processes. Further information is given by SELOMOV, (1964) a. o.

Possibilities of Calculating the Lattice Energy

Only for non-parametric diagonal lattices of heteropolar combinations derived from simple non-deformed ions (DOERFFEL, 1956).

1. BORN's formula (1919)

$$J = -\frac{AN_L z^2 e_0}{a}\left(1 - \frac{1}{m}\right)$$

In which A is Madelung's constant (to be determined for any one type of lattice; for instance see EUCKEN, 1949; KAPUSTINSKII, 1933), N_L is the Loschmidt number, i.e. the number of molecules per mole $= (6.0227 \pm 0.011) \cdot 10^{23}$ mole^{-1}, z is the ionic valence, e_0 is the elementary electrical charge $= 4.803 \cdot 10^{-10}$ (cgs)$_{st}$, a is a measure of the shortest interionic distance (lattice constant), m is Born's exponent of repulsion.

2. BORN-MAYER equation (1932)

$$U = N_L\left[-\frac{Ae_0^2}{a} + B\left(e^{-a/m'}\right) - \frac{C}{a^6} - \frac{D}{a^8} + \varepsilon\right]$$

where $B\left(e^{-a/m'}\right)$ is Unsöld's two-constant exponential expression, m' is the empiric term of repulsion (0.345), C/a^6 is the proportion of the van der Waals forces, D/a^8 are the dipole-quadrupole forces, z represents the zero of energy.

This equation can only be applied to highly symmetrical lattices and calculating it is rather time-consuming.

3. KAPUSTINSKII's equations (1933 and 1943)

Elimination of the Madelung number by combining it with the Loschmidt number and the electric elementary quantum from Born's equation and conversion into terms of the caloric system of measurement (thus, Born's formula can be applied to combinations with any type of lattice):

$$U = -256.1\frac{n z_K z_A}{r_K + r_A} \quad (1933),$$

$$U = -287.2\frac{n z_K z_A}{r_K + r_A}\left(1 - \frac{0.345}{r_K + r_A}\right) \quad (1943),$$

where n is the number of ions in a molecule, z_K, z_A represent the charge (valence) of cations and anions, r_K, r_A are the radii of cations and anions.

4. Methods of calculating have been developed on the basis of the theory of electron gas by SARKISOV (1954)

The equations given in 1. and 3. lead in many cases to results which are correct under certain conditions only because they assume ions to be rigid-elastic spheres.

Empiric Determination of the Lattice Energy

1. HABER-BORN Cycle (1919); in DOERFFEL (1956).

Heat of formation Q of a substance decomposed into individual partial amounts W_i (in which the lattice energy is included)

$$Q = \Sigma W_i.$$

105

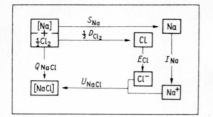

Fig. 18   HABER-BORN cycle (for NaCl)

Q Heat of formation; S Heat of sublimation; D Energy of dissociation; I Ionization energy; E Electron affinity; U Lattice energy

For NaCl (cf. Fig. 18), for example we obtain:

$$Q_{\mathrm{NaCl}} = S_{\mathrm{Na}} + \frac{1}{2} D_{\mathrm{Cl_2}} + I_{\mathrm{Na}} + E_{\mathrm{Cl}} + U_{\mathrm{NaCl}} ,$$

$$U_{\mathrm{NaCl}} = Q_{\mathrm{NaCl}} - S_{\mathrm{Na}} - \frac{1}{2} D_{\mathrm{Cl_2}} - I_{\mathrm{Na}} - E_{\mathrm{Cl}}$$

where Q is the heat of formation, S is the heat of sublimation, D is the energy of dissociation, I is the ionization energy, E is the electron affinity, U is the lattice energy.

Although, in a strict sense, this cycle is only applicable to heteropolar bonds, it can also be extended to substances with non-polar bonds (DOERFFEL, 1956, and others). The lattice energies resulting from the cyclic process must be considered the most exact ones.
Limitation: The Haber-Born cycle frequently cannot be interpreted numerically, because the thermo-chemical values required are missing.

2. Direct experimental determination of the lattice energy by measuring ionic partial pressures, see MAYER (1930) or MAYER and HELMHOLTZ (1934); Lit. cf. DOERFFEL, 1956.

Calculation of the Lattice Energy on the Basis of "Crystallochemical Electronegativities" According to Kapustinskiǐ (1949)

Crystallochemical electronegativities are not identical with the electronegativities calculated by Pauling (Table 16). The lattice energy values determined according to this method for compounds of the AB type of formula, with ions which are not highly polarisable, compare well with the values determined according to the Haber-Born cycle. Further information is given in KAPUSTINSKIǏ (1949a, b) and DOERFFEL (1956).

3.2.2.2. Energy Coefficients EK and Paragene Concept by A. E. Fersman

Energy Coefficients

The energy coefficient, EK, is the proportion of energy contributed by any ion (thought at infinity) to the formation of a heteropolar combination. It is a quantity which is characteristic of any ion.

Calculation of the Energy Coefficients of Single Ions

    1. On the basis of known lattice energies
        a) The lattice energy values calculated on the basis of data which were determined experimentally were called "experimental energy coefficients" by Fersman. They should be considered the most reliable ones.

106

b) On the basis of Kapustinskiĭ's equation (if the thermo-chemical data required for the cyclic process are missing).

2. On the basis of approximation formulas

a) For anions and cations of a lower valence, the following holds: $EK = -\dfrac{z^2}{2r}$ (z valence, r ionic radius).
The magnitudes of the energy coefficients of ions with the same radius and the same valence are equal to each other. The energy coefficients increase with increasing valence and decreasing ionic radius.

b) For other cations, the following obtains:

$$EK = -\frac{0.75\,z^2}{2r}\,(r + 0.20)\,.$$

3. On the basis of the ionization potential

The energy coefficients of certain cations show the same magnitude as their ionization energies. This relation has also been used for the calculation of the energy coefficients of cations and of the lattice energies of the corresponding combinations.

Calculation of the Energy Coefficients of Complex Ions

$$EK = -\frac{z_A'^2}{2r_A'}$$

$z_A' = $ valence of the anion complex

$r_A' = r_x + 2r_y\,.$

Where:

r_A' is the radius of the anion complex, r_X is the radius of the central ion, r_Y is the radius of the ligands.

Only approximate values are obtained for the radii of anion complexes, because in reality, the rather considerable deformation of the ligands is not taken into account. The energy coefficients of anion complexes determined according to the above formula are always too low.

For calculating the lattice energy of heterodesmic combinations (e.g. $BaSO_4$) see SZÁDECZKY-KARDOSS (1958).

A certain parallelism between energy coefficients and the crystallization sequence can only be found with simple sulphides, oxides, and certain halides and the like (SZÁDECZKY-KARDOSS, 1958).

Determination of the Lattice Energy with the Help of the Energy Coefficient according to A. E. Fersman

The lattice energy of U any combination (for exceptions see: SZÁDECZKY-KARDOSS, 1955a, 1958) can be calculated quickly with the help of the energy coefficient, and with an accuracy that is sufficient for geochemical purposes:

$$U = -256.1\,(m\,EK_K + n\,EK_A)\,.$$

Where m, n are the numbers of cations und anions, respectively; 256.1 is the Kapustinskiĭ number. According to the principle of addition, the general formula for more complicated

107

combinations (e.g. for leucite $KAlSi_2O_6$) is as follows:

$$U = -256.1 \, (EK_{K^+} + EK_{Al^{3+}} + 2\,EK_{Si^{4+}} + 6\,EK_{O^{2-}}) \, .$$

The energy coefficients are values which are almost parallel to the ionic potentials (SZÁDECZKY-KARDOSS, 1955 a).

Determination of the Lattice Energies from Ionic Energies according to K. Doerffel (1956)

In principle, this is the same method of calculating as with Fersman's energy coefficients. However, in contrast with the latter, we obtain the energy contribution from any single ion directly in the caloric system of measurement (multiplication by Kapustinskiï's factor $K = 256.1$ is not necessary). For further information see DOERFFEL (1956) who, among other things, gives details on electron affinities, tables of ionic energies, examples of calculations, examples from geochemistry and analytic chemistry and values of radii of complex anions.

Paragene Concept

The paragene is a function which determines the position of a given combination in the paragenetic sequence, i.e. in the order in which minerals precipitate or crystallize.

One obtains the paragene of a combination by an addition of the paragenes of the ions, analogously to the determination of the lattice energy U by an addition of the energy coefficients of the ions.

At present, an exact mathematical formulation of paragenes is not yet possible; they were empirically determined from the precipitation order of minerals for various ions. The majority of paragenes correspond approximately with the energy coefficients (see Table 29). In several cases, where such factors as dissociation, concentration, polarization, etc. are taken into account, paragenes characterize the actual order of the sequences of crystallization in question much better than the values of the lattice energy (SAUKOV, 1953, and others).

Qualification: These empirically determined paragenes only apply to the minerals of the paragenesis for which they were determined. The same minerals from different associations may show different paragenes (due to the different concentration of matter, etc.)! Paragenes are not generally applicable and are rarely used.

Table 29. Energy Coefficients and Paragenes according to FERSMAN (1935, 1937)

1. The energy coefficients (EK) are already multiplied [kcal/mole] by their proportionality factor $(K = 256.1)$ and are given without taking the minus sign into consideration. The values in brackets were newly calculated by DOERFFEL (1956);
2. EK from SAUKOV (1953);
   Paragenes from SAUKOV (1953).

Ions	EK 1.	EK 2.	Empiric Paragenes (FERSMAN)	Ions	EK 1.	EK 2.	Empiric Paragenes (FERSMAN)
Ag^+	154	0.60	0.6	$(AlO_4)^{5-}$	1,025	4.0	6
Au^+	166	0.65		$(AsO_4)^{3-}$	392	1.53	
Al^{3+}	1,270	4.95	7.5				
As^{3-}	680	2.65	2.5	B^{3+}	1,535 (1,620)	6.00	
As^{3+}	(1,200)			Ba^{2+}	346	1.53	1.2
As^{5+}	3,890	15.20		Be^{2+}	678	2.65	2

Table 29 continued

Ions	EK 1.	EK 2.	Empiric Paragenes (FERSMAN)	Ions	EK 1.	EK 2.	Empiric Paragenes (FERSMAN)
Bi^{3+}	(1,030)			K^+	92	0.36	0.35
Bi^{5+}	3,050	11.90					
Br^-	57	0.22	0.2	La^{3+}	918	3.58	
				Li^+	141	0.55	0.3
$(BO_3)^{3-}$	430	1.68		Lu^{3+}	(1,100)	3.98	4
$(BeO_4)^{6-}$	1,357	6.0					
				Mg^{2+}	538	2.10	4
C^{4+}	3,124 (3,400)	12.20	12	Mn^{2+}	512	2.00	3
Ca^{2+}	448	1.75	2	Mn^{4+}	2,330	9.10	
Cd^{2+}	512	2.00	1.8	Mo^{4+}	2,180	8.50	9
Ce^{3+}	(940)			Mo^{6+}	(4,760)		
Cl^-	64	0.25	0.3				
Co^{2+}	551	2.15	5	$(MoO_4)^{2-}$	149	0.58	1
Cr^{3+}	1,220	4.75	8	$(MgO_4)^{6-}$	1,358	5.30	
Cr^{6+}	(6,020)						
Cs^+	77	0.30	0.2	N^{5+}	5,060 (5,640)		
Cu^+	179	0.70	0.7	N^{3-}	920	3.60	
Cu^{2+}	538	2.10	2	Na^+	115	0.45	0.5
$(ClO_4)^-$	54	0.21		Nb^{5+}	3,485	13.60	
CN^-	64	0.25		Nd^{3+}	(965)		
$(CO_3)^{2-}$	200	0.78	0.5	Ni^{2+}	558	2.18	8
$(CrO_4)^{2-}$	172	0.67	1				
$(CrO_3)^{2-}$	192			$(NO_3)^-$	49	0.19	0.2
				$(NH_4)^+$	95	0.37	
Dy^{3+}	(1,015)						
				O^{2-}	397	1.55	1.5
Er^{3+}	(1,038)			Os^{4+}	2,280	8.90	9
Eu^{3+}	(972)						
				OH^-	95	0.37	
F^-	95	0.37	0.4				
Fe^{2+}	543	2.12	6	P^{3-}	690	2.70	
Fe^{3+}	1,320	5.15		P^{5+}	3,560	15.50	
Fr^+	(75)			Pb^{2+}	423	1.65	1.5
				Pb^{4+}	2,035	7.95	
Ga^{3+}	1,385	5.41	2	Pd^{2+}	(540?)		
Ge^{4+}	2,700	10.53		Pm^{3+}	(968?)		
Gd^{3+}	(986)			Pr^{3+}	(952)		
				Pt^{2+}	(521)		
H^+	282	1.10					
H^-	82	0.32	0.1	$(PO_4)^{3-}$	384	1.50	2
Hf^{4+}	2,000	7.81	8				
Hg^+	238	0.93	0.2	Ra^{2+}	(357)		
Hg^{2+}	538	2.10	0.2	Rb^+	84	0.33	0.3
Ho^{3+}	(1,030)			Re^{4+}	2,280	8.90	
				Re^{7+}	7,200	28.10	9
In^{3+}	1,115	4.35		Rn^{3+}	(1,275)		
Ir^{4+}	2,290	8.93	9	Rn^{4+}	2,330	9.10	9
I^-	46	0.18	0.1				
				S^{2-}	295	1.15	1.3
$(IO_3)^-$	36	0.14		S^{6+}	(5,810)		
$(IO_4)^{5-}$	855						

Table 29 continued

Ions	1.	EK 2.	Empiric Paragenes (Fersman)	Ions	1.	EK 2.	Empiric Paragenes (Fersman)
Sb^{3-}	590	2.30		Tl^+	108	0.42	0.4
Sb^{3+}	(951)			Tl^{3+}	885	3.45	
Sb^{5+}	3,135	12.25		Tm^{3+}	1,038		
Sc^{3+}	1,190	4.65	4				
Se^{2-}	282	1.10	1	$(TiO_4)^{4-}$	628	2.45	
Si^{4+}	2,200	8.60	10				
Sm^{3+}	(972)			U^{4+}	1,795	7.00	6
Sn^{2+}	(432)			U^{6+}	4,270		
Sn^{4+}	2,025	7.90	8				
Sr^{2+}	384	1.50	1.5	V^{3+}	1,360	5.32	
				V^{5+}	3,870	15.12	5
SH^-	59	0.23					
$(SO_4)^{2-}$	174	0.68	0.5	W^{6+}	4,956 (4,275)	19.35	
$(SeO_4)^{2-}$	167						
$(SiO_4)^{4-}$	704	2.75	4	$(WO_4)^{2-}$	146	0.57	2
Ta^{5+}	3,485	13.60		Y^{3+}	1,010	3.95	4
Tb^{3+}	(1,000)			Yb^{3+}	(1,071)		
Te^{2-}	243	0.95	0.9				
Th^{4+}	1,742	6.80	7	Zn^{2+}	565	2.20	2
Ti^{3+}	1,190	4.65	5	Zr^{4+}	2,003	7.85	8
Ti^{4+}	2,170	8.40	9	$(ZrO_4)^{4-}$	589	2.30	

3.2.2.3. Potential Values as an Expression of the Binding Forces

An exact thermodynamic calculation of geochemical phenomena and processes by is impossible (because only approximations are obtained).

The most important questions of theoretical geochemistry can be settled by easily computable potential values (compound potential, etc.). In contrast with energy calculations, the advantage of computations of potentials as used in geochemistry seems to be the fact that they are easier to handle because they start from the determination of the binding forces, and not from the energy content.

Concepts such as anionic potential, atomic potential and compound potential were coined (Szádeczky-Kardoss, 1954a, 1955a, b, 1958; Grassely, 1958; Kliburszky, 1958), analogously to the Cartledge ionic potentials (cf. 2.6.4.).

For instance, the ionic potential exercises a decisive influence on the formation of chemical compounds in an aqueous medium, e.g. in sedimentary processes (Goldschmidt, 1937), also in magmatic processes (Table 30): (mobility series of ions according to Korzhinskii, 1950 further Szádeczky-Kardoss, 1954b; also in Ringwood, 1955). In the sedimentary field, the separation of elements occurs in the sequence of dropping ionic potentials.

Metasomatic-metamorphic mobility series (according to Korzhinskii)

H, C, S, Cl, K, Na, O, Ca, Mg, Fe, Si, P, Al, Ti.

Decreasing mobility ⎯⎯⎯⎯⎯⎯⎯⎯⎯→

←⎯⎯⎯⎯⎯⎯⎯⎯Decreasing ionic potentials

"Energy index" according to Gruner (1950) as a measure used when a mineral crystallizes out (compare also Szádeczky-Kardoss, 1955).

ation	Φ	Behaviour in Magma	Cation	Φ	Behaviour in Magma
s+	0.60		Sc^{3+}	3.72	
b+	0.68		Th^{4+}	3.92	
l+	0.68		Sb^{3+}	3.96	mostly occurring as
.+	0.75		V^{3+}	4.05	free ions
g+	0.79		U^{4+}	4.15	
Ta+	1.03	occupy vacancies as	Fe^{3+}	4.68	
u+	1.04	free ions,	Cr^{3+}	4.77	
i+	1.47	do not form complexes			
Ba2+	1.49		Ga^{3+}	4.83	
b2+	1.67		Zr^{4+}	5.08	
r2+	1.79		Hf^{4+}	5.12	
Mg2+	1.82		Sn^{4+}	5.64	
a2+	2.02		Be^{2+}	5.72	complex-forming ions:
d2+	2.06		Al^{3+}	5.88	complexes can com-
n2+	2.16		Ti^{4+}	5.88	bine into lattices, the
Mn2+	2.50		Ta^{5+}	7.35	stability of the com-
La3+	2.64		Nb^{5+}	7.5	plexes decreases with
anthanides	2.64—3.48		Ge^{4+}	7.56	decreasing cationic
			V^{5+}	8.50	potential
n2+	2.70	at a high $\dfrac{Si + Al}{O}$ ratio	Si^{4+}	9.52	
e2+	2.70		Mo^{6+}	9.66	
Co2+	2.78	of the magma in	W^{6+}	9.66	
Cu2+	2.78	vacancies, at a low	As^{5+}	10.85	
Ni2+	3.00	$\dfrac{Si + Al}{O}$ ratio form-	B^{3+}	13.05	
Mg2+	3.04		P^{5+}	14.30	
Bi3+	3.12	ing lattices	S^{6+}	19.98	
			C^{4+}	25.00	

Bibliography

Born, M.: Eine thermochemische Anwendung der Gitterenergie. Verh. dtsch. physik. Ges. 21 (1919), 13, 679

Born, M., and J. E. Mayer: Z. Physik. 75 (1932), 15

Doerffel, K.: Das System der Ionenenergien — Ein Näherungsverfahren zum Berechnen von Gitterenergien heteropolarer Verbindungen. Freiberger Forschungshefte, C 20 (1956), 1—104

Fersman, A. E.: Use of the EK-values for the Determination of the Lattice Energy. Dokl. Akad. Nauk SSSR, Dept. Math. and Nat. Sc. 1935, 1426 (Russian)

Fersman, A. E.: Energy Indices in Geochemistry. Dokl. Akad. Nauk SSSR 2 (1935), 266

Fersman, A. E.: The EK-System. Dokl. Akad. Nauk SSSR 2 (1935), 564—566 (Russian)

Gopal, R.: Relation between Lattice Energy and Melting Points of Some Crystalline Substances. Z. anorg. allg. Chem. 278 (1955), 43

Grimm, G. G., and K. F. Herzfeld: Die chemische Valenz der Metalle als Energiefrage. Z. Physik 19 (1923), 141

Grassely, G.: Die Veränderlichkeit der Komplex-Anionenpotentiale in anisodesmischen und mesodesmischen Strukturen. Acta geol. Acad. Sci. hung. 5 (1959), 293—311

Grassely, G.: Confrontation of Complex Anionic Potentials and Compound Potentials with Thermochemical Data. Acta geol. Acad. Sci. hung. 9 (1965) 3/4, 329—338

GRUNER, J. W.: An Attempt to Average Silicates in the Order of Reaction Energies at Relatively Low Temperatures. Amer. Mineralogist *35* (1950), 137—148

HABER, F.: Verh. dtsch. physik. Ges. *21* (1919), 750

HOPPE, R.: Über eine neue einfache Methode zur Berechnung von MADELUNG-Faktoren. Z. anorg. allg. Chem. *283* (1956), 196

HUSH, N. S., and M. H. L. PRYCE: Influence of the Crystalfield Potential on Interionic Separation in Salts of Divalent Iron-group Ions. J. chem. Physics *28* (1958), 244

KAPUSTINSKIĬ, A. F.: A Universal Equation for the Lattice Energy of Ion Crystals. J. allg. Chem. *13* (1943), 492 (Russian)

KAPUSTINSKIĬ, A. F.: Allgemeine Formel für die Gitterenergie von Kristallen beliebiger Struktur. Z. physik. Chem., Abt. B, *22* (1933), 257—260

KAPUSTINSKIĬ, A. F.: Dokl. Akad. Nauk SSSR. LXVII (1949), No. 3, 467; No. 4, 663 (Russian)

KLIBURZKY, B.: Die physikalischen Grundlagen der geochemischen Potentialberechnung. Acta geol. Acad. Sci. hung. *5* (1958), 313—321

KRAFT, M.: Die Ausscheidungsfolge der Erzmineralien auf der Lagerstätte Freiberg/Brand in Abhängigkeit von der Gitterenergie. Geologie 1959, 303—314

LEBEDEV, V. I.: Die geoenergetische Theorie A. E. FERSMANS und ihre Entwicklung in einem Vierteljahrhundert. Freiberger Forschungsh., Ausg. C *79* (1960), 172

LEMMLEIN, G. G.: Sequence of Precipitating Silicates from the Magmatic Melt and their Crystal Energies. Dokl. Akad. Nauk SSSR, *I* (1936), 33 (Russian)

LEUTWEIN, F., and K. DOERFFEL: Über einige Verfahren zur theoretischen Klärung geochemischer Prozesse, unter besonderer Berücksichtigung der Gitterenergie. Geologie *5* (1956), 65—100

SELOMOV B. B.: An Experiment on the Energetic Analysis of Bowen's Reaction Series. Geokhimiya, *12* (1964), 1313—1317 (Russian)

SZÁDECZKY-KARDOSS, E.: Vorläufiges über Anionenpotentiale und Verbindungspotentiale. Acta geol. Acad. Sci. hung. *2* (1954a), 285—298

SZÁDECZKY-KARDOSS, E.: Studien über die geochemische Migration der Elemente, Part III (Über die Rolle der Oxydationsgrade, der Ionenwichten und der Ionenpotentiale in der Gesteinsmetamorphose). Acta geol. Acad. Sci. hung. *2* (1954b), 269—283

SZÁDECZKY-KARDOSS, E.: Das Verbindungspotential und seine Beziehungen zum Schmelzpunkt und zur Härte. Acta geol. Acad. Sci. hung. *3* (1955a), 115—161

SZÁDECZKY-KARDOSS, E.: Bemerkungen zu einer Arbeit von F. LEUTWEIN and K. DOERFFEL. Acta geol. Acad. Sci. hung. *5* (1958), 359—380

SZÁDECZKY-KARDOSS, E., and G. GRASSELY: On the Present Stage of Development of the Potential Concept in Geochemistry. Acta geol. Acad. Sci. hung. *9* (1965) 3/4, 313—328

TAUSSON, L. W.: Zur Frage der Energetik der heterovalenten Isomorphie in Silikaten. Sowjetwissenschaften, naturwiss. Abt. *1* (1950), 196

UNSÖLD, A.: Beiträge zur Quantenmechanik der Atome. Ann. Physique *82* (1927), 355

URUSOV, V. S.: On the Use of the Idea about the Crystal Lattice Energy. Geokhimiya *5* (1965), 551—555 (Russian)

3.2.3. Isomorphism Relations and Diadochy

3.2.3.1. Isomorphism

Isomorphism is said to exist if isostructural types of crystals form mixed crystals (isomorphous miscibility).

Originally, MITSCHERLICH (1819/1821) defined isomorphism in terms of substances of different chemical composition but having the same or a very similar crystalline form (further information is given in textbooks of crystal chemistry).

Isomorphous relations result from

1. crystallogeometric relations (i.e., "type"-relations)

   a) isotypical or isostructural: types of crystals having the same space group; analogy between the chemical summation formulas, lattice dimensions, and occupation of the same sites of atoms, e.g. NaCl, KCl, PbS. The absolute size of the structural unit and the character of bond are unimportant;

   b) homeotypic: structures which are no longer isotypical, but still show to a high degree a structural-geometric affinity (e.g. sphalerite-wurtzite, spinel-corundum);

   c) heterotypic: structures which do not exhibit any resemblance to each other.

2. the capability of forming mixed crystals (miscibility)

   a) formation of mixed crystals by isotypical crystal types = isomorphous miscibility;

   b) formation of mixed crystals by homotypic crystal types = homomorphous miscibility;

   c) formation of mixed crystals by heterotypic crystal types = heteromorphous miscibility.

   Structurally, a distinction can be made between the miscibility patterns shown in Table 31.

Miscibility without a gap

The two components (A and B) are miscible in any proportions.

Miscibility gaps

One component can take up only a certain percentage of the other component, i.e., in certain regions between A and B there are no mixed crystals. At higher temperatures usually complete miscibility (e.g. $KAlSi_3O_8$-orthoclase and $NaAlSi_3O_8$-albite). Pressure has minor influence.

Segregation

Mixed crystals which are homogeneous at higher temperatures break up into the end-member components on cooling down (e.g. perthitic alkali feldspars), especially in minerals where the structural units that replace each other show greater differences in radii (e.g. K 1.33 Å and Na 0.97 Å).

In the case of *isomorphous miscibility*, especially the size of the structural units that replace one another must be equal or similar. Deviations of up to 15% from the smaller particle are possible.

Thus, in such mixture series, changes occur in accordance with the properties of the two final end-members A and B and with their percentage in the mixture (cf. Vegard's rule; VEGARD, 1921/1927) with respect to:

1. lattice constant of the mixed crystal, Figs. 46 and 47,

2. refraction of light (or reflection factor; Fig. 19),

3. density,

4. hardness (e.g. micro-hardness test) etc.

Table 31. Possibilities of Mixed Crystal Formation (compiled according to LAVES, 1939, LEUTWEIN and SOMMER-KULASZEWSKI, 1960, and KLEBER, 1963)

Mixed Crystal Formation by	Types of Mixed Crystals	isovalent lattice sites are completely occupied	Method of Substitution	Examples of Miscibility
Simple substitution	isotypical homotypic heterotypic		n structural units of crystal A replaced by n structural units of crystal B	statistically irregular distribution of the foreign structural units (in heteropolar and homeopolar mixed crystals Mg^{2+} by Fe^{2+} in olivine $(Mg, Fe)SiO_4$
				regular distribution of the foreign atoms (superlattice structure; in metallic mixed crystals). intermediary or metallic phases, artificial alloys, natural Pt-alloys, Cu-arsenides, tellurides
Coupled substitution			diadochy between differently charged ion types, preserving the charge neutrality	Na^+ and Si^{4+} by Ca^{2+} and Al^{3+} in the plagioclases; many rock-forming silicates
Addition substitution	homotypic heterotypic		number of the occupied isovalent sites of the lattice is greater than in the pure phase	system —Al_2O_3-$MgAl_2O_4$ (spinel), Au in pyrite and arsenopyrite lattice; minerals with crystal water 1. coordination water: H_2O molecules occupy certain sites in the lattice (hydrates) 2. structural or zeolitic water: H_2O molecules statistically distributed to lattice voids (zeolites). Lattice expansion or swelling (montmorillonite)
Subtraction substitution			number of the occupied isovalent sites of the lattice is smaller as compared with the pure phase	between $NaAlSiO_4$ (nepheline) and SiO_2 (quartz)
Division substitution	heterotypic		a stoichiometrically conditioned number of structural units is distributed to a considerably greater number of sites of the lattice	system $AgBr$-$CuBr$

The determination of these physical properties permits conclusions on the chemical composition of a solid solution (dependence between refraction of light and chemical composition of rockforming silicates see, for instance, in TRÖGER, 1952).

It should be noted that isotypicality and isomorphism are related to the crystal as a whole (STRUNZ, 1961).

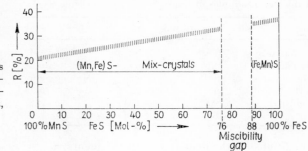

Fig. 19 Dependence of reflection values hatched; R [%]) on the chemical composition of MnS-FeS mixed crystals (sulphide inclusions in steel; after TROJER, 1962)

3.2.3.2. Diadochy

Diadochy ("vicariism of components"; FUCHS, 1815) or isomorphic substitutability (MITSCHERLICH, 1819/1821; GROTH, 1874); substitutability of individual structural units without regard to the "Type" relations of the crystal species (STRUNZ, 1961). Only the individual structural units (atoms, ions etc.) are examined for their substitutability.

Diadochy is one of the most important geochemical migration factors!

The isomorphic substitutability of certain ions for one another (these are diadochic) is a phenomenon occurring in almost all minerals. Besides the main elements included in the mineral formula (e.g. in the mixed crystals), other elements always occur which are present in traces (about $<1\%$) and which are not included in the formula (the so-called trace elements).

As to the trace elements, camouflage and capture or admission (these phenomena are called endocryptic by Fersman) play an important role:

1. Camouflage

The elements replacing one another have the same charge and similar radii (Table 32).

Table 32. Examples of the Camouflage of Trace Elements

In the Mineral	element	r [Å]	concealed behind	r [Å]
Zircon	Hf^{4+}	0.78	Zr^{4+}	0.79
Olivine	Ni^{2+}	0.69	Mg^{2+}	0.66
Al-silicates	Ga^{3+}	0.62	Al^{3+}	0.51
Silicates and quartz	Ge^{4+}	0.53	Si^{4+}	0.42

2. Capturing and Admission

The elements replacing one another have different charges but similar radii.

Capturing: A trace element having a higher charge is captured by a main element having a lower charge (Table 33).

Admission: A trace element having a lower charge replaces a main element having a higher charge. In this way, the lattice energy of the crystal is reduced (e.g. Mg^{2+} partly replaced by Li^+ or 0^{2-} by F^-).

Table 33. Examples of Capturing of Trace Elements

In the Mineral	is captured	r [Å]	by	r [Å]
Monazite	Th^{4+}	1.02	Ce^{3+}	1.07
Ti minerals	Nb^{5+}	0.69	Ti^{4+}	0.68
Mg silicates	Sc^{3+}	0.81	Mg^{2+}	0.66
Yttrofluorite	Y^{3+}	0.92	Ca^{2+}	0.99
Feldspar	Sr^{2+}	1.12 ⎫	K^+	1.33
	Ba^{2+}	1.34 ⎭		

Conditions for the diadochy of two elements

    1. Goldschmidt's rules (1937), established for minerals of magmatic origin

        a) The difference in ionic radii must not exceed 15 % (related to the smaller radius).

        b) Of two ions having the same charge, but different radii, the ion with the smaller radius is preferably incorporated.

        c) Of two ions having similar radii, but different charges, the ion with the higher charge is preferably incorporated.

    2. FERSMAN's rule (1937/1939)

    Elements can replace one another in crystal lattices if they belong to one and the same lattice plane and if their ionic or atomic radii resemble each other (from SAUKOV, 1953).

However, these rules cover only part of the factors which govern an isomorphic ability to substitute; among other things, the following properties are of importance (cf. SHAW, 1953; RINGWOOD, 1955; VENDEL, 1955/1958; TAUSON, 1958; AHRENS, 1964):

Polarization properties. Cations with a high-polarizing effect, for example, are concentrated in minerals where anions occur which are highly polarizable. Cations having a low polarizing effect are frequently found together with anions which are only polarizable to a low degree (DEVORE, 1955; RAMBERG, 1955).

Electronegativity (FYFE, 1951; RAMBERG, 1952); "reduced electronegativity rate" and correlation between electronegativity and polarization (VENDEL); connection between electronegativity and normal redox potential as well as the relation to the ionization potential, cf. SZÁDECZKY-KARDOSS (1955); Table 36.

Ionization potential. Elements with a pronounced inclination to diadochy possess an approximately equal ionization potential (GOLDSCHMIDT, 1937). The ionization potential of cations is the best measure of their polarizing effect (DEVORE, 1955). For ionization potential and substitutability see VENDEL (1955); Table 34.

Ionic potential (or valence; SZÁDECZKY-KARDOSS, 1952) plays a role especially in sedimentary processes (Section 8.4.).

To sum up, the following crystallographic conditions for diadochy can be given (STRUNZ, 1961):

    1. similar atomic or ionic radii,

    2. similar bond character within the specific type of crystal,

    3. maintenance of electrostatic equilibrium.

Above all, diadochy exists between the elements whose atomic numbers differ by 1, 8, $(1 + 8)$, 18, $(1 + 18)$, 32, and $(1 + 32)$ (e.g. diagonal relation of elements in the periodic system). Mostly one element is substituted for another one only if the lattice energy is increased as a consequence. Ions having a smaller radius or a higher charge are preferably incorporated (cfl. Goldschmidt's rules); rule concerning oriented (polar) diadochy (not dependent on concentration, FERSMAN in SAUKOV, 1953; DOERFFEL, 1956; LEBEDEV, 1961); Table 35.

Table 34. Comparison of the Ionization Potential of a Few Elements

Cation	Ionization Potential [eV]		Cation	Ionization Potential [eV]	Difference ΔI [%]	
Si^{4+}	44.9	← ——— →	Ge^{4+}	45.5	1	
K^+	4.34	← ——— →	Rb^+	4.17	4	
Fe^{2+}	16.5	← ——— →	Co^{2+}	17.4	5	
Co^{2+}	17.4	← ——— →	Ni^{2+}	18.2	5	
Al^{3+}	28.3	← ——— →	Ga^{3+}	≈30.6	8	decreasing tendency to diadochy
Fe^{2+}	16.5	← ——— →	Zn^{2+}	18.0	9	
Mg^{2+}	15.0	← ——— →	Fe^{2+}	16.5	10	
Fe^{2+}	16.5	← ——— →	Ni^{2+}	18.2	10	
Ba^{2+}	10.0	← ——— →	Sr^{2+}	11.0	10	
Mg^{2+}	15.0	← ——— →	Co^{2+}	17.4	16	
Mg^{2+}	15.0	← ——— →	Zn^{2+}	18.0	20	
Mg^{2+}	15.0	← ——— →	Ni^{2+}	18.2	21	
Rb^+	4.17	← ——— →	Tl^+	6.12	45	
Na^+	5.14	← ——— →	Cu^+	7.72	48	

Rule of thumb: The greater the difference I between two elements, the smaller their capability of mutual substitution

Table 35. Isomorphic Ability to Substitute of Elements with Gain in Energy (from SAUKOV, 1953; ionic radii according to GOLDSCHMIDT; for EK see Section 3.2.2.2.)

Ion that is replaced	r [Å]	EK	by	r [Å]	EK	ΔEK (gain)
Ca^{2+}	1.06	1.75	Y^{3+}	1.06	3.95	+2.20
Mg^{2+}	0.78	2.10	Sc^{3+}	0.83	4.65	+2.55
Fe^{2+}	0.83	2.12	Sc^{3+}	0.83	4.55	+2.53
Sc^{3+}	0.83	4.65	Zr^{4+}, Hf^{4+}	0.87	7.85	+3.20
Ti^{4+}	0.64	8.40	Nb^{5+}, Ta^{5+}	0.69	13.60	+5.20
K^+	1.33	0.36	Pb^{2+}	1.32	1.65	+1.29
Na^+	0.98	0.45	Ca^{2+}	1.06	1.75	+1.30

The diadochic behaviour of elements is distinctly dependent on temperature and pressure (hence on the formation conditions of the minerals; cf. VERNADSKIĬ in SAUKOV, 1953):

Within the magmatic range
(T and p high)
Within the metamorphic range   decrease in the isomorphic
(T medium and p high)   miscibility and inclination
Within the weathering range   to diadochy
(T and p low)

117

Table 36. Influence of Electronegativity in the Diadochic Incorporation of Elements within the Magmatic Range (from (RINGWOOD, 1955)

a) Of two cations having approximately the same ionic radius, the cation having the lower electronegativity (boldface) is preferably incorporated.

b) In the case of approximately the same electronegativity and valence of two cations, the cation having the smaller ionic radius (boldface) is preferably incorporated.

Cation	Ionic Radius [Å]	Electronegativity	Cation	Ionic Radius [Å]	Electronegativity
a)			**Ca^{2+}**	0.99	1.0
Na$^+$	0.97	0.9	Cd^{2+}	0.97	1.5
Cu$^+$	0.99	1.8	**Ca^{2+}**	0.99	1.0
Fe^{2+}	0.74	1.65	Hg^{2+}	1.10	1.9
Cu^{2+}	0.72	2.0	**Fe^{2+}**	0.74	1.65
Fe^{2+}	0.74	1.65	Sb^{3+}	0.76	1.8
Zn^{2+}	0.74	1.7	**Ca^{2+}**	0.99	1.0
K$^+$	1.33	0.8	Bi^{3+}	0.96	1.8
Pb^{2+}	1.20	1.6			
Ca^{2+}	0.99	1.0	b)		
Sn^{2+}	0.93	1.65	**Ni^{2+}**	0.69	1.7
V^{3+}	0.74	1.35	Fe^{2+}	0.74	1.65
Fe^{3+}	0.64	1.8	**Co^{2+}**	0.72	1.7
Cr^{3+}	0.63	1.6	Fe^{2+}	0.74	1.65
Fe^{3+}	0.64	1.8	**K$^+$**	1.33	0.8
Sc^{3+}	0.81	1.3	Rb$^+$	1.47	0.8
Fe^{2+}	0.74	1.65	**Mg^{2+}**	0.66	1.2
Ca^{2+}	0.99	1.0	Li$^+$	0.68	1.0
(Lanthanides)$^{3+}$	0.85—1.14	1.05—1.2	**Ba^{2+}**	1.34	0.85
K$^+$	1.33	0.8	K$^+$	1.33	0.8
Ag$^+$	1.26	1.8			

Bibliography

AHRENS, L. H.: The Use of the Ionization Potentials. Geochim. cosmochim. Acta [London] *3* (1953), 1—29

AHRENS, L. H.: The Significance of the Chemical Bond for Controlling the Geochemical Distribution of the Elements — Part I, Phys. Chem. of the Earth *5* (1964), 1—54

ANFILIGOV, V. N., B. I. BELOV, and JU. P. TRUSHIN: On the Behaviour of Isomorphous Impurities in the Cocrystallization Process in Open Systems. Geokhimiya *2* (1966), 246—249 (Russian)

DEVORE, G. W.: Crystal Growth and the Distribution of Elements. J. Geology *63* (1955) 471—494

DEVORE, G. W.: The Role of Adsorption in the Fractionation and Distribution of Elements. J. Geology *63* (1955), 159—190

DOERFFEL, K.: Das System der Ionenenergien. Freiberger Forschungsh., C *20* (1956), 1—104

FYFE, W. S.: Isomorphism and Bond Type. Amer. Mineralogist *36* (1951), 538

GOLDSCHMIDT, V. M.: The Principles of Distribution of Chemical Elements in Minerals and Rocks, J. chem. Soc. [London] (1937), 655—673

GOLDSCHMIDT, V. M.: Crystal Chemistry and Geochemistry, Chem. Products Chem. News *7* (1944), 1—6

HALLA, F.: Kristallchemie und Kristallphysik metallischer Werkstoffe, 3rd ed. Johann Ambrosius Barth, Leipzig 1957

HUSH, N. S., and M. H. L. PRYCE: Radii of Transition Ions in Crystal Fields. J. Chem. Physics *26* (1937), 143

ČIRKINSKIĬ, V. A.: Influence of Temperature on the Limits of Isomorphic Miscibility. Geokhimiya (1965), 406—413 (Russian)

ČIRKINSKIĬ, V. A.: On Some Regularities in the Behaviour of Isomorphous Mixtures Under Pressure. Geokhimiya *3* (1966), 303—311 (Russian)

KLEBER, W.: Kristallchemie. Math.-nat. Bibliothek No. 36. B. G. Teubner, Leipzig 1963

KRETZ, R.: The Distribution of Certain Elements Among Coexisting Calcic Pyroxenes, Calcic Amphiboles and Biotite in Skarns. Geochim. cosmochim. Acta [London] *20* (1960), 161

KORITNIG, S.: Das Reflexionsvermögen opaker Mischkristallreihen. Neues Jb. Miner., Mh. *8* (1964), 25—231

LAVES, F.: Elektrochem. angew. physik. Chem. *45* (1939), 7

LEBEDEV, V. I.: On the Laws of Isomorphism of Magmatic Crystallisation. Acta geol. Acad. Sci. Hung. *7* (1961), 47—52

LEBEDEV, V. I.: Energetic Fundamentals of the Analysis of Geochemical Processes. Univ. Leningrad 1957 (Russian)

LEUTWEIN, F., and CH. SOMMER-KULASZEWSKI: Allgemeine Mineralogie. Bergakademie Freiberg, Fernstudium, 4th ed. VEB Deutscher Verlag der Wissenschaften, Berlin 1960

NOCKOLDS, S. R., and R. L. MITCHELL: The Geochemistry of Some Caledonian Plutonic Rocks: A Study in the Relationship between the Major and Trace Elements of Igneous Rocks and their Minerals. Trans. Roy. Soc. Edinburgh *61* (1948), 533

NOCKOLDS, S. R., and R. S. ALLEN: The Geochemistry of Some Igneous Rock Series. Geochim. cosmochim. Acta [London] *4* (1953), 105 (Part I); *5* (1954) 245 (Part II): *9* (1956), 34 (Part III)

PERCHUK, L. L., and A. S. PAVLENKO: Temperature Influence upon the Distribution of the Isomorphous Components among Coexisting Minerals of Alkaline Rocks. Geokhimiya *9*(1967), 1063—1082 (Russian)

PYATENKO, JU. A.: On the Isomorphism of Atoms and Some of its Mineralogical Consequences. Geokhimiya (1965), 414—420 (Russian)

RAMBERG, H., and G. DEVORE: The Distribution of Fe^{++} and Mg^{++} in Coexisting Olivines and Pyroxenes. — J. Geology *59* (1951), 193—210

RAMBERG, H.: Chemical Bonds and Distribution of Cations in Silicates. J. Geology *60* (1952), 331 to 355

RINGWOOD, A. E.: The Principles Governing Trace Element Distribution during Magmatic Crystallization. Geochim. cosmochim. Acta [London] *7* (1955), 189—202 (Part I), 242—254 (Part II)

RYABCHIKOV, I. D., V. V. SHCHERBINA: The Influence of Pressure on Isomorphous Replacement in Minerals. Geokhimiya *10* (1965), 1207—1211 (Russian)

SHCHERBINA, V. V.: Interconnection of Mineral Formation and Isomorphism. Geokhimiya *3* (1965), 259—268 (Russian)

SHCHERBINA, V. V.: Dependence of Isomorphous Replacement on Pressure. Geokhimiya *5* (1965), 544—550 (Russian)

SHAW, D. M.: The Camouflage Principle and Trace Element Distribution in Magmatic Minerals. J. Geology *61* (1953), 142—151

SOBOLEV, B. P., et al: Isomorphous Replacements Indice of Refractory Systems in Moderate Range (300—500°) of Temperature. Geokhimiya *6* (1966), 634—649 (Russian)

SZÁDECZKY-KARDOSS, E.: Über zwei neue Wertigkeitsregeln und die geochemische Gruppierung der Elemente. Acta geol. Acad. Sci. hung. *1* (1952), 1—4

SZÁDECZKY-KARDOSS, E.: Das Verbindungspotential und seine Beziehungen zum Schmelzpunkt und zur Härte. Acta geol. Acad. Sci. hung. *3* (1955), 115—161

STRUNZ, H.: Die für den geochemischen Lebenslauf der Elemente verantwortlichen Faktoren. Acta geol. Acad. Sci. Hung. *7* (1961), 53—55

TAUSON, L. V.: The Influence of the Structure of Minerals on the Isomorphous Replacement in Silicates of the Eruptiva. Geokhimiya *8* (1958), 735 (Russian)

TROJER, F.: Reflexionsmessungen in der Mikroskopie hüttenmännischer Produkte. Berg- und hüttenmänn. Mh. *107* (1962), 33—39

TRÖGER, W. E.: Tabellen zur optischen Bestimmung der gesteinsbildenden Minerale. Schweizerbart, Stuttgart 1952

URUSOV, V. S.: Influence of Difference of Component Sizes on Limits of Isovalent Isomorphic Replacements. Geokhimiya *9* (1968), 1033—1043 (Russian)

Vegard, L.: Vidensk. Skrifter, Oslo, (1921) 6 and (1927) 4

Vendel, M.: Die Substituierbarkeit der Ionen und Atome vom geochemischen Gesichtspunkt aus, II. Acta geol. Acad. Sci. hung. 5 (1958), 381—433

Wager, L. R., and R. L. Mitchell: The Distribution of Trace Elements during Strong Fractionation of Basic Magma. Geochim. cosmochim. Acta [London] 1 (1951), 129—208

Wickmann, F. E.: Some Aspects of Geochemistry of Igneous Rocks and of Differentiation by Crystallization. Geol. Fören. Stockholm Förh. 65 (1943), 371—396

3.2.4. Radioactivity

Radioactive isotopes possess the property of spontaneously, i.e. without external cause, transforming their atomic nuclei into other types of atoms (having other properties than the original atoms).

Due to an excess of neutrons or protons, radioactive nuclei are unstable, they change into stable nuclei by the emission of particles (see for instance Section 2.4., Figs. 4, 5, 6, 7). The following law holds for this disintegration:

$$dN = -\lambda N \, dt \quad \text{or} \quad N = N_0 \, e^{-\lambda t},$$

where N is the number of atoms present at a time, N_0 is the number of active atoms present at time zero $(t = 0)$, λ is the decay constant, number of disintegrations per number of active nuclei and per unit time.

λ is a measure of the speed of decay. Another measure is the half-life $T\,{}^1/_2$, i.e. the time required for disintegration of one-half of the active nuclei of a sample of a radioactive substance; for this the following applies:

$$\lambda = \frac{0.693}{T_{1/_2}}.$$

Of the most important disintegration processes and types of emission, the following are mentioned here:

1. *Emission of an α-particle* $({}^4_2\text{He})$

$$^A_Z K_\text{a} \rightarrow {}^{A-4}_{Z-2} K_\text{e} + \alpha,$$

where K_a and K_e are the original type of nucleus and the end type of nucleus, A is the mass number, Z is the nuclear charge number.

2. *β⁻-radiation*. Emission of electrons in the case of an excess of neutrons:

$$^A_Z K_\text{a} \rightarrow {}^A_{Z+1} K_\text{e} + \beta^-.$$

3. *β⁺-radiation*. Emission of positrons in the case of an excess of protons:

$$^A_Z K_\text{a} \rightarrow {}^A_{Z-1} K_\text{e} + \beta^+.$$

4. *γ-radiation*. Transition of a nucleus from a higher to a lower energy by emission of one photon (γ-particle, electromagnetic radiation):

$$^A_Z K_\text{a} \rightarrow {}^A_Z K_\text{e} + \gamma.$$

α-decay and β-decay are frequently associated with γ-radiation.

5. *K-capture.* Passing of an electron from the *K*-shell (in the case of a few artificial nuclei also from the *L*-shell) into the nucleus (while the electron envelope emits characteristic X-rays):

$$\beta^- + {}^A_Z K_a \to {}^A_{Z-1} K_e,$$

e.g. in ^{40}K, ^{138}La, ^{176}Lu. This nuclear process is similar to positron emission.

6. *Nuclear fission.* The division of a heavy nucleus into two new nuclei of approximately the same size which emit further β-particles.
Spontaneous fission occurs in ^{238}U having the high half-life of $8 \cdot 10^{15}$a (fossil track dating).
A more frequent occurrence is fission induced by bombardments with neutrons, protons, deuterons or α-particles. Spallation is due to a bombardment with highly energetic particles (cosmic radiation); small nuclei, deuterons, tritium "peel off" from a larger nucleus, leaving behind only one heavy nucleus. The exposure ages, of meteorites for instance, are based on this process.

Bibliography (see textbooks of nuclear physics and Section 4.15.)

3.3. External Migration Factors

The external migration factors, i.e. the circumstances to which atoms and their combinations are exposed, vary with the geological development. This change at the same time causes a change in the laws of distribution of the chemical elements in the course of the earth's history (e.g. changes in the migration of elements with the progressing development of organisms).
The evolution of external migration factors and their varying influence on the migration of elements is the subject of "Historical Geochemistry" (bibliography given in SAUKOV, 1961 and also p. 18).
Below, only a few important external migration factors are dealt with.

Bibliography

SAUKOV, A. A.: Evolution der Migrationsfaktoren der Elemente in der Historischen Geologie. Wiss. Techn. Inform.-Dienst, Zentr. Geol. Inst. Berlin *2* (1961), 6, 14—33
SAUKOV, A. A.: Historicism in Geochemistry. Akad. Nauk SSSR, Moscow 1963 (Russian)
VOÏTKEVICH, G. V.: Changes in the Chemical Composition of the Earth during the Development of the Geological Formations. Priroda *40* (1951), 4, 28—37 (Russian)

3.3.1. Gravitational Energy

Gravitation is one of the most important indirect factors of migration. According to Newton's law of gravitation (see below), any matter is subject to a force which is directly proportional to its mass and indirectly proportional to the square of the distance.

The measured acceleration due to gravity is vectorially derived from the gravity acceleration b_0 caused by the total mass of the earth, the centrifugal acceleration q produced by the
earth's rotation and varying attractive forces b_1 resulting from the variable position of the
larger celestial bodies (only moon and sun are measurable).
The gravity acceleration b_0 constitutes the main part of the measured acceleration, the centrifugal acceleration reaches a maximum of $3.4^0/_{00}$ of the measured acceleration at the equator
and is equal to 0 at the poles; the temporal variation due to positional changes of the celestial
bodies may be neglected (this change only amounts to $\leqq \dfrac{g}{40 \cdot 10^6}$).

The effect of gravitation on geological processes, and thus on the migration of the elements,
can be proved in many cases or can be assumed. A few examples:

1. The differentiation of the earth's material in its early stage; formation of the geospheres.
2. Magmatic crystallization differentiation, particularly evident in basic magmas.
3. The denudation of parts of the surface of the earth; this applies especially to those
   surface areas which protrude from the level of the normal sphaeroid (mountains).
4. Accumulation of light gases in the atmosphere of the earth.
5. Several tectonic processes (e.g. creep. diapirism, rifts).

If gravitation were the only migration factor, the whole earth would be arranged in such a
way that the heaviest atoms are at the lowest and the lightest atoms at the highest level.
In as much as this is not the case, at least in many parts of the earth's crust, it is obvious that
a number of other migration factors counteract gravitation. This applies especially to the
heat of the earth's interior, the radioactive processes, and physico-chemical properties of
compounds (e.g. their volatility).

Some Definitions and Data

Newton's law of universal gravitation:

$$K = k_0 \frac{m_1 m_2}{r^2}$$

where $k_0 = 6.67 \cdot 10^{-8} \dfrac{\text{cm}^3}{\text{g s}^2}$ is the universal gravitational constant.
"Two masses m_1 and m_2 attract each other with a force K which is directly proportional to the
product of their masses and inversely proportional to the square of the distance r between
them" (formulation HAALCK, 1953).
The potential W of the gravitational force at a point, differentiated in any direction, results
(according to HAALCK, 1953) in the component of the gravitational force at this point in this
direction.

Potential or Force Function of Gravitation:

$$W = K^2 \int \frac{dm}{e},$$

where:

e is the linear absolute distance of the mass element from P; dm is the mass element of the
mass M.

122

y summation of the second differential quotient of P, we obtain *Laplace's equation* (valid way from the attracting masses):

$$\frac{\partial^2 W}{\partial x^2} + \frac{\partial^2 W}{\partial y^2} + \frac{\partial^2 W}{\partial z^2} = 0 .$$

, y and z are coordinates in the following system of coordinates:

-axis; surface normal of the equipotential surface (positive, reckoned upwards),

-axis; geographic eastward direction,

-axis; geographic northward direction.

'or W, as the potential to gravity on the surface of the earth, the term $2\omega^2$ is added to the ight of the Laplace equation, ω being the angular velocity due to the earth's rotation. *Equipotential surfaces* are normal to the lines of force at each point of the surface. On these urfaces, the potential is independent of position, hence, equipotential surfaces. A higher ravitational force means an increased set of equipotential surfaces and vice versa.

Geoid is defined as the equipotential surface of oceans at rest.

Normalsphaeroid is the shape which the earth would assume if it were liquid (ellipsoid of evolution).

The differences between geoid and ellipsoid of revolution are so small that the earth is considered to be an ellipsoid of revolution flattened at the poles (ellipticity: $\frac{1}{297}$).

Reference ellipsoid is defined as an ellipsoid which, for a limited part of the earth's crust, best onforms to a geoid.

Radii of the international reference ellipsoid:

equatorial semidiameter $a = 6378.388$ km

polar semidiameter $b = 6356.912$ km $\varDelta r \approx 21$ km.

Normal gravity is the regular distribution of gravity on the earth which is assumed to be an ellipsoid of revolution. Thus, it is a function of the geographic latitude and is given by the nternational gravity formula:

$= 978.0490 \, (1 + 0.005\,2884 \sin^2\beta - 0.000\,0059 \sin^2 2\beta) \, \text{Gal} .$

Table 37. Standard Gravity Values at Different Latitudes (GASSMANN-WEBER, 1960)

Latitude β	Standard Gravity γ [mGal]	Gravity Increment Q_β [mGal/km]
0°	978,049.0	0
15°	978,394.0	0.406
30°	979,337.8	0.704
45°	980,629.4	0.812
46°	980,719.7	0.812
47°	980,809.8	0.810
48°	980,899.8	0.808
60°	981,923.9	0.704
75°	982,873.4	0.406
90°	983,221.3	0

The increment Q_β of gravity when migrating from one point of latitude β for 1 km towards the north or south is:

$$Q_\beta = 0.812 \sin 2\beta \, \frac{\mathrm{m \, Gal}}{\mathrm{km}} .$$

Magnitude of g away from the earth:

$\frac{1}{100}$ of the standard value in terms of a distance of 10 radii of the earth (extreme limit of the *van Allen belt of radiation*, see Fig. 21);

$\frac{1}{3600}$ of the standard value in terms of a distance of 60 radii of the earth (lunar distance). Per increment in altitude of 100 m, g decreases by 0.03085 cm/s² or about $\frac{1}{30\,000}$ of g near the earth surface.

Accuracy of measurement of the absolute gravity measurement: $\frac{1}{10^8} \, g$.

The gravity values from the surface towards the core of the earth are shown in Fig. 20.

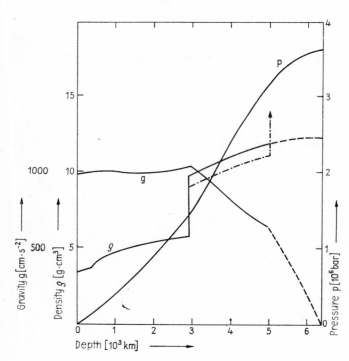

Fig. 20 Density, pressure and gravity distribution in the earth (after BULLEN; from HIERSEMANN, 1956)

Bibliography

GRANT, F. S., and G. F. WEST: Interpretation Theory in Applied Geophysics. Mc Graw Hill Book Co., New York 1965
RUNCORN, S. K. (ed.): International Dictionary of Geophysics. Oxford 1967

3.2. Solar Energy and Cosmic Energy

he energy radiated from the sun to the earth is of extraordinary importance to the processes king place on the surface of the earth. A survey of extraterrestrial sources of energy hich produce an effect on the earth is shown in Table 38.

our latitudes, the heat supply by solar radiation about $3.2 \cdot 10^{-2}$ cal/cm² s, totalling approximately ,2 · 10^{16} cal/s. It is about 5,000 to 10,000 times the agnitude of the terrestrial heat flow from the interr of the earth through the surface of the earth. lowever, the depth of penetration of the annual ariations in temperature only amounts to about 20) 30 m, that of the daily variations about 1 to m.

'he maximum amount of heat energy incident upon e earth and its atmosphere is derived from

$$ V = Cr^2 \sin h . $$

Vhere: C is the solar constant $= 1.94$ cal/cm² min, r is the apparent solar radius, h the sun's ltitude.

t is estimated that about 2,000 tons of meteoric mass reach the earth daily. 'he screening effect produced by the magnetic belt round the earth on the supply of cosmic ncluding solar energy is shown in Fig. 21.

Table 38. Extraterrestrial Sources of Energy Producing an Effect on the Earth

Sources	erg/s
Sun	$1.76 \cdot 10^{24}$
Full moon	$3.09 \cdot 10^{19}$
Light of sky at night	$2.61 \cdot 10^{17}$
Cosmic rays	$1.69 \cdot 10^{17}$
Meteors	$1.44 \cdot 10^{17}$

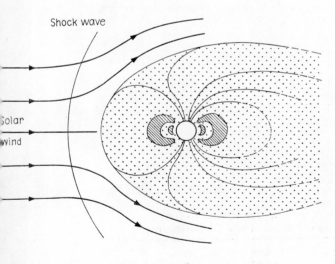

Fig. 21 Schematic representation of the magnetosphere

Due to the solar wind, the terrestrial magnetic field is restricted to the closed space (dotted). The trapped radiation belts (hatched) are embedded in this space. The outer boundary of the magnetosphere is called magnetopause. A shock wave is formed at the side facing the sun (illustration and text by courtesy of Prof. Dr.W. KERTZ, Braunschweig)

3.3.3. Radioactive Processes

t is assumed that the energy present in the earth and continuously emitted into outer space s only partly a residual energy of our planet, but for the major part (75 to 80 % according to Jrry) is produced by radioactive processes and the influence of gravity (e.g. gravitational ontraction, see 3.3.1.).

Table 39. Amounts of Heat Generated in the Interior of the Earth from its Origin by Radioactive Processes (from Lyubimova, 1963)

Author	Age of the Earth [10^9 a]	Sum of the Radiogenetic Energy [10^{38} erg]	Remarks
Urry (1949)	3	2.3	whole earth
Voïtkevich (1961)	5	1.9	whole earth
	6	3.7	whole earth
Vinogradov (1960)	4.5	1.1—1.5	earth's mantle
Urey (1952)	4.5	0.6	earth's mantle
Lyubimova (1958)	4.0	1.6	whole earth
	4.5	2.2	whole earth
	5.0	2.8	whole earth
	6.0	5.4	whole earth
	4.5	2.4	whole earth
McDonald (1959)	4.5	1.4	whole earth
Levin, Maera (1961)	4.5	1.2—1.5	whole earth

Estimates of the energy supplied by radioactive processes during the earth's existence from the beginning are shown in Table 39.
The amount of heat presently generated by the most important nuclides is shown in Table 40.

Nuclide	[cal/a]
^{238}U	$171.8 \cdot 10^{18}$
^{235}U	$7.14 \cdot 10^{18}$
^{232}Th	$173.57 \cdot 10^{18}$
^{40}K	$73.80 \cdot 10^{18}$
Sum	$4.26 \cdot 10^{20}$

Table 40. Amount of Heat Presently Generated by Radioactive Elements in the Earth (according to Voïtkevich, 1956, from Saukov, 1961)

It is assumed that heat generation by radioactive processes varied considerably during the earth's history. A survey is given in Table 41.

Table 41. Change in Heat Generation by Radioactive Processes in the Course of the Development of the Earth (according to Zvyagintsev, from Schössler and Schwarzlose, 1959)

Element	Heat Emission [10^{16} cal/h] today	$2 \cdot 10^9$ years ago	$3 \cdot 10^9$ years ago	$4 \cdot 10^9$ years ago
U	21.26	29.42	34.06	39.03
AcU	0.89	6.33	17.34	43.43
Th	20.27	22.78	24.12	29.13
K	4.70	19.46	79.53	80.82
Total	47.12	77.99	155.05	188.41

The distribution of radioactive elements in the various rock types is shown in Table 42.

Table 42. Contents of Radioactive Elements and Radioactive Generation of Heat in Various Rock Types (according to several authors, from SCHÖSSLER and SCHWARZLOSE, 1959)

Type of Rock	Ra [10^{-12} g/g]	Th [10^{-6} g/g]	K [10^{-2} g/g]	Heat generation [10^{-6} cal/g year]			Sum
				Ra	Th	K	
Acid magmatites	1.37	13.0	2.8	3.0	2.6	0.2	5.8
Intermed. magmatites	0.51	4.4	2.0	1.1	0.9	0.2	2.2
Basic magmatites	0.38	3.9	1.4	0.8	0.8	0.1	1.7
Sediments	0.57	3.3	2.0	1.2	0.7	0.2	2.1

Bibliography

KABRAKHMANOV, T.: On the Migration of Radioactive Chemical Elements in Nature. Trudy Inst. klimencheskoĭ eksperiment. chirurgy Akad. Nauk Kazakh. SSR 4 (1958), 159—164 (Russian)

LEE, W. H. K.: Terrestrial Heat Flow. Geophysical Monographie Series 8, Amer. Geophys. Union 1965

LYUBIMOVA, E. A.: The Origin of the Innerplanetary Heat. Khimiya Zemnoĭ Kory, Vol. I, pp. 30—38, Izd. Akad. Nauk SSSR, Moscow 1963 (Russian)

SCHÖSSLER, K., and J. SCHWARZLOSE: Geophysikalische Wärmeflußmessungen. Freiberger Forschungsh. C 75. Akademie-Verlag, Berlin 1959

3.3.4. Heat Content and Heat Emission of the Earth (Heat Flow)

The inhomogeneous distribution of heat in the earth and the heat balance with regard to the earth's surroundings can be of importance to geochemical considerations and computations.
Fig. 22 shows the earth's initial temperature distribution due to melting point behaviour calculated by several authors.
Fig. 23 shows the temperature distribution in the earth's crust down to a depth of 50 km, while Fig. 24 shows the estimates and calculations with respect to the earth's mantle to a depth of 600 km.
The majority of authors assumes temperatures in the earth's core of 3000 to 4000 °C and more (cf. Fig. 25).
In as much as the lavas are the only substances which we can subject to measurements of the temperatures of deepcrustal and subcrustal products, Table 43 gives a few of such values.

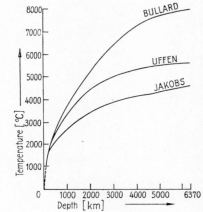

Fig. 22 Initial temperature distribution in the earth according to several authors (from SCHÖSSLER and SCHWARZLOSE, 1959)

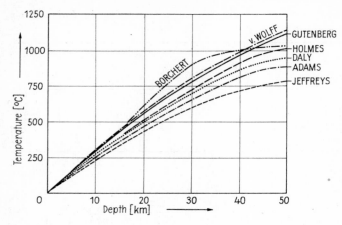

Fig. 23 Temperature distribution in the earth's crust according to several authors (from SCHÖSSLER and SCHWARZLOSE, 1959)

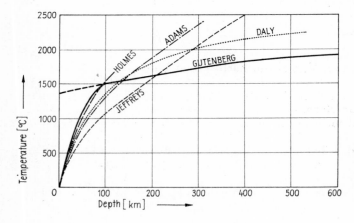

Fig. 24 Temperature distribution in the earth's mantle according to several authors (from SCHÖSSLER and SCHWARZLOSE, 1959)

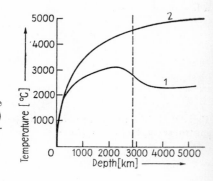

Fig. 25 Distribution of temperature in the earth (at the present time) due to radioactivity (after LYUBIMOVA, 1963) Curve 1: Assuming an iron core; Curve 2: Assuming a silicate core

128

Table 43. Maximum Temperatures of Some Lavas
(from GUTENBERG, in SCHÖSSLER and SCHWARZLOSE, 1959)

Place	Temperature [°C]	Author
Kilauea	1,185	DAY and SHEPHERD
Kilauea	1,200	JAGGAR
Vesuvius (1904)	1,100	BRUN
Vesuvius (1913)	1,200	PERRET
Stromboli (1901)	1,150	BRUN
Etna (1892)	1,060	BARTOLI
Etna (1892)	1,300	PHILIPP
Sakura-jima (1914)	1,048	KOTÓ
Oshima	1,200—1,300	TSUBOI

Further information about the earth's heat content and heat economy is derived from heat flow measurements in the accessible parts of the upper crust of the earth. They also give a clue to the emission of heat to outer space. Table 44 shows a few data from several continents.

Table 44. Regional Mean Value of Heat Flow (from LEE and UYEDA, 1965, and others)

Geological Feature	Number of Values	Mean and Standard Deviation [μcal/cm^2 sec]
Precambrian Shields	26	0.92 ± 0.17
South African Shield	5	1.03 ± 0.13
Ukrainian Shield	3	0.69 ± 0.07
Canadian Shield	10	0.88 ± 0.13
Post-Precambrian non-orogenic areas	23	1.54 ± 0.38
Interior Lowlands of North America	8	1.25 ± 0.18
Paleozoic orogenic areas	21	1.23 ± 0.40
Appalachian system	12	1.04 ± 0.23
Great Britain	7	1.31 ± 0.38
German Democratic Republic	10	1.40 ± 0.45
Mesozoic-Cenozoic orogenic areas	19	1.92 ± 0.49
Alpine System	10	2.09 ± 0.38
Cordilleran System	9	1.73 ± 0.53
Cenozoic volcanic areas (excluding geothermal areas)	11	2.16 ± 0.46
Atlantic Ocean	206	1.16 ± 0.33
Indian Ocean	210	1.53 ± 0.85
Pacific Ocean	497	1.80 ± 1.13
Ocean Trenches	21	0.99 ± 0.61
Ocean Basins	273	1.28 ± 0.53
Ocean Ridges	338	1.82 ± 1.56

As to mean value of all measurements made so far, a value of about $1.5 \pm 10\% \mu$cal/cm^2 sec can be given. Since, today, the energy comes mainly from the earth's crust (radioactive sources), the heat flow on continents and sea-bottoms is of the same order of magnitude. The lowest values occur mostly in old crystalline shields, the highest values in sialic orogenic zones.

Bibliography

BULLARD, E. C.: The Transfer of Heat from the Core of the Earth. Monthly Notices Roy. Astronom. Soc., Geophys. Suppl. *6* (1950), 36—41

DONALDSON, J. G.: Temperature Gradients in the Upper Layers of the Earth's Crust Due to Convective Water Flows. J. Geophys. Res. *67* (1962), 9, 3449—3459

GUTENBERG, B.: Internal Constitution of the Earth, 2nd ed. Dover Publications, Inc., New York 1951

HERZEN, R. P. VON, and A. E. MAXWELL: Measurements of Heat Flow at the Preliminary Mohole Site of Mexico. J. Geophys. Res. *69* (1964), 4, 741—748

HIERSEMANN, L.: Geologisch-geophysikalische Theorien über den Aufbau und die Dynamik der Erdkruste. Freiberger Forschungsh. C *24* (1956)

JACOBS, J. A., and D. W. ALLAN: Temperature and Heat Flow within the Earth. Proc. Roy. Soc. Canada *48* (1954), 33—39

LEE, W. H. K., and S. UYEDA: Review of Heat Flow Data. In W. H. K. LEE (ed.): Terrestrial Heat Flow. Geoph. Monogr. Ser. N. *8* (1965), 87—190

LYUBIMOVA, E. A.: On the Evolution Conditions of Magmatism and the Role of Volcanism in Relation to the Thermic Conditions in the Earth's Crust. J. Phys. Earth, Tokyo *8* (1960), 2, 17—21

LYUBIMOVA, E. A.: The Origin of the Inner Planetary Heat. Khimiya Zemnoǐ Kory, Vol. I, 30—38. Moscow: Izd. Akad. Nauk SSSR 1963 (Russian)

PEKERIS, C. L.: Thermal Convection in the Interior of the Earth. Monthly Notices Roy. Astronom. Soc. Geophys. Suppl. *3* (1935), 343—367

RINGWOOD, A. E.: Some Aspects of the Thermal Evolution of the Earth. Geochim. cosmochim. Acta *20* (1960), No. 3/4

SCHÖSSLER, K., and J. SCHWARZLOSE: Geophysikalische Wärmeflußmessungen. Freiberger Forschungsh. C *75*, 1959 120 pp.

SCHWINNER, R.: Über den Wärmehaushalt des Erdballes. Gerlands Beitr. Geophysik *58* (1942), 234—296

SIMMONS, G., and K. HORAI: Heat Flow Data 2. J. Geoph. Rs. *73* (1968) 20, 6608—6629

UFFEN, R. J.: A Method of Estimating the Melting Point Gradient in the Earth's Mantle. Trans. Amer. Geophys. Union *33* (1952), 893—896

URRY: Thermal History of the Earth. Trans. Amer. Geophys. Union *30* (1949), 171—180

ZVYAGINTSEV, O.: Radioactivity and Thermal Economy of the Earth. Priroda *12* (1951), 93—104 (Russian)

3.3.5. Dependence of Mineralogical-geochemical Systems on pTC

The dependence of mineral formation, mineral properties and thus of geochemical migration on pressure, temperature and concentration is one of the most important fields of geochemistry and has a bearing on all other geoscientific branches. A complete representation of this field within the scope of this book is impossible, only a few problems can be pointed out. Two important special cases of the influence of concentration on geochemical systems are dealt with in Sections 3.3.6. and 3.3.7. Therefore, emphasis in the following discussion is placed on examples which demonstrate the influence of pressure and temperature. They focus on three complexes of questions:

1. Within which pressure-temperature range is a given mineral stable?

2. Which minerals can form a stable paragenesis under certain conditions of pressure and temperature?

3. How do solubility and miscibility change under the influence of pressure and temperature?

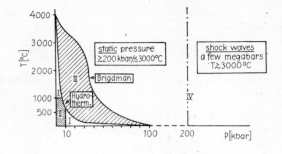

Fig. 26   The experimental pressure-temperature field within reach (state of 1965; from
NEUHAUS, 1965)

I   Technically utilised p-T range
II p-T field of the classical hydrothermal
    method
II p-T range acquired by BRIDGMAN up to 1950
V   Field of the static high-pressure high-tem-
    perature methods

To solve these problems, laboratory investigations into relatively simple systems have proved useful, besides investigations into the usually very complicated natural conditions. The theoretical foundations of this field of research are for the greater part borrowed from thermodynamics. Concepts and laws which are of importance in this connexion are defined in Section 3.3.5.1., others have already been dealt with in Section 2.10. Full details about these subjects are given in textbooks of physical chemistry.

The fact that properties of geochemical systems (e.g. miscibility, stability of minerals, etc.) are dependent on pressure and temperature, enable us to use them as geothermometers or geobarometers, provided the numerical relations between temperature and pressure, and the property in question are known.

Examples of frequently used geothermometers are given on page 140

The pressure/temperature range which is at present accessible for experiments is shown in Fig. 26. Information about the pressure/temperature conditions in the interior of the earth are given in Sections 3.3.4 and 3.3.1.1.

The high pressure/high-temperature methods (range IV in Fig. 26) have become very important to investigations into the structure and composition of the earth's mantle and the earth's core. The pressure-coordination and pressure-homology rule advanced by NEUHAUS (1966) enable substantiated predictions on properties of matter in p-T ranges which presently are not accessible to experiments. An illustration on the structure of the earth developed on this basis is given in Fig. 49 (p. 217).

The results of this field of research are of great importance not only to geosciences but also to technology (e.g. synthesis of diamond).

3.3.5.1. Some Concepts and Laws

System: In thermodynamics, a system is defined as a quantity of substance whose properties can be completely described by certain quantities (pressure, temperature, mass, chemical composition) and which is separated from its surrounding by any physical or imaginary walls or boundaries. In geochemistry, such systems are, for example, a hydrothermal solution or a rock. Frequently, this therm is also used for a combination of certain elements or compounds (e. g. the Fe-S-O system or $MgO-Al_2O_3-SiO_2$ system), where not only the elements mentioned or the compounds themselves are covered, but also all combinations between them. Systems are grouped as follows, depending on their interchanges with their surroundings:

isolated or enclosed systems, have neither interchanges of energy nor interchanges of substance with their surroundings,

closed systems, may have interchange of energy but no interchange of substances across their boundaries, *open systems,* may have interchanges of energy and substances with their surroundings. Within the scope of the geosciences, most systems are open.

According to other viewpoints, systems are grouped into:

homogeneous systems; their macroscopic properties (e.g. density, refractive index, hardness) are the same in all parts,

heterogeneous systems; at certain boundary surfaces, sudden changes in the macroscopic properties occur, the parts of the system separated by the boundary surfaces can generally be separated mechanically.
Examples of homogeneous systems: minerals (diamond), water, ionic solutions.
Examples of heterogeneous systems: rocks (granite), water-ice mixture, a saline solution with a solid phase at the bottom.

Phase: The homogeneous parts of a heterogeneous system are called phases. A phase possesses the same macroscopic properties in all of its parts and is separated from any other phase by boundary surfaces (e.g. quartz in granite). A homogeneous system has only one phase: A genuine solution consists of only one phase (differences only in the molecular range). A grouping in the colloidal range is uncertain.
Since all gases are completely miscible, only one gas phase exists in any case.

State of equilibrium: The state of equilibrium of a system is characterised by the fact that it remains unchanged for any period of time, unless it is subject to external influences, and that no processes take place which are macroscopically detectable. Naturally, reactions take place in the molecular order of magnitude, however, the number of reactions and back reactions is equal (dynamic equilibrium in contrast with the statical mechanic equilibrium). Reactions in this sense include not only the chemical processes proper, but also evaporation and condensation processes, transitions from one modification to another one, etc. A system in a state of equilibrium must respond to a change in the external conditions (pressure, temperature) with a change in the substantial or phase composition. When the system returns to the initial values of pressure and temperature, the original composition must be regained. Quantitatively, the state of equilibrium can be characterized with the help of the free enthalpy G (see 2.10). Equilibrium is given if $dG = 0$ and G is at a minimum.

Stable equilibrium: In this state of equilibrium, the system is, under the given conditions, stable with respect to all other states.

Metastable equilibrium: The system is in a state which is stable with respect to immediately neighbouring states, but not with respect to all other states. (G is higher than in a stable state.) In nature, metastable equilibria occur frequently (e.g. supercooled liquids).

Phase diagrams:

Phase relations in a system are usually represented graphically. The representation of the influences of pressure, temperature and concentration on one and the same graph is only possible for binary systems (three-dimensional). For a representation of complex systems other than binary systems, n-dimensional spaces would be required. That is why representations projected into a plane are often used (*p-T*, *T-C* and *C-C* diagrams, less frequently *p-C* diagrams).

Tie lines: On an equilibrium phase diagram, lines joining coexisting minerals or phases.

132

Activity:

All thermodynamic laws which contain concentrations as variable quantities (e.g. law of mass action, Nernst distribution law) in a strict sense only apply to the borderline case of infinitely great dilutions (no interaction between the solved particles). To maintain the form of the laws also for real conditions, activities a are introduced in the place of concentrations C. The conversion of the two quantities is obtained by $a = f \cdot C$, where f, the activity coefficient, is dependent on the temperature, pressure and activities of all other components. For low concentrations, a satisfactory approximation can frequently be obtained when assuming that $f = 1$, i.e. $a = C$. Information about the determination of the activity coefficient is given in textbooks of chemical thermodynamics.

Le Chatelier Principle:

This principle permits qualitative statements of the direction of changes in equilibria of stable systems. It may be worded as follows: If a system which is in a state of equilibrium is subjected to a constraint by changing the external conditions (p, T, C), the equilibrium is modified in such a way that the constraint is partially annulled. If, for instance, the pressure in a system consisting of two phases of different density is increased, the equilibrium is modified in such a way that the formation of the phase of higher density is encouraged (volume reduction).

Clausius-Clapeyron equation:

This equation establishes quantitative relation between pressure and temperature in the case of phase transformations (evaporation, melting, passing from one polymorphous or allotropic modification to another one) in unary systems (under the condition of a stable equilibrium). Its general form is:

$$\frac{dp}{dT} = \frac{L}{T(V_2 - V_1)}$$

where:

dp = change in the transformation pressure
dT = change in the transformation temperature
L = latent heat of transformation per mole
T = temperature $[°K]$
V_2 = molar volume of the phase developing by the absorption of heat
V_1 = molar volume of the phase developing by the emission of heat

Here, the numerical example of the melting equilibrium of water at $0\,°C$:

$$V_2 - V_1 = -1.64\ \text{cm}^3, \qquad L = 59270\ \text{cm}^3 \cdot \text{atm/mole},$$

$$T = 273.16\,°K,$$

$$\frac{dp}{dT} = -132\ \text{atm/degree}$$

With an increase in pressure of 132 atm, the melting point decreases by $1\,°C$.
In such computations, the dependence of L, V_2 and V_1 on p and T must be taken into account.

Nernst distribution law:

This law describes the distribution of a substance between two phases in a state of equilibrium. It was derived for liquid phases, but may also be applied to other states of aggregation (e.g. distribution of a trace element between two coexistent minerals or between solution and mineral separated from the solution).

For real conditions, it is written as follows:

$$\frac{a_1}{a_2} = K$$

where $a_1 =$ activity of the given substance in phase 1
$a_2 =$ activity of the given substance in phase 2
$K =$ coefficient of distribution, it has a constant value for the given system irrespective of the absolute value of a; this value is temperature-dependent.

This relation only holds if the dissolved substance is present in the same form in both phases (e.g. in both phases ionic, but not ionic in one phase and molecular in the other one). For dilute solutions, the concentrations can be used in place of the activities.

Gibbs' phase rule:

A mathematical expression which shows the conditions of equilibrium in a system as a relation between the number of components Ko, the number of phases P, and the degrees of freedom F. This phase rule holds if all components can cross the phase boundaries independently of each other. The formular expression is:

$$P + F = Ko + 2$$

where:

$P =$ number of phases
$F =$ number of degrees of freedom; the quantities (pressure, temperature, concentration of components) which can be freely selected within certain limits without causing a phase to disappear are called degrees of freedom.
$Ko =$ number of components; they are defined as the chemical components which are indispensable to build up the system.

The particular significance of the phase rule to the geosciences is that it offers the possibility of determining the maximum number of minerals which may occur in a given system, for example a certain rock. Since P and T are not constant in mineral-forming processes, the following inequality can be given:

$$P \leq Ko$$

This relation is called the mineralogical phase rule (GOLDSCHMIDT, 1912). If further so-called mobile components (H_2O, CO_2) have taken part in the formation of a system and have escaped later, the maximum number of phases is then reduced; the following holds according to Korzhinskiĭ

$$P \leq Ko - n$$

where $n =$ the number of the mobile components.

The determination of the number of components, which is difficult in more complicated systems, is facilitated by the following form of the mineralogical phase rule, as suggested by Buchheim (1964):

$$P \leqq B + Q - R$$

where

B = total number of all types of molecules occurring in the given system
Q = number of types of molecules which take part in several reactions in the system
R = number of reactions which take place in the given system.

Law of mass action:

This important law describes the relation between the concentrations or activities of the parent products and the end products of a chemical reaction for the case of equilibrium. For a reaction

$$aA + bB \rightleftharpoons cC + dD \, ,$$

the law of mass action is

$$\frac{[C]^c \cdot [D]^d}{[A]^a \cdot [B]^b} = K \, .$$

The square brackets represent the activities or concentrations of the substances. The equilibrium constant K is dependent on temperature and, in reactions where different numbers of gaseous co-reactants occur on both sides of the equation, also distinctly on pressure. The activities of pure solid or liquid reacting agents are taken as 1, hence, they do not appear in the law of mass action. Details on the experimental determination and calculation of K should be looked up in textbooks of physical chemistry.

Bibliography

BARTH, T. F. W.: Theoretical Petrology. I. Wiley and Sons, New York-London 1962
BUCHHEIM, W.: Die mineralogische Phasenregel bei Berücksichtigung chemischer Reaktionen. Geologie *13* (1964), 6/7, 836
EITEL, W.: Physikalische Chemie der Silikate. Univ. Chicago Press, 1954
GARRELS, R. M., and C. I. CHRIST: Solutions, Minerals and Equilibria. Harper and Row, New York 1965
GOLDSMITH, J. R.: Metastability and Hangovers in Crystals. Geochim. cosmochim. Acta *31* (1967), 913—919
GORBUNOV, L. V.: Empiric Dependence of Mole Volumes of Crystalline Matters on Entropy and Enthalpy During Thermal Expansion. Geokhimiya (1968), 3, 364—368 (Russian)
KADIK, A. A., and N. I. KHITAROV: Influence of Pressure on the Mass-transfer between Magmatic Melt and External Medium's Water. Geokhimiya (1965), 5, 507—518 (Russian)
KERN, R., and A. WEISBROD: Thermodynamique de Base pour Minéralogistes, Petrographes et Géologues. Masson Cie, Paris 1964

Khodakovskiĭ, I. L.: Thermodynamics of Aqueous Solutions of Electrolytes at Heightened Temperatures (Ion Entropy in Aqueous Solutions at Heightened Temperatures). Geokhimiya (1969) 1, 57—63 (Russian)

Kirkinskiĭ, V. A.: The Influence of Pressure on the Limits of Solid Solution. Geokhimiya (1965), 5, 534—543 (Russian)

Korzhinskiĭ, D. S.: Physicochemical Basis of the Analysis of the Paragenesis of Minerals, Traduc. Consultants Bureau, Inc., New York 1959

Kullerud, G., and H. S. Yoder: Pyrite Stability Relations in the Fe-S-System. Econ. Geol. *54* (1959), *4*, 533—572

Levin, E. M., C. R. Robbins and H. F. Mc Murdie: Phase Diagrams for Ceramists. The American Ceramic Society, Ohio 1964

Letnikov, F. A.: Isobar Potentials of the Evolution of Minerals (Chemical Affinity) and Their Use in Geochemistry. Izd-vo "Nedra", Moscow 1965 (Russian)

Naumov, G. B., B. N. Ryzhenko, and I. L. Khodakovskiĭ: Thermodynamics of Aqueous Solutions of Electrolytes at Heightened Temperatures (Thermodynamic Calculations in a Single Hydrogen Scale of Mineral Equilibria with Participation of the Hydrous Phase). Geokhimiya (1968), 7, 795—805 (Russian)

Neuhaus, A.: Die moderne Hochdruck-Hochtemperatur-Forschung und ihre geochemisch-petrologischen Aspekte. Freiberger Forschungsh. C *210·* (1966), 113—131

Neuhaus, A.: Über Phasen- und Materiezustände in den tieferen und tiefsten Erdzonen (Ergebnisse der modernen Hochdruck-Hochtemperatur-Forschung zum geochemischen Erdbild). Geol. Rdsch. *57* (1968), 3, 972—1001

Ovchinnikov, L. N., and N. F. Chelishchev: On the Statistical Estimation of Physical Conditions of Mineral Formation. Geokhimiya (1967), 11, 1328—1335 (Russian)

Ostapenko, G. T., and L. N. Khetsikov: Equations of Matter Transitions from one Phase to Another at an Isochoric Temperature Change in Closed Liquid-Vapour System. Geokhimiya (1968), 4, 485—488 (Russian)

Polkanov, A. A.: Physicochemical Tendencies of the Formation of Granite. Khimiya Zemnoĭ Kory, Vol. I, Izv. Akad. Nauk SSSR, Moscow 1963 (Russian)

Ramberg, H.: The Origin of Metamorphic and Metasomatic Rocks. Univ. Chicago Press, Chicago 1952

Rosenfeld, J. L.: The Contamination-reaction Rule. Amer. J. Sci. *259* (1961), 1—23

Schreyer, W.: Möglichkeiten und Grenzen der experimentellen Petrologie. Ber. geol. Ges. GDR *10* (1965), 2, 181—200

Szádeczky-Kardoss, E.: Sinn und Anwendung mineralogisch-geochemischer Modelle. Gesteinsmetamorphose bei verschiedenen Druckarten. Ber. Deutsch. Ges. geol. Wiss. *13* (1968), 43—58

Thompson, J. B., Jr.: The Thermodynamic Basis for the Mineral Facies Concept. Amer. J. Sci. *253* (1955), 65—103

Toropov, N. A., and V. P. Barsakovskiĭ: High-temperature Chemistry of Silicates and Other Oxygen Systems. Izd. Akad. Nauk SSSR, Moscow-Leningrad 1963 (Russian)

Voronov, A. N.: Use of Entropy for the Determination of Information Sufficiency in Geochemistry. Geokhimiya (1968), 11, 1404—1406 (Russian)

Weill, D. F., and W. S. Fyfe: A Discussion of the Korzhinsky and Thompson Treatment of Thermodynamic Equilibrium in Open Systems. Geochim. cosmochim. Acta *28* (1964), 565—576

Wehrenberg, J. P., and A. Silverman: Studies of Base Metal Diffusion in Experimental and Natural Systems. Econ. Geol. *60* (1965), 2, 317—350

Winkler, H. G. F.: Petrogenesis of Metamorphic Rocks. Springer, Berlin/Heidelberg/New York 1965

Zharikov, V. A.: Thermodynamical Characteristic of Irreversible Natural Processes. Geokhimiya (1965), 10, 1191—1206 (Russian)

Annual Report of the Director. Geophys. Laborat. Carnegie Institution 1965—1966

Group of authors: Melting and Transformation Point in Oxide and Silicate Systems at Low Pressure. Geol. Soc. Amer., Mem. 97, (1966), 301—322

3.3.5.2. Examples of the Dependence of Mineralogical-geochemical Systems on p and T

The following examples illustrate the complex of questions dealt with in Section 3.3.5. The stability ranges of some minerals are shown in Figs. 27, 28, 29, and 30. The possible paragenesis of minerals as a function of T and p is shown in these Figures and represented in another manner (only a certain temperature is taken into consideration) in Figs. 31 and 32.

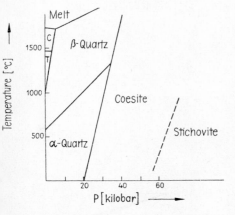

Fig. 27 Stability ranges of the most important SiO_2 varieties (C cristobalite; T tridymite)

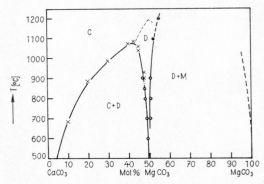

Fig. 28 The phase diagram of the $CaCO_3$-$MgCO_3$ system.

C = calcitic phase, D = dolomitic phase, M = magnesite-rich phase. The region in which dolomite is observed as an ordered compound is delimited by a dashed line. (Reproduced from GOLDSMITH and HEARD 1961; from GOLDSMITH, 1967)

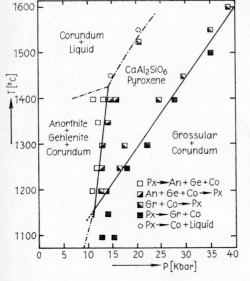

Fig. 29 The stability field of $CaAl_2SiO_6$ pyroxene (from HAYS, 1966)

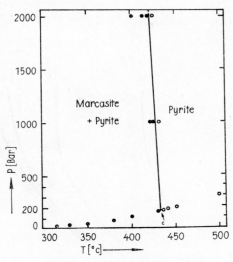

Fig. 30 Stability boundary between marcasite and pyrite in the Fe-S-OH system (from KULLERUD, 1966)

Finally, examples of the dependence of solubility and miscibility are given in Figs. 33, 34, 35, 36. It should be noted that, in Fig. 35 $k > 1$ means an enrichment of the given element in the the precipitate related to the mother liquor, whereas $k < 1$ means a depletion.

Fig. 31 Phase relations in the Fe-S-O system below 675 °C (from KULLERUD, 1953)

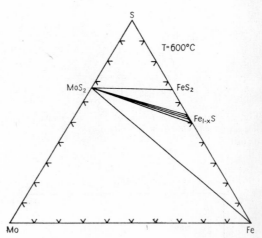

Fig. 32 Phase relations in the Fe-Mo-S system at 600 °C (from KULLERUD, 1966)

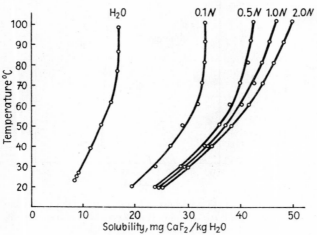

Fig. 33 Solubility of fluorite in NaCl solutions (from STRÜBEL, 1965)

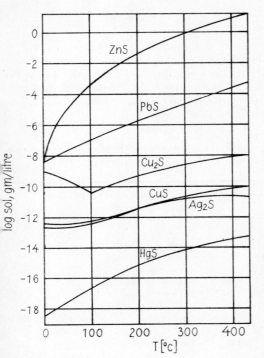

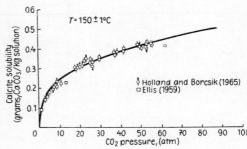

Fig. 34 Solubility of several metallic sulphides in aqueous solution at $pH = 7$, calculated by VERHOOGEN (1938) up to 400 °C. This is the calculated concentration of simple metallic ion only on condition that the ratio of total sulphur to metallic ions in the solution is the same as in the solid sulphide solute (from SMITH, 1963)

Fig. 35 The solubility of calcite in water as a function of the CO_2 pressure in the vapor phase at 150 °C (from HOLLAND and BORCSIK, 1965)

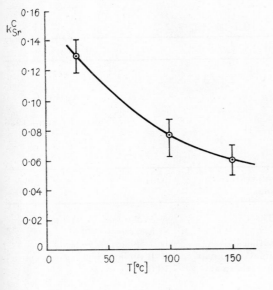

Fig. 36 The variation of the distribution coefficient k_{Sr}^{Ca} for strontium in calcite as a function of temperature (from BODINE et al., 1965)

$$k_{Sr}^{Ca} = \frac{\dfrac{c_{Sr(s)}}{c_{Ca(s)}}}{\dfrac{c_{Sr(aq)}}{c_{Ca(aq)}}}$$

$c_{Sr(s)}$ = Concentration of strontium in calcite
$c_{Ca(s)}$ = Concentration of calcium in calcite
$c_{Sr(aq)}$ = Concentration of strontium in the mother liquor of calcite
$c_{Ca(aq)}$ = Concentration of calcium in the mother liquor of calcite

3.3.5.3. Geological Thermometers

To study the course of geological processes, geological thermometers (barometers, etc.) are required. The thermometric statement is obtained via phase transformations, trace element contents, physical phenomena, oxygen-isotope exchange equilibria (see 10.2.), etc. In the case of some ore minerals, the statements given in Table 45 are possible. The bibliography given below should be used as an introduction into the problems and contains further examples.

Table 45. Ore Minerals as Useful Geothermometers
(data from INGERSON, 1955, BARTON and SKINNER 1967; L. BAUMANN and P. BEUGE, pers. inform.)

1. *Transformation Products*

Mineral	Formula	High-temperature modification	T [°C]	Low-temperature modification	Remarks
Argentite	Ag_2S	low Argentite (cubic)	176	Acanthite (orthorhombic)	
Hessite	Ag_2Te	intermediate H. (cubic)	145	low Hessite (orthorhombic)	
Naumannite	Ag_2Se	cubic	133	orthorhombic	
Schapbachite	$AgBiS_2$	cubic	210	orthorhombic	
Chalcosite	Cu_2S	hexagonal	104	orthorhombic	
Digenite	Cu_9S_5	high Digenite	76	Digenite	in presence of Covellite
		high Digenite	120	Digenite	in presence of Bornite
Pyrrhotite	FeS	β-Pyrrhotite	138	α-Pyrrhotite	KULLERUD,
		γ-Pyrrhotite	325	β-Pyrrhotite	YODER, 1959
Bravoit	$(Ni, Fe)S_2$	Pyrite + vaesite	137	Bravoit	CLARK, KULLERUD, 1963
Sulfur	S	monoclinic	95	orthorhombic	

2. *Melting point*

Mineral	Formula	Melting point T [°C]	Remarks
Bismuth	Bi	271, 5	KLEMENT a. o. 1963
Sulfur	S (monocl.)	115,2	WEST 1959
	S (orthorhomb.)	110,2	
Selenium	Se	217	STULL, SINKE 1956

3. *Decomposition of Solid Solutions*

Host Mineral	Dissociation Mineral	Temperature Range	Author
Silver	Dyscrasite	± 250 °C	A. WESTGREN, G. HÄGG, S. ERICSSON, 1929
Gold	Copper	katathermal	D. L. SCHOLTZ, 1936
Gold	Bismuth	pegmat.-pneumat. to katathermal	P. RAMDOHR, F. BUSCHENDORF, 1931

Table 45 continued

Host Mineral	Dissociation Mineral	Temperature Range	Author
Iron	Cohenite	$\pm 900\,°C$	LÖFQUIST and C. BENEDIKS, 1941
Platinum	Iridium	magmatic	P. RAMDOHR, 1960
Arsenic	Stibarsenic	epithermal	P. E. WRETBLAD, 1951
Stibarsenic	Arsenic	epithermal	P. E. WRETBLAD, 1951
Stibarsenic	Antimony	epithermal	J. ORCEL, G. RIVERA PLAZA, 1928
Antimony	Stibarsenic	epithermal	J. ORCEL, G. RIVERA PLAZA, 1928
Tellurobismuthite	Tetradymite	mesothermal	T. WATANABE, 1933 et al.
Tetradymite	Tellurobismuthite	mesothermal	T. WATANABE, 1933
Chalcosite	Neodigenite	$300° — 78\,°C$	W. KURZ, 1936
Chalcosite	Stromeyerite	$300\,°C$	G. M. SCHWARTZ, 1935
Chalcosite	Bornite	$170\,°C$	G. M. SCHWARTZ, 1935
Neodigenite	Chalcosite	$300° — 78\,°C$	N. W. BUERGER, 1941,
Neodigenite	Covelline	epithermal	P. RAMDOHR, 1943
Neodigenite	Bornite	$175\,°C$	P. RAMDOHR, 1955
Bornite	Chalcopyrite	$300\,°C$	A. SUGAKI and N. YAMAE, 1950
Bornite	Chalcopyrite	$475\,°C$	G. M. SCHWARTZ, 1931
Bornite	Chalcosite	$225\,°C$	G. M. SCHWARTZ, 1928
Bornite	Tetrahedrite	$275\,°C$	A. B. EDWARDS, 1946
Bornite	Tetrahedrite	$300\,°C$	H. SCHNEIDERHÖHN, 1922
Pentlandite	Pyrrhotite	liquid-magmatic	W. H. NEWHOUSE, 1927
Pentlandite	Linneite	liquid-magmatic	V. O. VÄHÄTALO, 1953
Sphalerite	Pyrrhotite	mesothermal	O. W. OELSNER, 1929
Sphalerite	Pyrrhotite	$<580\,°C$	H. E. USDOVSKI, 1967
Sphalerite	Chalcopyrite	$650\,°C$	G. M. SCHWARTZ, 1931b
Sphalerite	Chalcopyrite	$350—400\,°C$	N. W. BUERGER, 1934
Sphalerite	Chalcopyrite	$550\,°C$	H. BORCHERT, 1934
Sphalerite	Cubanite	kata- to mesothermal	J. W. GRUNER, 1929
Sphalerite	Stannite	mesothermal	O. W. OELSNER, 1929
Chalcopyrite	Chalcosite	$480\,°C$	H. D. MERWIN and R. H. LOMBARD, 1937
Chalcopyrite	Bornite	$500\,°C$	H. BORCHERT, 1934
Chalcopyrite	Chalcopyrrhotite	$255—450\,°C$	H. BORCHERT, 1934
Chalcopyrite	Sphalerite	$350—400\,°C$	A. SUGAKI and N. YAMAE, 1952
Chalcopyrite	Stannite	$\pm 500\,°C$	G. M. SCHWARTZ, 1931
Chalcopyrite	Pyrrhotite	$\pm 250\,°C$	H. BORCHERT, 1934
Chalcopyrite	Valleriite	$225\,°C$	H. BORCHERT, 1934
Chalcopyrite	Cubanite	$450\,°C$	G. M. SCHWARTZ, 1927
Chalcopyrite	Cubanite	$250—300\,°C$	J. W. GRUNER, 1929
Chalcopyrite	Tetrahedrite	$500\,°C$	A. B. EDWARDS, 1946
Chalcopyrrhotite	Pyrrhotite	$550\,°C$	H. BORCHERT, 1934
Stannite	Sphalerite	highly mesothermal	F. AHLFELD, 1934
Stannite	Chalcopyrite	$<250\,°C$	P. RAMDOHR, 1944
Stannite	Cubanite	$<250\,°C$	P. RAMDOHR, 1944, et al.
Pyrrhotite	Pentlandite	$\pm 600\,°C$	G. M. SCHWARTZ, 1931
Pyrrhotite	Pentlandite	$425\,°C$	R. L. HEWITT, 1938
Pyrrhotite	Chalcopyrite	$255\,°C$	H. BORCHERT, 1934
Pyrrhotite	Chalcopyrite	$\pm 600\,°C$	R. L. HEWITT, 1938
Pyrrhotite	Chalcopyrrhotite	$350—550\,°C$	H. BORCHERT, 1934
Nickelite	Breithauptite	epithermal	P. RAMDOHR, 1960
Cubanite	Chalcopyrite	$250—300\,°C$	G. M. SCHWARTZ, 1927

Table 45 continued

Host Mineral	Dissociation Mineral	Temperature Range	Author
Cubanite	Chalcopyrite	235 °C	H. Borchert, 1934
Cubanite	Chalcopyrite	400 °C	V. Ross, 1954
Galena	Tetradymite	mesothermal	T. Watanabe, 1933
			O. Ödman, 1941
Galena	Argentite	mesothermal	H. Schneiderhöhn, 1922
Galena	Schapbachite	215 °C	E. P. Chapman and
			R. E. Stevens, 1933
Galena	Polybasite	mesothermal	W. L. Whitehead, 1916
Galena	Freibergite	mesothermal	H. V. Warren, 1932
			H. Schneiderhöhn, 1922
Altaite	Aguilarite	hydrothermal	F. L. Stillwell, 1926
Alabandite	Pyrrhotite	katathermal	G. A. Thiel, 1924
			P. Ramdohr, 1960
Schapbachite	Galena	≈225 °C	P. Ramdohr, 1938, E. P. Chapman, and R. E. Stevens, 1933
Bismuthinite	Emplectite	epithermal	P. Ramdohr, 1960, et al.
Linneite	Chalcopyrite	hydrothermal	P. Ramdohr, 1960, et al.
Spinel	Magnetite, Ilmenite	magmatic	A. Maucher, 1939
Magnetite	Ilmenite	700—400 °C	P. Ramdohr, 1926
Magnetite	Ulvöspinel	600 °C	P. Ramdohr, 1926, et al.
Magnetite	Ulvöspinel	750 °C	Kawai, Naoto, 1954
Magnetite	Hematite	>1,000 °C	P. Ramdohr, 1960
Magnetite	Ilmenite	600—700 °C	P. Ramdohr, 1926
Magnetite	Ilmenite	800 °C	G. M. Schwartz, 1931 b
Magnetite	Spinel	800 °C	P. Ramdohr, 1926
Magnetite	Spinel	1,000 °C	G. M. Schwartz, 1931 b
Magnetite	Franklinite	magmatic	P. Ramdohr, 1960
Magnetite	Geikielite	magmatic	P. Ramdohr, 1926, 1939
Chromite	Hematite	magmatic	P. Ramdohr, 1924, 1950
Chromite	Ilmenite	magmatic	P. Ramdohr, 1924, 1950
Franklinite	Gahnite	pegmat.-pneumat.	H. Ries and Bowen, 1922,
Franklinite	Zincite, Ilmenite	pegmat.-pneumat.	H. Ries and W. Bowen, 1922
Franklinite	Hausmannite	pegmat.-pneumat.	E. Mason, 1943, 1947
Franklinite	Magnetite	pegmat.-pneumat.	P. Ramdohr, 1960
Corundum	Hematite	hydrothermal	J. M. Bray, 1939
Hematite	Ilmenite	600 °C	P. Ramdohr, 1926, 1939
Ilmenite	Hematite	600—700 °C	P. Ramdohr, 1926, 1939
Ilmenite	Magnetite	700—400 °C	P. Ramdohr, 1926, 1939
Ilmenite	Rutile	hydrothermal	A. B. Edwards, 1949
Ilmenite	Corundum	magmatic	P. Ramdohr, 1960
Rutile	Ilmenite	hydrothermal	P. Ramdohr, 1926
			A. B. Edwards, 1949
Rutile	Hematite	hydrothermal	P. Ramdohr, 1960
Cassiterite	Tapiolite	pegmat.-pneumat.	P. Ramdohr, 1935
Olivine	Magnetite	magmatic	P. Ramdohr, 1960
Olivine	Ilmenite	magmatic	P. Ramdohr, 1960

Bibliography

BARTH, T. F. W.: The Feldspar Geologic Thermometers. Norsk geol. T. *42* (1962), 2, 330—339

BARNES, H. L. (Edit.): Geochemistry of Hydrothermal Ore Deposits. Holt, Rinehardt and Winston, Inc. New York a. o. 1967

BARTON, P. B., and B. J. SKINNER: Sulfide Mineral Stabilities. In H. L. Barnes: Geochemistry of Hydrothermal Ore Deposits, New York etc. 1967

HELLNER, E., and R. EULER: Hydrothermale und röntgenographische Untersuchungen an gesteins-bildenden Mineralen, I. Über ein geologisches Thermometer auf Grund von Untersuchungen an Bio-titen. Geochim. cosmochim. Acta *12* (1957), 47—56

HOLLAND, H. D., and M. BORCSIK: On the Solution and Deposition of Calcite in Hydrothermal Systems: Symposium, Problems of Postmagmatic Ore Deposition, Prague, 2 (1965) 364—374

HOLLAND, H. D.: Gangue Minerals in Hydrothermal Deposits: in BARNES (Geochemistry of Hydro-thermal Ore Deposits) New York a.o. 1967

INGERSON, E.: Geologic Thermometry in the Crust of the Earth. Geol. Soc. America, Spec. Paper No. 62 (1955), 465—488

INGERSON, E.: Methods and Problems of Geologic Thermometry. Econ. Geol. *50* (1955), 1, 341—410

MACDIARMID, R. A.: The Application of Thermoluminescence to Geothermometry. Econ. Geol. *58* (1963), 1218—1228

OFTEDAHL, I.: Scandium in Biotite as a Geologic Thermometer. Norsk geol. T. *23* (1943), 202

PARK, CH. F., and R. A. MACDIARMID: Ore Deposits. W. H. Freeman and Comp., San Francisco and London 1964

STRÜBEL, G.: Quantitative Untersuchungen über die hydrothermale Löslichkeit von Flußspat (CaF_2). N. Jb. Miner (1965), 3. 83—95

DEN TEX, E.: Gefügekundliche und geothermometrische Hinweise auf die Herkunft von Lherzolith-knollen aus Basalten. N. Jb. Miner. Mh. 1962, 9/10, 225—237

3.3.6. Redox Potential Eh

The redox potential (reduction-oxidation potential) is a measure of the degree of oxidation or reduction in a reversible oxidation-reduction system. The term redox potential is a misnomer because this quantity actually is not a potential but an electromotive force (*emf*), that is to say, a potential difference. The redox potential of a system is the *emf* of that galvanic element which consists of the redox system into which a noble metal electrode is dipped and the normal hydrogen electrode ((Pt) H_2 (p = 1 atm)/H^+ (a=1), 25 °C). The dependence on con-centration (or, more precise, on activity) of the redox potential is described by the Nernst equation:

$$Eh = Eh_o + \frac{RT}{nF} \cdot \ln \frac{[Ox]}{[Red]} \quad \text{[Volt]}$$

where:

Eh = redox potential at the given concentration, temperature, and pressure conditions [Volt]
Eh_0 = redox potential under standard conditions ($T = 25$ °C, $p = 1$ atm, activity = 1) [Volt]
R = gas constant
T = absolute temperature [°K]
n = number of electrons which are exchanged in the redox process
F = Faraday number = 96,484 coulomb

After introducing the constants and at standard conditions, the relation is reduced to

$$Eh = Eh_o + \frac{0.059}{n} \cdot \lg K \quad \text{[Volt]} \qquad K = \frac{[Ox]}{[Red[}$$

where K is the equilibrium constant of Red $\rightleftharpoons$ Ox $+ e^-$

In publications, Eh_0 is frequently called normal or standard potential. Eh_0 values of some important redox reactions are given in Table 46. When the Eh_0 values of the metals are arranged in the order of their magnitudes, the electromotive series of metals is obtained. For data on their geochemical significance see, among other publications, Section 8.4., oxidation zone. Further information in LATIMER (1952), SATO (1960), TISCHENDORF and UNGETHÜM (1964) and textbooks of electrochemistry.

The redox potential may depend on the pH value and will then vary by ~ 0.06 volt per pH unit on an average. It is the higher, the greater the degree of oxidation in the redox system and the lower, the greater the degree of reduction (see Fig. 38).

Table 46. Geochemically Interesting Redox Potentials of Elements, Ions and Combinations in Acid Solution and at Normal Conditions. (The Potential Values Marked by an Asterisk Apply to Alkaline Solution.)

Element	Reaction	Normal Potential or Redox Potential [Volt]
Ag	$Ag^+ + 1e \rightleftharpoons Ag$	$+0.799$
	$Ag^{2+} + 1e \rightleftharpoons Ag^+$	$+1.98$
	$Ag_2O + 2H^+ + 2e \rightleftharpoons 2Ag + H_2O$	$+1.173$
	$AgCl + 1e \rightleftharpoons Ag + Cl^-$	$+0.222$
	$Ag_2SO_4 + 2e \rightleftharpoons 2Ag + (SO_4)^{2-}$	$+0.653$
	$Ag_2S + 2e \rightleftharpoons 2Ag + S^{2-}$	$-0.69\star$
	$Ag_2O + H_2O + 2e \rightleftharpoons 2Ag + 2(OH)^-$	$+0.344\star$
Al	$Al^{3+} + 3e \rightleftharpoons Al$	-1.66
As	$HAsO_2 + 3H^+ + 3e \rightleftharpoons As + 2H_2O$	$+0.247$
	$As_2O_3 + 6H^+ + 6e \rightleftharpoons 2As^+ + 3H_2O$	$+0.234$
	$2H_3AsO_4 + 4H^+ + 4e \rightleftharpoons As_2O_3 + 5H_2O$	$+0.577$
	$(HAsO_4)^{2-} + 3H^+ + 2e \rightleftharpoons (AsO_2)^- + 2H_2O$	$+0.565$
	$As + 3H^+ + 3e \rightleftharpoons AsH_3 + 3H_2O$	-0.545
Au	$Au^+ + 1e \rightleftharpoons Au$	about $+1.68$
	$Au^{3+} + 3e \rightleftharpoons Au$	$+1.42$
	$(AuO_2)^- + 4H^+ + 3e \rightleftharpoons Au + 2H_2O$	$+1.603$
	$Au(OH)_3 + 3H^+ + 3e \rightleftharpoons Au + 3H_2O$	$+1.45$
	$Au_2O_3 + 6H^+ + 6e \rightleftharpoons 2Au + 3H_2O$	$+1.364$
B	$H_3BO_3 + 3H^+ + 3e \rightleftharpoons B + 3H_2O$	-0.87
Ba	$Ba^{2+} + 2e \rightleftharpoons Ba$	-2.90
	$Ba(OH)_2 .8H_2O + 2e \rightleftharpoons Ba + 8H_2O + 2(OH)^-$	$-2.97\star$
Be	$Be^{2+} + 2e \rightleftharpoons Be$	-1.85
Bi	$(BiO)^+ + 2H^+ + 3e \rightleftharpoons Bi + H_2O$	$+0.32$

Table 46 continued

Element	Reaction	Normal Potential or Redox Potential [Volt]
Br	$Br_2 + 2e \rightleftharpoons 2\,Br^-$	$+1.066$
	$HBrO + H^+ + 1e \rightleftharpoons \frac{1}{2}\,Br_2 + H_2O$	$+1.59$
Ca	$Ca^{2+} + 2e \rightleftharpoons Ca$	-2.87
	$Ca(OH)_2 + 2e \rightleftharpoons Ca + 2(OH)^-$	$-3.03^\star$
Cd	$Cd^{2+} + 2e \rightleftharpoons Cd$	$+0.403$
	$CdS + 2e \rightleftharpoons Cd + S^{2-}$	$-1.21^\star$
	$Cd(OH)_2 + 2e \rightleftharpoons Cd + 2(OH)^-$	$+0.809$
Ce	$Ce^{3+} + 3e \rightleftharpoons Ce$	-2.48
	$Ce^{4+} + 1e \rightleftharpoons Ce^{3+}$	$+1.61$
Cl	$Cl_2 + 2e \rightleftharpoons 2\,Cl^-$	$+1.359$
	$HClO + H^+ + 1e \rightleftharpoons \frac{1}{2}\,Cl_2 + H_2O$	$+1.63$
Co	$Co^{2+} + 2e \rightleftharpoons Co$	-0.277
	$Co^{3+} + 1e \rightleftharpoons Co^{2+}$	$+1.82$
	$Co(OH)_2 + 2H^+ + 2e \rightleftharpoons Co + 2H_2O$	$+0.097^\star$
	$Co(OH)_3 + H^+ + 1e \rightleftharpoons Co(OH)_2 + H_2O$	$+1.026^\star$
Cr	$Cr^{3+} + 3e \rightleftharpoons Cr$	-0.74
	$Cr^{3+} + 1e \rightleftharpoons Cr^{2+}$	-0.41
	$(Cr_2O_7)^{2-} + 14H^+ + 6e \rightleftharpoons 2\,Cr^3 + 7H_2O$	$+1.33$
Cs	$Cs^+ + 1e \rightleftharpoons Cs$	-2.923
Cu	$Cu^+ + 1e \rightleftharpoons Cu$	$+0.521$
	$Cu^{2+} + 2e \rightleftharpoons Cu$	$+0.337$
	$Cu^{2+} + 1e \rightleftharpoons Cu^+$	$+0.153$
	$Cu_2O + 2H^+ + 2e \rightleftharpoons 2\,Cu + H_2O$	$+0.470$
	$2\,Cu(OH)_2 + 2H^+ + 2e \rightleftharpoons Cu_2O + 3H_2O$	$+0.747$
	$CuCl + 1e \rightleftharpoons Cu + Cl^-$	$+0.137$
	$Cu(NH_3)_2 + 1e \rightleftharpoons Cu + 2NH_3$	$+0.12^\star$
Eu	$Eu^{3+} + 1e \rightleftharpoons Eu^{2+}$	-0.43
F	$F_2 + 2e \rightleftharpoons 2F^-$	$+2.65$
	$F_2O + 2H^+ + 4e \rightleftharpoons 2F^- + H_2O$	$+2.1$
Fe	$Fe^{2+} + 2e \rightleftharpoons Fe$	-0.44
	$Fe^{3+} + 1e \rightleftharpoons Fe^{2+}$	$+0.771$
	$Fe(OH)_2 + 2H^+ + 2e \rightleftharpoons Fe + 2H_2O$	-0.048
	$Fe(OH)_3 + H^+ + 1e \rightleftharpoons Fe(OH)_2 + H_2O$	$+0.272$
	$Fe(OH)_2 + 2e \rightleftharpoons Fe + 2(OH)^-$	$-0.877^\star$
	$Fe(OH)_3 + 1e \rightleftharpoons Fe(OH)_2 + (OH)^-$	$-0.56^\star$
Ga	$Ga^{3+} + 3e \rightleftharpoons Ga$	-0.53

Table 46 continued

Element	Reaction	Normal Potential or Redox Potential [Volt]
Gd	$Gd^{3+} + 3e \rightleftharpoons Gd$	-2.40
Ge	$GeO_2 + 4H^+ + 4e \rightleftharpoons Ge + H_2O$	$+0.15$
H	$H^+ + 1e \rightleftharpoons \frac{1}{2}H_2$	± 0.000
	$\frac{1}{2}H_2 + 1e \rightleftharpoons H^-$	-2.25
	$O_2 + 4H^+ + 4e \rightleftharpoons 2H_2O$	$+1.229$
Hf	$Hf^{4+} + 4e \rightleftharpoons Hf$	-1.7
Hg	$Hg_2^{2+} + 2e \rightleftharpoons 2Hg$	$+0.789$
	$2Hg^{2+} + 2e \rightleftharpoons Hg_2^{2+}$	$+0.920$
I	$I_2 + 2e \rightleftharpoons 2I^-$	$+0.536$
	$HIO + H^+ + 1e \rightleftharpoons \frac{1}{2}I_2 + H_2O$	$+1.45$
In	$In^{3+} + 3e \rightleftharpoons In$	-0.342
K	$K^+ + 1e \rightleftharpoons K$	-2.925
La	$La^{3+} + 3e \rightleftharpoons La$	-2.52
Li	$Li^+ + 1e \rightleftharpoons Li$	-3.045
Mg	$Mg^{2+} + 2e \rightleftharpoons Mg$	-2.37
	$Mg(OH)_2 + 2e \rightleftharpoons Mg + 2(OH)^-$	$-2.69\star$
Mn	$Mn^{2+} + 2e \rightleftharpoons Mn$	-1.182
	$Mn^{3+} + 1e \rightleftharpoons Mn^{2+}$	$+1.510$
	$MnO_2 + 4H^+ + 2e \rightleftharpoons Mn^{2+} + 2H_2O$	$+1.230$
	$MnO_2 + 2H^+ + 2e \rightleftharpoons Mn(OH)_2$	$+0.776$
	$(MnO_4)^- + 1e \rightleftharpoons (MnO_4)^{2-}$	$+0.56$
	$(MnO_4)^- + 8H^+ + 5e \rightleftharpoons Mn^{2+} + 4H_2O$	$+1.51$
	$Mn(OH)_2 + 2H^+ + 2e \rightleftharpoons Mn + 2H_2O$	-0.727
	$Mn(OH)_2 + 2e \rightleftharpoons Mn + 2(OH)^-$	-1.55
	$Mn(OH)_3 + 1e \rightleftharpoons Mn(OH)_2 + (OH)^-$	$+0.1$
Mo	$Mo^{3+} + 3e \rightleftharpoons Mo$	about -0.2
	$Mo^{6+} + 1e \rightleftharpoons Mo^{5+}$	$+0.78$
	$MoO_3 + 2H^+ + 2e \rightleftharpoons MoO_2 + H_2O$	$+0.261$
	$(MoO_4)^{2-} + 4H^+ + 2e \rightleftharpoons MoO_2 + 2H_2O$	$+0.258$
Na	$Na^+ + 1e \rightleftharpoons Na$	-2.714
Nb	$Nb^{3+} + 3e \rightleftharpoons Nb$	about -1.1
	$Nb_2O_5 + 10H^+ + 10e \rightleftharpoons 2Nb + 5H_2O$	-0.65
Nd	$Nd^{3+} + 3e \rightleftharpoons Nd$	-2.44

ement	Reaction	Normal Potential or Redox Potential [Volt]
i	$Ni^{2+} + 2e \rightleftharpoons Ni$	—0.250
	$Ni(OH)_2 + 2H^+ + 2e \rightleftharpoons Ni + 2H_2O$	+0.110
	$NiO_2 + 4H^+ + 2e \rightleftharpoons Ni^{2+} + 2H_2O$	+1.680
	$NiO_2 + 2H^+ + 2e \rightleftharpoons Ni(OH)_2$	+1.318
	$O_3 + 2H^+ + 2e \rightleftharpoons O_2 + H_2O$	+2.07
s	$OsO_4 + 8H^+ + 8e \rightleftharpoons Os + 4H_2O$	+0.85
	$H_3PO_2 + H^+ + 1e \rightleftharpoons P + H_2O$	—0.51
b	$Pb^{2+} + 2e \rightleftharpoons Pb$	—0.126
	$PbCl_2 + 2e \rightleftharpoons Pb + 2Cl^-$	+0.27
	$PbO_2 + 4H^+ + 2e \rightleftharpoons Pb^{2+} + 2H_2O$	+1.455
	$PbS + 2e \rightleftharpoons Pb + S^{2-}$	—0.95
	$PbO + 2H^+ + 2e \rightleftharpoons Pb + H_2O$	+0.248
	$PbO_2 + 2H^+ + 2e \rightleftharpoons PbO + H_2O$	+1.110
d	$Pd^{2+} + 2e \rightleftharpoons Pd$	—0.987
t	$Pt^{2+} + 2e \rightleftharpoons Pt$	about +1.2
	$Pt(OH)_2 + 2H^+ + 2e \rightleftharpoons Pt + 2H_2O$	+0.98
	$PtS + 2H^+ + 2e \rightleftharpoons Pt + H_2S$	—0.33
Rb	$Rb^+ + 1e \rightleftharpoons Rb$	—2.925
Rh	$Rh^{3+} + 3e \rightleftharpoons Rh$	about +0.8
	$S + 2e \rightleftharpoons S^{2-}$	—0.48*
	$H_2SO_3 + 4H^+ + 4e \rightleftharpoons S + 3H_2O$	—0.529
b	$Sb^{3+} + 3e \rightleftharpoons Sb$	+0.24
	$Sb_2O_3 + 6H^+ + 6e \rightleftharpoons 2Sb + 3H_2O$	+0.152
c	$Sc^{3+} + 3e \rightleftharpoons Sc$	—2.08
e	$Se + 2e \rightleftharpoons Se^{2-}$	—0.78*
	$H_2SeO_3 + 4H^+ + 4e \rightleftharpoons Se + 3H_2O$	+0.74
i	$SiO_2 + 4H^+ + 4e \rightleftharpoons Si + 2H_2O$	—0.86
	$(SiF_6)^{2-} + 4e \rightleftharpoons Si + 6F^-$	—1.2
Sm	$Sm^{3+} + 3e \rightleftharpoons Sm$	—2.41
Sn	$Sn^{2+} + 2e \rightleftharpoons Sn$	—0.136
	$Sn^{4+} + 2e \rightleftharpoons Sn^{2+}$	+0.15
	$(SnF_6)^{2-} + 4e \rightleftharpoons Sn + 6F^-$	—0.25
Sr	$Sr^{2+} + 2e \rightleftharpoons Sr$	—2.89
	$Sr(OH)_2 + 8H_2O + 2e \rightleftharpoons Sr + 2(OH)^- + 8H_2O$	—2.99*

Table 46 continued

Element	Reaction	Normal Potential or Redox Potential [Volt]
Ta	$Ta_2O_5 + 10\,H^+ + 10\,e \rightleftharpoons 2\,Ta + 5\,H_2O$	—0.81
Te	$Te^{2+} + 2\,e \rightleftharpoons Te$	—1.14*
	$Te + 2\,H^+ + 2\,e \rightleftharpoons H_2Te$	—0.72
	$TeO_2 + 4\,H^+ + 4\,e \rightleftharpoons Te + 2H_2O$	+0.529
Th	$Th^{4+} + 4\,e \rightleftharpoons Th$	—1.90
Ti	$Ti^{2+} + 2\,e \rightleftharpoons Ti$	—1.63
	$Ti^{3+} + 1\,e \rightleftharpoons Ti^{2+}$	about —0.37
	$(TiO)^{2+} + 2\,H^+ + 4\,e \rightleftharpoons Ti + H_2O$	* —0.89
	$(TiO)^{2+} + 2\,H^+ + 1\,e \rightleftharpoons Ti^{3+} + H_2O$	+0.1
	$(TiF_6)^{2-} + 4\,e \rightleftharpoons Ti + 6\,F^-$	—1.19
Tl	$Tl^+ + 1\,e \rightleftharpoons Tl$	—0.336
	$Tl^{3+} + 2\,e \rightleftharpoons Tl^+$	+1.25
U	$U^{3+} + 3\,e \rightleftharpoons U$	—1.80
	$U^{4+} + 1\,e \rightleftharpoons U^{3+}$	—0.61
	$(UO)_2^{2+} + 1\,e \rightleftharpoons (UO_2)^+$	+0.05
	$(UO_2)^{2+} + 4\,H^+ + 2\,e \rightleftharpoons U^{4+} + 2\,H_2O$	+0.334
V	$V^{2+} + 2\,e \rightleftharpoons V$	about —1.18
	$V^{3+} + 1\,e \rightleftharpoons V^{2+}$	—0.255
	$(VOH_4)^+ + 4\,H^+ + 5\,e \rightleftharpoons V + 4\,H_2O$	* —0.253
	$(VO)^{2+} + 2\,H^+ + 1\,e \rightleftharpoons V^{3+} + H_2O$	+0.361
W	$WO_3 + 6\,H^+ + 6\,e \rightleftharpoons W + 3\,H_2O$	—0.09
Y	$Y^{3+} + 3\,e \rightleftharpoons Y$	—2.37
Zn	$Zn^{2+} + 2\,e \rightleftharpoons Zn$	—0.763
	$ZnS + 2\,e \rightleftharpoons Zn + S^{2-}$	—1.44
	$(ZnO_2)^{2-} + 2\,H_2O + 2\,e \rightleftharpoons Zn + 4(OH)^-$	—1.216
	$ZnCO_3 + 2\,e \rightleftharpoons Zn + (CO_3)^{2-}$	—1.06
	$ZnO + 2\,H^+ + 2\,e \rightleftharpoons Zn + H_2O$	—0.438
Zr	$Zr^{4+} + 4\,e \rightleftharpoons Zr$	—1.53

A great number of chemical elements have more than one valence. In a redox system, the valence of an element can be changed into a higher or a lower one. This changes the solubility of the element or element combination. As a consequence, the migration capability can be considerably increased or reduced and a dispersion or concentration may occur in the geochemical cycle. The influence of the redox potential is known to be highest in sedimentary processes.

Table 47. List of the Most Important Element Valences of Geochemical Interest
(bold face: most important valence; numbers in brackets: less important)

Symbol	Atomic Number	Valence
Ac	89	**+3**
Ag	47	**0** **+1** +2
Al	13	(+1) **+3**
As	33	−3 0 +3 **+5**
Au	79	**0** **+1** (+2) +3
B	5	**+3**
Ba	56	**+2**
Be	4	**+2**
Bi	83	−3 0 **+3** +5
Br	35	**−1** +1 +3 +5 +7
C	6	**−4** 0 +2 **+4**
Ca	20	**+2**
Cd	48	**+2**
Ce	58	**+3** +4
Cl	17	**−1** 0 +1 +3 +5 +7
Co	27	(+1) **+2** **+3** +4
Cr	24	+2 **+3** **+6**
Cs	55	**+1**
Cu	29	**0** +1 **+2** (+3)
Dy	66	**+3**
Er	68	**+3**
Eu	63	+2 **+3**
F	9	**−1**
Fe	26	0 **+2** **+3** (+6)
Ga	31	+1 +2 **+3**
Gd	64	**+3**
Ge	32	−4 +2 **+4**
H	1	(−1) 0 **+1**
Hf	72	**+4**
Hg	80	**0** +1 **+2**
Ho	67	**+3**
I	53	**−1** (+1) (+3) **+5** (+7)
In	49	+1 +2 **+3**
Ir	77	**0** +2 +3 +4 (+6)
K	19	**+1**
La	57	**+3**
Li	3	**+1**
Lu	71	**+3**
Mg	12	**+2**
Mn	25	**+2** +3 **+4** (+5) (+6) (+7)
Mo	42	(+2) +3 **+4** (+5) **+6**
N	7	**−3** **0** (+1) +2 +3 +4 +5

Symbol	Atomic Number	Valence
Na	11	**+1**
Nb	41	(+2) +3 (+4) **+5**
Nd	60	**+3**
Ni	28	+1 **+2** +3
O	8	**−2** **0**
Os	76	**0** +2 +3 +4 (+6)
P	15	−3 +3 (+4) **+5**
Pa	91	(+4) **+5**
Pb	82	**0** **+2** +4
Pd	46	0 **+2** (+3) +4
Pm	61	**+3**
Po	84	**+2** +6
Pr	59	**+3** +4
Pt	78	0 **+2** (+3) +4 (+6)
Ra	88	**+2**
Rb	37	**+1**
Re	75	(+1) (+2) (+3) (+4) (+5) **+6** **+7**
Rh	45	0 +2 **+3** +4 (+6)
Ru	44	**0** +2 +3 (+4) (+6)
S	16	**−2** **0** (+3) +4 (+5) **+6**
Sb	52	−3 **+3** +4 +5
Sc	21	**+3**
Se	34	**−2** 0 +4 (+6)
Si	14	−4 (+2) **+4**
Sm	62	+2 **+3**
Sn	50	+2 **+4**
Sr	38	**+2**
Ta	73	(+3) (+4) **+5**
Tb	65	**+3** +4
Te	52	**−2** +4 (+6)
Th	90	(+2) (+3) **+4**
Ti	22	+2 +3 **+4**
Tl	81	**+1** +3
Tm	69	**+3**
U	92	+3 **+4** +5 **+6**
V	23	+2 +3 **+4** **+5**
W	74	+2 +3 +4 +5 **+6**
Y	39	**+3**
Yb	70	+2 **+3**
Zn	30	**+2**
Zr	40	(+3) **+4**

An excellent indicator of the degree of oxidation of sediments and magmatites is iron which occurs frequently and in two valences. Its degree of oxidation is indicative of the valences which other elements may have (Table 48).

Table 48. The Occurrence of Elements with Respect to the Degree of Oxidation of Iron (according to SAUKOV, 1953)

In the presence of	may occur	cannot occur
Fe^{2+}	$Ti(3^+, 4^+)$, $V(3^+, 4^+)$, $Cr(3^+)$, $Mo(5^+, 6^+)$, $Mn(2^+)$, $Co(2^+)$, $Ni(2^+)$, $W(6^+)$, $Pb(2^+)$	$V(5^+)$, $Cr(6^+)$, $Mn(4^+)$, $Co(3^+)$, $Ni(3^+)$, $Pb(4^+)$, $NO_3(1^-)$
Fe^{3+}	$Ti(4^+)$, $V(4^+, 5^+)$, $Cr(3^+, 6^+)$, $Mo(5^+, 6^+)$, $Mn(2^+, 4^+)$, $Co(2^+, 3^+)$, $Ni(2^+, 3^+)$, $Pb(2^+, 4^+)$, $NO_3(1^-)$	$Ti(3^+)$, $V(3^+)$, $U(4^+)$

Further applications of the redox potential are demonstrated in Figs. 37, 38, 41—43, and in Section 8.

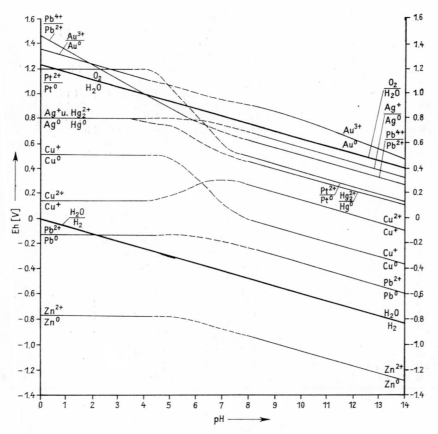

Fig. 37 Behaviour of the reduction-oxidation potential (Eh) of several valencies of the elements lead, gold, copper, platinum, mercury, silver, and zinc as a function of the hydrogen ion concentration (pH) under normal conditions (298.16 °K, 1 atm.); from TISCHENDORF and UNGETHÜM, 1964

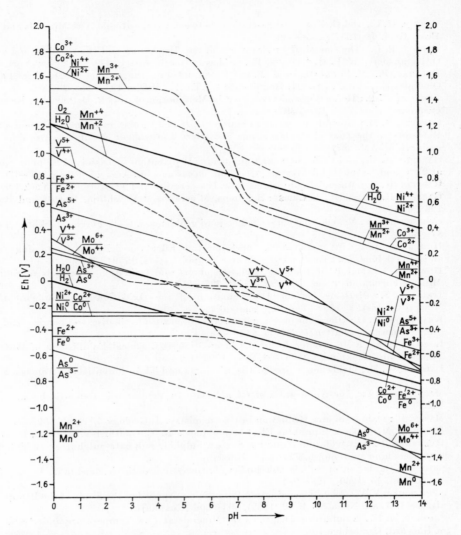

Fig. 38 Behaviour of the reduction-oxidation potential (*Eh*) of several valencies of the elements arsenic, iron, cobalt, manganese, nickel, and vanadium as a function of the hydrogen ion concentration (*pH*) at normal conditions (298.16 °K, 1 atm.); from TISCHENDORF and UNGETHÜM, 1964

Bibliography

BAAS-BECKING, L. G. M., I. R. KAPLAN, and D. MOORE: Limits of the Natural Environments in Terms of *p*H and Oxidation-reduction Potentials. J. Geology *68* (1960), 243—248

BARDOSSY, G., and M. BOD: A New Method to Characterize the State of Oxidation of Rocks. Acta geol. Acad. Sci. hung. *7* (1962), 29—35

BARNES, H. L., and G. KULLERUD: Equilibria in Sulfur-containing Aqueous Solutions, in the System Fe—S—O, and Their Correlation during Ore Deposition. Econ. Geol. *56* (1961), 648—688

BARNES, H. L., and G. KULLERUD: Relations between Composition of Ore Minerals and Ore Solutions. Econ. Geol. *52* (1957), 825—830

BARTON, P. B.: The Chemical Environment of Ore Deposition and the Problem of Low-temperature Ore Transport, in: P. H. ABELSON, Researches in Geochemistry, 279—300. New York 1959

DELAHAY, P., M. POURBAIX, and F. VAN RYSSELBERGHE: Diagramme d'Equilibre Potential-pH de Quelques Éléments. C. R. III. Réunion du C.I.T.C.E., Milano 1952, 15—29

EUGSTER, H. P.: Reduction and Oxidation in Metamorphism, in: P. H. ABELSON, Researches in Geochemistry, New York 1959, 397—426

GARRELS, R. M.: Mineral Species as Function of pH and Oxidation-reduction Potentials, with Special Reference to the Zone of Oxidation and Secondary Enrichment of Sulphide Ore Deposits. Geochim. cosmochim. Acta *5* (1954), 153—168

GARRELS, R. M., and C. R. NAESER: Equilibrium Distribution of Dissolved Sulphur Species in Water at 25 °C and 1 atm. Total Pressure, Geochim. et cosmochim. Acta *15* (1958), 113—130

GARRELS. R, M.: Mineral Equilibria at Low Temperature and Pressure. New York 1960

GARRELS, R. M., and C. L. CHRIST: Solutions, Minerals and Equilibria. Harper and Row, New York 1965, 450 p.

GOLDSCHMIDT, V. M.: Oksydasjon og reduksjon i geokjemien. Geol. Fören. Stockholm Förh. *65* (1943), 84—85

HEM, J. D., and M. W. SKOUGSTAD: Coprecipitation Effects in Solutions Containing Ferrous, Ferric, and Cupric Ions. U.S. Geol. Survey, Water Supply Papers, 1459-E (1960), 95—110

HUBER, N. K., and R. M. GARRELS: Relation of pH and Oxidation Potentials to Sedimentary Iron Mineral Formation. Econ. Geol. *48* (1953), 337—357

ITKINA, J. S.: Methods for the Determination of the Oxidation-reduction Potential in Rocks. Mineral Oil Inst. *2* (1952), 84—91 (Russian)

KRAUSKOPF, K. B.: Übersicht über moderne Ansichten zur physikalischen Chemie erzbildender Lösungen. Naturwissenschaften *48* (1961), 411—445

LISITSIN, A. K.: A Method for $Eh-pH$ Determination at Chemical Equilibrium of Water with Rocks and Minerals. Geokhimiya (1967), 8, 994—1002 (Russian)

LATIMER, W. M.: Oxidation States of the Elements and Their Potentials in Aqueous Solutions. New York 1952

MACHAMER, J. F.: The Application of $Eh-pH$ Data to Metasomatic Deposits. Econ. Geol. *54* (1958), 1122—1125

MASON, B.: Oxidation and Reduction in Geochemistry. J. Geology *57* (1949), 62—72

POURBAIX, M.: Atlas d'Equilibres Electrochimiques à 25 °C. Gauthier-Villars Cie., Paris 1963

RYZHENKO, B. N.: Oxidation-reduction System Sulphide-Sulphate-Sulphur at Heightened Temperature. Geokhimiya (1966), 6, 726—728 (Russian)

SATO, M.: Oxidation of Sulfide Ore Bodies, 1. Geochemical Environments in Terms of Eh and pH. Econ. Geol. *55* (1960), 928—961

SHCHERBINA, V. V.: Oxidation-reduction Potentials as Applied to the Study of the Paragenesis of Minerals. Dokl. Akad. Nauk SSSR *22* (1939), 503—506 (Russian)

SCHMIDT, H. H.: Equilibrium Diagrams for Minerals at Low Temperature and Pressure. Geol. Club of Harvard, Cambridge 1962

SKRIPCHENKO, N. S.: Oxidation-reduction Conditions of the Medium as a Cause of the Primary Zone Formation in a Few Copper Deposits. Geokhimiya 1963, 402—409 (Russian)

SOLOMIN, G. A.: Methods of Determining Eh and pH in Sedimentary Rocks. Consultants Bureau, Plenum Press, New York, 1965, (Translation from the Russian)

SZÁDECZKY-KARDOSS, E.: Über zwei Wertigkeitsregeln der Geochemie und die geochemische Gruppierung der Elemente. Acta geol. Acad. Sci. hung. *1* (1952), 231—268

TISCHENDORF, G., and H. UNGETHÜM: Über die Bedeutung des Reduktions-Oxydations-Potentials (Eh) und der Wasserstoffionenkonzentration (pH) für Geochemie und Lagerstättenkunde. Geologie *13* (1964), 2, 125—158

TISCHENDORF, G., and H. UNGETHÜM: Zur Anwendung von $Eh-pH$-Beziehungen in der geologischen Praxis. Z. f. angew. Geologie, Berlin *11* (1965), 57—65

ZOBELL, C. E.: Studies on Redox Potential of Marine Sediments. Bull. Amer. Assoc. Petroleum Geologists *30* (1946), 477—513

3.7. Hydrogen Ion Concentration and *pH* Value

The hydrogen ion concentration is indicative of the acidity or alkalinity of an aqueous solution. Its negative logarithm is known as the hydrogen exponent *pH* or *pH* value. A simplified survey of the relations in question is given in Fig. 39.

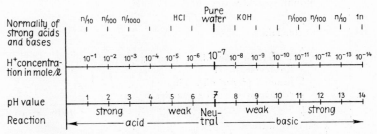

Fig. 39   Relations between the normality of the hydrogen ion concentration and the *pH* value in basic and acid aqueous solutions (schematically)

The solubility and thus the migration capability of many elements and their combinations in aqueous solutions are largely dependent on the *pH* value. The most widely known conditions relative to hydroxides are given in Table 49.

Table 49. Beginning of the Hydroxide Precipitation of Some Elements as a Function of the *pH*-Value (according to RANKAMA-SAHAMA, 1949, supplemented)

Ion	*pH*-Value	Ion	*pH*-Value	Ion	*pH*-Value
Ti^{4+}	1.4—1.6	Al^{3+}	4.1—5	Ni^{2+}	$\leqq 6.7$
Fe^{3+}	2—3	Fe^{2+}	5.1—5.5	Co^{2+}	$\leqq 6.8$
Sn^{2+}	2—3	Cr^{3+}	5.5	Hg^{2+}	7.9
Zr^{4+}	3	Cu^{2+}	$\leqq 5.4$	Mn^{2+}	8—8.8
Th^{4+}	3.5	Zn^{2+}	5.2—6	Mg^{2+}	10—10.5
In^{2+}	3.7	Pb^{2+}	6	Ca^{2+}	11

Fig. 40 and Table 50 give information about further relations, especially in the sedimentary field. For further data and additional publications see Redox Potential (3.3.6) and Geochemistry of Sediments (8.5).
Figs. 41, 42 and 43 give information about the stability relations of a few important combinations in given systems as functions of *pH* and *Eh*.

Bibliography

ALEKSANDROV, I. V.: On the Acid-basic Interaction in Bulk Activity Coefficients. Geokhimiya (1967), 9, 1117—1120 (Russian)
ATKINS, W. R.: Some Geochemical Applications of Measurements of Hydrogen Ion Concentration. Sci. Proc. Roy. Dublin Soc. *19* (1930), 455
BATES, R. G.: Electrometric *pH* Determinations (Theory and Practice). John Wiley and Sons Inc., New York. Chapman and Hall Ltd., London 1954
CHAĬNIKOV, B. I.: On the Problem of Using *pH* Suspensions of Minerals in Geology. Geokhimiya (1965), 11, 1372 (Russian)

Table 50. *pH* Values in Natural Waters and Sediments (combined on the basis of data from SAUKOV, 1953, RANKAMA-SAHAMA, 1949, and information from WÜNSCHE, Freiberg)

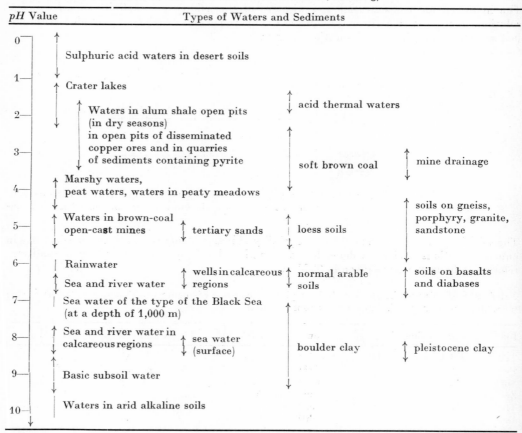

pH Value	Types of Waters and Sediments

Sulphuric acid waters in desert soils

Crater lakes

acid thermal waters

Waters in alum shale open pits (in dry seasons) in open pits of disseminated copper ores and in quarries of sediments containing pyrite

soft brown coal — mine drainage

Marshy waters, peat waters, waters in peaty meadows

soils on gneiss, porphyry, granite, sandstone

Waters in brown-coal open-cast mines — tertiary sands — loess soils

Rainwater

Sea and river water — wells in calcareous regions — normal arable soils — soils on basalts and diabases

Sea water of the type of the Black Sea (at a depth of 1,000 m)

Sea and river water in calcareous regions — sea water (surface) — boulder clay — pleistocene clay

Basic subsoil water

Waters in arid alkaline soils

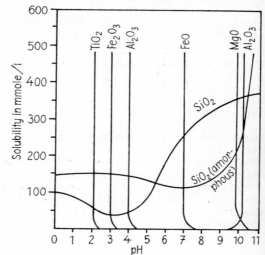

Fig. 40 Solubility of a few important oxides as a function of the *pH* value

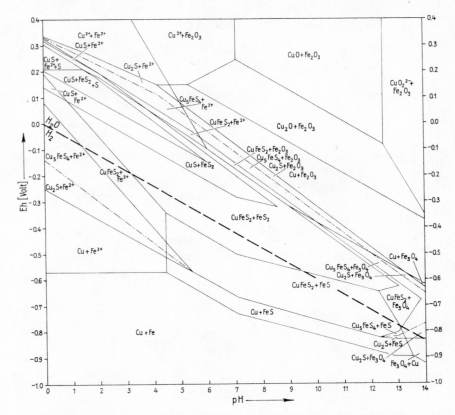

Fig. 41 Stability relations between some Cu, Fe, and Cu-Fe compounds in the Cu-Fe-S-H$_2$O system at 25 °C and a total pressure of 1 atm. Total activity: Σ Cu = 10^6, Σ Fe = 10^{-4}, Σ S = 10^{-2} (from TISCHENDORF and UNGETHÜM, 1965)

GLOGOCZOVSKY, J. J.: Possibilities of Determining the Physico-chemical Conditions of the Space of Sedimentation on the Basis of Geochemical Investigations. Nafta *11* (1959), 297—302 (Polish)

HUBER, N. K., and R. M. GARRELS: Relation of *pH* and Oxidation Potentials to Sedimentary Iron Mineral Formation. Econ. Geol. *48* (1953), 337—357

KHITAROV, N. J., and B. N. RYZHENKO: On the Assessment of the *pH* of Hydrothermal Solutions. Geokhimiya *12* (1963), 1152—1154 (Russian)

KORZHINSKIĬ, D. S.: Consideration of the Dependence of the Activity Coefficients on the Acidity of Solution. Geokhimiya (1968), 10, 1270—1272 (Russian)

KRUMBEIN, W. C., and R. M. GARRELS: Origin and Classification of Chemical Sediments in Terms of *pH* and Oxidation-reduction Potentials, J. Geology *60* (1952), 1—33

MARAKUSHEV, A. A.: Redox Processes in Connection with the Acidity Governing Post-magmatic Solutions. Geokhimiya *3* (1960), 214—222 (Russian)

RYZHENKO, B. N.: On the Possibility of Determining the Total Coefficient of the Acidity and Base Activity. Geokhimiya (1965), 5, 556—561 (Russian)

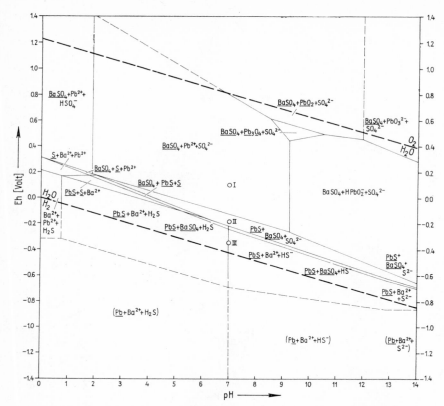

Fig. 42 Stability relations between some Ba, Pb, and S compounds in the Ba-Pb-S-H₂O system at 25 °C and a total pressure of 1 atm. Total activity: Σ Ba $= 10^{-4}$, Σ Pb $= 10^{-6}$, Σ S $= 10^{-2}$ (from TISCHENDORF and UNGETHÜM, 1965). The solid phases are underlined. For example, in the neutral range, either BaSO₄ (point I), BaSO₄ + PbS (point II) or PbS (point III) are precipitated, depending on the redox potential.

SCHALSCHA, E. B., H. APPELT, and A. SCHATZ: Effect of *pH* and Complexing Agents on Rock Materials. Geol. Soc. Amer. Progr. 1964. ann. Meet. (et: spec. Paper No. 82/1965), 174

SOLOMIN, A. G.: On the Determination of the Oxidation-reduction Potential and the *pH* of Sedimentary Rocks. Izd-vo "Nauka", Moscow 1964 (Russian)

URUSOV, V. S.: Acid-base Equilibria from the Point of View of Chemical Bond Polarity. Geokhimiya (1965), 10, 1186—1190 (Russian)

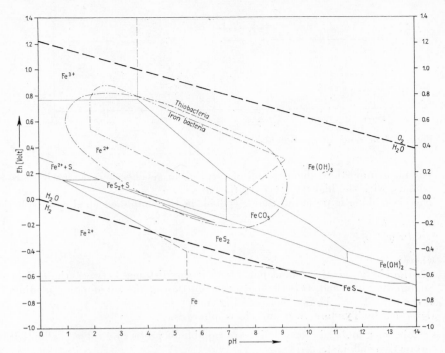

Fig. 43 Stability relations between some Fe compounds in the Fe-S-CO$_2$-H$_2$O system at 25 °C and a total pressure of 1 atm. Total activity: Σ Fe = 10^{-6}, Σ S = 10^{-2}, Σ [H$_2$CO$_3$] + [HC$_3^-$ + [C$_3^{2-}$] = 10^{-1} In addition, the *Eh-pH* characteristics of thiobacteria and iron bacteria are plotted on the graph according to BAAS-BECKING et al., (1960) (from TISCHENDORF and UNGETHÜM, 1965)

3.3.8. Sorption, Colloids, Ion Exchange

Sorption

General: The taking up of one substance by another. The following phenomena are subsumed under this generalised term: absorption, adsorption, chemosorption.

Of particular importance to geochemistry is *adsorption* which takes place at the surface of a solid (or a liquid) where particles of gaseous or dissolved substances are accumulated without any chemical reaction. Quantitatively determined by Langmuir's adsorption isotherm and Gibbs' adsorption equation.

Exchange adsorption is defined as the displacement of adsorbed substances from the surface of the adsorbent by other substances (cause: increased linkage force, higher concentration or higher pressure). Also known as *ion exchange*.

Colloids

Colloids are substances of any type within the order of magnitude from 1 to 100 μm (10 to 1000 Å) according to Ostwald, according to other authors also from 1 to 500 nm. They are positively or negatively charged. As to the classification of colloids see Fig. 44 and the bibliography.

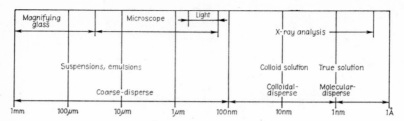

Fig. 44   The range of the colloids

The most important phenomenon of sorption in geochemistry is the ion adsorption by solid colloids, with the highest concentration in gels.

The electrokinetic potential ζ of a colloid, in the simplified ideal case (spherical form) is:

$$\zeta = \frac{e\,\delta}{Dr\,(r + \delta)},$$

where:

e is the charge on the colloid particle, δ is the distance between the colloid surface and its surrounding ion shell, D is the dielectric constant of the surrounding medium, r is the radius of the colloid and the ion shell (see Fig. 45).

Detection by Tyndall effect and electrophoresis.

Certain ions, molecules or colloids are capable of protecting solved colloids from flocculation (protective colloids).

Table 51. Increase in the Surface of a Cube with Increasing Decimal Division

Side Length of Cube	Number of Cubes	Total Surface [cm²]
1 cm	1	6
1 mm	10^3	60
100 μm	10^6	600
10 μm	10^9	6,000
1 μm	10^{12}	60,000
0.1 μm	10^{15}	600,000
0.001 μm	10^{18}	6,000,000

Most important phenomena within the colloidal range are:

1. Precipitation of solved colloids,
2. Flocculation of colloids by electrolytes,
3. Flocculation of colloids by other colloids of opposite charge,
4. Sorption and ion exchange by colloids.

The degree of sorption of ions by colloids, especially clay colloids, largely depends on the ionic potential (see Section 2.6.4.). The following relative adsorbability can be established: (cf. NOLL, 1931; JASMUND, 1954):

Monovalent ions: $Cs > Rb > K > Na > Li$.
Bivalent ions: $Mg > Ca > Sr > Ba$.

Positive Charge	Negative Charge
$Al(OH)_3$	metallic hydrosols (Au, Ag, Pt)
$Fe(OH)_3$	sulphide hydrosols (Pb and others)
$Cr(OH)_3$	S, As_2S_3, Sb_2S_3
$Cd(OH)_2$	MnO_2-hydrate
$Ce(OH)_4$	V_2O_5-hydrate
TiO_2-hydrate	SnO_2-hydrate
ZrO_2-hydrate	humus colloids
ThO_2-hydrate	silicic acid (SiO_2)
	clays

Table 52. Electric Charge of Natural Colloids or Sols (cf. Fig. 45)

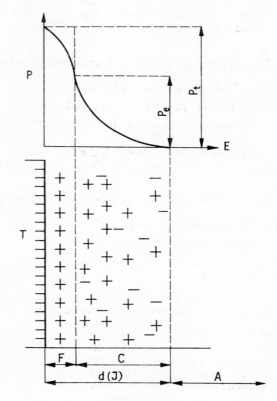

Fig. 45 Schematic representation of the electrokinetic potential of a colloid with negative charges on the surface of the particles (from FIEDLER and REISSIG, 1964)

T Particle surface; F Firmly adhering inner layer; C Loosely adhering outer layer with diffuse ionic distribution; d Electric double layer which also corresponds to the interior solution (I); A Exterior solution; E Distance from the particle surface; P_e Electrokinetic (zeta) potential; Pt Thermodynamic potential

Bibliography

CHUKHROV, F. V.: The Colloids in the Earth's Crust. Iz. Akad. Nauk SSSR Moscow 1955 (Russian)
EDELMANN, K.: Lehrbuch der Kolloidchemie, Vol. I (1962), Vol. II (1964). VEB Deutscher Verlag der Wissenschaften, Berlin
KUHN, A.: Kolloidchemisches Taschenbuch, 4th ed. Akademische Verlagsgesellschaft Geest & Portig, Leipzig 1953
LIPATEV, S. M.: Physikalische Chemie der Kolloide. Akademie-Verlag, Berlin 1953
NOLL, W.: Über die geochemische Rolle der Sorption. Chem. d. Erde 6 (1931), 552

WARD, A. G.: Colloids, their Properties and Applications. London 1946
WEISER, H. B.: A Text Book of Colloid Chemistry. New York 1949

Ion Exchange

The capability of clay minerals (due to their excess negative charges) to hold cations loosely so that they remain to be exchangeable.
Minerals of the ultramarine groups (sodalite, nosean, hauyne, lasurite) and zeolites also possess this property.
The ion-exchanging ability is given in milli-equivalent of interchangeable substance per 100 g of clay (m-e/100 g). For the various clay mineral groups, characteristic values are obtained which, however, differ largely (Table 53).

The property of exchanging ions is influenced by

1. Time allowed for a solution to act,

2. Ion concentration of the solution,

3. Granular size of the clay minerals.

The period of time after which the ion stock in a solution is no longer subject to changes (i.e. the time until the state of equilibrium is reached) is different for the individual clay minerals.

The adhesion or bond strength of the ion is dependent on

1. State of hydration,

2. Size of ions,

3. Valence.

Group of Clay Minerals	Ion-exchange Ability m-e/100 g
Kaolinite	3— 15
Sepiolite, Palygorskite	20— 30
Illite, Cihlorite	10— 40
Halloyste	5— 50
Montmorillonite	80—150
Vermiculite	100—150

Table 53. Ability of Clay Minerals to Exchange Ions

The selectivity, the different degree of adsorption of individual ions from solutions having extremely low cation concentrations ($<$1 per cent of the entire ability to exchange ions of the clay minerals) may be attributed to a different surface density of the negative charges on the groups of strata.
Studies of this problem have been conducted especially by SCHACHTSCHABEL, 1943; JASMUND and LANGE, 1966. Comprehensive bibliographies are given by DEUEL and HOSTETTLER, 1950.

Exchanger	Exchange Capacity [mval/100 g]
Kaolinite	3— 15
Montmorillonite	50—150
Nontronite	75— 80
Illite	10— 70
Vermiculite	100—150
Humic acids	100—500
Organic substances	up to 300

Table 54. Exchange Capacity of Clay Minerals, Humic Acids, and Organic Substances (from FIEDLER and REISSIG, 1964 and other authors)

Importance of Ion Exchange

In nature

Exchange of elements in formations due to weathering (nutrient cations, adherent to clay minerals in exchangeable form can be absorbed by plant roots; see Table 54).
Changes in structure and permeability of the soil (Ca^{++} and Mg^{++} ions improve the crumb structure, occupation by Na^+ causes compaction of the soil).
Change of the thixotropic properties in clays (clay minerals occupied by monovalent cations will yield suspensions which are more readily thixotropic than those of clay minerals occupied by alkaline earths or hydrogen).

Artificial utilisation

Change in the plastic properties (occupation by Ca^{++} improves and occupation by Na^+ reduces the ability to build up supporting frameworks in ceramic masses, thus improving or reducing the plasticity of these masses).
Production of acid-like properties (when occupied by H^+, clay minerals obtain acid-like properties; for example, this accounts for the effect of the H^+ montmorillonite catalyst in cracking crude oil).

Bibliography

DEUEL, H., and F. HOSTETTLER: Experimentia [Basle] *6* (1950), 445—456
JASMUND, K., and H. LANGE: Adsorption und Selektivität an Na-, K-, Ca-Kaoliniten und K-, Ca-Montmorilloniten mit radioaktiv markiertem Rubidium, Cäsium und Kobalt. Westdeutscher Verlag, Köln and Opladen 1966
WEISS, A.: Über das Kationenaustauschvermögen der Tonminerale I. Vergleich der Untersuchungsmethoden. Z. anorg. allg. Chem. *297* (1958), 232—256. II. Der Kationenaustausch bei den Mineralen der Glimmer-, Vermikulit- und Montmorillonit-Gruppe. Z. anorg. allg. Chem. 257—286. III. Der Kationenaustausch bei Kaolinit. Z. anorg. allg. Chem. *299* (1959), 92—120
SCHACHTSCHABEL, P.: Z. Bodenkunde und Pflanzenernährung *29* (1943), 213—219

4. Geochemical Practice and Testing Methods

4.1. Sampling and Preparation of Samples

Sampling

The way of sampling, the quantity and number of samples may differ, depending on the aim of the geochemical investigations. Since the sample is a part of a whole which must be representative of the whole, sampling should be carried out according to mathematical-statistical considerations, taking the geological conditions into account.

Sampling instructions regarding ores are given by OELSNER (1952), STAMMBERGER (1956); regarding rocks by HOENES (1957), regarding sediments by MÜLLER (1964); regarding waters by TRUEISEN (1959).

Preparation of Samples

This includes the drying of samples (especially in the case of organic materials, loose sediments and the like), the incineration (of organic materials), the removal of water-soluble salts and pore water (in the case of loose sediments and several saline rocks), the dissolving out of certain mineral components by means of acids (especially in the case of carbonate and silicate rocks), etc.; as to sedimentary rocks see MÜLLER (1964).

Instruments suitable for the reduction of samples, that is to say, for reducing the sample to the quantity required for the analysis, are described by MÜLLER (1964).

Besides the reduction of samples, a sizing of the sampled material is necessary especially when pure monomineral fractions are to be obtained from rocks and ores.

In the case of screen sizing minerals with a hardness >6 (MOHS), perlon netting should be used for the subsequent trace element analysis (when wire netting is used, there is the risk of introducing heavy metal in the form of abraded particles adhering to the material).

For the use of appropriate crushing machines, screening devices and other sizing methods which are particularly suitable for the handling of larger quantities of samples see text-books on mineral dressing (GRÜNDER, 1951; KIRCHBERG, 1953; SCHUBERT, 1964).

Preparation of Monomineral Fractions

Some methods of producing mineral concentrates which are of particular importance to geochemical investigations are given in Table 55. Almost all of them require relatively closely sized material to be treated.

The following main points govern the selection of the most favourable method of concentration (LANGE and WIEDEMANN, 1962):

The Problem to be Solved

If, for instance, minerals are to be subjected to trace element analyses, a chemical treatment of the sample, a separation by means of heavy liquids or by flotation should be avoided, unless the possible effect on the mineral sample is controlled.

Table 55. Survey of Some Methods of Producing Mineral Concentrates of Importance to Geochemical Investigations (cf. LANGE and WIEDEMANN, 1962; REHWALD, 1954; and textbooks on mineral dressing and preparation)

Method	Separation on the basis of the following mineral properties	Applicable[1] to grain sizes from...to... [mm]	Most favourable range of grain sizes [mm]	Test sample quantity	Remarks	References
Separation by heavy liquids (cf. Table 56)	Density (mainly minerals <4.2) (cf. Tables 57 and 58)	0.001—100	0.01—2	economical only if <100—200 g (depending on material)	sometimes poisonous; offensive smell; possibility of introducing elements due to adsorption or ion exchanges especially in the case of layered minerals. Clerici solution contains up to 0.1% of Pb. Separations with differences in density of up to 0.01 possible	TWENHOFEL and TYLER (1941), BONSTEDT-KUPLETSKAYA (1954), MÜLLER (1964), TH. v. WOLFF (1942). F. v. WOLFF and JÄGER (1930)
Micropanner	Density	a few micrometers up to a few millimetres		mg-quantities (up to about 50 mg)	economical especially in the case of homogeneous series samples	MÜLLER (1958 a, b)
Mechanised sifting troughs	(up to >5)					
Haultain Superpanner		about 0.03—2	0.04—0.2	1 g up to a few 10 g		GRÜNDER and GEYER (1957)
Laboratory table (e.g. WEDAG-Bochum)	Density and/or grain shape	0.05—2.5 (10)	heavy minerals >0.05 silicates >0.15 (close sizing favourable)	1 kg up to a few kg	mainly for pre-concentration	LANGE and WIEDEMANN (1962)
Table treatment Fast sweep table (e.g. model HUMBOLDT)				>10 kg		

Method	Separation on the basis of the following mineral properties	Applicable[1] to grain sizes from...to... [mm]	Most favourable range of grain sizes [mm]	Test sample quantity	Remarks	References
Laboratory magnetic separator (Type FRANTZ, COOK)	Susceptibility (cf. Table 60)	0.03 up to about 1	0.04—0.25 (close sizing favourable)	a few grams up to >100 g	separation of ferro-, para- and diamagnetic minerals	HESS (1959)
Magnetic separation						
Ring band separator (laboratory model)		0.06 up to about 5	(closely sized grain-size fractions favourable)	>100 g up to a few kg	separation of ferro- and paramagnetic minerals	LANGE and WIEDEMANN (1962)
Electrostatic separation	electric conductivity (etc.), dielectric constant	0.05—2.5	0.1—2.5 (close sizing favourable)	quantities of the order of g to kg	more liable to troubles than magnetic separation; e.g. finest inclusions affect the electric conductivity	DRUZHININ (1960), HOWELL and LICASTRO (1961)
Foam flotation	wettability	0.006—0.5	0.01—0.25	quantities of the order of g to kg	very time-consuming tests are necessary; possibility of introducing elements by flotation reagents	BUCKENHAM and ROGERS (1955), PETERSEN (1941)

[1] The grain size ranges may be different depending on the composition of the sample.

ample Quantity to be Put Through depends

1. on the required quantity of pure material,
2. on the contents of the interesting constituent of the rock or ore (main constituents — accessories) and
3. on the intergrowth conditions. In the case of a small grain size or close intergrowth, part of the rock or ore sample will remain non-disintegrated in the process of size reduction.

Grain Size

The various methods depend on grain size (e.g. REHWALD, 1954). The range of grain size within which the treatment has to take place is dependent on the degree of disintegration which, in turn, is governed by the conditions of intergrowth.

Differences in the Physical Properties — and also in the Chemical Properties —

between the component to be separated and the other components (e.g. differences in density or susceptibility).

If a monomineralic fraction is to be obtained from a rock or an ore, various separating methods have to be combined in most cases. In almost any case, the last "impurities" have to be removed by hand-picking with the help of the preparing lens or binocular (hand-picking needle; hand-picking apparatus MURTHY, 1957; HOPPE, 1960).

Table 56. Commonly Used Heavy Liquids (after TWENHOFEL and TYLER, 1941)

Heavy Liquids and Diluents	Density [g/cm³]	Temp. [°C]	Density	Melting Point [°C]	Boiling Point [°C]	Recovery
Bromoform, $CHBr_3$	2.87	20		9	151.2	
Acetone			0.79—2.87	— 94.6	56.5	washed out with water
Alcohol			0.79—2.87	—114	78.4	washed out with water
Antimony tribromide			2.49—3.64	94.2	280	distilled
Benzol			0.88—2.87	5.4	80.4	distilled
Carbon tetrachloride			1.58—2.87	— 23	76	distilled
Nitrobenzol			1.20—2.87	5.4	210	distilled
Acetylene tetrabromide, $C_2H_2Br_4$	2.96	20		0.1	126	
Acetone			0.79—2.96	— 94.6	56.5	washed out with water
Alcohol			0.79—2.96	—114	78.4	washed out with water
Benzol			0.88—2.96	5.4	80.4	distilled
Carbon tetrachloride			1.58—2.96	— 23	76	distilled
Nitrobenzol			1.20—2.96	5.4	210	distilled
Potassium-mercury-iodide (*Thoulet solution*; $KI + HgI_2$)	3.2	20				
Water			1.00—3.2	0.0	100	evaporated
Methylene iodide, CH_2I_2	3.2	20		5.7	180	
Acetone			0.79—3.32	— 94.6	56.5	distilled
Alcohol			0.79—3.32	—114	78.4	distilled
Benzol			0.88—3.32	5.4	80.4	distilled
Carbon tetrachloride			1.58—3.32	— 23	76	distilled

Table 56 continued

Heavy Liquids and Diluents	Density [g/cm³]	Temp. [°C]	Density	Melting Point [°C]	Boiling Point [°C]	Recovery
Antimony tribromide, SbBr₃	3.64	108		94.2	280	
Acetone			0.79—3.64	— 94.6	56.5	evaporated
Alcohol			0.79—3.64	—114	78.4	evaporated
Antimony trichloride			2.59—3.64	73.2	223.5	distilled
Carbon disulphide			1.25—3.64	—112.8	46.2	distilled
Thallium formiate,	4.95	105		94		
TlCO₂H	3.40	20				
Water			1.00—4.95		0.0	100	evaporated
Thallium-formiate-malonate,	5.10	100				
CH₂(TlCO₂)₂ Clerici solution	4.25	20				
Water			1.00—5.10		0.0	100	evaporated

Table 57. Density Values (up to $\varrho = 5$) of Important Rock-forming and Ore-forming Minerals (after TRÖGER, 1950)

Density [g/cm³]	Mineral	Density [g/cm³]	Mineral	Density [g/cm³]	Mineral
1.9 —2.2	Opal	2.89—2.93	Wollastonite	3.36—3.45	Vesuvianite
2.20—2.24	Graphite	2.90—3.00	Boracite	3.45—3.56	Chloritoid
2.31—2.33	Gypsum	2.90—3.07	Biotite	3.45—3.57	Titanite
2.44—2.48	Leucite	2.93—2.95	Aragonite	3.46—3.60	Hypersthene
2.54—2.59	Orthoclase	2.93—2.99	Anhydrite	3.50—3.59	Aegirine
2.54—2.60	Microcline	2.93—3.02	Zinnwaldite	3.50—3.80	Pyrope
2.55—2.62	Sanidine	2.95—3.05	Magnesite	3.53—3.65	Disthene
2.55—2.61	Anorthoclase	2.95—3.11	Ankerite	3.53—3.75	Grossularite
2.57—2.59	Kaolinite	2.98—3.06	Tremolite	3.55—3.58	Topas
2.57—2.60	Sodium orthoclase	3.05—3.50	Allanite (isotrope)	3.58—4.00	Spinel
2.57—2.66	Cordierite	3.06—3.35	Actinolite	3.70—4.20	Allanite
2.58—2.62	Chalcedony	3.06—3.40	Hornblende	3.71—3.77	Staurolite
2.60—2.65	Nepheline	3.07—3.15	Glaucophane	3.75—3.95	Andradite
2.61—2.63	Albite	3.08—3.24	Schorl	3.80—3.90	Siderite
2.63—2.65	Oligoclase	3.11—3.22	Andalusite	3.90—3.99	Celestite
2.64—2.66	Quartz	3.14—3.22	Apatite	3.91—4.02	Ludwigite
2.65—2.68	Andesine	3.15—3.20	Fluorite	3.94—4.09	Sphalerite
2.65—2.70	Glauconite	3.18—3.27	Enstatite	3.95—4.10	Corundum
2.66—2.78	Beryl	3.18—3.32	Sillimanite	4.02—4.25	Spessartite
2.68—2.71	Labradorite	3.20—3.32	Forsterite	4.05—4.35	Almandite
2.70—2.74	Calcite	3.20—4.35	Olivine	4.18—4.42	Rutile
2.71—2.75	Bytownite	3.22—3.31	Diopside	4.23—4.35	Fayalite
2.71—2.96	Phlogopite	3.24—3.36	Axinite	4.43—4.50	Barite
2.75—2.77	Anorthite	3.27—3.44	Bronzite	4.58—4.69	Pyrrhotite
2.77—2.93	Muscovite	3.29—3.33	Omphacite	4.68—4.86	Zircon
2.80—2.88	Lepidolite	3.31—3.40	diopsid. Augite	4.68—5.00	Ilmenite
2.86—2.95	Dolomite	3.35—3.38	Zoisite	4.70—5.08	Chromite

Table 58. Density Values of Important Saline Minerals

Density [g/cm³]	Mineral	Density [g/cm³]	Mineral	Density [g/cm³]	Mineral
1.602	Carnallite	2.201	Leonite	2.695	Vanthoffite
1.604	Bischofite	2.232	Astrakhanite	2.697	Glaserite
1.664	Tachhydrite	2.320	Gypsum	2.775	Polyhalite
1.987	Sylvite	2.374	Loeweite	2.828	Langbeinite
2.030	Schönite	2.573	Kieserite	2.850	Glauberite
2.132	Kainite	2.603	Syngenite	2.94—2.985	Anhydrite
2.168	Halite	2.673	Thenardite		

When small quantites of double sulphates are found, it is recommendable to concentrate them before separation by selective dissolution.

Table 59. Methods of Extracting Gases and Liquids from Inclusions in Minerals (from GOGUEL, 1963)

Treatment of the Host Mineral	Advantages	Disadvantages	Recommendable for the Extraction of ...
Dissolving	complete recovery	various troubles caused by solvents	noble gases and gases from salts
Decrepitation by freezing	simple procedure, trouble: A, B	explosive action only in the case of large liquid inclusions of a high degree of filling	large liquid inclusions of a high degree of filling
Decrepitation by heating		only medium and large inclusions will be opened, trouble: A, B	large inclusions in thermally constant minerals
Heating to more than 1,000 °C Melting	complete recovery of noble gases	trouble: A, B	nobel gases
Boring Crushing	trouble: A, B	only big inclusions will be opened	reactive gases from big inclusions
Reducing to powder	almost complete recovery, trouble: B	trouble: A	reactive gases from small inclusions

Trouble A: Decomposition of the minerals in heating (thermal) and in pulverising (mechanical); e.g. separation of hydroxidic water and carbonaceous carbon dioxide

Trouble B: Reaction of gases, e.g. $CO_2 + H_2 = CO + H_2O$
$4 H_2 + CO_2 = CH_4 + 2 H_2O$

Table 60. Survey of the Mass Susceptibility of Some Minerals (from LANGE and WIEDEMANN, 1962)

Magnetic Properties	Intensity $\chi \cdot 10^{-6}$	Mass Susceptibility of Various Minerals	
		$\chi \cdot 10^{-6}$	Mineral
Ferro magnetic	high >1,500	80,000—20,000	Magnetite
		6,100— 1,500	Pyrrhotite
paramagnetic	medium 300—50	271—113	Ilmenite
		124— 50	Almandite
			Olivine
		125— 53	Chromite
	low 100—10	120— 38 (13)	Amphiboles
		22— 13	Amphiboles with Fe (as FeO) + MnO = 19.9—7.5 %
		130— 26 (2)	Pyroxenes
		61— 15	Klinopyroxenes with Fe (as FeO)+MnO = 31—11 %
		73— 2	Orthopyroxenes with Fe (as FeO)+MnO = 48—3 %
		93— 51	Andradite
			Chlorite
		80— 23 (12)	Biotite
		50— 12	Biotite with Fe (as FeO) + MnO = 30.2—7.2 %
		56— 11	Pyrope (with Almandite)
		47— 13	Grossularite (with Andradite)
		31— 22	Tourmaline (black-brown)
		24	Epidote
		33— 6	Cordierite
		46— 9	Lepidolite
		19— 17	Olivine (10.3 % FeO)
	very low <10	(22)— 1	Titanite
		8— 5	Phlogopite
		4	Zinnwaldite
		2	Rutile (black-brown)
		6— 1	Muscovite
			Micaceous feldspars
			Zoisite
		9— 1	Tourmaline (green)
		1.12— 0.75	Tourmaline (pale pink)
		0.6	Feldspars
diamagnetic	up to —2.5	0.73 up to —0.17	Zircon
			Pyrite
			Rutile (reddish)
		—0.08	Cassiterite
		—0.28	Fluorite
		—0.30	Barite
		—0.34	Corundum
		—0.36 (—0.46)	Quartz
		—0.37	Orthoclase
		—0.37	Calcite
		—0.42	Topaz
		—2.46	Apatite

Bibliography

ADLER, R.: Zerkleinerung von Feststoffen bis zur kolloidalen Feinheit durch Wirbelstrommühlen. Chem. Technik *8* (1956), 485—486

BONSTEDT-KUPLETSKAYA, E. M.: Die Bestimmung des spezifischen Gewichts von Mineralen. Jena 1954

BORCHERT, H.: Einfluß der Lagerstättenverhältnisse auf die Probenahme. Erzmetall (1952), 297—303

BUCKENHAM, M. H., and J. ROGERS: Flotation of Quartz and Feldspar by Dodecylamin. Trans. Inst. Min. Met. *64* (1955), 11—33

BUERGER, M. J., and J. S. LUKECH: The Preparation of Oriented Polished Sections of Small Single Crystals. Amer. Mineralogist (1936), 667—669

DRUZHININ, J. P.: On a Few Peculiarities in the Separation of Quartz and Feldspar by Means of an Electrostatic Separator "T". Izv. Akad. Nauk SSSR, ser. geol. *10* (1960), 92—96 (Russian)

FREIER, R. K.: Wasseranalyse. Walter de Gruyter, Berlin 1964

GOGUEL, R.: Die chemische Zusammensetzung der in den Mineralen einiger Granite und ihrer Pegmatite eingeschlossenen Gase und Flüssigkeiten. Geochim. cosmochim. Acta *27* (1963), 155—181

GRÜNDER, W.: Aufbereitungskunde, Vol. II: Arbeitsmethoden im Aufbereitungslaboratorium. Hermann Hübener, Wilhelmshagen/Goslar 1957

GRÜNDER, W., and G. GEYER: Der Haultain-Superpanner. Erzmetall *10* (1957), 370—373

HAASE, L. W. et al.: Deutsche Einheitsverfahren zur Wasser-, Abwasser- und Schlammuntersuchung. Verlag Chemie, Weinheim/Bergstraße 1954

HESS, H. H.: Notes on Operation of Frantz Isodynamic Magnetic Separator. Prospectus on Frantz isodynamic magnetic separator. New York 1959

HOENES, D.: Mikroskopische Grundlagen der technischen Gesteinskunde, in: H. FREUND, Handbuch der Mikroskopie in der Technik, *4/1* (Umschau), Frankfurt/Main 1947

HOPPE, G.: Ein pneumatisches Auslesegerät für kleine Partikel. Z. angew. Geol. *6* (1960), 515—516

HOWELL, B. F., and P. H. LICASTRO: Dielectric Behaviour of Rocks and Minerals. Amer. Mineralogist *48* (1961), 269—288

KIRCHBERG, H.: Aufbereitung bergbaulicher Rohstoffe, Vol. I: Allgemeine Aufbereitung, Verfahren und Maschinen; Vol. II: Spezielle Aufbereitung, Wilhelm Gronau, Jena 1953

LANGE, H., and F. WIEDEMANN: Zur Gewinnung reiner Mineralfraktionen aus Gesteinen und den dabei möglichen Aussagen über die quantitative Zusammensetzung einzelner Gesteinstypen. Bergakademie *14* (1962), 431—439, 511—518

LAVERGNE, P. J.: Preparation of Geological Materials for Chemical and Spectrographic Analysis. Geol. Surv. Canada, Paper 65—78 (1965)

LUDWIG, G.: Methodisches zur Bestimmung von Kalisalzen unter dem Blickwinkel der Aufbereitung. Freiberger Forschungsh., A *120*, Berlin 1959

LYKKEN, W. H.: Mahlen und Klassieren im Bereich unter der Siebgrenze. Mining Enging. *9* (1957), 10, 1118—1120

MULLER, L. D.: The Micropanner — an Apparatus for the Gravity Concentration of Small Quantities of Materials. Bull Inst. Mining Metallurgy Trans. *68* (1958), 10, 1—7

MULLER, L. D.: Discussions and Contributions "The Micropanner — an Apparatus for the Gravity Concentration of Small Quantities of Materials". Bull. Inst. Mining Metallurgy Trans. *68* (1958), No. 625, 95—100

MÜLLER, G.: Methoden der Sedimentuntersuchung. Schweizerbart, Stuttgart 1964

MURTHY, M. V. N.: An Apparatus for Hand-Picking Mineral Grains. Amer. Mineralogist *42* (1957), 694—696

NEUMANN, H.: A Pulveriser for Mica and Micaceous Materials. Norsk Geol. Tidsskr. *36* (1956), 52

OELSNER, O.: Grundlagen zur Untersuchung und Bewertung von Erzlagerstätten. Jena 1952

REHWALD, G.: Die Anwendung erzmikroskopischer Untersuchungsmethoden bei Aufbereitung der Edelmetalle und der Nichteisenmetallerze, in: H. FREUND, Handbuch der Mikroskopie in der Technik, vol. II, part 2, p. 251, Frankfurt 1954

SCHEIBE, W.: Die Fein- und Feinstzerkleinerung in der Verfahrenstechnik. Bergakademie 1962, 476—482

SCHUBERT, H.: Aufbereitung fester mineralischer Rohstoffe Vol. I and II. VEB Deutscher Verlag für Grundstoffindustrie, Leipzig 1964—71

STAMMBERGER, F.: Einführung in die Berechnung von Lagerstättenvorräten fester mineralischer Rohstoffe. Akademie-Verlag, Berlin 1956

STRASHEIM et al.: A Possible Method of Pulverizing Mica and Other Floating Materials for Spectra Analysis. Appl. Spectroscopy *15* (1961), 177

TRUEISEN, C.: Vorschriften für die Entnahme und chemische Untersuchung von Wasserproben Bohrtechnik/Brunnenbau/Rohrleitungsbau *10* (1959), 1, 45—46

TWENHOFEL, W. H., and S. A. TYLER: Methods of Study of Sediments. New York/London 1941

v. WOLFF, F., and W. JÄGER: Die mechanische Phasenanalyse. Abderhaldens Handbuch der biologischen Arbeitsmethoden *10* (1930) 943—1022

v. WOLFF, TH.: Methodisches zur quantitativen Gesteins- und Mineraluntersuchung mit Hilfe der Phasenanalyse, ausgeführt am Beispiel der mafischen Komponenten des Eklogits von Silberbach Z. Kristallogr. Mineralog. Petrogr. Abt. B *54* (1942) 1—20

4.2. Chemical Analysis

The "classical" chemical analysis is still today one of the standard methods of geochemistry, especially for the determination of the main components of rocks and ores. Since it is not specific to geochemistry, however, and a special treatment would require much space, here only reference is made to the most important publications in this field.

Bibliography

BECKE, GOEHRING, and FLUCK: Einführung in die Theorie der quantitativen Analyse, 1st ed. Theodor Steinkopff, Dresden 1961

BILTZ-FISCHER: Ausführung quantitativer Analysen. 8th ed. Hirzel, Stuttgart 1960

BRUNCK, LISSNER, and SELTMANN: Quantitative. Analyse (Gravimetrie) 3rd ed. Theodor Steinkopff, Dresden 1962

CHARLOT, G., and D. BEZIER: Méthodes Modernes d'Analyse Quantitative Minérale. Paris 1949

ERDEY, L.: Gravimetric Analysis. Pergamon Press, Oxford 1963

GRILLOT, H. et al.: Methodes d'analyse quantitative appliquées aux Roches et aux Prélèvements de la prospection géochimique. Mem. BRGM, No. 30, Ed. Techn., Paris 1964

JAKOB: Chemische Analyse der Gesteine und silikatischen Mineralien. Birkhäuser, Basle 1952

JANDER-BLASIUS: Lehrbuch der analytischen und präparativen anorganischen Chemie, 4th ed. Hirzel, Stuttgart 1962

JANDER and JAHR: Maßanalyse, 7th ed. Walter de Gruyter, Berlin 1956

JEFFERY, P. E.: Chemical Methods of Rock Analysis. Pergamon Press, Oxford 1970

PRIBIL: Komplexometrie. Vol. I: Prinzipien und Grundbestimmung, 1963; Vol. II: Analyse von Legierungen, 1962; Vol. III: Analyse von anorganischen Rohstoffen und Erzeugnissen, 1962. VEB Deutscher Verlag für Grundstoffindustrie, Leipzig 1962

RÖDICKER: Analytische Chemie, Vol. I: Maßanalyse, 3rd ed. 1964; Vol. II: Gewichtsanalyse, 2nd ed. 1962. VEB Deutscher Verlag für Grundstoffindustrie, Leipzig

TREADWELL: Kurzes Lehrbuch der analytischen Chemie, Vol. I: Qualitative Analyse, 22nd ed. 1948; Vol. II: Quantitative Analyse, 11th ed. 1949, Franz Deuticke, Vienna

VICKERY, R. C.: Analytic Chemistry of the Rare Earths. Pergamon Press, New York 1961

VOLBORTH, A.: Elemental Analysis in Geochemistry, Part A: Major Elements. Elsevier, Amsterdam/London/New York 1969

WILSON, C. L., and B. W. WILSON (ed.): Comprehensive Analytical Chemistry, Vols 1 A and 1 B: Classical Analysis (General Methods of Analysis), Vol 1 C: Classical Analysis (Gravimetric and Titrimetric Determination of Elements), Vols 2 A and 2 B: Electroanalytical Methods. Elsevier, Amsterdam/London/New York 1959—1971

AHRENS, L. H.: A Note on the Relationship Between the Precision of Classical Methods of Rock Analysis and the Concentration of each Constituent. Mineralog. Mag. *30* (1954), 467

FAIRBAIRN, H. W. et al.: A Cooperative Investigation of Precision and Accuracy in Chemical, Spectrochemical and Modal Analysis of Silicate Rocks. U.S. Dep. Interior, geol. Surv., Bull. *980* (1951)

FAIRBAIRN, H. W.: Precision and Accuracy of Chemical Analysis of Silicate Rocks. Geochim. Cosmochim. Acta *4* (1953), 113

MERCY, E. L. P.: The Accuracy and Precision of "Rapid Methods" of Silicate Analysis. Geochim. Cosmochim. Acta *9* (1956) 161

THIERGÄRTNER, K.: Zur Genauigkeit und Richtigkeit von Silikatanalysen. Z. angew. Geol. (1965), H. 9

WASHINGTON, H. S.: Chemical Analysis of Igneous Rocks, Published from 1884 to 1900 with a Critical Discussion of the Character and Use of Analysis. U.S. Geol. Surv. Prof. Paper *14* (1903)

4.3. Chemical Microanalysis

The use of small and minute quantities of substances for analytical purposes is a growing demand in these days because of the necessity of saving time and material. This means detection by means of special reactions, primarily in the field of qualitative analysis (methods of spot analysis and precipitation of crystals).

Qualitative and quantitative analyses dealing with quantities of the following orders.

Semi-micromethod   *Micromethod*

100—10 mg of substance   10—0.1 mg

Bibliography

ACKERMANN, G.: Einführung in die qualitative anorganische Halbmikroanalyse. 5th ed. VEB Deutscher Verlag für Grundstoffindustrie, Leipzig 1968

ALIMARIN, J. P., and W. N. ARKHANGELSKAYA: Qualitative Halbmikroanalyse. 1st ed. VEB Deutscher Verlag der Wissenschaften, Berlin 1956

ALIMARIN, J. P., and B. J. FRID: Quantitative mikrochemische Analyse von Mineralen und Erzen. 1st ed. Theodor Steinkopff, Dresden 1965

DUVAL, C.: Traité de Micro-Analyse Minérale: Press, Scient. Intern., Paris 1954—1957

FEIGL, F.: Spot Tests. Elsevier, Amsterdam/London/New York/Princeton 1956 and 1958

HECHT, F., and M. K. ZACHEL: Handbuch der mikrochemischen Methoden, Vol. I to 3. Springer, Vienna 1954, 1955, 1961

KOCH, D. G., and G. A. KOCH-DEDIC: Handbuch der Spurenanalyse. Springer, Berlin/Göttingen/Heidelberg/New York 1964

MALISSA, H., BENEDETTI, and A. A. PICHLER: Monographien aus dem Gebiete der qualitativen Mikroanalyse, Vol. I: Anorganisch-qualitative Mikroanalyse. Springer, Berlin/Göttingen/Heidelberg 1958

MIKA, J.: Methoden der Mikromaßanalyse. 2nd ed. Ferdinand Enke, Stuttgart 1958

MILTON, R. F., and W. A. WATERS: Methods of Quantitative Microanalysis. 2nd ed. E. Arnold, London 1955

4.4. Photometry and Colorimetry

These belong to the most rapid and sensitive methods of chemical analysis. They can always be used when the presence of small quantities of a material can be detected on the basis of inherent colouring or colour reactions (colorimetry). It is a relative method, using calibration curves or reference substances. Measurement of absorption (photometry): Qualitative and quantitative determination of many chemical elements; great importance to geochemical investigations.

Dithizone Method

Dithizone (diphenylthiocarbazone $C_6H_5 \cdot N:N \cdot CS \cdot NH \cdot NH \cdot C_6H_5$) forms together with 18 different heavy-metal ions, the so-called "dithizone metals", in defined *pH* ranges in each case, inner complex salts (primary and secondary dithionates). 18 dithizone metals: Mn, Fe, Co, Ni, Cu, Zn, Pd, Ag, Cd, In, Sn, Pt, Au, Hg, Tl, Pb, Bi, Po.
For all methods, the use of heterogeneous solutions, especially of the liquid two-phase system water/chloroform (shaking-out reaction) has proved successful as a typical process technology for dithizone.
Basic determination methods are volumetric and colorimetric methods.
The dithizone method is particularly suitable for the determination of metal contents between 0.1 and 100 ppm. It is of great importance to the microanalysis and trace analysis (especially in geochemical prospecting!).

Bibliography

BOLTZ, D. F.: Colorimetric Determination of Nonmetals. Interscience Publ., New York 1958
HAWKES, H. E.: Dithizone Field Tests. Econ. Geol. *58* (1963), 4, 579
IWANTSCHEFF, G.: Das Dithizon und seine Anwendung in der Mikro- und Spurenanalyse. Verlag Chemie, Weinheim/Bergstraße 1958
KAKÁC, B., and J. V. ZDENEK: Handbuch der Kolorimetrie, 4 Volumes, 1st ed. VEB Gustav Fischer Verlag, Jena 1962
KORTÜM, G.: Anleitungen für die chemische Laboratoriumspraxis, Vol. II, Kolorimetrie, Photometrie und Spektrometrie, 4th ed. Springer, Berlin/Göttingen/Heidelberg 1962
LANGE, B.: Kolorimetrische Analyse. 5th ed. Verlag Chemie Weinheim/Bergstraße 1956
SANDELL, E. B.: Colorimetric Determination of Traces of Metals. Interscience Publisher Inc., New York 1950
SCHRÖN, W.: Anwendung der Dithizonchemie bei der geochemischen Prospektion. Z. angew. Geol. H. *8* (1960), 395—397

4.5. Chromatography and Ion Exchange

Paper Chromatography and *Gas Chromatography*

These methods are of increasing importance to geochemical analyses.
Paper chromatography primarily is a qualitative test method which allows the detection of quantities of a certain element of the order of 5—50 γ. Main field of application is that of organic chemistry; it is also used for the determination of anions and cations, even if they are chemically closely allied.

Ion Exchangers

Ion exchangers are solid or liquid substances which, in an electrolytic solution, can exchange ions against other ions. Depending on the type of charge of the exchangeable ions, a distinction is made between cation exchangers and anion exchangers. Separation and concentration of cations and anions which then are determined on the basis of other methods.

Bibliography

AYER, E.: Gaschromatographie. 2nd ed. Springer, Berlin/Göttingen/Heidelberg 1962

LASIUS, E.: Chromatographische Methoden in der analytischen und präparativen anorganischen hemie unter besonderer Berücksichtigung der Ionenaustauscher. Ferdinand Enke, Stuttgart 1958

RAMER, F.: Papierchromatographie. 5th ed. Verlag Chemie, Weinheim/Bergstraße 1962

ORFNER, K.: Ionenaustauscher. Walter de Gruyter, Berlin 1963

ORFNER, K.: Ionenaustauscher-Chromatographie. 1st ed. Akademie-Verlag, Berlin 1963

RIESSBACH, R.: Austauschadsorption in Theorie und Praxis. 1st ed. Akademie-Verlag, Berlin 1957

AIS, I. M., and K. MACEK: Handbuch der Papierchromatographie. Vol. 3. VEB Gustav Fischer Verg, Jena 1963

AISER, R.: Gaschromatographie. 2nd ed. Akademische Verlagsgesellschaft Geest & Portig, Leipzig 62

INSKENS, H. F., and L. STANGE: Praktikum der Papierchromatographie. Springer, Berlin/Göttingen/ eidelberg 1961

ACEK, K. et al.: Bibliography of Paper Chromatography 1957—1960 and Survey of Applications. ubl. House of the Czech. Acad. Sciences, Prague 1962

URNELL, J. H.: Progress in Gas Chromatography. Interscience (Wiley), New York 1968

ITSCHIE, S.: Chromatography in Geology. Elsevier Publ. Comp., Amsterdam/London/New York 64

ITSCHIE, S.: Gas Chromatography Abstracts, 1966, Elsevier Publ. Comp., Amsterdam/London/New ork 1967

roup of authors: Gaschromatographie, Fortschrittsberichte 1959. 1st ed. Akademie-Verlag, Berlin 61

.6. Potentiometry and Polarography

otentiometry is based on the measurement of an electrode potential; it enables to draw con-lusions as to the concentration conditions of the solution.

Applications

recipitation, complex-forming and neutralization reactions, oxidation and reduction pro-esses.

Advantages of Potentiometry

1. Carrying-out of simultaneous determinations;

2. Selective determinations, also possibility of titrations of turbid or intensely coloured solutions;

3. Possibility of using volumetric solutions for which no other useful indicator is known;

4. Carrying-out of microanalyses, because accuracy is higher than with "classical" methods.

As to polarography see Bibliography below.

Bibliography

JANDER, G., and K. F. JAHR: Maßanalyse. 7th ed. Walter de Gruyter, Berlin 1956, 240—285

KORTÜM, G.: Lehrbuch der Elektrochemie. 3rd ed. Verlag Chemie, Weinheim/Bergstraße 1962

MÜLLER, E.: Elektrochemische Maßanalyse. Theodor Steinkopff, Dresden/Leipzig 1942

Polarography

HEYROVSKÝ, J., and R. KALVODA: Oszillographische Polarographie mit Wechselstrom. Theoretische Grundlagen und praktische Anwendung. Theodor Steinkopff, Dresden 1960
HEYROVSKÝ, J., and P. ZUMAN: Einführung in die praktische Polarographie. VEB Verlag Technik Berlin 1959
HEYROVSKÝ, J., and J. KUTA: Grundlagen der Polarographie. Akademie-Verlag, Berlin 1965
KOLTHOFF, J. M., and J. J. LINGANE: Polarography, 2nd ed. New York
KRYUKOVA, T. M., S. I. SINYAKOVA, and T. V. AREFEVA: Polarographische Analyse. VEB Deutscher Verlag für Grundstoffindustrie, Leipzig 1964
SCHMIDT, H., and M. V. STACKELBERG: Die neuartigen polarographischen Methoden, ihr Prinzip und ihre Möglichkeiten. Verlag Chemie, Weinheim/Bergstraße 1962

4.7. Flame Spectrometry and Atomic Absorption Spectrometry

Flame Spectrometry

Characteristic spectral lines of certain elements are used for qualitative and quantitative analyses. The fundamentals are the same as those of the emission spectrometric analysis.

Advantages

1. Simultaneous determinations without chemical separation processes,
2. Little time required, suitable for serial analyses,
3. Small quantities of material required,
4. High accuracy,
5. Little expenditure of work.

Applications

Specifications for the determination of about 60 elements; particularly suitable for alkalies and alkaline earths. The applicability of flame spectrometry is restricted by the interaction of elements in the flame.

Atomic Absorption Spectrometry

Determination of about 50 main and trace elements and of a few isotopes in rocks, ores, minerals, waters, and organic matter. It is of increasing importance to geochemistry.

Bibliography

ANGINO, E. E., and G. K. BILLINGS: Atomic Absorption Spectrometry in Geology. Elsevier Publ. Comp., Amsterdam 1967
BERTHELAY, J. C.: Dosage d'éléments majeurs (Al, K, Na, Ca, Mg, Fe) dans les roches par spectro-photométrie d'absorption atomique. Ann. Fac. Sci. Univ. Clermont, No. 38, Clermont-Ferrand 1968
ELWELL, W. T., and J. A. F. GIDLEY: Atomic Absorption Spectrophotometry. Macmillan, New York 1962
FORNASERI, M., and A. PENTA: Determinazione del tallio neu silicati par fotometria di fiamma. La metallurgia italiana, *55* (1963) 8, 437—441
FORNASERI, M., and L. GRANDI: Flame Photometric Determination of Strontium in Silicates. Strontium Content of the Granite G-l and the Diabase W-l. Geochim. et Cosmochim. Acta, *19*, July (1960)
RUBESKA, I.: Absorptionsflammenphotometrie und ihre Verwendungsmöglichkeiten. Geol. Průzkum, ČSSR, *7* (1965), 3, 77—78

4.8. Emission Spectrographic Analysis

Spectrographic analysis is one of the most important analytical methods of geochemistry. It enables the qualitative and quantitative determination of chemical elements in rocks and minerals. Due to its high sensibility, it is used especially for trace element analysis.

The fundamentals of this method should be looked up in textbooks. Below, a few essential aspects of the qualitative and quantitative determination of elements in powder samples by emission spectrographic analysis are discussed. The information refers to analytical methods on the basis of carbon arc excitation within the wavelength range from 2,000 to 9,000 Å and photographic recording.

4.8.1. Spectrographs

The efficiency of the various spectrographs differs (dispersion and resolving power, working range, luminous intensity). The selection of the suitable spectrograph is dependent on the analytical problem to be solved (Table 61). The principal reference lines of geochemically important elements are given in Table 62.

A greater dispersion and a higher resolving power of a spectrograph (ensured especially by grating spectrographs) show two advantages:

a) Avoidance of line coincidences,
b) Increase in the detecting efficiency by thinning the background.

Table 61. Survey of the Spectrographs for Various Wavelength Ranges (Working Ranges) and the Required Type of Plate

Wavelength Range [Å]	Spectrographs	Type of Spectral Plate
1,650—2,000	Vacuum spectrographs (prism or grating spectrographs)	Schumann plates
2,000—3,500	Quartz-prism spectrographs (e.g. Zeiss Q-24, ISP-30, Hilger E 742)	ORWO Spectral-Blue
3,500—3,700	Range of transition from quartz- to glass-prism spectrographs	ORWO Spectral-Yellow
3,700—9,000	Glass-prism spectrographs (e.g. Zeiss Three-prism Spectrograph, ISP-51, ISP-67, Hilger E 744)	ORWO Spectral-Red ORWO Spectral-Total ORWO Infrared
2,000—9,000	Grating spectrographs (Zeiss PGS-2, DFS-8, Hilger E 543)	in accordance with the given wavelength range

4.8.2. Excitation

The following principal means of excitation are used: D.C. permanent arc, D.C. interruption arc, A.C. permanent arc, A.C. interruption arc.

Voltage: 220 V; current: 3 to 15 A controllable; arc ignition: primarily high-frequency.

In the spectroscopic analysis of solutions (rotating disk method) and a few other methods (carbon pellets, etc.) spark excitation is used.

The majority of the geochemically important elements are most efficiently excited in the arc; extremely easily excitable elements, such as alkaline metals, in very soft light sources (flames;

Table 62. Principal Reference Lines of Geochemically Important Elements (from Moenke-Blankenburg, 1962)

1,650—2,000 Å

Element	Lines
As	1,890.5
C	1,657.01
Hg	1,849.68
P	1,774.94
S	1,807.31
Se	1,960.91

2,000—3,500 Å

Element	Lines	Element	Lines
Ag	3,280.683; 3,382.891	Lu	2,615.42; 2,911.39; 3,077.60
As	2,349.84	Mn	2,576.104
Au	2,427.95; 2,675.95	Mo	3,170.347
B	2,496.778; 2,497.733	Ni	3,414.765; 3,492.956
Be	2,348.610; 3,130.416; 3,131.072	Os	2,909.061; 3,058.66
Bi	3,321.343; 2,897.975; 3,067.716	Pb	2,833.069
Cd	2,288.018; 3,261.057; 3,466.201	Pd	3,404.580; 3,421.24
Co	3,405.120; 3,453.505	Pt	2,659.454; 3,064.712
Cu	3,247.540; 3,273.962	Re	3,460.47
Er	3,372.750	Rh	3,434.893
Ga	2,943.637	Ru	3,436.737; 3,498.942
Gd	3,362.244; 3,422.466	Sb	2,528.535; 2,598.062; 2,877.915
Ge	2,651.178; 3,039.064; 3,269.494	Sn	2,839.989; 3,175.019
Hg	2,536.519	Ta	3,262.328; 2,714.674; 3,311.162
Ho	3,456.00	Te	2,142.75; 2,383.25; 2,385.76
In	3,256.09	Th	2,837.29
Ir	2,543.971; 3,133.321; 3,220.780	Tm	3,462.20
La	3,337.488	V	3,185.396
Li	3,232.61	Y	3,242.280
		Yb	3,289.37
		Zn	2,138.56; 3,345.020
		Zr	3,391.975

3,500—3,700 Å

Element	Lines
Dy	3,531.712
Er	3,692.652
Gd	3,646.196
Ni	3,524.541
Pb	3,683.471
Rh	3,692.357
Sm	3,568.258; 3,592.595; 3,609.484
Tb	3,509.17; 3,676.35
Tl	3,519.24
Yb	3,694.203
Y	3,710.290

>3,700 Å

Element	Lines	Element	Lines	Element	Lines
Ba	4,554.042; 4,934.086; 5,535.551	Hg	4,358.35	Pr	4,222.98; 4,225.327
Bi	4,722.552	Ho	3,748.17	Rb	4,201.851; 4,215.556; 7,800.227; 7,947.60
Ce	4,012.388; 4,133.800; 4,186.599; 4,222.599	In	3,891.02; 4,511.323	Re	4,889.17
La	4,296.680	K	4,044.140; 4,047.201; 7,664.907; 7,698.979	Sc	3,911.810; 4,023.688; 4,246.829
Cs	4,555.355; 4,593.177; 8,521.11; 8,943.50	La	3,949.106; 3,988.518	Sm	4,296.750; 4,424.342
Cr	4,254.346; 4,274.803; 4,289.721	Li	4,333.734; 4,602.863; 6,103.642; 6,707.844	Sr	4,077.714; 4,607.331
Dy	4,000.454; 4,045.983; 4,211.719	Lu	4,518.57	Tb	4,278.51; 4,326.48
Er	3,906.316; 4,007.967	Mn	4,030.755; 4,033.073; 4,034.490	Th	4,019.137
Eu	3,819.66; 3,907.110	Mo	3,798.252; 3,902.963	Tl	3,775.72; 5,350.46
Ga	4,205.046; 4,594.02	Nb	4,058.938; 4,079.729	Tm	3,717.92
Gd	4,032.982; 4,172.056; 4,251.736; 4,262.095	Nd	4,012.250; 4,061.085	U	3,761.333; 3,761.917; 4,241.669; 4,244.372
		Pb	4,247.367; 4,303.573	V	4,379.238
		Pr	4,057.820; 3,908.431; 4,179.422	W	4,008.753; 4,294.614; 4,302.108
				Y	4,374.935
				Yb	3,987.994
				Zn	4,810.534
				Zr	4,687.803

ee flame spectrometry) and elements which are difficult to excite in hard light sources (spark xcitation, electric gas discharge).

Note: The excitation energy of elements is dependent on their ionization potential (see ection 2.5.). Elements with a high ionization potential ($I > 10$) are difficult to excite.

.8.3. Spectral Carbon Products

Mostly high purity graphite is used as electrode material; a distinction is made between graphite" and "carbon", depending on whether or not an adjustment of the crystallites e.g. in graphite) has taken place. These structural differences produce an effect especially on

Table 63. Quality of Spectral Carbon Products

Type	Structure	Impurities	Form
Ultra Carbon Corporation, Bay City, Michigan			
U-1, U-2, UF4S, U7, ST, ST, U-5	Graphite resp Carbon (U-5)	<5 ppm	R, D, P, PP
VEB Elektrokohle Berlin-Lichtenberg			
T 0	Graphite	—	R
T 1	Graphite	B	R,D
T 2	Graphite	B, Si (Cu, Fe, Mg)	R,D
T 3	Graphite	B, Ca, Ti (Si, Cu, Fe, Mg)	R,D
EK-0	Graphite	as T 0	PP
EK-1	Graphite	as T 1	PP
EK-2	Graphite	as T 2	P
EK-3	Graphite	as T 3	PP
Electrocarbon Topolčany (ČSSR)			
SU	Graphite	Si, Fe, Al (Mg, Ti)	R,D, PP, P
SW	Carbon	Si, Fe, Al (Mg, Ti)	R
Ringsdorf GmbH, W. Germany			
RW-0	Graphite	(Mg Si)	R
RW-I	Graphite	B (Mg? Si?, Ti?, V?)	R, D
RW-II	Carbon	B (Ca, Si, Ti, V, Mg)	R
RW-III	Graphite	B (Ca, Mg, Si, Ti, V, Fe?, Cu?)	R
RW-A	Graphite	as RW-I	PP
RW-B	Graphite	as RW-I	P
RW-C	Graphite	as RW-III	PP
RW-D	Graphite	as RW-0	PP
Tokai electrode MFG. Co., LTD Tokio, Japan			
Glass Carbon GC 305	Glass	Al, Fe, Si (Mg, Cu, B)	R, D

Explanation:

R   Spectral carbon rods
D   Spectral carbon disks
P   Spectral carbon powder, non-pressable
PP Spectral carbon powder, pressable
( ) Impurities which occur only occasionally

thermal conductivity and burning characteristics of the material. "Carbon" possesses better burning characteristics and a lower thermal conductivity than "graphite". A more steady carbon arc will be sustained already when combining a "carbon electrode" with a "graphite electrode". Table 63 gives a few data of purity and properties of various spectral carbon products. Pressable graphite powder is used for the production of carbon pellets (i.e. pressed electrodes).

4.8.4. Chemicals

For spectrochemical purposes, chemicals of the highest degree of purity must be used. Chemicals of the "p.a." purity degree are usually not suitable.
"Spectropure" chemicals; e.g. from Messrs. Johnson Matthey and Co., London; Politechnika Slaska, Gliwice. Occasionally minerals also prove to be spectropure.

4.8.5. Quantitative Analyses

The accuracy of quantitative analyses is essentially determined by the evaporation and excitation processes in the light source. Plate faults, interpretation errors, etc. are of minor importance.
The principle of the quantitative spectral analysis is a comparison of samples with exactly defined contents (standard mixtures) and samples of the analysis. A precondition of such a comparison is the fact that the evaporating and excitation processes in the individual tests are identical. Compliance with this requirement is given if samples of the same type (similar chemical composition of the matrix, similar structure) are excited by means of an appropriate excitation method (total evaporation of the sample in a 10-A d.c. continuous arc). Dissimilar samples can be transformed into a homogenous condition by the addition of spectrochemical agents (cf. SCHROLL 1963) or by melting by means of a flux (cf. DÜMECKE et al. 1958).
In the event of an inappropriate excitation, a non-uniform composition of the rock or mineral samples can effect different temperatures of the arc plasm which affect the line intensities. For example, alkalies may reduce the arc temperature and thus cause systematic errors in the analysis (cf. SCHRÖN 1964). Perturbation effects in the excitation of the sample can be avoided by the use of the addition method developed by GATTERER (1943).

Calibration

Perfect homogeneity of samples and standard mixtures can rarely be achieved in geochemistry because of the diversity of the materials to be examined. Standard mixtures are used which are similar in chemical composition and differ partly in structure. It should be emphasised, however, that the use of unsuitable standard mixtures may lead to considerable systematic errors which, occasionally, exceed one order of magnitude. Considerably simpler and more accurate is the calibration in the spectroscopy of solutions because in this case the requirement of "homogeneity of samples and standard mixtures" can conveniently be complied with. Information about the preparation of standard mixtures is given by MYERS et al. (1961); SCHRÖN (1964).
It is recommendable to check the calibration by means of standard rocks and minerals (cf. Section 4. 16).

Limits of Detection

In Table 64, the chemical elements are arranged according to their geochemical frequency (Clarke-value) and the spectrochemical limit of detection.

Table 64. Spectrochemical Limit of Detection (DC Arc) and Geochemical Frequency (Clarke-value) of Some Elements (from Murata, 1958)

Element	Frequency ppm	Limit of Detection ppm	Ratio of Frequency to Limit of Detection	Element	Frequency ppm	Limit of Detection ppm	Ratio of Frequency to Limit of Detection
Cr	200	1	200	Dy	4	30	0.1
Mn	1,000	10	100	Ho	1	30	0.03
Cu	70	1	70	Er	2	10	0.2
Li	65	1	65	Tm	0.2	10	0.02
V	150	3	50	Lu	0.8	30	0.03
Sr	450	10	45	Ge	1	10	0.1
Ba	430	10	43	Cs	3	50	0.06
Rb	280	10	28	Tl	1.3	50	0.03
Zr	220	10	22	Bi	0.2	10	0.02
Ni	100	5	20	Ag	0.02	1	0.02
Co	40	5	8	Th	10	500	0.02
Be	6	1	6	Hf	5	300	0.02
Sc	5	1	5	In	0.1	10	0.01
Y	40	10	4	Sb	1	100	0.01
Yb	3	1	3	U	4	500	0.008
Ga	15	10	1.5	Ta	2	400	0.005
Pb	15	10	1.5	Cd	0.2	50	0.004
B	10	10	1	As	2	500	0.004
La	20	30	0.7	W	1	500	0.002
Mo	2	5	0.4	Hg	0.5	500	0.001
Nb	20	50	0.4	Pd	0.01	3	0.003
Zn	80	200	0.4	Os	0.05	500	0.000,1
Sn	6	20	0.3	Pt	0.005	10	0.000,5
Ce	40	200	0.2	Rh	0.001	3	0.000,3
Pr	5	700	0.007	Ir	0.001	100	0.000,01
Nd	20	100	0.2	Ru	0.005	80	0.000,06
Sm	6	30	0.2	Au	0.001	30	0.000,03
Eu	1	30	0.03	Re	0.001	50	0.000,02
Gd	6	30	0.2	Se	0.1	10,000	0.000,01
Tb	1	40	0.02	Te	0.002	1,000	0.000,002

Elements with a ratio of ≥ 1 can be determined spectrochemically in the majority of silicate rocks. The other elements (ratio < 1) usually can only be determined if they are concentrated (naturally or artificially) to a certain degree.

In the case of spark excitation, the efficiency of detection is generally lower than in the case of arc excitation. The efficiency also varies considerably with variations in the "inductivity" of the spark.

Elements with an ionization potential up to 9 eV can be detected in the arc with high efficiency when the cathode-glow layer method (according to Mannkopf and Peters 1931) is used.

Accuracy

In the results of spectral analysis, a distinction has to be made between incidental (scattering) and systematic errors. The incidental error can be easily determined and usually is stated in the form of a variation coefficient (relative standard deviation). In trace element determinations in the arc it is anything between 5 to 30 % (occasionally up to 50 % and more), in general 20 %, depending on the excitation technique, type of sample, element, and concentration. In flame spectrometry, solution analysis, and other special procedures, relative standard deviations below 5 % can be achieved.

Systematic errors can be determined with the help of geochemical standards (standard rocks and minerals)or by outside checking. Systematic errors can be very great, even if the incidental errors are small (i.e. in spite of a good reproducibility of the results). Therefore, both the incidental error and the systematic error should be included in the statement of the results.

4.8.6. Statement of Method

When stating the results of spectral analyses, information about the method of analysis is essential. For reporting these details, the following scheme is suggested:

1. Aim and efficiency of the method
2. Calibration: Production of the synthetic standard mixtures, use of geochemical standard rocks and minerals
3. Preparation of samples
4. Electrodes: Material, form of the electrodes, electrode gap length
5. Type of excitation: Exciting apparatus and electrical data
6. Light guide to the spectrograph
7. Spectrograph: Type, efficiency, width of slit (In the case of grating spectrographs, in addition data of blaze and spectral order.)
8. Photographic material
9. Exposure time (perhaps pre-sparking time)
10. Evaluation of spectra, evaluation instruments
11. Limits of detection and working range
12. Analytic lines and reference lines
13. Variation coefficient
14. Systematic errors
15. Bibliography

Bibliography

Books

AHRENS, L. H., and S. R. TAYLOR: Spectrochemical Analysis, 2nd ed. Addison-Wesley Publ. Comp., Inc., USA 1961

ASTM-Committee: Methods for Emission Spectrochemical Analysis. ASTM, Philadelphia 1960

BOROVIK-ROMANOVA, T. F. et al.: The Spectrometric Determination of Rare and Dispersed Elements in Minerals, Rock, Soils, Plant, and Natural Waters. Izd. Akad. Nauk SSSR, Moscow 1962 (Russian)

BRODE, W. R.: Chemical Spectroscopy, 2nd ed. John Wiley & Sons, New York 1952

GERLACH, W., and E. SCHWEITZER: Die chemische Emissions-Spektralanalyse, Vol. I: Grundlagen und Methoden. L. Voss, Leipzig 1930

GERLACH, WA., and WE. GERLACH: Die chemische Emissions-Spektralanalyse, Vol. II: Anwendung in Medizin, Chemie und Mineralogie, 2nd ed. L. Voss, Leipzig 1933

HARRISON, G. R., R. C. LORD, and J. R. LOOFBOUROW: Practical Spectroscopy. Prentice-Hall, New York 1949

MAXWELL, J.: Rock and Mineral Analysis. Inters., Wiley, New York 1968

KAISER, FISCHER, LAQUA et al.: Spektrochemische Analyse, in: O. PROSKE, Analyse der Metalle, 2nd Vol. Betriebsanalysen, 2nd Part, 547—618. Springer-Verlag, Berlin/Göttingen/Heidelberg 1953

MOENKE, H.: Spektralanalyse von Mineralien und Gesteinen, 1st ed. 3—102 and 159—197. Akademische Verlagsgesellschaft Geest & Portig, Leipzig 1962

MOENKE, H., and L. MOENKE: Einführung in die Laser-Mikro-Emissionsspektralanalyse. 2nd ed. Akad. Verlagsgesellschaft Geest & Portig, Leipzig 1968

MORITZ, H.: Spektrochemische Betriebsanalyse, 2nd ed. (G. JANDER: Die chemische Analyse, 43.) Ferdinand Enke, Stuttgart 1956

NACHTRIEB, N. H.: Principles and Practise of Spectrochemical Analysis. Mc Graw Hill, New York 1950

RUSANOV, A. K.: Quantitative Spectrum Analysis of Rare and Dispersed Elements. State Sc Techn. Publ. f. Geol. Lit., Moscow 1960 (Russian)

SCHELLER, H.: Einführung in die angewandte spektrochemische Analyse, 3rd ed. VEB Verlag Technik, Berlin 1960

SCHRÖN, W., and L. ROST: Atom-Spektralanalyse. Dtsch. Verl. f. Grundstoffindustrie, Leipzig 1969

SEITH, W., and K. RUTHARDT: Chemische Spektralanalyse, 5th ed., Springer, Berlin/Göttingen/Heidelberg 1958

SMALES, A. A., and L. R. WAGER: Methods in Geochemistry. Interscience Publ. New York 1960

TÖRÖK, T., and K. ZIMMER: Quantitative Evaluation of Spectrograms by Means of l-Transformation. Akad. Kiadó, Budapest 1972

Publication of Papers Read at International Meetings

VIII. Colloquium Spectroscopicum Internationale 1959 Luzern. Sauerländer & Co., Arau, 1960

IX. Colloquium Spectroscopicum Internationale 1961 in Lyon, I—III. Publication du Groupement pour l'avancement des Méthodes Spectrographiques, Paris 1962 (Publ. GAMS)

XII. Colloquium Spectroscopicum Internationale 1965 in Exeter, England. Hilger and Watts, Ltd., London 1965

Tables and Atlases

GATTERER, A., and J. JUNKES: Atlas der Restlinien, 1st to 3rd Vol. Specola Vatikana, Rome 1945, 1947, 1949

GERLACH, W., and E. RIEDL: Die chemische Emissionsspektralanalyse, Vol. III: Tabellen zur qualitativen Analyse 3rd ed. Johann Ambrosius Barth, Leipzig 1949

HARRISON, G. R.: Wavelength Tables with Intensities in Arc, Spark or Discharge Tube of more than 100,000 Spectrum Lines Between 10,000 and 2,000 A. John Wiley & Sons, New York 1939

KALININ: Spectrum Atlas for Glass Spectrographs, 2nd ed., Alma Ata 1960 (Russian) (German edition in preparation)

KAYSER, H., and R. RITSCHL: Tabelle der Hauptlinien der Linienspektren aller Elemente, nach Wellenlängen geordnet. Springer, Berlin/Göttingen/Heidelberg 1939

KOLENKO, L. J.: Atlas of Spectrum Lines for the Analysis of Ores and Minerals by Means of a Diffraction Spectrograph. Izd-vo "Nauka", Moscow 1967 (Russian)

KROONEN, J., and D. VADER: Line Interference in Emission Spectrographic Analysis. Elsevier Amsterdam/London/New York 1963

MEGGERS, CORLISS, and SCRIBNER: Tables of Spectral-line Intensities, Washington 1962

MEGGERS, CORLISS, and SCRIBNER: Relative Intensities for the Arc Spectra of Seventy Elements. Spectrochim. Acta [London] *17* (1961), 1137—1172

SAIDEL, A. N., V. K. PROKOFEV, and S. M. RAISKI: Spektraltabellen, VEB Verlag Technik, Berlin 1955.

N.N. Atlas "Q-24 Eisenspektrum" von 4555—2227 A. VEB Gustav Fischer Verlag, Jena 1958

Important Special Works

BIRKS, F. T.: The Application of the Hollow Cathode Source to Spectrographic Analysis. Spectrochim. Acta [London] *6* (1954), 169—179

BLACKBURN, W. H., Y. I. A. PELLETIER and W. H. DENNEN: Spectrochemical Determinations in Garnets Using a Laser Microprobe. Applied Spectroscopy 22 (1968), 278—283

BROOKS, R. R., L. H. AHRENS, and S. R. TAYLOR: The Determination of Trace Elements in Silicate Rocks by a Combined Spectrochemical Anion Exchange Procedure. Geochim. cosmochim. Acta 18 (1960), 162—175

DANIELSON, A., G. SUNDKVIST et al.: The Tape Machine. Spectrochim. Acta 1959, 122 (Part I), 126 (Part II), and 134 (Part III)

DOERFFEL, K., and R. GEYER: Untersuchungen zur spektrochemischen Anregung pulverförmiger Stoffe. Z. analyt. Chemie 200 (1964), 6, 411

DOERFFEL, K., and J. LICHTNER: Spektrometrische Lösungsanalyse mit dem kaskadenstabilisierten Lichtbogen. Spectrochim. Acta 22 (1966), 7, 1245—1252

DÜMECKE, G., GERSÖNE, and WIEGMANN: Über ein spektrochemisches Analysenverfahren für Silikate und verwandte oxidische Minerale. Silikattechnik 9 (1958), 488—491. (Determination of Main Elements)

EHRLICH, G., and R. GERBATSCH: Untersuchungen zur Anwendung des Eichzusatzverfahrens. Z. anal. Chem., Munich 209 (1965), 35—46

FRED, M., N. H. NACHTRIEB, and F. S. TONKINS: Spectrochemical Analysis by the Copper Spark Method. J. opt. Soc. America 37 (1947), 279

GATTERER, A.: Quantitative Bestimmung kleinster Beträge von Zusatzelementen in einem Grundelement, das spektralrein nicht erhältlich ist. Spectrochim Acta, 1 (1943), 6, 513—531 (Addition Method)

GURNEY, J. J., and A. J. ERLANK: DC Arc Spectrographic Technique for Determination of Trace Amounts of Li, Rb and Cs in Silicate Rocks. Anal. Chem. 38 (1966), 1836—1939

HEGEMANN, F.: Methodische Untersuchungen zur quantitativen spektrochemischen Mineralanalyse. Fortschr. Mineralog. 35 (1957), 56—59. (Summary of 40 Publications on Spectrochemical Mineral Analysis Conducted by the Research Centre for Geochemistry, TH Munich)

HOLDT, G.: Zur Anwendung des Streudiagramms in der Spektralanalyse. Z. anal. Chem., Munich 209 (1965), 46—58

KAISER, H.: Der Einfluß des Untergrundes auf die Gestalt spektrochemischer Eichkurven. Spectrochim. Acta, Rome 3 (1948), 297—319 (Background Correction)

KAISER, H.: Bemerkungen zur spektrochemischen Spurenanalyse. Erzmetall 5 (1952), 138—141

KAISER, H.: Zur Definition der Nachweisgrenze, der Garantiegrenze u. der dabei benutzten Begriffe. Z. anal. Chem., Munich 216 (1966), 80—94

KRINBERG, I. A.: Spectrum Analysis in Geology and Geochemistry. Materials of the 2nd Siberian Conf. on Spectrosc., Moscow 1967 (Russian)

MANNKOPFF, R., and CL. PETERS: Über quantitative Spektralanalyse mit Hilfe der negativen Glimmschicht im Lichtbogen. Z. Physik 70 (1931), 4444 (Glow Layer Effect)

MARGOSHES, M., and B. F. SCRIBNER: The Plasma Jet as a Spectroscopic Source. Spectrochim. Acta, 15 (1959), 138—145

MOENKE, H., and L. MOENKE: Der Arbeitsbereich des Laser-Mikro-Spektralanalysators LMA-1. Bergakademie 18 (1966), 697—699

MURATA, K. J.: Spectrochemical Analysis for Trace Elements in Geological Material, ASTM, Spec. Techn. Publ., Philadelphia 221 (1958), 67—79

MYERS, A. T., R. G. HAVENS, and P. J. DUNTON: A Spectrochemical Method for the Semiquantitative Analysis of Rocks, Minerals, and Ores. Geol. Surv. Bull. 1084-I, Washington 1961, 205—229

ONDRICK, C. W., N. H. SUHR and J. U. MEDLING: Investigation into Increasing the Precision of a Rotating Disk — Solution Technique for the Analysis of Silicates. Applied Spectroscopy 23 (1969), 111—115

PRICE, W.J.: Spectrographic Analysis of Complex with Particular Reference to Slag and Ores. Spectrochim. Acta 6 (1953), 26—38

RESHETINA, T. C.: Quantitative Spectrum Analysis of the Main Components of Intrusive Rocks. Izv. Akad. Nauk SSSR, Ser. Fiz. 23 (1959), 1150—1151 (Russian)

ROST, L.: Die Lösungsspektralanalytische Bestimmung von Gesteinshauptkomponenten. Bergakademie 17 (1965), 263

RUNGE, E. F., R. W. MINCK, and F. R. BRYAN: Spectrochemical Analysis Using a Pulsed Laser Source. Spectrochim. Acta 20 (1964), 733—736

182

RUSANOV, A. K., and VOROBEV: Uniform Blowing of Powder into the Arc Plasma in Spectral Analysis. Zav. Labor. *30* (1964), 41 (Russian)

SCHROLL, E.: Über die Anwendung thermochemischer Reaktionen in der emissionsspektrographischen Spurenanalyse und ihre Bedeutung für den Carriereffekt. Z. analyt. Chem. *198* (1963), 40—55 (Carrier-Effect, Spectrochemical Buffer)

SCHRÖN, W.: Der Einfluß des Matrixeffektes bei der Eichung von emissionsspektrochemischen Verfahren für die Bestimmung von Spurenelementen in Silikatgesteinen und -mineralen. Silikattechnik *15* (1964), 281—288

SCHRÖN, W., and S. RÄPKE: Ein verbessertes Gerät zum Formen von Spektralkohlestäben. Ber. dtsch. Ges. geol. Wiss., B *11* (1966), 389—390

SCHWANDER, H.: Neue Anwendungen zur quantitativen spektrochemischen Bestimmung der Hauptkomponenten in Silikatgesteinen und -mineralen. Schweiz. mineralog. petrogr. Mitt. *40* (1960), 289 to 311 (Stabilisation of the d. c. Permanent Arc, Main Elements of Silicates)

SHAW, D. M., O. I. JOENSUU, and L. H. AHRENS: A Double-arc Method for Spectrochemical Analysis of Geological Materials. Spectrochim. Acta *4* (1950), 233

STALLWOOD, B. J.: Air Cooled Electrodes for the Spectrochemical Analysis of Powders. J. Opt. Soc. America *44* (1954), 171—176

TAYLOR, B. L., and F. T. BIRKS: A Flexible Computer Program for Calculations in Emission — Spectrographic Analysis. The Analyst *96* (1971), Nr. 1148, 753

TENNANT, W. C., and S. K. FELLOWS: Spectrochemical Determination of Trace Amounts of Rare Earths. Geochim. Cosmochim. Acta *31* (1967), 1473—1478

TENNANT, W. C., and J. R. SEWELL: Direct Reading Spectrochemical Determination of Trace Elements in Silicates — Incorporating Automatic Background and Matrix Corrections. Geochim. Cosmochim. Acta *33* (1969), 640—645

THIERS, R. E., and B. L. VALLEE: The Effect of Noble Gas Atmospheres on the Characteristics of the DC Arc. Spectrochim. Acta, Special volume 1956, 179

THOMPSON, G. v., and D. C. BANKSTON: Sample Contamination from Grinding and Sieving Determined by Emission Spectroscopy. Applied Spectroscopy *24* (1970), 210—219

YOUNG, L. G.: Emission Spectroscopy of Solutions. Analyst *87* (1962), 6.

4.9. Infrared-spectrometry

Infrared-spectrometry is a method of molecular spectroscopy. The infrared absorption bands of organic and inorganic materials are suitable for the determination of kinds of chemical linkages, molecular parameters, for constitution determination, and for purity tests, qualitative and quantitative analyses.

Due to the construction of commercial spectrophotometers for infrared, infrared-spectrometry has made much headway in organic chemistry in the last two decades. Recently, while preparation techniques for methods of powder analysis (potassium bromide moulding technique) have been improved, infrared-spectrometry has also become of importance to mineralogy and geochemistry (MOENKE 1962).

It is mainly applied to mineral diagnosis; structurally similar minerals show absorption bands whose form and position are indicative of the fact that they are allied (structure groups). By way of trial, it has also been used in petrography (phase analysis of rocks and ores).

Bibliography

BRÜGEL, W.: Einführung in die Ultrarot-Spektroskopie, 3rd ed. Theodor Steinkopff, Darmstadt 1962

CROSS, A. D.: An Introduction to Practical Infra-red Spectroscopy. Butterworth, London 1960

GÖRLICH, P., H. MOENKE, and L. MOENKE-BLANKENBURG: Ultrarot-Spektralphotometrie als Hilfsmittel in Mineralogie und Geologie. Jenaer Jb. *1* (1959), 154

JANCKE, H.: Über den Stand der Spektralphotometrie im infraroten Spektralbereich. Exp. Techn. Phys. *1* (1953), 157—173

MOENKE, H.: Spektralanalyse von Mineralien und Gesteinen. 1st ed., 103—212. Akademische Verlags-
gesellschaft Geest & Portig, Leipzig 1962
MOENKE, H.: Entwicklung, Stand und Möglichkeiten der Ultrarotspektralphotometrie von Mineralien.
Fortschr. Mineralog. *40* (1963), 76—123
MOENKE, H.: Mineralspektren. Akademie-Verlag, Berlin 1962
PLYUSNINA, I. I.: The Infrared Spectra of Silicates. Izd-vo Mosc. Univ., Moscow 1967 (Russian)
N. N.: Neue mineralogische Untersuchungsmethoden. Freiberger Forschungsh. C *172* (1964)

4.10. X-ray Spectral Analysis and Electron Microprobe Analysis

When mineral samples are exposed to an intensive radiation by X-rays, the elements are
excited to emit their characteristic radiation (X-ray fluorescence analysis). The elements of
mineral samples can also be excited directly by cathode rays to emit their characteristic
radiation (X-ray spectral analysis with primary excitation). This characteristic radiation is
spectrally dispersed by means of analysers (various single-crystals). The wavelengths and
radiation components are recorded by means of detectors. All elements have a characteristic
radiation which consists of a few spectral lines of certain wavelengths. A representation is
given, for instance, by LANDOLDT-BOERNSTEIN; Springer Verlag 1950—1955.

Application

1. Quantitative determination of all elements with an atomic number over 11. For elements whose atomic number is below 23, vacuum spectrographs are required.
2. A quantitative determination of elements is rendered difficult by the matrix effect-
   therefore, use of calibration samples (outer standard), inner standards, diluents or mix-
   ing the sample with powerful absorbers is necessary. The elements can be identified
   within a wide range of concentration, whereas contents of a few ppm and less can only
   be detected in the most favourable cases.
3. A rapid determination of individual elements in the field by X-ray fluorescence analysis is possible by means of a *portable radioisotope X-ray fluorescence analyser*. Certain
   elements are excited to emit their characteristic radiation by radioactive preparations.
   The fluorescent radiation of an element, isolated by filters, is recorded by a detector.
4. With the help of the *electron microprobe analysis*, the element content (atomic number
   over 4) of mineral samples can be determined, within the micrometer range, both
   quantitatively and qualitatively on the basis of the X-ray spectral analysis. Besides
   the qualitative and quantitative element determination of a point of the sample
   ($\sim 1 \, \mu m^2$), the electron microprobe analysis enables the recording of the contents of
   individual elements along an adjustable line or within an adjustable range on the
   surface of the sample (so-called scanning records).

Bibliography

Books

ADLER, I.: X-ray Emission Spectrography in Geology. Elsevier Publ. Com., Amsterdam-London-
New York 1966
BIRKS, L. S.: X-Ray-Spectrochemical Analysis. Elsevier Publ. Com., Amsterdam-London-New York
1959
BIRKS, L. S.: Electronprobe Microanalysis. Wiley, New York, London 1963
BLOCHIN, M. A.: Methoden der Röntgenspektralanalyse. B. G. Teubner, Leipzig 1963

LIEBHAFSKY, H. A., H. G. PFEIFFER, E. H. WINSLOW et al.: X-ray Absorption and Emission in Analytical Chemistry. John Wiley & Sons, New York 1960

ZALISSA, H.: Elektronenstrahl — Mikroanalyse. Springer-Verlag, Wien-New York 1960

LEGLER, F.: Einführung in die Physik der Röntgen- und Gammastrahlen. Karl-Thiemig, Munich 1967

MÜLLER, R.: Spektrochemische Analyse mit Röntgenfluoreszenz. Oldenburg/Munich/Vienna 1967

NAGEL, K.: Tabellen zur Röntgen-Emissions- und Absorptions-Analyse. Springer, Berlin/Göttingen/ Heidelberg 1959

Special Works

ADAMS, J. A. S., and P. GASPARINI: Gamma-Ray Spectrometry of Rocks. Elsevier, Amsterdam 1970

ADLER, I.: Some Applications of X-ray Fluorescence Spectrography to Mineralogical Problems. . Washington Acad. Sci. *48* (1958), No. 4, 135 f.

BALL, D. F.: Rapid Analyses for Some Major Elements in Powdered Rock by X-ray Fluorescence Spectrography. Analyst *90* (1965), 258—265

BOWIE, S. H. U., A. G. DARNEY, and J. R. RHODES: Portable Radioisotope X-ray Fluorescence Analyser. Extract from Transaction of the Institute of Mining and Metallurgy *74* (1964/65), 7 f.

CARL, H. F., and W. J. CAMPBELL: The Fluorescent X-ray Spectrographic Analysis of Minerals. Symposium of Fluorescent X-ray Spectrographic Analysis. ASTM Spec. Techn. Publ. No. 157 (1954), 63—68

CARTER, F. G.: X-ray Fluorescence Analysis of Light Elements, Aluminium-Chromium. Appl. Spectroscopy *16* (1962), No. 5, 159—163

CHODOS, A. A., and C. G. ENGEL: Fluorescent X-ray Spectrographic Analyses of Amphibolite Rocks. Amer. Mineralogist *46* (1961), No. 1/2, 120—133

CLAISSE, F., and G. SAMSON: Heterogeneity Effects in X-ray Analysis. Adv. in X-ray Analysis *5* (1962), 335—354

DE VRIES, J. L.: Einführung in die Methode und Anwendung der Röntgenspektralanalyse. Berichte der I. Informationstagung der C.H.F. Müller G.m.b.H. Hamburg in Darmstadt (1961)

FUCHS, H., and R. SCHINDLER: Die Röntgenfluoreszenzspektralanalyse und Beispiele ihrer Anwendung in der Geologie. Z. angew. Geol. *10* (1964), 5, 267—271

GUILLEMIN, C., and M. CAPITANT: Utilisation de la Microsonde Electronique de Castaing pour des Etudes Mineralogiques. Int. Geol. Congr., Copenhague, Session XXI, Part *21* (1960), 201 f.

HARRIS, D. C.. and E. J. BROOKER: X-Ray Spectrographic Analysis of Minute Mineral Samplex. Canalian Mineralogist 8 (1966), 471—480

HOLLAND, J. G., and E. I. HAMILTON: Mass Absorption Corrections in X-ray Fluorescence Analysis of Natural Igneous Rocks and their Metamorphosed Equivalents. Spectrochim. Acta *21* (1965), 206—208

HOWER, J.: Matrix Corrections in X-ray Spectrographic Trace Element Analysis of Rocks and Minerals. Amer. Mineralogist *44* (1959), 19—32

KARTTUNEN, J. O., and W. R. HARMON: Determination of U in Ores and in Solution Using a Portable Non-Dispersive X-Ray Spectrograph. Spectrochim. Acta *24* B (1969), 301—311

KRAUSE, H.: Analytische und röntgenographische Untersuchungen natürlicher Zinkblenden. Neues Jb. Mineralog. *97* (1962), 2, 143—164

LEAKE, B. E., G. L. HENDRY and A. KEMP: The Chemical Analysis of Rock Powders by Automatic X-Ray Fluoreszence. Chem. Geol. *5* (1969), 7—86

LIEBHAFSKY, H. A., E. H. WINSLOW, and H. PFEIFFER: X-ray Absorption and Emission. Analyt. Chem. *32* (1960), 240 R—248 R

MACK, M., and N. SPIELBERG: Statistical Factors in X-ray Intensity Measurements. Spectrochim. Acta London *12* (1958), 2/3, 169—178

MEYER, J. W.: Determination of Iron, Calcium and Silicon in Calcium Silicates by X-ray Fluorescence. Analytic. Chem. *33* (1961), 6, 692—696

NORRISH, K., and J. T. HUTTON: An Accurate X-Ray Spectrographic Method for the Analysis of a Wide Range of Geological Samples. Geochim. Cosmochim. Acta *33* (1969), 431—453

OTTEMANN, J.: Quantitative Röntgenfluoreszenzanalyse auf Blei, Kupfer und Zink in Gesteinen, insbesondere im Kupferschiefer. Z. angew. Geol. *6* (1960), 10, 496—502

PARRISH, W.: X-ray Spectrochemical Analysis. Philips' Techn. Rev. *17* (1959), 269 f.

PHILIBERT, J.: The Castaing "Mikrosonde" in Metallurgical and Mineralogical Research. Journ. of the Inst. of Metals *90* (1961), 241

REYNOLDS, R. C.: Matrix Corrections in Trace Element Analysis by X-ray Fluorescence: Estimation of the Mass Absorption Coefficient by Compton Scattering. Amer. Mineralogist *48* (1963), 9/10, 1133—1143

ROSÈ, H. J., I. ADLER, and F. J. FLANAGAN: X-ray Fluorescence Analysis of the Light Elements in Rocks and Minerals. Appl. Spectroscopy *17* (1963), 4, 81—85

ROTTER, R.: Untersuchungen über die Anwendungsmöglichkeit der Röntgenspektralanalyse als Meß- und Kontrollmethode bei der Aufbereitung von Erzen in der ČSSR. Bergakademie *16* (1964), 1, 25—32

SALMON, M. L., and J. P. BLACKLEDGE: A Review of Proved Mineral Analyses by Fluorescent X-ray Spectrography. Norelco Reporter (1956), 68—73

SHERMAN, J.: The Correlation between Fluorescent X-ray Intensity and Chemical Composition. Symposium of Fluorescent X-ray Spectrographic Analysis-ASTM Spec. techn. publ. No. 157 (1954), 27—33

SIEMENS, H.: Bestimmungen von Elementen geringer Konzentration im Bleiglanz mit Hilfe der Röntgenfluoreszenzanalyse. Erzmetall *15* (1962), 9, 463—471

SILBER, A., and H. BLAAS: Röntgenfluoreszenzanalyse von Schlacken. Berg- und Hüttenmänn. Mh. *109* (1964), 7, 237—241

STRASHEIM, A., and M. P. BRANDT: A Quantitative X-ray Fluorescence Method of Analysis for Geological Samples Using a Correction Technique for the Matrix Effects. Spectrochim. Acta 23 B (1967), 183—196

STUMPFEL, E. F.: Some New Platinoid-rich Minerals, Identified with the Electron Microanalyse. Min. Mag. *23* (1961), 254

TÖGEL, K.: Über die Röntgenfluoreszenzanalyse im Vakuum auf Elemente der Ordnungszahlen 12 —22. Siemens-Z. *34* (1960), 726—733

TRAILL, R. J., and G. R. LACHANCE: A New Approach to X-ray Spectrochemical Analysis. Geol. Surv. Canada, Paper 64—57 (1965)

VOLBORTH, A.: Dual Grinding and X-ray Analysis of All Major Oxides in Rocks to Obtain True Composition. Applied Spectroscopy *19* (1965), 1—7

WEDEPOHL, K.-H.: Die Anwendung der Röntgenfluoreszenz-Spektralanalyse zu geochemischen Untersuchungen. Fortschr. Mineralog. *37* (1959), 94—97

4.11. X-ray Analysis

X-ray analysis is used for mineral phase identification and in this form indirectly serves geochemical investigation. Its main applications for geochemical purposes are as follows:

1. Qualitative determination of all crystalline minerals by comparing the d-values found with d-value surveys (e.g. ASTM data file). Sample quantity required 1—30 mg.

2. Semi-quantitative determination of the chemical composition of mixed crystals on the basis of changes in lattice plane spacings which depend on the composition (Figs 46 and 47).

3. Qualitative phase identification of mineral assemblages by measuring the intensities of the diffraction lines, using calibration samples. Increase in accuracy by the addition of reference substances (inner standard method) or by the determination of the mass extinction coefficient of the samples.

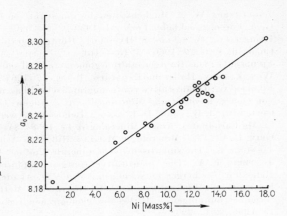

g. 46 Relation between Ni content and ttice constant a_0 [kX] in skutterudite (from LLE, 1964)

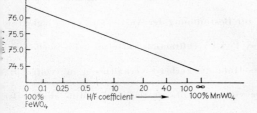

Fig. 47 Dependence of the H/F coefficient of the (Mn, Fe) WO$_4$ mixed-crystal diffraction angle (171) in the case of CuK$_{\alpha 1}$ radiation (from STARKE, 1959)

ibliography

ooks

zaroff, L. V., and M. J. Buerger: The Powder Method in X-ray Crystallography. McGraw-Hil ew York/Toronto/London 1958

rindley, G. W.: The X-ray Identification and Crystal Structures of Clay Minerals. Mineralogical ociety 1951

rown, G.: The X-ray Identification and Crystal Structures of Clay Minerals. Mineralogical Society Clay Minerals Group), London 1961

locker, R.: Materialprüfung mit Röntgenstrahlen unter besonderer Berücksichtigung der Röntgen- etallkunde. 4th ed. Springer, Berlin/Göttingen/Heidelberg 1958

uinier, A.: X-ray Diffraction. Elsevier Publ. Comp., Amsterdam-London-New York 1963

Klug, H. P., and L. E. Alexander: X-ray Diffraction Procedures for Polycrystalline and Amorphous aterials. John Wiley & Sons, New York 1954

eff, H.: Grundlagen und Anwendung der Röntgen-Feinstruktur-Analyse. 2nd ed., R. Oldenbourg, unich 1962

eiser, H. S., H. P. Rooksby, and A. J. C. Wilson: X-ray Diffraction by Polycrystalline Materials. he Institute of Physics, London 1955

pecial Works

Brehler, B.: Über die Messung von Massenschwächungskoeffizienten bei der Phasenanalyse mit dem ählrohrgoniometer. Beitr. Mineralog. Petrogr. 6 (1957), 52—58

Bunge, H. J., and A. G. Herrmann: Eine röntgenographische Methode zur Bestimmung des in An- ydrit isomorph eingebauten Strontiums. N. Jb. Miner. Mh. Jg. 1960, No. 7/8, 186—189

Engelhardt, W. v.: Über die Möglichkeit der quantitativen Phasenanalyse von Tonen mit Röntgen- trahlen. Z. Kristallogr. 106 (1955), 6, 430—459

ENGELHARDT, W. V.: Möglichkeiten der quantitativen Phasenanalyse von Tonen mit Röntgenstrahlen. Geol. Fören. Stockholm Förh. *81* (1959), 374—388

ENGELHARDT, W. V., and S. HAUSSÜHL: Röntgenographische Phasenanalyse grobkristalliner Gesteine. Chem. d. Erde *20* (1960), 3, 155—161

FIEDLER, G.: Das Röntgenzählrohrgoniometer und seine Anwendung in Mineralogie und Petrographie. Wiss. Z. Univ. Halle, math.-naturw. Reihe, *13* (1964), 3, 227—240

FIEDLER, G.: Quantitative röntgenographische Phasenanalyse von Mineralen. Part I: Bestimmmung von Quarz, Kaolinit und Montmorillonit. Z. angew. Geol. *12* (1966), 130—135

GOLDSMITH, J. R., and D. L. GRAF: Relation between Lattice Constants and Composition of th Ca-Mg-Carbonates. Amer. Mineralogist *43* (1958), 84—101

GRIM, R. E.: Clay Mineralogy. McGraw-Hill, New York/Toronto/London 1953

JASMUND, K.: Die silikatischen Tonminerale. 2nd ed. Verlag Chemie, Weinheim/Bergstraße 1955

JUMPERTZ, E. A.: Die quantitative röntgenographische Gemenge-Analyse mit monochromatische Reflexion. Freiberger Forschungsh., ed. C *121*. Akademie-Verlag, Berlin 1962

KRANZ, G., J. WIEGMANN, and C.-H. HORTE: Zur Methodik mineralanalytischer Untersuchungen be Serienbestimmungen, z. B. bei der geologischen Erkundung von Kaolinlagerstätten. Silikattechnik *17* (1966), 291—293

KRAUSE, H.: Analytische und röntgenographische Untersuchungen natürlicher Zinkblenden. N. Jb Miner. Abh. *97*, No. 2, 143—164

OVERKOTT, E.: Eine röntgenographische Methode zur Bestimmung der An-Gehalte von Plagioklasen N. Jb. Miner. Mh. 1958, 113—120

SMITH, J. R., and H. S. YODER: Variations in X-ray Powder Diffraction Patterns of Plagioclase Feld spars. Amer. Mineralogist *41* (1956), 632—647

STARKE, R.: Röntgenographische Bestimmung des Hübnerit-Ferberit-Verhältnisses in Wolframiten Bergakademie *11* (1959), 1, 23—25

STARKE, R.: Die Anwendung der Röntgenanalyse von Pulverpräparaten in der Mineralogie. Berg akademie *12* (1960), No. 9, 495—498

STARKE, R., and H. HAUPT: Beitrag zur quantitativen Bestimmung der Mineralgehalte von Tonen und Tongesteinen. Ber. deutsch. Ges. Geol. Wiss. B, *13* (1968), 515—516

STARKE, R., and D. RÜHLICKE: Eine Methode zur flammenphotometrischen und röntgenographischer Bestimmung von Strontium und Kalzium im Baryt. Bergakademie *13* (1961), 7, 505—511

STARKE, R., and J. RENTZSCH: Quantitative röntgenographische Phasenanalyse erzführender Karbonat gesteine. Bergakademie *13* (1961), 12, 755—757

TÖLLE, H.: Chemismus und genetische Stellung des Skutterudits etc. Freib. Forschungsh. C *171* Leipzig 1964

WIEGMANN, J., and G. KRANZ: Beitrag zur quantitativen Mineralanalyse von Tonen, Kaolinen und ähnlichen Gesteinen mittels röntgenographischer Methoden. Ber. dtsch. keram. Ges. *38* (1961), 7 294—302

YODER, H. S., and TH. SAHAMA: Olivine X-ray Determinative Curve. Amer. Mineralogist *42* (1957) 475 f.

UNGER, H.: Quantitative röntgenographische Phasenanalyse von Mineralen. Part II: Die Bestim mung von Tonmineralverhältnissen. Z. f. angew. Geol. *12* (1966), 197—198

N. N.: X-ray Analysis of Mineral Raw Materials. Periodical. Gosgeoltekhizdat, Moscow 1962 ff. (Russian)

Books of Tables

BERRY, L. G., and R. M. THOMPSON: X-ray Powder Data for Ore Minerals: The Peacock Atlas. The Geological Society of America, Memoir 85, New York 1962

DETTMAR, H. R., and H. KIRCHNER: Tabellen zur Auswertung der Röntgendiagramme von Pulvern Verlag Chemie, Weinheim/Bergstraße 1956

HANAWALT, J. D., H. W. RINN, and L. K. FREVEL: Chemical Analysis by X-ray Diffraction. Ind Engng. Chem., Analyt. Edit. *10* (1938), 457—512

MIKHEEV, V. J.: X-ray Identification of Minerals. Gosgeoltekhizdat, Moscow 1957 (Russian)

MOLLOY, M. W., and P. F. KERR: Diffractometer Patterns of A.P.I. Reference Clay Minerals. Amer. Mineralogist *46* (1961), 583—605

URBAN, H.: Röntgenkartei zur Bestimmung von Ton- und Sedimentmineralien. Gustav-Feller-Nottuln, Kettwig/Ruhr 1954

National Bureau of Standards Circular 539: Standard X-ray Diffraction Powder Patterns. Washington: U.S Department of Commerce 1953 ff. (to be supplemented)

American Society for Testing Materials: X-ray Powder Data File. Philadelphia: American Society for Testing Materials, 1950 ff., (to be supplemented)

American Society for Testing Materials: Index to the X-ray Powder Data File. Philadelphia: American Society for Testing Materials, 1958 ff., (to be supplemented)

4.12. Differential Thermal Analysis

Phase transformations by means of time/temperature curves during the heating of the system at a uniform rate. Sample and comparison substance are heated in an oven. The temperature difference between sample and inert substance is measured by means of a differential thermocouple. A second simple thermocouple measures the temperature in the sample. Phase transformations, decomposition phenomena and oxidation are indicated by irregularities in the behaviour of the differential thermo-curve; after the transformation, the temperature difference is equalized in any case.

This method is primarily used for clay minerals:

1. Qualitative determination of the individual clay minerals concurrently; also used for other substances, e.g. calcite, dolomite, quartz,

2. Quantitative determination in general by planimetering the individual endothermic and exothermic peaks. For this purpose, comparison curves are required.

Bibliography

CLAUSNITZER, H., and D. RÜHLICKE: Ein Gerät zur Differentialthermoanalyse. Bergakademie 15 (1963), 96—100

COLL, W. F., and N. M. ROWLAND: An Abnormal Effect in Differential Thermal Analysis of Clay Minerals. Amer. Mineralog. 46 (1961), 304—312. Referat: Z. angew. Geol. 2 (1962), 106—107

TSVETKOV, A. Y., and G. O. PILOYAN: Modern Trends in the Development of Differential Thermal Analysis. Izv. Akad. Nauk SSSR, Moscow 1965, 159—236 (Russian)

DREES, W.: Quantitative differential-thermoanalytische Untersuchungen an quarzhaltigen Staubproben. Bergakademie 14 (1962), 518—523

FAUST, G. T.: Thermal Analysis of Quartz and its Use in Calibration in Thermal Analysis Studies. Amer. Mineralogist 33 (1948), 5/6, 337/345

HORTE, C. H., and J. WIEGMANN: Fragen zur Auswertung und Dokumentation von Meßergebnissen der Differentialthermoanalyse (DTA). Ber. Deutsch. Ges. Geol. Wiss., B, 11 (1966), 2, 239—245

KIRSCH, H.: Die Anwendung der DTA bei der Kohleuntersuchung. Brennstoff-Chemie 38 (1957), 5/6, 87

KÖHLER, A., and P. WIEDEN: Bemerkungen zu chemischen Gesteinsanalysen. Mineralog. Petrogr. Mitt. 4 (1954), 431—439

KOPP, O. C., and P. F. KERR: Differential Thermal Analysis of Sulfides and Arsenides. Amer. Mineralogist 42 (1957), 7/8, 445

LEHMANN, H., and R. FAHN: DTA-Untersuchung an Kaolin. Tonind.-Ztg. 79 (1955), 3—5

LEHMANN, H. et al.: Die DTA (Monographie). Tonind.-Ztg., Supplement I

MACKENZIE, R. C.: The Differential Thermal Investigation of Clays. Mineralog. Society, London 1957

MORITA, H., and H. M. RICE: DTA von organischen Substanzen. Analytic. Chem. 27 (1955), 336—339

PILOYAN, G. O.: On the Quantitative Identification of a Few Minerals in Natural and Artificial Mixtures with the Help of Differential Thermal Analysis. Izv. Akad. Nauk SSSR, ser. geol. *6* (1963),
24—42 (Russian)

SMOTHERS, W. I., and VAO CHIANG: Differential Thermal Analysis, Theory and Practice. Chemical
Publishing Co. Inc., New York 1958

STOLL, H.: Die DTA, Meßverfahren und Meßgeräte. Silikattechnik *4* (1953), 3, 123—126

STONE, R. L.: Differential Thermal Analysis of Kaolin Group Minerals under Controlled Partial Pressures of H_2O. J. Amer. Ceram. Soc. *35* (1952), 4, 90—99

4.13. Isotope Analysis

Purpose of Isotope Analysis

Investigation of geological samples by measuring the composition of stable and unstable
isotopes on the basis of natural variations in isotopic abundances or the changes in isotopic
abundances in the course of geological events.

The isotopic abundances of elements occurring in nature show normally (in one and the same
element) only insignificant variations. They are shown in Table 13.

Possibilities of Expressing Isotopic Abundances

Isotopic abundances can be expressed in atomic percentages (the sum of the isotopic abundances of all considered isotopes of one element is equal to 100 per cent). Frequently, the ratio
of the two most abundant isotopes is used, e.g. $^{32}S/^{34}S$ or its reciprocal value $^{34}S/^{32}S = R$.
The expression mentioned at first is more easily retained (the values are >1).

In precision measurements, the sample (Sa) is compared with a standard (St). The resulting
δ-values $[^0/_{00}]$ express the relative deviation of the $^{34}S/^{32}S$ ratio from the standard, e.g.

$$\delta^{34} S = \frac{(R_{Sa} - R_{St})}{R_{St}} \cdot 1000 , \qquad 1/R_{St} = 22.22 .$$

Exception: For the deuterium/hydrogen ratio, the δ-value is given in per cent. For several
elements, more than one standard are in use (e.g. A and B). Conversions are based on the
formula:

$$\delta_{(x,B)} = \delta_{(x,A)} \cdot (1 + 10^{-3}\delta_{(A,B)}) + \delta_{(A,B)} ,$$

where:

$\delta_{(x,A)}$ known δ-value of a sample related to standard A,
$\delta_{(x,B)}$ δ-value of the same sample to be calculated, related to standard B,
$\delta_{(A,B)}$ δ-value of standard A related to standard B.

Table 65 shows some standards.

Possibilities of Determining the Isotopic Composition

As a general principle, any physical property which depends on the nuclear mass can be
utilized for measuring the isotopic composition. There are the following methods of determination:

    1. Mass spectrometry
      a) classical methods
      b) more modern methods (omegatron, favitron, time of flight mass spectrometer, etc.)

2. Emission spectral analysis
3. Infrared absorption
4. Density measurements
5. Refraction of light
6. Katharometry (measuring the thermal conductivity)

lthough many new methods of isotope analysis have been developed, the classical mass
ectrometry has remained superior to other methods, especially with regard to precision
otope analyses, the diversity of the samples, resolving power, and the like. On the other hand,
is method involves high expenditures for apparatus because of the ultra-high vacuum
chniques required (poor background) and highly stabilized electronics employed.
eterminations of the relative isotopic abundances, e.g. $^{32}S/^{34}S$ or $^{16}O/^{18}O$ in the case of the
tural isotopic composition, can be carried out at an accuracy of about $0.3^0/_{00}$ provided the
mple to be analysed is a gas. In the event of lower isotopic ratios, the accuracy decreases
cordingly. In these cases, either the single collector or double collector compensation method
used.
ass spectrometric isotope analyses of minute quantities of gas (up to 10^{-10} cm^3) can be
rried out with the help of the static method. In these cases, one mass spectrometer may be
mployed only for one special measuring problem for many years.

able 65. Isotope Standards

arbon	Ratio $^{12}C/^{13}C$	Ratio $^{13}C/^{12}C \cdot 10^5$	δ-value [$^0/_{00}$] δ (^{13}C)
O_2 from *Belemnitella americana*, Creta- eous, P-D-Formation, South Carolina (U.S.A.), Chicago (P.D.B.) Standard O_2 from limestone, ura Solnhofen, S-W-Germany	88.99	1,123.72	0
) Niers Standard	89.05	1,123.00	—0.64
) Nat. Bur. Stand. No. 20	89.08	1,122.53	—1.06
O_2 from calcite-marble, Gheiba, Valle li Peccia Switzerland P. Baertschi (Basle) tandard	88.74	1,126.63	+2.77

)xygen	Ratio $^{16}O^{18}O/^{16}O^{16}O \cdot 10^5$	Ratio $^{18}O/^{16}O$	δ-value [$^0/_{00}$] δ (^{18}O)
Chicago (P.D.B.) Standard	415.80		0
Niers Standard	413.97		—4.40
Nat. Bur. Stand. No. 20	414.08		—4.14
P. Baertschis (Basel) Standard	415.13		—1.62
SMOW (Standard-Mean-Ocean-Water)		$(1,993.4 \pm 2.5)\,10^{-6}$	—0.22

Sulphur	Ratio $^{32}S/^{34}S$		
Troilite phase of the Fe meteorite (octahedrite) from the Cañon Diablo	22.21 (also 22.220)		

Note: The values of carbon and oxygen are quoted from the works by CRAIG (1957, 1961). The δ-
values are related to the Chicago (PDB)-Standard; in the case of oxygen, they do not represent the
isotopic compositions of the given carbonate or water, but those of the CO_2 gas formed of the carbonate on the basis of the standard method or of the CO_2 which has attained equilibrium in the
isotopic exchange reaction with the given water at 25.2 °C.

Solid substances can be measured by means of a thermal or oven ion source, ensuring an accuracy of about 1%. In many cases, the substances required are of the order of 1 µg, especially if a multiplier is used for pre-amplification. In recent time, compensation methods and comparisons with standards have been employed; the error could be reduced to anything below 1⁰/₀₀, though considerably larger quantities of substances were required.

Bibliography

ALEKSANDRUK, V. M. et al.: On the Accuracy of Spectral Methods Used for the Determination of the Isotopic Composition of Elements. Geokhimiya (1967), 5, 637—640 (Russian)
BEYNON, J. H.: Mass Spectrometry and its Application to Organic Chemistry. Elsevier, Amsterdam/London/New York/Princeton 1960
BIRKENFELD, H., G. HAASE, and H. ZAHN: Massenspektrometrische Isotopenanalyse. VEB Deutscher Verlag der Wissenschaften, Berlin 1962
BRÜNNEE, C., and H. VOSHAGE: Massenspektrometrie. Verlag Karl Thiemig, Munich 1964
CRAIG, H.: Isotopic Standards for Carbon and Oxygen and Correction Factors for Mass-spectrometric Analysis of Carbon Dioxide. Geochim. Cosmochim. Acta 12 (1957), 133—149
EWALD, H., and H. HINTENBERGER: Methoden und Anwendungen der Massenspektroskopie. Verlag Chemie, Weinheim/Bergstraße 1953
INGHRAM, M. G., and R. J. HAYDEN: A Handbook on Mass Spectrometry. Nuclear Science Series Rep. Nr. 14. Nat. Acad. of Sci. — Nat. Res. Council, Washington, D.C., 1954
KIENITZ, H.: Massenspektrometrie. Verlag Chemie, Weinheim/Bergstraße 1967
Standard for Reporting Concentrations of Deuterium and Oxygen-18 in Natural Waters. Science 133 (1961), 1833—1834

4.14. Neutron-activation Analysis

Stable atomic nuclei can change into an active state if they absorb an elementary particle, e.g. a neutron. This phenomenon is the basis of a highly efficient detecting method for traces of an element, the neutron-activation analysis. A sample is irradiated with high-speed or slow neutrons so that one of the isotopes of the element to be detected is transformed, to a certain degree, into a radioactive isotope. The increase in radioactive nuclei due to irradiation by neutrons is proportional to the neutron flux, the capture cross-section, and the number of nuclei to be activated. Usually, the active isotope will be chemically separated from the irradiated sample and then its activity will be determined.

The concentration to be found can be determined with the help of a known comparison sample which is subjected to the same treatment; a comparison of the activities of both samples then gives the result. Irradiation by neutrons can favourably be carried through until saturation (i.e. the number of nuclei newly formed by irradiation by neutrons per unit of time is equal to the number of nuclei decaying per unit of time), however, this is not possible in many cases because of the long half-life. Impurities which are introduced by non-radioactive chemicals during the chemical procedures do not impair the measuring result.

The applicability of this method differs according to circumstances. Since, however, for irradiation by neutrons in reactors a neutron flux of about 10^{12} neutrons per sec and cm^2 has been available, this method compares well with a mass-spectrometric isotope dilution analysis as far as sensitivity and accuracy are concerned. Table 66 gives a survey of the efficiency of the neutron-activation analysis.

lement	Efficiency per one Gram of the Sample
u, Dy	10^{-12}
o, In, Ir, Lu, Mn, Re, Sm	10^{-11}—10^{-10}
b, As, Br, Cu, Ga, Au, I, a, Pd, Sc, Na, Pr, Ta, Tb, u, W, Yb	10^{-10}—10^{-9}
a, Ca, Os, Cl, Co, Er, Gd, e, Hf, Ni, Os, P, K, Pb, e, I, Zn	10^{-9}—10^{-8}
e, Cr, Hg, Mo, Sr, Te, TI, d, Pt, Ru, Ag, Sn, Zr	10^{-8}—10^{-7}
i, Ca, Fe, S, Si	10^{-7}—10^{-6}

ibliography

IBBONS, D., B. A. LOVERIDGE, and R. J. MILLETT: Radioactivation Analysis. A Bibliography. Atomic nergy Research Establ. (Gt.Brit.) I/R-2208

AITSEV, E. J., et al.: Einige Möglichkeiten der Anwendung kernphysikalischer Analyseverfahren bei eochemischen Sucharbeiten auf Erzlagerstätten. Z. angew. Geol. *11* (1965), 4, 180—183

MALES, A. A., D. MAPPER, and A. J. WOOD: The Determination, by Radioactivation, of Small Quanti-ies of Nickel, Cobalt and Copper in Rocks, Marine Sediments and Meteorites. Analyst *82* (1957), 75

.15. Measuring Radioactivity

)etermination of the activity which is caused by radioactive elements, e.g. U, Th and their isintegration products, and ^{40}K, ^{87}Rb, ^{14}C and ^{3}T. For radioactivity see Section 3.3.3. The nethod excels because of its high efficiency.

able 67. Use of Various Detectors for the Determination of the Individual Type of Radiation

ype of Detector	Type of Radiation				Cosmic Rays
	α	β	γ	n	
ieiger-Müller counter tube	+	+	+	+	+
'roportional counter tube	+	+	—	+	—
onisation chamber	+	+	+	+	+
icintillation counter with — gases	+	+	—	+	—
icintillation counter with — solutions	+	+	+	+	+
icintillation counter with — plastic substances	+	+	+	+	+
icintillation counter with — crystals	+	+	+	+	+
'ilm emulsions	+	+	+	+	+
Wilson and diffusion chamber	+	+	—	+	+
ipark counter	+	—	—	+	+
iecondary emission counter	+	+	—	—	—
Irystal counter	+	+	+	+	—
Ihemical dosimeter	—	+	+	+	—
Ialorimetric detectors	+	+	+	—	—
Bubble chamber	+	+	—	—	+

There are the following applications for geochemical purposes:

1. Quantitative determination of the total activity of a sample (e.g. measurement by means of the Geiger-Müller counter tube, proportional counter tube, scintillation counter).
2. Quantitative measurement of the given type of radiation (α, β or γ; e.g. with specific scintillators, Geiger-Müller counter tube with appropriate absorbers).
3. Quantitative (or semi-quantitative) determination of the various radioactive elements (γ-spectrometry).
4. Localisation of radioactive substances in an ore or rock sample, etc. (by autoradiography).

Bibliography

BRODA, E., and T. SCHÖNFELD: Die technischen Anwendungen der Radioaktivität. VEB Verlag Technik, Berlin, Porta-Verlag, Munich 1956
CORK, J. M.: Radioactivity and Nuclear Physics. Van Nostrand, New York 1950
FRIEDLÄNDER, G., and J. W. KENNEDY: Nuclear and Radiochemistry. John Wiley and Sons, New York 1960
HAISSINSKY, M., and J. P. ADLOFF: Radiochemical Survey of the Elements.
Elsevier, Amsterdam/London/New York 1965
HERFORTH, L., and H. KOCH: Radiophysikalisches und radiochemisches Grundpraktikum, 2nd ed. VEB Deutscher Verlag der Wissenschaften, Berlin 1962
HEVESY, E., and F. PANETH: Lehrbuch der Radioaktivität, 2nd ed. Leipzig 1931
KMENT, V., and A. KUHN: Technik des Messens radioaktiver Strahlung, 2nd ed. Akademische Verlagsgesellschaft Geest & Portig, Leipzig 1963
SHARPE, J.: Nuclear Radiation Detectors. Methuen, London. John Wiley and Sons, New York 1955
TAYLOR, D.: Methoden und Geräte zur Messung radioaktiver Substanzen. VEB Verlag Technik, Berlin 1958

4.16. Geochemical Standard Samples

For checking analytic procedures, numerous laboratories in various countries use rocks and minerals which are analyzed for main and trace elements (geochemical standard samples).
The first standards used on an international scale are the standard rocks Granite G-1 and Diabase W-1 recommended by FAIRBAIRN et al. (1951). Meanwhile, the number of standards has increased.
Table 68 gives a survey of some of such standard samples.
Table 69 lists the main components and trace element contents of some standard rocks.

Table 68. Geochemical Standard Samples

Standard Rock		Institution
Granite	GM	Zentrales Geologisches Institut Berlin
Basalt	BM	Zentrales Geologisches Institut Berlin
Argillite	TB	Zentrales Geologisches Institut Berlin
Limestone	KH	Zentrales Geologisches Institut Berlin
Anhydrite	AN	Zentrales Geologisches Institut Berlin
Diabase	W-1	U.S. Geological Survey

Standard Rock		Institution
Granite	G-1	U.S. Geological Survey
Granite	G-2	U.S. Geological Survey
Granodiorite	GSP-1	U.S. Geological Survey
Andesite	AGV-1	U.S. Geological Survey
Peridotite	PCC-1	U.S. Geological Survey
Dunite	DTS-1	U.S. Geological Survey
Basalt	BCR-1	U.S. Geological Survey
Nepheline Syenite	STM-1	U.S. Geological Survey
Syenite-rock 1		Canadian Association for Applied Spectroscopy
Sulfide-ore 1		Canadian Association for Applied Spectroscopy
Granite	GA	Centre de Recherches Pétrographiques et Géochimiques (Nancy)
Granite	GH	Centre de Recherches Pétrographiques et Géochimiques (Nancy)
Granite	GR	Centre de Recherches Pétrographiques et Géochimiques (Nancy)
Basalt	BR	Centre de Recherches Pétrographiques et Géochimiques (Nancy)
Tonalite	T-1	Geological Survey of Tanganyika
Granite	1771	Geological Survey of Great Britain
Granite	G-B	People's Republic of Bulgaria
Nephelinsyenite (Chibin Massif)		Leningrad University, Chair for Mineralogy

Table 69. Main Constituents and Trace Element Contents of a Few Standard Rocks

Element	GM	BM	TB	KH	G-1	W-1	Syenite-rock 1	Sulfide-ore 1
A. Main Constituents in %								
SiO_2	73.55	49.65	60.30	8.60	72.64	52.64	59.45	34.52
TiO_2	0.22	1.15	0.93	0.14	0.26	1.07	0.49	0.81
Al_2O_3	13.45	16.20	20.60	2.30	14.04	14.85	9.58	9.46
Fe_2O_3	2.00[1]	9.60[1]	6.90[1]	0.90[1]	0.87	1.41	2.27	32.70[1]
FeO	1.15	7.30	5.45	0.30	0.98	8.74	5.44	—
MnO	0.04	0.14	0.05	0.09	0.03	0.16	0.40	0.11
MgO	0.35	7.45	1.90	0.70	0.41	6.62	4.07	3.93
CaO	1.00	6.35	0.30	47.7	1.39	10.96	10.32	3.96
Na_2O	3.80	4.70	1.30	0.10	3.32	2.07	3.24	1.03
K_2O	4.75	0.20	3.85	0.40	5.45	0.64	2.75	0.63
H_2O^+	0.34	3.60	3.70	0.90	0.34	0.53	0.42	2.91
CO_2	0.26	1.35	0.15	37.5	0.08	0.05	0.36	—
P_2O_5	0.07	0.11	0.10	0.11	0.09	0.14	0.20	0.10
B. Trace Elements in ppm								
Ag	—	—	—	—	0.04	0.06	1.6	3.9
As	4	11	11	(1)	0.7	2.2	10.4	424
Au	(∼0.05)	(∼0.005)	(∼0.004)	—	0.005	0.005	(<20)	0.11
B	20	(15)	70	—	1.5	17	107	23
Ba	300	180	680	42	1,220	180	273	227
Be	4	2	3.5	(1)	3	0.8	25	(1)
Bi	—	—	—	—	0.2	—	(<7)	5
Cd	—	—	—	—	0.06	0.3	1.5	(<20)
Ce	70	35	120	25	200	25	593	(<200)

Table 69 continued

Element	GM	BM	TB	KH	G-1	W-1	Syenite-rock 1	Sulfide-ore 1
Cl	—	—	—	—	50	—	—	—
Co	4	33	16	6	2.4	50	19	546
Cr	9	115	70	22	22	120	58	368
Cs	4	—	6	(25)	1.5	1.1	0.9	—
Cu	14	48	60	9	13	110	25	8,291
Eu	(<1)	(1.2)	—	(<1)	0.8	—	(<500)	—
F	700	240	725	590	700	250	0.17	200
Ga	15	14	22	4	20	17	18	13
Ge	1.6	1.5	2.5	0.32	1.1	1.6	1.7	0.7
Hf	—	—	—	—	7.6	—	(<100)	(<100)
In	0.02	0.03	0.07	0.01	0.025	0.055	0.037	0.14
Ir	(0.01)	(0.005)	(0.02)	—	0.006	—	(<100)	(<100)
La	(36)	(11)	(72)	(11)	120	20	243	(<20)
Li	55	70	113	8	24	12	126	10
Mo	1.5	0.7	(0.7)	0.3	7	(0.7)	4.1	8.3
N	—	—	—	—	7	12	—	—
Nb	20	—	(15)	—	20	10	146	(<10)
Nd	(30)	(18)	(63)	(15)	55	17	325	(<100)
Ni	8	55	40	20	1—2	78	42	13,103
Pb	30	11	9	8	49	8	516	248
Pd	—	—	—	—	0.001	0.02	(<3)	0.18
Rb	270	10	180	25	220	22	204	119
Re	—	—	—	—	0.000,7	0.000,7	(<50)	(<50)
S	600	570	750	1,000	(175)	130	400	120,400
Sb	0.54	2.5	3.3	—	0.4	1.1	0.7	0.2
Sc	5.3	34	15	(<1)	3	34	14	22
Sm	(4)	(5)	(7)	(3)	11	5	(<100)	(<100)
Sn	4.5	2	5	—	4	3	12	11
Sr	132	218	160	560	250	180	320	110
Ta	—	—	—	—	1.6	0.7	(<200)	(<200)
Th	37	8	25	10	51	2.3	1,324	3.4
Tl	(0.6)	(0.2)	(2)	(0.1)	1.3	0.15	1.8	(<50)
U	(7)	(2)	(3)	(2)	3.9	0.49	2,400	1.7
V	10	180	110	26	16	240	87	192
W	2	1	3	(1)	0.4	0.45	(<100)	(<100)
Y	29	28	41	13	13	25	447	21
Yb	3.4	5	5	2	2	4	73	8.6
Zn	40	100	100	8	45	82	200	298
Zr	150	110	170	35	210	100	3,048	108

GM: Granite GM, Zentrales Geologisches Institut Berlin (unpublished values)
BM: Basalt BM, Zentrales Geologisches Institut Berlin (unpublished values)
TB: Shale TB, Zentrales Geologisches Institut Berlin (unpublished values)
KH: Limestone KH, Zentrales Geologisches Institut Berlin (unpublished values)
G-1: Granite G-1, U.S. Geological Survey (values according to FLEISCHER 1965)
W-1:Granite W-1, U.S. Geological Survey (values according to FLEISCHER 1965)
Syenite-rock 1, Canadian Association for Applied Spectroscopy (values according to WEBER 1965)
Sulfide-ore 1, Canadian Association for Applied Spectroscopy (values according to WEBER 1965)
The data given by the authors have been partly supplemented.

[1]) $\sum$ Fe as Fe_2O_3

Bibliography

ALEKSIEV, E., and R. BOYADJIEVA: Content of Rare Earths in the Standard Igneous Rocks G-1, W-1 and G-B. Geochim. Cosmochim. Acta *30* (1966), 511—513

BRUNFELT, A. O., and E. STEINNES: Instrumental Neutron-activation Analysis of "Standard Rocks". Geochim. Cosmochim. Acta *30* (1966), 921—928

CARMICHAEL, I. S. E., J. HAMPEL, and R. N. JACK: Analytical Data on the U.S.G.S. Standard Rocks. Chem. Geol. *3* (1968), 59—64

CHAYES, F.: Modal Composition of USGS Reference Sample G-2. Geochim. Cosmochim. Acta *31* (1967), 463—464

FAIRBAIRN, H. W., et al.: A Cooperative Investigation of Precision and Accuracy in Chemical, Spectrochemical and Modal Analysis of Silicate Rocks. U.S. Dep. Interior, Geol. Surv., Bull. 980, Washington 1951 (G-1 and W-1)

FLANAGAN, F. J.: U. S. Geological Survey Silicate Rock Standards. Geochim. Cosmochim. Acta *31* (1967), 289—308

FLANAGAN, F. J., and M. E. GWYN: Sources of Geochemical Standards. Geochim. Cosmochim. Acta *1* (1967), 1211—1213

FLEISCHER, M.: Summary of New Data on Rock Samples G-1 and W-1. 1962—1965. Geochim. Cosmochim. Acta *29* (1965), 1263—1283

FLEISCHER, M., and R. E. STEVENS: Summary of New Data on Rock Samples G-1 and W-1. Geochim. Cosmochim. Acta *26* (1962), 525—543

FRIESE, G., and H. GRASSMANN: Die Standardgesteinsproben des ZGI. 4. Mitteilung: Diskussion der Gehalte an einigen Hauptkomponenten auf Grund neuer Analysen. Z. angew. Geol. *13* (1967), 473

FUCHS, H., et al.: Spurenelementgehalte der Standardgesteine des Zentralen Geologischen Institutes, Berlin. Ber. deutsch. Ges. geol. Wiss. B, *11* (1966), 109—113

GRASSMANN, H.: Standardgesteinsproben, WTI 8 des ZGI Berlin (1962), 18—23

GRASSMANN, H.: Die Standardgesteinsproben des ZGI. Z. Angew. Geol. *10* (1964), 10, 555—557

GRASSMANN, H.: Die Standardgesteinsproben des ZGI, Ergebnisse der chem. Analysierung auf Hauptkomponenten. Z. angew. Geol. *12* (1966), 368—378

HUANG, W. H., and W. D. HOHNS: The Chlorine and Fluorine Contents of Geochemical Standards. Geochim. Cosmochim. Acta *31* (1967), 597—602

INGAMELLS, C. O., and N. H. SUHR: Chemical and Spectrochemical Analysis of Standard Silicate Samples. Geochim. Cosmochim. Acta *27* (1963), 897—910

KUKHARENKO, A. A., et al.: Clarke Values of the Chibine Alkali-Massive. Zap. vses. miner. obshch. vtor. ser. 1968, 133—149 (Russian)

MORGAN, J. W., and K. S. HEIER: Uranium, Thorium and Potassium in Six USGS Standard Rocks. Earth Planet. Sci., Lett. *1* (1966), 4, 158—160

ROUBAULT, M., H. DE LA ROCHE, and K. GEVINDARAJU: L'Analyse des Roches Silicatées par Spectrométrie Photo-Electrique en Quantométrie ARL et Son Contrôle par des Roches Etalons. Sciences de la Terre *IX* (1962—1963), 4, 339—371, Nancy 1964

ROUBAULT, M., H. DE LA ROCHE, and K. GOVINDARAJU: Rapport sur Quartere Roches Etalons Géochimiques: Granites GR, GA, GH et Basalte BR. Sciences de la terre *XI* (1966), 1, 105—121, Nancy 1966

SCHINDLER, R.: Die Standardgesteinsproben des ZGI. Erste zusammenfassende Betrachtung der Spurenelementanalysierung der Standardgesteine des ZGI, Berlin und ihre Vergleiche mit den amerikanischen Gesteinen G-1 und W-1. Z. angew. Geol. *12* (1966), 4, 188—196

SCHRÖN, W., and W. KLEMM: Erfahrungen mit Standardproben bei der quantitativen spektrochemischen Spurenanalyse. Z. angew. Geol. *14* (1968), 537—541

STEVENS, R. E., et al.: Second Report on a Cooperative Investigation of the Composition of Two Silicate Rocks. U.S. Dep. Interior, Geol. Surv., Bull. 1113, Washington 1960 (G-1 and W-1)

N.N.: Report of Nonmetallic Standard Committee Canadian Association for Applied Spectroscopy. App. Spectroscopy *15* (1961), 159 (Syenite-rock-l and Sulfide-ore-l)

TAYLOR, R. S., and P. KOLBE: Geochemical Standards. Geochim. cosmochim. Acta *28* (1964), 447—454

WEBBER, G. R.: Second Report of Analytical Data for CAAS Syenite and Sulphide Standards. Geochim. cosmochim. acta *29* (1965), 4, 229—248

5. Representation and Mathematical Processing of Geochemical Data

5.1. Representation of Geochemical Information

Information of different contents accrues from geochemical studies. It is represented in the form of descriptions, numbers or graphs. The representations are distinguished from each other by their information content, degree of uniqueness of the expression used, vividness, possibility of further processing, etc., and all of them are of fundamental importance with respect to their intended purpose.

Non-graphical representations

Texts: for the communication of geochemical facts;
Tables: condensed representation, primarily of numbers assigned to geochemical facts;
Punched cards, holes of different shapes: these can easily be sorted and re-sorted according to given characteristics;
Punched cards and punched tapes prepared by special machines to record the information required by electronic data processing equipment.

Graphical representations

Vivid forms of documentation of results used for communication purposes (teaching) and as analytical material in studies etc.

a) Representation of two properties

Properties (features, characteristics) are always a geochemical component, a second geochemical component (Figs. 58, 59, 66, 69, 88), the frequency (abundance) of the occurrence of a component (histograms; Figs. 48, 70), ordinal numbers (numbers of samples), a space or time coordinate (Figs. 79, 85, 87, 89), other factors (Figs. 4, 8, 19, 37, 46, 126, 134). This also applies to paragraphs b) und c).
Form of the diagrams: rectangular system of coordinates (column, line, correlation diagrams), circles divided into sectors (Fig. 53) and the like, nomograms.

b) Representation of three properties

Form of the diagrams: rectangular system of coordinates with a double axis (e.g. columns of different components side by side), triangular diagrams, cartograms for one component (maps with isolines or symbols), block diagrams in two-dimensional projection, and the like, nomograms.

c) Representation of more than three properties

The vividness rapidly decreases with increasing number of features represented (e.g. figure diagrams and n-dimensional petrochemical diagrams). Cartograms showing a few features by the use of dots, symbols (Fig. 119) or representations side by side instead of one upon another are useful.

2. Obtaining of Geochemical Data

important aspects of the gathering of geochemical information which can be reasonably and correctly analyzed and interpreted are as follows:

  a) Kind and number of data must meet the requirements of the specified mathematical evaluation processes (especially for statistical purposes!).
  b) Reasonable sampling, i.e. optimum number and mass of samples for the determination of spatial geochemical regularities or laws, suitable location of sampling points in the field or of the geological bodies, for the determination of random-statistical regularities, statistical random sampling (no preference to be given to "typical" material), the set of samples taken should be representative of the object under investigation.
  c) Avoidance of losses of information and economical restriction in the case of unnecessary excess supply of information.
  d) It should be possible to check that the data are free from errors; for this purpose, instructions regarding analyses, limits of detection and determination, accuracy and precision must be given.
  e) Statistical homogeneity of the data in the case of statistical work.
  f) For computer processing, the data must be acquired in a form that is legible by the processing equipment, unless the so-called on-line presentation is used.

5.3. Reduction of the Number of Features

Frequently, many characteristics of the same material are determined. For the major part, they are causally interrelated by the geochemical conditions of formation. If the evaluation is not deliberately focussed on a few characteristics (e.g. value of contents and the geological position; renouncing the treatment of other properties), either a method suitable for the observation of many features must be used (correlation, factor analysis) or a reduction of characteristics must be achieved by the formulation of mathematical operating figures. This is done by

  a) formation of ratios or proportionality factors
  b) simple addition of characteristics suitable for this operation
  c) calculation of specific characteristics (petrochemistry)
  d) statistical methods (see bibliography)

However, this always involves losses of information in differing degrees; on the other hand, concealed relations can be detected in this way and thus new information acquired.

5.4. Mathematical Processing (without statistics)

Non-statistical mathematical methods of processing geochemical data are used if any datum is considered to be a numerical value free from random variations. Any value is considered significant and representative of a defined geological subject (sample, development, bore, even rock bodies). The evaluation is based on strictly functional relations: calculation of normative rock compositions, of mineral formulas, thermodynamic equilibria, reaction equations, age conditions, etc.

5.5. Statistical Processing

The data are assumed to be a statistical set. Only a random statistical sample represents a certain geological object, the individual data being subject to random variations which are of no interest in most cases.

Table 70 gives a survey of the statistical methods which are most widely used in geochemistry. Table 71 shows the statistical distributions and parameters which are commonly used in geochemistry.

Either single methods only or several methods subject to a common interpretation of partial results are employed. In all statistical procedures, the results must be checked for their significance by appropriate tests. In many statistical methods and tests, the frequency distribution of the data must be taken into account.

For mathematical processing (statistically or functionally), many programs for electronic data processing are available in the international programming languages.

Bibliography

a) *Comprehensive Works and Textbooks-*

CRAMER, H.: Mathematical Methods of Statistics. Princeton University Press, Princeton (1947)

DOERFFEL, K.: Statistik in der analytischen Chemie. VEB Deutsch. Verlag für Grundstoffind. Leipzig (1966)

ECKSCHLAGER, K.: Fehler bei chemischen Analysen. Akad. Verlagsgesellschaft Geest & Portig, Leipzig (1964)

FISZ, M.: Wahrscheinlichkeitsrechnungen und mathematische Statistik. 5th ed. Deutsch. Verl. Wissensch. Berlin (1970)

GRIFFITH, J. C.: Scientific Method in Analysis of Sediments. McGraw-Hill Book Co. New York etc (1967)

KOCH, J. G. S., and R. F. LINK: Statistical Analysis of Geological Data, John Wiley & Sons New York/London, Vol. 1 (1970), Vol. 2 (1971)

KRUMBEIN, W. C., and F. A. GRAYBILL: An introduction to statistical Models in Geology. McGraw-Hill Book Co. New York etc. (1965) (Russian transl.: KRUMBEIN, U., and F. GRAYBILL: Statistical Models in the Geology. Mir, Moscow (1969))

LINDER, A.: Planen und Auswerten von Versuchen. Birkhäuser Verlag, Basle/Stuttgart (1959)

MARSAL, D.: Statistische Methoden für Erdwissenschaftler. E. Schweizerbart'sche Verlagsbuchh. Stuttgart (1957)

MILLER, R. L., and J. S. KAHN: Statistical Analysis in the Geological Sciences. John Wiley & Sons New York/London (1962) (Russian transl. Mir, Moscow (1965))

RODIONOV, D. A.: The Distribution Functions of the Element and Mineral Contents in Eruptive Rocks. Nauka Moscow (1964) (Russian)

SHARAPOV, I. P.: Application of Mathematical Statistics in the Geology. Nedra Moscow (1965) (Russian)

SMITH, F. G.: Geological Data Processing-Using FORTRAN. IV. Harper & Row Publishers New York/London (1966)

THIERGÄRTNER, H.: Grundprobleme der statistischen Behandlung geochemischer Daten. Freiberger Forschungshefte C 237 Leipzig (1968)

VISTELIUS, A. B.: Studies in Mathematical Geology. Consultants Bureau New York (1967)

Inst. Geoch. SO.AN.SSSR: Mathematical Methods of Geochemical Studies. Nauka Moscow (1966) (collection of articles, in Russian)

b) *Selected Special Papers*

AGTERBERG, F. P.: Methods of Trend Surface Analysis. Quart. Colorado School Mines 59 (1964) 4 A, 111—130

AGTERBERG, F. P.: The Use of multivariate Markov Schemes in Petrology. Journ. Geol. 74 (1966) 764—785

Table 70. Survey of Important Statistical Methods Used in Geochemistry

Type of variation to be analyzed	of the magnitude of one variable	of the magnitude of several variables	according to one spatial dimension	according to two spatial dimensions	according to three spatial dimensions	according to time	according to other factors
Analysis of the random portion $\varepsilon\ (0, \sigma^2)$	statistical random distributions and their parameters (estimations)	statistical random distributions comparisons between them					
	mixed distributions	discriminant analysis			MATHERON model		variance, factor, vector analysis
			elimination and statistical analysis of the residual components of trend analyses				
Analysis of the (oriented) trend portion (regional) and anomalie (local)	correlation regression		spatial one-dimensional trend analysis auto-correlation iterations	trend surface analysis	spatial three-dimensional trend analysis	time-trend analysis, auto-correlation iterations	variance, factor, vector analysis with interpretation
			cycles etc.			cycles etc.	

Table 71. Statistical Distributions and Statistical Parameters Frequently Used in Geochemistry

Group	Distribution/Parameter	Estimated Value from Random Samples
Distributions	normal, Poisson, log-normal, binominal, Pearson, Weibull	
Mean values (mean)	arithmetic	$x_{ar} = \dfrac{1}{n} \sum\limits_{i=1}^{n} x_i$
	weighted arithmetic	
	geometric	$x_{geom} = \sqrt[n]{\prod\limits_{i=1}^{n} x_i}$
		$\log x_{geom} = \dfrac{1}{n} \sum\limits_{i=1}^{n} \log x_i$
	density (mode) central value (median) mean of the log-normal distribution	x - value of true clusters in a histogram
Measures of dispersion (variance)	standard deviation	$s = \sqrt{\dfrac{1}{n^{-1}} \sum\limits_{i=1}^{n} (x_i - x_{ar})^2}$
	range	$w = x_{max} - x_{min}$
	mean deviation	$e = \dfrac{1}{n} \sum\limits_{i=1}^{n} \left| x_{ar} - x_i \right|$
	coefficient of variation (relative measure)	$v = \dfrac{s}{x_{ar}}$
Skewness (asymmetry)	3rd central moment	
Kurtosis	4th central moment	
Correlation coefficient	linear correlation of two variables	$r = \dfrac{\sum\limits_{i} x_i y_i - \dfrac{1}{n} \sum\limits_{i} x_i \sum\limits_{i} y_i}{(n-1) s_x s_y}$
	linear partial correlation coefficient linear multiple correlation coefficient non-linear correlation coefficient	
Regression equation	linear regression of two variables	$y = a + bx$ with $a = y_{ar} - bx_{ar}$
		$b = \dfrac{\sum\limits_{i} x_i y_i - \dfrac{1}{n} \sum\limits_{i} x_i \sum\limits_{i} y_i}{(n-1) s_x^2}$
	partial regression	

Notation:

$_i$ individual values of the geochemical feature
$_i$ individual values of another geochemical feature
     number of individual values to be taken into account
$_{max}$ maximum individual value
$_{min}$ minimum individual value
$_x$, s_y standard deviation of the x and/or y values

AHRENS, L. H.: Element Distribution in Specific Igneous Rocks. VIII. Geochim. Cosmochim. Acta *30* 1966) 1, 109—122

BONDARENKO, V. N.: On the Possibility Solution of the Problem on Classifying Vulcans – Plutonic Formations. Sov. Geologiya *11* (1968) 4, 148—153 (Russian)

CAMERON, E. M.: A Computer Program for Factor Analysis of Geochemical and Other Data. Geol. Surv. Canada Paper (1967) No. 34

COTIADI, E. E., et al.: Methodologische Fragen der Einführung mathematischer Methoden und von Elektronenrechenmaschinen in die Praxis der geologischen Untersuchung. Z. angew. Geol. Berlin *13* 1967) 4, 194—197

ANOVICI, V., et al.: La Répartition Lognormale en Géologie. Stud. cerc. geol., geof., geogr.; ser.geol. *14* 1969) 1, 41—61

GY, P.: Variography. Quart. Colorado School Mines *59* (1964) 4 B, 693—711

HARTMANN, L.: Die Anwendung elektronischer Rechenautomaten für die Auswertung geologischer Daten. Z. angew. Geol. Berlin *12* (1966) 10, 536—544

KANCEL', A. V., et al.: On the Possibilities of the Typisation and Prognosis of Ore bodies and Deposits Based on Mathematical Characteristics of the Ores. Izv. AN SSSR, ser.geol. *5* (1967), 31—39 (Russian)

KÖHLER, K.: Beiträge zur Theorie der Probenahme. Freiberger Forschungshefte C 214 Leipzig (1966)

KRUMBEIN, W. C.: The General Linear Model in Map Preparation and Analysis. Computer Contrib. 1967) 12, 38—44

KUTOLIN, V. A.: Statistical Petrochemical Criteria for Classifying of Basalts and Dolerites. Doklady AN SSSR *178* (1968) 2, 434—437 (Russian)

MATHERON, G.: Les Variables Régionalisées et leur Estimation. Masson et Cie.éd.Paris (1965)

MERRIAM, D. F.(ed.): Computer Programs for Multivariate Analysis in Geology. Computer Contrib. 1968) 20

MIDDLETON, G. V.: The Origin of the Lognormal Frequency Distribution in Sediments. Voprosy natem, geologii. Nauka Leningrad (1968), 37—45 (Russian)

PÄLCHEN, W., and H. THIERGÄRTNER: Anwendung der Trendanalyse zum Studium des Aufbaus von Granitoidkörpern. Geologie Berlin *19* (1970) 1, 56—71

PERPY, K. jr.: A Computer Program for Representation of Mineral Chemical Analysis in Terms of End Member Molecules. Contrib. to Geology *7* (1968) 1, 7—14

RODIONOV, D. A.: Statistical Methods of Delimitation of Geologic Objects by the Complex of Features. Nedra Moscow (1968) (Russian)

SHAW, D. M.: Principles of Machine Computation of Spectrochemical Analysis. Canad. Spectroscopy *10* (1965) 1, 3—6

THIERGÄRTNER, H., and I. KOCH: Varianzanalyse und Duncantest zur Gliederung von gruppierten Daten. Z. angew. Geol. Berlin *15* (1969) 5, 253—256

VISTELIUS, A. B.: On the Checking of the Theoretical Backgrounds of the Stochastical Models in Concrete Geological Investigations. 23rd Intern. Geol. Congr., Prague (1968), vol. 13, 153—161

WHITTEN, E. H. T.: Models in the Geochemical Study of Rock Units. Quart. Colorado School Mines *59* (1964) 4 A, 149—168

WHITTEN, E. H. T.: On the Dispersion of Some Features of Granitic Rocks. Voprosy matem. geologii. Nauka Leningrad (1968), 240—252

6. Distribution of the Elements in the Cosmos and in Meteorites

Extraterrestrial matter (solid, liquid, gaseous) can be grouped in stellar matter, interstellar matter, and matter of our solar system. Naturally, the composition of our solar system is of greatest interest to us because we may assume that, in all probability, its entire matter (sun, planets, matter of comets and meteorites, gaseous atmosphere of the sun and of the planets) has a common origin. It is also assumed that the meteorites come exclusively from within the range of our solar system.

The knowledge of the chemical composition of cosmic matter, especially of our solar system, is of great importance to the solution of many problems. In particular, it enables us to draw conclusions as to the development of, and interrelations between, the solar systems and the origin and development of the terrestrial matter. It can be stated that there is no difference in the chemism of the matter of the world which is accessible to measurements.

The chemical composition of extraterrestrial matter is studied mainly in three ways, namely,

1. by spectral analysis of luminous matter
2. by examining meteorites
3. by examining lunar rock samples.

6.1. The Elements in the Cosmos and in the Sun

Table 72 shows the relative abundance of the chemical elements in the cosmos and in the sun.

Fig. 48 shows the regularities of the elementary distribution.

Table 72. Abundance of the Elements in Cosmos
(from Suess and Urey, 1956, and Cameron, 1959) and in the Sun
(from Aller, 1961); atomic concentration related to Si $= 10^6$

Element	Cosmos		Sun
	Suess-Urey	Cameron	Aller
1 H	$4.00 \cdot 10^{10}$	$2.50 \cdot 10^{10}$	$3.16 \cdot 10^{10}$
2 He	$3.08 \cdot 10^9$	$3.80 \cdot 10^9$	—
3 Li	100	100	$2.8819 \cdot 10^{-1}$
4 Be	20	20	$7.239 \cdot 10^0$
5 B	24	24	$1.58 \cdot 10^3$
6 C	$3.5 \cdot 10^6$	$9.3 \cdot 10^6$	$1.658 \cdot 10^7$
7 N	$6.6 \cdot 10^6$	$2.4 \cdot 10^6$	$3.107 \cdot 10^6$
8 O	$2.15 \cdot 10^7$	$2.5 \cdot 10^7$	$2.881 \cdot 10^7$
9 F	1,600	1,600	—
10 Ne	$8.6 \cdot 10^6$	$8.0 \cdot 10^5$	—

Table 72 continued

Element		Cosmos		Sun
		SUESS-UREY	CAMERON	ALLER
1	Na	$4.38 \cdot 10^4$	$4.38 \cdot 10^4$	$3.512 \cdot 10^4$
2	Mg	$9.12 \cdot 10^5$	$9.12 \cdot 10^5$	$7.937 \cdot 10^5$
3	Al	$9.48 \cdot 10^4$	$9.48 \cdot 10^4$	$5.008 \cdot 10^4$
4	Si	$1.00 \cdot 10^6$	$1.00 \cdot 10^6$	1.10^6
5	P	$1.00 \cdot 10^4$	$1.00 \cdot 10^4$	$6.914 \cdot 10^3$
6	S	$3.75 \cdot 10^5$	$3.75 \cdot 10^5$	$6.304 \cdot 10^5$
7	Cl	8,850	2,610	—
8	Ar	$1.5 \cdot 10^5$	$1.5 \cdot 10^5$	—
9	K	3,160	3,160	$1.583 \cdot 10^3$
20	Ca	$4.90 \cdot 10^4$	$4.90 \cdot 10^4$	$4.465 \cdot 10^3$
21	Sc	28	28	$2.087 \cdot 10^1$
22	Ti	2,440	1,680	$1.512 \cdot 10^3$
23	V	220	220	$1.583 \cdot 10^2$
24	Cr	7,800	7,800	$7.239 \cdot 10^3$
25	Mn	6,850	6,850	$2.509 \cdot 10^3$
26	Fe	$6.00 \cdot 10^5$	$8.50 \cdot 10^4$	$1.173 \cdot 10^5$
27	Co	1,800	1,800	$1.379 \cdot 10^3$
28	Ni	$2.74 \cdot 10^4$	$2.74 \cdot 10^4$	$4.465 \cdot 10^4$
29	Cu	212	212	$3.463 \cdot 10^3$
30	Zn	486	202	$7.937 \cdot 10^2$
31	Ga	11.4	9.05	$7.239 \cdot 10^0$
32	Ge	50.5	25.3	$6.162 \cdot 10^1$
33	As	4.0	1.70	—
34	Se	67.6	18.8	—
35	Br	13.4	3.95	—
36	Kr	51.3	42.0	—
37	Rb	6.5	6.50	$9.543 \cdot 10^0$
38	Sr	18.9	61.0	$1.257 \cdot 10^1$
39	Y	8.9	8.9	$5.618 \cdot 10^0$
40	Zr	54.5	14.2	$5.365 \cdot 10^0$
41	Nb	1.00	0.81	$2.816 \cdot 10^0$
42	Mo	2.42	2.42	$2.509 \cdot 10^0$
43	Tc	—	—	—
44	Ru	1.49	0.87	$8.506 \cdot 10^{-1}$
45	Rh	0.214	0.15	$1.904 \cdot 10^{-1}$
46	Pd	0.675	0.675	$5.125 \cdot 10^{-1}$
47	Ag	0.26	0.26	$4.360 \cdot 10^{-2}$
48	Cd	0.89	0.89	$9.113 \cdot 10^{-1}$
49	In	0.11	0.11	$4.566 \cdot 10^{-1}$
50	Sn	1.33	1.33	$1.095 \cdot 10^0$
51	Sb	0.246	0.227	$2.752 \cdot 10^0$
52	Te	4.67	2.91	—
53	I	0.80	0.60	—
54	Xe	4.0	3.35	—
55	Cs	0.456	0.456	—
56	Ba	3.66	3.66	$3.978 \cdot 10^0$
57	La	2.00	0.50	—
58	Ce	2.26	0.575	—
59	Pr	0.40	0.23	—
60	Nd	1.44	0.874	—

Table 72 continued

| Element | | Cosmos | | Sun |
		SUESS-UREY	CAMERON	ALLER
61	Pm	—	—	—
62	Sm	0.664	0.238	—
63	Eu	0.187	0.115	—
64	Gd	0.684	0.516	—
65	Tb	0.095,6	0.090	—
66	Dy	0.556	0.665	—
67	Ho	0.118	0.18	—
68	Er	0.316	0.583	—
69	Tm	0.031,8	0.090	—
70	Yb	0.220	0.393	$1.070 \cdot 10^0$
71	Lu	0.050	0.035,8	—
72	Hf	0.438	0.113	—
73	Ta	0.065	0.015	—
74	W	0.49	0.105	—
75	Re	0.135	0.054	—
76	Os	1.00	0.64	—
77	Ir	0.821	0.494	—
78	Pt	1.625	1.28	—
79	Au	0.145	0.145	—
80	Hg	0.284	0.408	—
81	Tl	0.108	0.31	—
82	Pb	0.47	21.7	$6.756 \cdot 10^{-1}$
83	Bi	0.144	0.3	—
90	Th	—	0.027	—
92	U	—	0.007,8	—

From Fig. 48 the following conclusions can be drawn:

1. The abundance of an element in the cosmos decreases considerably with increasing atomic number to a certain extent (up to the element with $Z = 40$; that is to say, up to the middle of the 5th period). Then the heavier elements exhibit relatively constant values. Exceptions are, above all, Li, Be, B (with relatively low contents) and Fe and Ni (with high contents).

2. Elements with even atomic numbers show higher contents than the elements (neighbouring in the periodic system) having odd atomic numbers (Oddo-Harkin's rule; cf. Section 3.2.).

6.2. Composition of Meteorites and Moon Basalt

Until some time back the meteorites were the only direct evidence of extraterrestrial matter. Their chemical, and especially their mineralogical compositions are of particular interest. Table 73 shows a survey of the classification of meteorites. Table 74 gives qualitative and quantitative statements on the mineral composition, whereas Tables 66 to 79 show the average contents of elements in meteorites and moon basalt (Tab. 78), and the gases in meteorites. For a few years, analyses of samples of lunar rocks have become available. First statements on their chemical composition are given in Table 78.

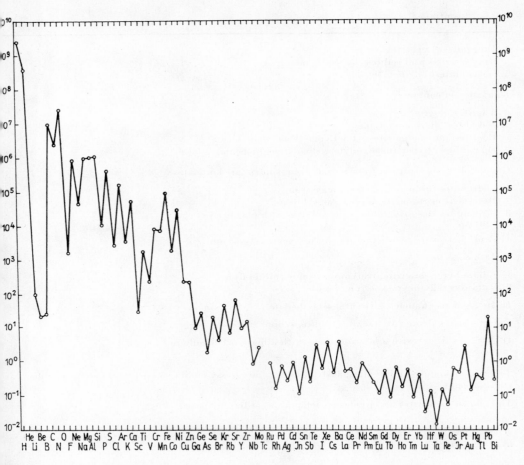

Fig. 48   Abundance of the elements in the cosmos (atomic concentrations related to $Si = 10^6$): values after CAMERON, 1959

Table 73. Classification of the Meteorites (from HEIDE, 1957)

I. Stony meteorites (silicate portion $>$ metallic portion)

A. Chondrites

1. Enstatite-chondrites (enstatite, little nickel-iron)
2. Bronzite-olivine chondrites (bronzite, olivine, little nickel-iron)
3. Hypersthene-chondrites (hypersthene, olivine, little nickel-iron)

B. Achondrites

low calcium content:

1. Aubrites (enstatite)
2. Ureyite (clinobronzite, olivine)

Table 73 continued

3. Diogenites (hypersthene)
4. Amphoterites and rodites (hypersthene, olivine)
5. Chassignites (olivine)

high calcium content:

6. Angrites (augite)
7. Nakhlites (diopside, olivine)
8. Eucrites and shergottites (clinohypersthene, anorthite)
9. Howardites (hypersthene, olivine, clinohypersthene, anorthite)

C. Siderolithes (transition from rock to iron, silicate predominates)

1. Lodromites (bronzite, olivine, nickel-iron)
2. Mesosiderites (bronzite, olivine, nickel-iron)
3. Grahamites (bronzite, olivine, nickel-iron with plagioclase)

II. Iron meteorites (consisting mainly or exclusively of iron)

A. Lithosiderites

1. Siderophyres (bronzite, tridymite, troilite, nickel-iron)
2. Pallasites (olivine, nickel-iron, troilite)

B. Hexaedrites (kamacite, troilite, schreibersite)

C. Octaedrites (kamacite, taenite, troilite, schreibersite)

D. Ataxites (those with a low and those with a high nickel content)

Note
Tectites (moldavites, billitonites, australites, etc.) are glassy bodies whose origin and development are still uncertain.

The known meteorites are divided as follows (according to Moore, 1962):

chondrites   51 per cent
achondrites   4.2 per cent
iron meteorites   44.8 per cent

Table 74. Minerals Found in Meteorites (according to Mason, from Moore, 1962, and supplements)

Mineral	Formula	Mineral	Formula
Nickel-iron	(Fe, Ni)	Calcite	$CaCO_3$
Kamacite	(with about 6% Ni)	Ilmenite	$FeTiO_3$
Taenite	(with about 13—48% Ni)	Rutile	TiO_2
Copper	Cu	Magnetite	Fe_3O_4
Gold	Au	Chromite	$FeCr_2O_4$
Diamond	C	Spinel	$MgAl_2O_4$
Graphite	C	Quartz	SiO_2
Cliftonite	C	Tridymite	SiO_2
Sulphur	S	Cristobalite	SiO_2
Moissanite	SiC	Apatite	$Ca_5(PO_4)_3Cl$
Cohenite	Fe_3C	Merrillite *)	$Na_2Ca_3(PO_4)_2O$
Schreibersite *)	$(Fe, Ni)_3P$	Farringtonite *)	$Mg_3(PO_4)_2$
Osbornite *)	TiN	Stanfieldite	P-mineral

Mineral	Formula	Mineral	Formula
Troilite	FeS	Epsomite	$MgSO_4 \cdot 7\,H_2O$
Oldhamite	CaS	Zircon	$ZrSiO_4$
Alabandite	MnS	Olivine	$(Mg, Fe)_2SiO_4$
Sphalerite	ZnS	Orthopyroxene	$(Mg, Fe)SiO_3$
Chalcopyrrhotine	$CuFeS_2$?	Clinopyroxene	$(Ca, Mg, Fe)SiO_3$
Djerfisherite	Alk.-Cu-Fe-Sulphide	Ureyite	$NaCrSi_2O_6$
Pentlandite	$(Fe, Ni)_9S_8$	Merrihueite	$(K, Na)_2(Mg,Fe)_5Si_{12}O_{30}$
Daubreelithe *)	$FeCr_2S_4$	Richterite	Na-tremolite
Lawrencite *)	$FeCl_2$	Plagioclase	$(Na, Ca)\,(Al, Si)_4O_8$
Magnesite	$MgCO_3$	Serpentine	$Mg_6Si_4O_{10}(OH)_8$

*) Minerals marked by an asterisk have not been found on the earth.

Table 75. Modal Content (Important Minerals) of a Few Types of Meteorites [volume-%] (from HEIDE, 1957, and other authors)

Minerals	Meteoritic Iron	Pallasites	Meso-siderites	Chondrites	Achondrites free from feldspar	Achondrites containing feldspar
Nickel-iron	98.3	50	45	9	0.6	(0.5)
Olivine	—	48	1.5	45	15	3
Pyroxene	—	—	31	25—30	70	45
Feldspar	—	—	16	11	10	50
Troilite	0.12	0.3	3	6	1	0.6
Schreibersite	1.12	0.2	2.5	(—)	(—)	(—)

Table 76. Mean Chemical Composition of Meteorites Compared with that of Eruptive Rocks (from HEIDE, 1957 and 1964)

Element	Iron Meteorites	Stony Meteorites	Chondrites	Eruptive Rocks
O	—	35.71	34.84	46.60
Fe	89.70	23.31	25.07	5.00
Si	—	18.07	17.78	27.72
Mg	—	13.67	14.38	2.00
S	0.08	1.80	2.09	0.05
Ca	—	1.73	1.39	3.63
Ni	9.10	1.53	1.34	0.08
Al	—	1.52	1.32	8.13
Na	—	0.65	0.68	2.83
Cr	—	0.32	0.25	0.02
K	—	0.17	0.084	2.59
C	0.12	0.15	0.1	0.03
Co	0.62	0.12	0.08	0.002
P	0.18	0.11	0.05	0.12
Ti	—	0.11	0.066	0.44

Table 77. Gases in Meteorites (from HEIDE, 1957; volume-%)

	CO$_2$	CO	CH$_4$	H$_2$	N$_2$	total
Stony meteorites	3.77	0.24	0.20	0.50	0.09	4.80
Iron meteorites	0.21	0.67	0.02	1.67	0.24	2.81

Table 78. Chemical Composition of Tectites (TAYLOR and SACHS, 1964),
Chondrites and Achondrites (VINOGRADOV, 1958) and Moon Basalt [Mass %]

	Tectites (Australite)	Chondrites (Silicate phase)	Achondrites (free from feldspar)	Moon Basalt (Mean Comp. Apollo 11)
SiO$_2$	73.31	47.0	52.56	41
TiO$_2$	0.41	0.14	0.12	11
Al$_2$O$_3$	11.68	3.09	1.09	9
Fe$_2$O$_3$	0.60 ⎱	15.40	11.45	19
FeO	4.03 ⎰			
MgO	2.02	29.48	30.47	8
CaO	3.55	2.41	1.20	11
Na$_2$O	1.25	1.2	0.36	0.5
K$_2$O	2.26	0.21	0.11	

Table 79. Distribution of Elements in Meteorites and Meteoritic Phases [ppm]

Symbol	Nickel-iron (HEIDE)	Troilite (HEIDE)	Chondrite (VINOGRADOV)	Silicate (HEIDE)	Meteorites (total) (HEIDE)	(RANKAMA and SAHAMA)
Ag	4	18	0.094	—	2	2.0
Al	40	—	13,000	17,400 ± 1,400	13,391	13,800
Ar	—	—	—	present	—	present?
As	360	1,020	0.3	20	149	present
Au	4	0.5	0.17	—	0.65	0.65
B	—	—	2	3	2.3	1.5
Ba	—	—	6	8	6.15	6.9
Be	—	—	3.6	1	0.8	1
Bi	0.5	2	0.003	0.02	0.25	present
Br	1	—	0.5	25	19.4	20
C	1,200	—	400	400	492	300
Ca	500	—	14,000	19,700 ± 2,000	15,231	13,300
Cd	8	30	0.1	1.6	4.8	present
Ce	—	—	0.5	2.5	1.9	1.77
Cl	present	—	70	900	692	1,000 ± 1,500
Co	6,300 ± 200	100	800	206 ± 86	1,135	1,200
Cr	300	1,200	2,500	3,450 ± 590	2,792	3,340
Cs	—	—	0.1	0.1	0.08	0.08
Cu	310	500	100	1.6	87.4	170
Dy	—	—	0.35	2.5	1.9	1.80
Er	—	—	0.2	2.1	1.6	1.48
Eu	—	—	0.08	0.33	0.25	0.25

Table 79 continued

Symbol	Nickel-iron (Heide)	Troilite (Heide)	Chondrite (Vinogradov)	Silicate (Heide)	Meteorites (total) (Heide)	Meteorites (total) (Rankama and Sahama)
	—	—	28	40	30.8	28
Fe	907,800 ± 2,600	611,000	250,000	156,400 ± 5,400	306,969	288,000
Ga	36	0.5	3	0.5	5.96	4.2
Gd	—	—	0.4	2	1.5	1.42
Ge	500	30	10	5	83	79
H	—	—	—	630 ± 150	485	present
He	present	—	—	present	present	present
Hf	—	—	0.5	1	0.8	1.6
Hg	—	0.2	3	<0.01	0.02	present
Ho	—	—	0.07	0.72	0.55	0.51
	0.6	—	0.04	1.25	1.1	1
In	0.5	0.5	0.001	—	0.1	0.15
Ir	4	0.4	0.48	—	0.65	0.65
K	—	—	850	1,990 ± 200	1,531	1,540
La	—	—	0.3	2.2	1.7	1.58
Li	—	—	3	3	2.3	4
Lu	—	—	0.035	0.65	0.5	0.46
Mg	320	—	140,000	158,200 ± 3,000	121,742	123,000
Mn	300	460	2,000	2,960 ± 670	2,359	2,080
Mo	16.6	11	0.6	2.5	5.3	5.3
N	present	—	1	0.9	0.7	present
Na	—	—	7,000	7,790 ± 530	5,992	5,950
Nb	0.2	—	0.3	0.5	0.4	0.41
Nd	—	—	0.6	3.7	2.85	2.59
Ni	85,900 ± 2,400	1,000	13,500	1,380 ± 240	14,354	15,680
O	—	—	350,000	410,200 ± 9,500	315,538	323,000
Os	8	9	0.5	—	1.9	1.92
P	1,800	3,000?	500	700	1,046	1,050
Pb	56	20	0.2	2	11.7	11
Pd	9	2	1	—	1.5	1.54
Pr	—	—	0.1	1	0.8	0.75
Pt	20	2	2	0.083	3.3	3.25
Rb	—	—	5	4.5	3.5	3.5
Re	0.008,2	0.001	$8 \cdot 10^{-4}$	0.008	0.007,5	0.002,0
Rh	5	0.4	0.19	—	0.8	0.80
Ru	20	9	1	—	2.2	2.23
S	present	347,633	20,000	—	26,740	21,200
Sb	2	7.8	0.1	0.1	0.98	present
Sc	—	—	6	5	3.8	4
Se	3	100	10	13	18	7
Si	40	—	180,000	205,700 ± 1,800	158,237	163,000
Sm	—	—	0.2	1.3	1	0.95
Sn	100	15	1	5	20.4	20
Sr	—	—	10	26	20	20
Ta	<0.06	—	0.02	<0.38	0.3	0.30
Tb	—	—	0.05	0.64	0.5	0.45

Table 79 continued

Symbol	Nickel-iron (HEIDE)	Troilite (HEIDE)	Chondrite (VINOGRADOV)	Silicate (HEIDE)	Meteorites (total) (HEIDE)	(RANKAMA and SAHAMA)
Te	—	1.3	0.5	—	0.1	0.1
Th	0.04	—	0.04	2	1.54	0.8
Ti	100	—	500	1,800	1,400	1,320
Tl	—	0.3	0.001	0.15	0.15	present
Tm	—	—	0.04	0.38	0.3	0.26
U	0.007	—	0.015	0.4	0.31	0.36
V	6	—	70	50	39.4	39
W	8.1	—	0.15	18	15.1	15
Y	—	—	0.8	7	5.4	4.72
Yb	—	—	0.2	2	1.5	1.42
Zn	115	1,520	50	3.4	137	138
Zr	8	—	30	100	78.2	73

Bibliography

ADLER, I., and J. TROMBKA: Geochemical Exploration of the Moon and the Planets. Springer 1970

AHRENS, L. H., R. A. EDGE, and S. R. TAYLOR: The Uniformity of Concentration of Lithophile Elements in Chondrites — with Particular Reference to Cs. Geochim. Cosmochim. Acta 20 (1960), 260—272

ALLER, L. H.: The Abundance of the Elements. New York 1961

ALLER, L. H.: Abundances of Elements in Stars and Nebulae. AFOSR 3024, Technical Note No. 2

ALLER, L. H.: The Abundance of Elements in the Solar Atmosphere. Chapter I in: Advances in Astronomy and Astrophysics. Vol. 3. Academic Press 1965

ANDERS, E., and G. G. GOLES: Theories on the Origin of Meteorites. J. Chem. Educat. 38 (1961), 58—66

ANNELL, C., and A. HELZ: Emission Spectrographic Determination of Trace Elements in Lunar Samples. Science 167 (1970), 521—523

BAEDECKER, P. A., and W. D. EHMANN: The Distribution of Some Noble Metals in Meteorites and Natural Materials. Geochim. Cosmochim. Acta 29 (1965), 329—342

CAMERON, A. G. W.: The Origin of the Elements. Phys. Chem. of the Earth 3 (1959), 199—223

CAMERON, A. G. W.: A Revised Table of Abundance of the Elements. Astrophysic. J. 129 (1959), 676—699

COHEN, A. J.: Germanium Contents of Tectites and Other Natural Glasses. Implications Concerning the Origin of Tectites. Internat. Geol. Congr.-Rep. 21, Sess. North, 1960, Copenhague, P. I. Geochem. Cycles, 30—39

CRAIG, H., S. L. MILLER, and G. J. WASSERBURG: Isotopic and Cosmic Chemistry. North-Holland Publ. Com., Amsterdam 1964

DOOD, R. T., and W. R. VAN SCHMUS: Merrihueite, A New Alkali-Ferromagnesian Silicate from the Mezö-Madaras Chondrite. Science, 149 (1965), 3687, 972—964

EASTON, A. J., and J. F. LOVERING: The Analysis of Chondritic Meteorites. Geochim. Cosmochim. Acta, 27 (1963), 7, 753—767

EHMANN, W. D.: The Abundance of Mercury in Meteorites and Rocks by Neutron Activation Analysis. Geochim. Cosmochim. Acta 31 (1967), 357—376

FISHER, D. E.: The Aluminium Content of Chondritic Meteorites as Determined by Activation Analysis. Geochim. cosmochim. Acta 28 (1964), 743—749

FOUCHÉ, K. F., and A. A. SMALES: The Distribution of Trace Elements in Chondritic Meteorites. Chem. Geol. 2 (1967), 2, 105—134

OWLER, W. A.: The Origin of the Elements. Proc. Nat. Acad. Sci. *52* (1964), 524—548

RONDEL, C., and C. KLEIN: Ureyite, $NaCrSi_2O_6$: A New Meteoritic Pyroxene. Science, *149* (1965), 685, 742—744

AMOV, G.: History of the Universe. Science, *158* (1967), 3802, 766—769

IBBONS, J. H., and R. L. MACKLIN: Neutron Capture and Stellar Synthesis of Heavy Elements. cience, *156* (1967), 3778, 1039—1049

OLDBERG, L., E. A. MÜLLER, and L. H. ALLER: The Abundance of the Elements in the Solar Atmophere. Astrophysic. J. Suppl. *45* (1960), Vol. 5, 1—138

OLDSCHMIDT, V. M.: Geochemische Verteilungsgesetze und kosmische Häufigkeit der Elemente. aturwissenschaften *47—49* (1930), 999—1013

OLES, G. G., and F. ANDERS: Abundances of Iodine, Tellurium and Uranium in Meteorites. Geochim. osmochim. Acta *26* (1962), 723—738

REENLAND, L., and I. E. LOVERING: Minor and Trace Element Abundances in Chondritic Meteorites. eochim. Cosmochim. Acta *29* (1965), 8, 821—858

IASKIN, L. A., and F. A. FREY: Dispersed and Not-so-Rare Earths. The Relative Abundance of the lements Reflect the Earth's Geochemical Evolution from Primordial Matter. Science *152* (1966), 720, 299—314

IEIDE, F.: Kleine Meteoritenkunde, 2nd ed. Springer, Berlin/Göttingen/Heidelberg 1957

IEY, M. H.: Catalogue of Meteorites (3rd edition). British Museum, London 1966

IINTENBERGER, H., H. WEBER et al.: Rare Gases, Hydrogen and Nitrogen Concentrations and Isotopic omposition of Lunar Material. Science *167* (1970), 543—545

KRINOV, E. L.: Principles of Meteorites. Pergamon Press, London 1960

EVIN, B., S. V. KOZLOVSKAYA, and A. G. STARKOVA: The Average Chemical Composition of Meteorites. eochim. Cosmochim. Acta, *13* (1958), 76

IPSCHÜTZ, M. B., and E. ANDERS: The Record in the Meteorites — IV. Geochim. cosmochim. Acta *4* (1961), 83—105

OVERING, J. F., W. NICHIPONK, A. CHODOS, and H. BROWN: The Distribution of Gallium, Germanium, obalt, Chromium, and Copper in Iron and Stony-iron Meteorites in Relation to Nickel Content and tructure. Geochim. cosmochim. Acta *11* (1957), 263—278

MARVIN, U. B., and C. KLEIN: Meteoritic Zircon. Science *146* (1964), 3646, 919—920

MASON, B.: Meteorites. 274 pp. John Wiley and Sons, New York 1962

MASON, B.: Geochemistry and Meteorites. Geochim. Cosmochim. Acta *30* (1966), 365—374

MASON, B., and W. G. MELSON: Comparison of Lunar Rocks with Basalts and Stony Meteorites. Proc. Apollo 11 Lunar Sci. Conf. *1*, 661—671. Geochim. Cosmochim. Acta *34* (1970), Suppl. Vol. 1

MOORE, C. B., and CH. LEWIS: Carbon Abundances in Chondritic Meteorites. Science *149* (1965), 3681, 317—318

MOORE, C. B.: Researches on Meteorites. John Wiley & Sons, New York/London 1960

PEPIN, R. O., and P. SIGNER: Primordial Rare Gases in Meteorites. Science *149* (1965), 3681, 253—265

REED, G. W., et al.: Determination of Concentration of Heavy Elements in Meteorites by Activation Analysis. Geochim. cosmochim. Acta *20* (1960), 122—140

REES, M. J., and W. L. W. SARGENT: Composition and Origin of Cosmic Rays. Nature *219* (1968), 5158, 1005—1009

RINGWOOD, A. E.: Cohenite as a Pressure Indicator in Iron Meteorites. Geochim. cosmochim. Acta *20* (1960), 155—157

RINGWOOD, A. E.: Chemical and Genetic Relationship Among Meteorites. Geochim. cosmochim. Acta, *24* (1961), 159—197

RINGWOOD, A. E., and E. ESSENE: Petro Genesis of Apollo 11 Basalts, Internal Constitution and Origin of the Moon. Proc. Apollo 11 Lunar Sci. Conf. *1*. Geochim. Cosmochim. Acta *34* (1970), Suppl. Vol. 1

RUNCORN, S. K. (ed.): The Application of Modern Physics to the Earth and Planetary Interiors. J. Wiley and Sons, New York/London 1969

SCHMITT, R. A. et al.: Abundance of the Fourteen Rare-earth Elements, Scandium and Yttrium in Meteoritic and Terrestrial Matter. Geochim. cosmochim. Acta *27* (1963), 577—622

SCHMITT, R. A., H. WAKITA and P. REY: Abundance of 30 Elements in Lunar Rock, Soil and Core Samples. Science *167* (1970), 512—515

STARIK, I. E., E. V. SOBOTOVICH, and M. M. SHATS: On the Question of the Origin of Meteorites and Tectites. Geokhimiya *3* (1963), 245—252 (Russian)

STUDIER, M. H., R. HAYATSU, and E. ANDERS: Origin of Organic Matter in Early Solar System, I Hydrocarbons. Geochim. cosmochim. Acta *32* (1968), 2, 151—173

TAYLOR, S. R., and M. SACHS: Geochemical Evidence for the Origin of Australites. Geochim. cosmochim. Acta *28* (1964), 235—264

TURKEVICH, A. L., E. J. FRANZGROTE, and J. H. PATTERSON: Chemical Analysis of the Moon at the Surveyor VII Landing Site, Science *162* (1968), 3849, 117—119

UREY, H. C.: The Abundance of the Elements. Physic. Rev. *88* (1952), 248—252

UREY, H. C.: The Planets, their Origin and Development. Yale Univ. Press, New Haven, Conn., 1952

UREY, H. C.: Biological Material in Meteorites: A Review. Science *151* (1966), 3707, 157—166

VINOGRADOV, A. P.: Meteorites and the Earth's Crust (Geochemistry of Isotopes). Congress for the Peaceful Use of Atomic Energy, Sess. e.-7b, P 2523, Geneva (1958), 255—269

VINOGRADOV, A. P.: Atomic Abundance of the Chemical Elements in the Sun and in Stony Meteorites. Geokhimiya *4* (1962), 291—295 (Russian)

VOIGT, H. H.: Anomale Elementhäufigkeiten im Kosmos. Naturwissenschaften *44* (1957), 501—504

VOROB'EV, G. G.: On the Chemical Composition of Tectites in Connection with the Problem of their Origin. Geokhimiya *5* (1960), 427—442 (Russian)

WÄNKE, H., R. RIEDER et al.: Major and Trace Elements in Lunar Material. Proc. Apollo 11 Lunar Sci. Conf. *2*, 1719—1727. Geochim. Cosmochim. Acta *34* (1970), Suppl. Vol. 1

WILDT, R.: Cosmochemistry. Scientia (Milano) 4. Ser. *34* (1940), 85—90

WILK, H. B.: The Chemical Composition of Some Stony Meteorites. Geochim. cosmochim. Acta *9* (1956), 279

ZÄHRINGER, J.: Rare Gases in Stony Meteorites. Geochim. Cosmochim. Acta *32* (1968), 2, 209—237

Note:

Additional information is given in the omnibus volume:

AHRENS, L. H. (Edit.): Origin and Distribution of the Elements. Pergamon Press, Oxford and New York 1968

7.1. Geochemical Structure of the Earth

One of the most difficult problems of geochemistry is to ascertain the chemical composition of the whole earth. The physical inhomogeneities are relatively well known from geophysics, especially due to seismology, whereas the substantial interpretation of the deeper regions of the earth is largely hypothetical.

A basis discovery is the shell-structure of the earth, determined by changes in physical and chemical parameters. The term "shells" was introduced by VERNADSKIĬ, the synonym "geophase" by MURRAY (1910). The structural model of the deeper regions of the earth suggested by GOLDSCMIDT (1922/1933) as a result of investigations on meteorites and comparisons of blast-furnace processes (MORITZ and CISSARZ, 1930) has been widely favoured. A survey of the most important hypotheses of the material structure of the earth is given in Fig. 49.

The following geospheres have been identified:

Atmosphere: gaseous envelope surrounding the earth, uppermost geosphere (mainly the gases in the mass of air surrounding the earth, but also in the hydrosphere and uppermost lithosphere; see Section 7.5.4. and 8.6.).

Biosphere: (according to de LAMARCK, about 1800, and SUESS, 1875): (the sphere in which life can exist; living beings together with their environment (see Section 7.5.5. and 8.7.).

Hydrosphere: (according to SUESS, 1875): the aqueous envelope of the earth (mainly water of the oceans, while fresh water, rain, snow, ice, etc. are of minor importance; detailed discussion in Sections 7.5.3. and 8.5.).

Lithosphere: (according to SUESS, 1875): silicate crust of the earth (sedimentary rocks, metamorphous rocks, igneous rocks; the silicates become more and more basic with increasing distance from the earth's surface towards the interior; full details are given in Sections 7.4. and 7.5.).

Chalcosphere: (according to GOLDSCHMIDT, 1922/1933): oxide-sulphide shell (roughly corresponds to the troilite phase of meteorites, contains more Fe-oxides, however).

Siderosphere: (according to GOLDSCHMIDT, 1922/1933): nickel-iron core of the earth (roughly corresponds to the composition of the iron meteorites).

Recently J. BARRELL postulates the existence of the *Astenosphere*: the shell of the upper mantle lying beneath the more rigid lithosphere in which crustal plates are embedded, isostatic readjustments are supposed to be effected and in which magmas are generated.

The boundary regions between the upper geospheres are not sharply defined, but interlinked. The division of the earth into the earth's core, mantle and crust is superimposed on this structure, the Gutenberg discontinuity (at a depth of 2,900 km) and the Mohorovičič discontinuity (at a depth of 15 to 30 km on an average) being considered the boundaries.

Rough surveys of the composition of the main structural units of the earth are given in Table 80, Table 81 and Fig. 49.

Table 80. Most Important Substantial Data of the Various Geospheres

Geosphere	Physical State	Main Constituents	Thickness
Atmosphere	gaseous	N_2, O_2 inert gases and other gases	total atmosphere 100 km: troposphere 12—15 km stratosphere $\sim$ 30 km mesosphere $\sim$ 35 km
Biosphere	solid, partly liquid	C, H, O, N, S, H_2O, in-organic skeleton material	about 1 km, frequently much more
Hydrosphere	liquid, rarely solid	salt-water and fresh water, snow, ice, solved salt (chlorides, sulphates, etc.)	about 4 km
Lithosphere (earth's crust)	solid	silicate rocks (so-called sial) upper region acid: Si, Al, alkalis, OH lower region basic: Si, Al, Ca, Mg, Fe	10—80 km average about 30—40 km
Lithosphere and chalcosphere (earth's mantle)	solid	basic to ultrabasic silicate rocks (so-called sima); in the lower part probably with sulphides. Si, Mg, Fe, Ca, Al, S	about 2,900 km
Siderosphere (earth's core)	solid and/or liquid	Ni-Fe alloys	3,470 km

Table 81. Chemical Composition of the Earth's Crust, Mantle, Core, and Total Earth (from MASON, 1966; data in mass-% of the elements or elementary oxides)

Element	Core	Mantle + Crust	Crust	Total Earth	Elementary Oxides	Mantle + Crust	Mantle
O		43.7	45.4	29.53			
Si		22.5	25.8	15.20	SiO_2	48.09	43.06
Mg		18.8	3.1	12.70	MgO	31.15	31.32
Fe	86.3	9.88	6.5	34.63	FeO	12.71	6.66[1]
Ca		1.67	6.0	1.13	CaO	2.32	2.65
Al		1.60	8.1	1.09	Al_2O_3	3.02	3.99
Na		0.84	2.2	0.57	Na_2O	1.13	0.61
Cr		0.38		0.26	Cr_2O_3	0.55	0.42
Mn		0.33	0.2	0.22	MnO	0.43	0.13
P		0.14	0.1	0.10	P_2O_5	0.34	0.08
K		0.11	1.6	0.07	K_2O	0.13	0.22
Ti		0.08	1.0	0.05	TiO_2	0.13	0.58
Ni	7.28			2.39	NiO		0.39
Co	0.40			0.13	CoO		0.02
S	5.96			1.93	H_2O^+		0.21
Sum	99.94	100.03	100.0	100.0		100.0	100.0

[1] +1.66% Fe_2O_3

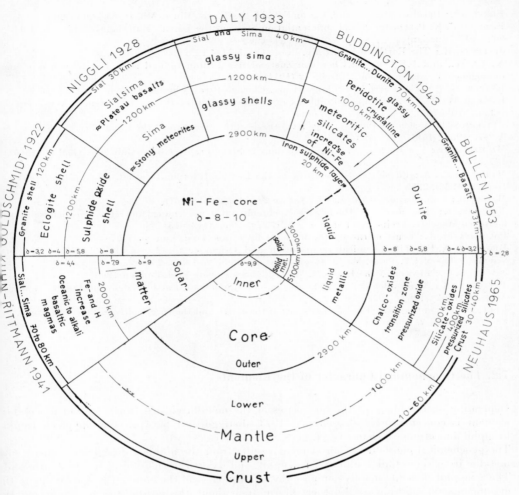

Fig. 49   Schematic representation of the most important hypotheses of the earth's structure

Bibliography

Borchert, H.: Chemismus und Petrologie der Erdschalen sowie Entstehung und Ausgestaltung der wichtigsten Diskontinuitäten der Erdkruste. N. Jb. Miner. Mh. (1962), 7/8, 143—163

Buddington, A. F.: Some Petrological Concepts and the Interior of the Earth. Amer. Mineralogist 28 (1943), 119—140

Bullen, K. E.: An Introduction to the Theory of Seismology. Cambridge University Press, Cambridge 1953

Chirvinsky, P. N.: On the Chemism and Mineralogical Composition of the Peridotite-Dunite Stratum of the Earth. Zap. Vsesoyuzn. min. Obs. SSSR 85 (1956), 397—401 (Russian)

Daly, R. A.: Igneous Rocks and the Depths of the Earth. McGraw Hill Book Comp., New York/ London 1933

ERNST, TH.: Basaltmagmenentstehung und Peridotit. N. Jb. Miner. Mh. (1963), 203—205

FLORENSKIĬ, K. P.: On the Initial Differentiation Stage of the Earth's Material. Geokhimiya (1965), 8, 909—917 (Russian)

JEFFREYS, H.: The Earth, 3rd ed. Univ. Press, Cambridge 1952

KUHN, W., and A.RITTMANN: Über den Zustand des Erdinnern und seine Entstehung aus einem homogenen Urzustand. Geol. Rdsch. 32 (1941), 215—245

KUIPER, G. P.: The Earth as a Planet. Univ. Chicago Press, Chicago 1954

LATIMER, W. M.: Astrochemical Problems in the Formation of the Earth. Science, New York 112 (1950), 101—104

LYTTLETON, R. A.: On the Phase-change Hypothesis of the Structure of the Earth. Proc. Roy. Soc., Ser. A 287 (1965), 1411, 471—493

MACHAIRAS, G.: Les Roches Leur Equilibre dans la Structure du Globe. Bull. Soc. Géol. 7 (1965), 1, 187—194

MAGNITSKY, V. A.: Physics and Chemistry of the Earth's Interior. Internat. Upper Mantle Project, Report (1965), 2, 85—87

MASON, B.: Composition of the Earth. Nature 211 (1966), 616—618

NEUHAUS, A.: Die moderne Hochdruck-Hochtemperatur-Forschung und ihre geochemisch-petrologischen Aspekte. Freiberger Forschungsh. C 210 (1966), 113—131

POLDERVAART, A.: Crust of the Earth. 3rd ed. Oliver and Boyd, London 1957

RINGWOOD, A. E.: The Chemical Composition and Origin of the Earth. In "Advances in Earth Science". The M.I.T. Press, Massachusetts Inst. Technol. 1 (1966), 287—356

STACEY, F. D.: Physics of the Earth. J. Wiley and Sons, New York/London 1969

TAYLOR, S. R.: Trace Element Abundances and the Chondritic Earth Model. Geochim. Cosmochim. Acta 28 (1964), 1989—1998

UREY, H. C.: On the Early Chemical History of the Earth and the Origin of Life. Proc. nat. Acad. Sci. USA 38 (1952), 351

7.2. The Geochemical Character of the Elements

Depending on the phase of the meteorites or the metallurgical products in which a certain element is concentrated, GOLDSCHMIDT (1923) distinguished between siderophil, chalcophil, lithophil and atmophil elements (cf. geospheres in 7.1.).

The geochemical character of an element results from the affinity of a given element with metallic iron (siderophil), sulphur (chalcophil) and oxygen (formation of silicates: lithophil). The atmophil elements mainly compose the atmosphere and the biophil elements build up the organic matter (biosphere). Further information about the possibilities of classifying the elements in geochemical groups is given in Tables 82 and 83. It should be noted that there are several other geochemical classifications, e.g. classifications by NIGGLI (1928; endogeosphere and exogeosphere elements), WASHINGTON (1920; two groups), VERNADSKIĬ (1922; three groups, 1930; six groups), BERG (1929; eight groups), SAVARITSKIĬ (1944; ten groups), and by other experts.

The geochemical character of an element is largely dependent on the electronic configuration of its atoms; this is why it is closely related with the position of the elements in the periodic system (Fig. 10 in Section 2.6.).

The dependence of the geochemical character of an element on the heat of formation ($\Delta H°$) of the oxides in question was shown by BROWN and PATTERSON (1948).

The different ions of an element can possess a different geochemical character (dependence of the valence, see Table 84). According to SZÁDECZKY-KARDOSS (1952), that geochemical character is to be attributed to any element which should be assigned to its most abundant and stable ion.

218

Table 82. Geochemical Grouping of the Elements by Various Authors

WASHINGTON (1920)	GOLDSCHMIDT (1923)	RANKAMA-SAHAMA (1950)	SZÁDECZKY-KARDOSS (1952)
metallogenic — the elements primarily in form: sulphides, selenides, tellurides, arsenides, antimonides, bromides, iodides; native	siderophil — concentrated in the siderosphere; soluble in Fe-melt; native		siderophil – especially in the metallic phase
	chalcophil — pronounced affinity for S; concentrated in sulphide-oxide shell (chalcosphere)	sulphophil — in particular: sulphides, selenides, tellurides, arsenides, antimonides, inter-metallic combinations, native, etc., no O(F, Cl) combinations, (mainly covalent or metallic bonds)	sulpho – chalcophil especially sulphide – (troilite) – phase
			oxychalcophil in oxides and sulphides
petrogenic — the elements in oxides, silicates, fluorides, chlorides, etc.	lithophil — pronounced affinity for oxygen; primarily in the silicate crust (lithosphere)	oxyphil — primarily combined with O (F, Cl); oxides, silicates, phosphates, carbonates, nitrates, borates, sulphates, etc. (mainly ionic bonds)	lithophil in a narrow sense especially in the main crystallisation
			pegmatophil a) transition elements b) typically pegmatophil
			sedimentophil easily volatile; chemically active "agents minéralisateurs"
	atmophil — typical elements of the atmosphere		atmophil easily volatile, inert
	biophil — typical elements of the biosphere		

Table 83. Classification of the Elements According to Their Geochemical Character

V. M. Goldschmidt (1923)					E. Szádeczky-Kardoss (1952)						
siderophil	Au				siderophil	Au					
	Ge	Sn	(Pb)								
	C	P	(As)								
	Mo	(W)									
	Re										
	Fe	Co	Ni			Fe	Co	Ni			
	Ru	Rh	Pd			Ru	Rh	Pd			
	Os	Ir	Pt			Os	Ir	Pt			
chalcophil	Cu	Ag			sulpho-	Cu	Ag				
	Zn	Cd	Hg		chalcophil	Zn(?)	Cd(?)	Hg			
	Ga	In	Tl								
	(Ge)	(Sn)	Pb								
	As	Sb	Bi			As	Sb	S	Se	Te	P(?)
	(Mo)										
	S	Se	Te		oxy-	Ge	Sn	Pb			
	Fe	(Co)	(Ni)		chalcophil	Ga	In	Tl			
	(Ru)	(Pd)	(Pt)			Zn(?)	Cd(?)				P(?)
lithophil	Li	Na	K	Rb	Cs	lithophil	Li(?)	Na	K	Rb	Cs
	Fr					in a narrow					
	Be	Mg	Ca	Sr	Ba	sense	Be(?)	Mg	Ca	Sr	Ba
	Ra										
	(Zn)	(Cd)									
	B	Al	Sc	Y			Al	Si			
	La	Cs	Pr	Nd	Sm						
	Eu	Gd	Tb	Dy	Ho	pegmatophil	Ti, V, Zr, Mn ↑ transition				
	Er	Tm	Yb	Lu			↓ elements				
	Ac	Th	Pa	U	Np		Sc	Y			
	Pu	Am	Cm				La—Lu				
	Ga	(In)	(Tl)				Th	U	Hf(?)		
	C	Si	Ti	Zr	Hf		Nb	Ta	W	Mo	
	(Ge)	(Sn)	(Pb)								
	V	Nb	Ta			sedimentophil	B				
	P	As					C				
	O	Cr	W	Mn			(N)				
	(Fe)	(Co)	(Ni)				F	Cl	Br(?)	I(?)	
	H	F	Cl	Br	I						
atmophil	H	C	N			atmophil	H(?)	N(?)			
	O	I	Hg				He	Ne	Ar		
	He	Ne	Ar				Kr	Xe	Rn		
	Kr	Xe	Rn								
biophil	H	C	N	O	P						
	(Na)	(Mg)	S	Cl							
	(K)	(Ca)	(Fe)								
	(B)	(F)	(Si)								
	(Mn)	(Cu)	(I)								

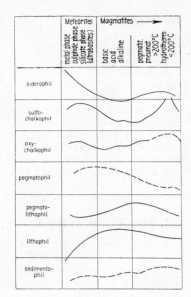

Fig. 50 The relative concentrations of the individual geochemical groups of elements in meteorites and magmatic formations (after Szádeczky-Kardoss, 1952)

Table 84. Changes in the Geochemical Character with an Increase in the Electropositive Charge (from Szádeczky-Kardoss, 1952)

siderophil ⟵―――――⟶	chalcophil ⟵―――――⟶	lithophil
Ag	Ag⁺	
As³⁻		
As (su)¹)	As³⁺ (su-o)	As⁵⁺ (o)
C⁴⁻	←C→	C⁴⁺ (o)
Co	Co²⁺	Co³⁺
Cu	Cu¹⁺	←Cu²⁺
Fe	Fe²⁺ (su-o)	Fe³⁺ (o)
Ge	Ge²⁺ (su)	Ge⁴⁺ (o)
(Mn)	Mn²⁺	
	Mn³⁺ (o)	Mn⁴⁺ (o)
Mo	Mo⁴⁺	Mo⁵⁺
		Mo⁶⁺ (o)
		N⁵⁺ (o)
N		
Ni	Ni²⁺	
P³⁻	←P→	P⁵⁺ (o)
(Pb)	Pb²⁺ (o)	Pb⁴⁺ (o)
Pt	Pt⁴⁺	
	S²⁻	S⁶⁺ (o)
	S	
(Sb³⁻)	Sb³⁻	
	Sb	Sb³⁺ Sb⁵⁺ (o)
	Se²⁻	Se⁶⁺
	Se	
(Si)		Si⁴⁺ (o)
	V²⁺	V³⁺
		V⁴⁺
		V⁵⁺

¹) (su) = sulphophil, (o) = oxyphil;   atoms or ions in brackets are questionable in the given group

221

Bibliography

AHRENS, L. H.: The Significance of the Chemical Bond for Controlling the Geochemical Distribution of the Elements, Part I, Phys. Chem. of the Earth 5 (1964), 1—54

GOLDSCHMIDT, V. M.: Geochemische Verteilungsgesetze der Elemente, Videnskapsselskapets Skrifter I, mat.-naturw. Cl. 3 (1923)

GOLDSCHMIDT, V. M.: Geochemische Verteilungsgesetze der Elemente. Die Mengenverhältnisse der Elemente und der Atomarten. Videnskapsselskapets Skrifter I, mat.-naturw. Cl. 4, 1937

KHOMYAKOV, A. P.: On the Role of Chemical and Crystallochemical Factors in the Distribution of Rare Earths. Geokhimiya (1967), 2, 197—205 (Russian)

KOSTOV, I.: A Geochemical Classification of the Minerals. Omnibus volume: Probl. Geokhim. Izd-vo "Nauka", Moscow 1965 (Russian)

SHCHERBAKOV, YU. G.: Classification Géochimique des Éléments. Dokl. Akad. Nauk 164 (1965), 4, 917 — 920. B.R.G.M. Paris, Trad. S.I.G. (1965), 4909

SZÁDECZKY-KARDOSS, E.: Über zwei neue Wertigkeitsregeln der Geochemie und die geochemische Gruppierung der Elemente. Acta geol. Acad. Sci. hung. 1 (1952), 231—268

7.3. The Mantle of the Earth and its Relation to the Earth's Crust

According to modern geophysical and geochemical investigations and considerations, it is highly probable that at most the upper 1000 km of the earth were subject to a true substantial differentiation, whereas the deeper regions owe their different physical parameters to phase transitions. From this follows that the exploration of the upper mantle and its relation to the earth's crust is of particular scientific and practical importance. A great part of this research work has been carried out since 1962 within the scope of the international upper mantle project (UMP).

The determination of the composition of the upper mantle is made possible in particular:

1. directly by thorough investigations into the material brought to the surface of the earth by geological processes (above all basic and ultrabasic igneous rocks)
2. indirectly by laboratory studies of the conditions of the state of silicate matter under pT- conditions which resemble those in the upper mantle (work conducted by RINGWOOD and other authors)
3. indirectly by comparisons with meteoritic substance (the "pyrolite" of the upper mantle, for instance, is compared with the chondrites containing C).

Table 81 gives information about the assumed composition of the upper mantle. The assumed relations between upper mantle and the earth's crust are shown in Fig. 51.

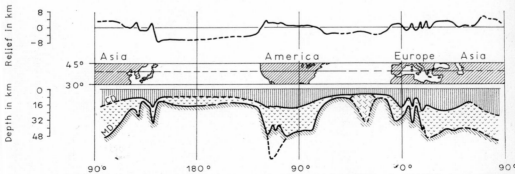

Fig. 51 Conrad discontinuity (CD) and Mohorovičić discontinuity (MD) along 40° latitude (in the upper part of the illustration as an explanation of the morphologic and geographic conditions), after Militzer and Porstendorfer 1968

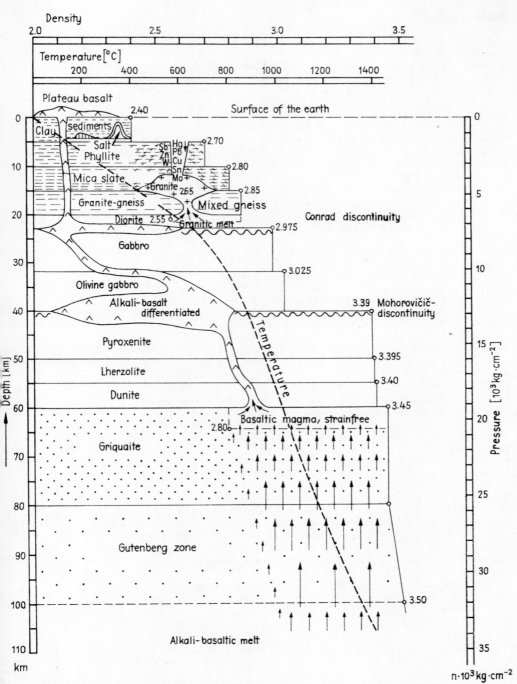

Fig. 52   Material structure of the lithosphere and origin of magmas (from BORCHERT, 1962)

Bibliography

ANDERSON, D. L.: Recent Evidence Concerning the Structure and Composition of the Earth's Mantle. Phys. Chem. Earth *6* (1965), 1—131

ANDERSON, D. L.: Phase Changes in the Upper Mantle. Science *157* (1967), 3793, 1165—1173

BELOUSOV, V. V.: The Upper Mantle and Its Influence on the Development of the Earth's Crust, "Upper Mantle Project". The Upper Mantle Symp., New Delhi 1964, 9—13, Copenhague 1965

BIRCH, F.: Density and Composition of Mantle and Core. Journ. Geophys. Res. *69* (1964)

BORCHERT, H.: Vulkanismus und Oberer Erdmantel in ihrer Beziehung zum äußeren Erdkern und zur Geotektonik. Bull. Geofis. Teorica ed Appl. *IX* (1967), 35, 194—213

BOTT, M.: The Interior of the Earth. St. Martin's Press, New York 1970

DEMENITSKAYA, R. M.: The Earth's Crust and Mantle. Izd-vo "Nedra", Moscow 1967 (Russian)

ESKOLA, P.: Wie ist die Anordnung der äußeren Erdsphären nach der Dichte zustande gekommen? Geol. Rdsch. *27* (1936), 61—73

GALAKTIONOV, V. A., and N. V. BELOV: The Mohorovičič Discontinuity and the Structure of Silicates. Geokhimiya (1967), 12, 1411—1417 (Russian)

GASKELL, T. F. (Edit.): The Earth's Mantle. Acad. Press. London-New York 1967

GORSHKOV, G. S.: On the Relations of Volcanism and the Upper Mantle. Bull. Vulcanol. Ital. *28* (1965), 159—167

HARRIS, P. G., and J. A. ROWELL: Some Geochemical Aspects of the Mohorovičič Discontinuity. J. Geophys. Res. *65* (1960), 2443—2459

HART, P. J. (ed.): The Earth's Crust and Upper Mantle. AGU, Washington 1970

JUAN, V. C.: Earth's Crust and Mantle. Proc. Geol. Soc. China (1965), 9, 3—8

KIRKINSKIĬ, V. A.: Mineralogy of the Principal Chemical Components of the Earth's Mantle. Geokhimiya (1967), 1, 65—74 (Russian)

MASUDA, A.: Genesis of the Earth's Crust and the Structure within the Mantle: A Unified Chemical Theory. Chem. Geol. *1* (1966), 135 —163

MASUDA, A., and Y. MATSUI: The Difference in Lanthanide Abundance Pattern between the Crust and the Chondrite and Its Possible Meaning to the Genesis of Crust and Mantle. Geochim. Cosmochim. Acta *30* (1966), 239—250

MATSUI, Y.: Structure of the Earth's Mantle Derived from the Abundance of the Elements. J. geol. Soc. Jap. *71* (1965), 843, 582—594

MATSUMOTO, T.: On the Classification of the Mantle Facies and on the Differences in the Primary Magmas. J. Geol. Soc. Jap. *71* (1965), 843, 619—632 (Japanese)

MERCY, E. L. P.: Geochemistry of the Mantle. Chapter 13 in: Gaskell. The Earth's Mantle, Acad. Press, London/New York 1967

MIYASHIRO, A.: Oxidation and Reduction in the Earth's Crust. Geochim. cosmochim. Acta, *28* (1964), 5

NAYDENOV, B. M., and V. V. CHERDYNTSEV: The Evolution of the Earth's Crust According to Data on the Isotopes of Mineral Lead. Dokl. Akad. Nauk SSSR *172* (1967), 3, 688—691 (Russian)

NICHOLLS, G. D.: Petrological and Geochemical Evidence for Convection in the Earth's Mantle. Philos. Trans. R. Soc. London, Ser. A *258* (1965), 1088, 168—179

PAPKE, K.-H.: Die Mohorovičič-Diskontinuität. Geologie, Berlin (1967) *16* Supplement 57, 1—127

RINGWOOD, A. E.: The Constitution of the Mantle I: Thermodynamics of the Olivinespinel Transition. Geochim. Cosmochim. Acta *13* (1958), 303—321

RINGWOOD, A.: Mineralogical Constitution of the Deep Mantle. J. Geophys. Res. *67* (1962), 4005 to 4010

RITSEMA, A. R. (ed.): The Upper Mantle. Elsevier, Amsterdam 1972

RITTMANN, A.: Zur geochemischen Entwicklung der prägeologischen Lithosphäre. Schweiz. Min.-petr. Mitt. *28* (1948), 36—48

SCHREYER, W.: Zur mineralogischen Konstitution des Erdmantels. Naturw. Rdsch. *19* (1966), 184 to 189

STILLE, H.: Zur Frage der Herkunft der Magmen. Abh. Preuss. Akad. Wiss., Math.-nat. Cl. *19* (1939)

VINOGRADOV, A. P., and A. A. YAROSHEVSKIĬ: On the Physical Parameters of Zone Melting in the Earth's Mantle. Geokhimiya (1965), 7, 779—790 (Russian)

7.4. Distribution of the Elements in the Earth's Crust

The determination of the distribution of the elements in the earth's crust is the most important task of geochemistry for theoretical and practical reasons. Using the knowledge and data gathered in this field, geochemistry takes part in the solution of almost all geological problems, especially those of petrogenesis and of the formation of ore deposits.

7.4.1. Fundamentals and Methods of Calculation

Many scientists devoted themselves to the investigation of the quantitative distribution of the chemical elements in the earth's crust, above all CLARKE (1889) and, in collaboration with him, WASHINGTON (1924), further VOGT (1898 and 1931), VERNADSKYIĬ (1925 to 1930), HEVESY (1932 to 1934), I. and W. NODDACK (1930 to 1934), GOLDSCHMIDT (1922 to 1935), FERSMAN (1933 to 1939), WICKMANN (1954), VINOGRADOV (for the last time in 1962), BROTZEN (1966) and RONOV (1967).

Fersman introduced the term "Clarke" (in appreciation of the merits of the American chemist F. W. CLARKE). The "Clarke" or "Clarke-Value" of an element means the average abundance of a chemical element in the earth's crust. The most important measures used are [mass-%], [g/t] or [ppm] and [atom-%]. Recently, so-called "regional Clarkes" have been determined of regional geological units and "local Clarkes" of smaller geological objects.

The Clarkes determined by various research workers differ more or less from each other (especially those of the trace elements) because of

a) the different methods of analysis used,
b) the different conceptions about the composition of the earth's crust,
c) the various ways of obtaining mean values.

A few authors include in the earth's crust not only the lithosphere, but also parts of the hydrosphere, the atmosphere (CLARKE and WASHINGTON, 1924) and of the biosphere (VERNADSKIĬ, 1934), whereas others (e.g. VINOGRADOV, 1962) only take the lithosphere into consideration. According to Clarke and Washington, the earth's crust is composed as follows thickness of the earth's crust: 10 miles ≈ 16 km):

Lithosphere 93.06%
Hydrosphere 6.91%
Atmosphere 0.03%

The possible composition of the lithosphere is shown in Table 85.

Table 85. Calculated Composition of the Lithosphere

Type of Rock	CLARKE and WASHINGTON	v. ENGELHARDT (1936)
Magmatites	95%	95%
Shales	4%	4%
Sandstones	0.75%	0.6%
Carbonates (limestone + dolomite)	0.25%	0.4%

The Clarkes of a few elements resulting from given combinations are shown in Table 86.

Table 86. Average Contents of the 9 Most Abundant Chemical Elements in Certain Geospheres or Rock Groups [mass-%] Calculated by CLARKE and WASHINGTON (1924)

Element	I.	II.	III.	IV.
O	49.52	48.08	46.71	46.59
Si	25.75	26.72	27.69	27.72
Al	7.51	7.79	8.07	8.13
Fe	4.70	4.87	5.05	5.01
Ca	3.39	3.52	3.65	3.63
Na	2.64	2.69	2.75	2.85
K	2.40	2.49	2.58	2.60
Mg	1.94	2.01	2.08	2.09
H	0.88	0.51	0.14	0.13

The analytical values apply to the following combinations:

I. 10-miles earth's crust: lithosphere, hydrosphere and atmosphere
II. 20-miles earth's crust: lithosphere, hydrosphere and atmosphere
III. 10-miles earth's crust: lithosphere (magmatites and sedimentites)
IV. 10-miles earth's crust: lithosphere (only magmatites)

In general, only the mean composition of eruptive rocks is taken into consideration for determining Clarkes.

There are different ways of obtaining mean values:

1. Arithmetic mean of a great number of individual analyses; e.g. CLARKE (1924) determined the mean from about 6,000 analyses of the most different types of rock.
2. Determination of the individual elements in an artificial mixture of the most abundant eruptive rocks (e.g. B.I. and W. NODDACK, 1930, 1934; HEVESY, 1930 and 1932; GOLDSCHMIDT, 1938; VOGT, 1931).
3. Calculation on the basis of mean contents of elements in individual types of rock of the lithosphere, taking into consideration their relative abundance (Table 87).

Table 87. The Average Composition of the Earth's Crust Assumed by Various Authors (cf. FLEISCHER and CHAO, 1960)

Author	Average Composition of the Earth's Crust	
POLDERVAART (1955) (up to 40 km depth)	granodiorite	40.8 mass-%
	diorite and andesite	10.3 mass-%
	basalt and tholeiite	48.9 mass-%
VINOGRADOV (1962) (applies especially to continental regions of the earth)	acid: basic rock = 2:1	
SMITH (taking the ocean basins into consideration)	acid: basic rock = 1:2	

he importance of the relative abundance of the individual types of rock in the lithosphere
as underlined especially by DALY (1914), SEDERHOLM (1925), and VOGT (1931) (Table 88).
elow there follows the relative distribution of the sedimentary rocks in the lithosphere
'able 89) to supplement Table 88.

able 88. Distribution in percent of the Most Important Igneous Rocks in the Lithosphere
rom FLEISCHER and CHAO, 1960)

	I.[1]) VOGT (1931)	II.	III. v. ENGELHARDT (1936)	IV. DALY (1914)
ranite	50	60	34	36.1
harnockite	—	—	5	—
uartzmonzonite and Granodiorite	10	9	—	36.6 (grano-diorite)
uartzdiorite and Diorite	8	6	39 (diorite)	3.2
abbro	18	15	18	22.8
northosite	4	3		
yroxenite and Peridotite	0.5	0.25	0.25	0.5
lkali-lime Syenite	3	2	3 (syenite)	0.6
onzonite	4	3	—	—
epheline Syenite	1	0.5	0.5	0.2
ordmarkite and Pulaskite	1	1	—	—
ssexite and the like	0.5	0.25	—	—
lkali-Gabbro	—	—	0.25	(0.04)

I. on the basis of the conditions given in Norway; II. and III. rough estimate; IV. the data given by
)ALY which were summarised by FLEISCHER and CHAO (1960).

able 89. Relative Distribution of Sedimentary Rocks in the Lithosphere

	CLARKE (1924)	Data from v. ENGELHARDT (1936)
rgillaceous shales, clays	80 %	79.6 %
andstones	15 %	11.9 %
imestones	} 5 %	8.1 %
olomites		0.4 %

The quantitative distribution of the chemical elements in nature and hence in the earth's crust
s rather uneven, though it follows certain laws.
This pronounced uneven distribution is clearly mirrored by the so-called *decadic system*, a
epresentation introduced by FERSMAN and shown in Table 91 in condensed form. The regu-
arities of this uneven distribution are caused by the internal and external migration factors
and are due to the atomic structure of the chemical elements and their positions in the
eriodic system. The relations which can be derived from *Niggli-Sonder's rule* are a conse-
uence of this dependence (Table 90).

Table 90. Periodicity of the Atomic Abundance (Niggli-Sonder's Rule; according to Saukov)

Most abundant elements of a series	O	Si	Ca	Fe	Rb	Zr	Sr	Y	Sn	Ba	W	Pb	Th
Atomic numbers	8	14	20	26	36	40	38	39	50	56	74	82	90
Difference of atomic numbers		6	6	6	6	12	12	12	6	6	18	8	8

(18)

Table 91. "Decadic System" of Distribution of the Chemical Elements (according to Vernadskii)
(the Clarke values used are rounded off)

100—10%	10—1%	1—0.1%	1,000—100 ppm	100—10 ppm	10—1 ppm	1—0.1 ppm	0.1—0.01 ppm	0.01—0.001 ppm	<0.001 ppm
O 47.0	Al 8.1	Ti 0.5	P 930	V 90	Gd 9	Lu 0.8	Hg 0.08	Pt 0.005	Re
Si 29.1	Fe 4.7	Mn 0.1	H 700	Cr 83	Sm 9	Sb 0.5	Ag 0.07	Au 0.004	Xe
	Ca 3.0		F 660	Zn 83	U 4	I 0.4	Se 0.05	Te 0.001	Ra
	K 2.5		Ba 650	Ce 70	Ta 3	In 0.3	Ar 0.04	He 0.001	Rn
	Na 2.5		S 470	Ni 58	Hf 3	Tm 0.3	Pd 0.01	Re 0.001	
	Mg 1.9		Sr 340	Cu 47	Sn 3	Cd 0.1	Bi 0.01	Ir 0.001	
			C 230	Li 32	Br 3			Rh 0.001	
			Cl 170	Y 29	Cs 3			Os 0.001	
			Zr 170	Nb 20	Be 2			Ru 0.001	
			Rb 150	Ga 19	As 2				
				N 19	Eu 1				
				Co 18	Ge 1				
				Pb 16	Tl 1				
				Th 13	Mo 1				
				B 12	W 1				
				and others	and others	and others			
Σ 76.1%	22.7%	0.6%	4,670 ppm	685 ppm	~70 ppm	~3 ppm	~0.3 ppm	~0.02 ppm	<0.001 ppm

ibliography

ROTZEN, O.: The Average Igneous Rock and the Geochemical Balance. Geochim. Cosmochim. Acta, 7 (1966), 873—868

ROTZEN, O.: Geochemical Ranking of Rocks. Sver. Geol. Und. Serv. C. 617, Stockholm 1966 (With bles containing 600 rock compositions)

ORRENS, C. W.: Die geochemische Bilanz. Naturwissenschaften 35 (1948), 7—12

AY, F. H.: The Chemical Elements in Nature. 372 pp. Harrap & Co., London 1948

OLDSCHMIDT, V. M.: The Principles of Distribution of Chemical Elements in Minerals and Rocks. J. hem. Soc., London (1937), March. 655—673

EIER, K. S., and I. A. S. ADAMS: Concentration of Radioactive Elements in Deep Crustal Material. eochim. Cosmochim. Acta 29 (1965), 53—61

EIER, K. S.: Metamorphism and the Chemical Differentiation of the Crust. Geol. Fören Stockholm örhdlg. 521 (1965), 249—256

ORN, M. K., and J. A. ADAMS: Computer-derived Geochemical Balances and Element Abundances. eochim. Cosmochim. Acta 30 (1966), 279—298

AMBERT, I. B., and K. S. HEIER: The Vertical Distribution of Uranium, Thorium and Potassium in ne Continental Crust. Geochim. Cosmochim. Acta 31 (1967), 377—390

ARKER, R. L.: Composition of the Earth's Crust. — Geol. Surv. Prof. Pap. 440-D. Washington 1967, ol. 1041

ONOV, A. V., and A. A. YAROSHEVSKIĬ: The Chemical Structure of the Earth's Crust. Geokhimiya 1967), 11, 1285—1309 (Russian)

HCHERBAKOV, YU. G.: Periodicity of Background Ratios and the Geochemical Evolution of the Crust Dokl. Akad. Nauk 161 (1965), 2, 451—454 (Russian)

ERNADSKIĬ, V. I.: The Parageneses of the Chemical Elements in the Earth's Crust. Selected Works, ol. 1, Izd. AN SSSR, Moscow 1954 (Russian)

VICKMAN, F. E.: The "Total" Amount of Sediments and the Composition of the "Average Igneous Rocks". Geochim. Cosmochim. Acta 5 (1954), 97—110

.4.2. Average Abundance of Chemical Elements in the Lithosphere

The abundance of chemical elements in the whole lithosphere (Clarke values) is given in Table 92 and in Figs. 53 to 55. Tables 93 and 94 show the data used for this survey. However, he information contained in the Clarke value is insufficient for the solution of many petro-

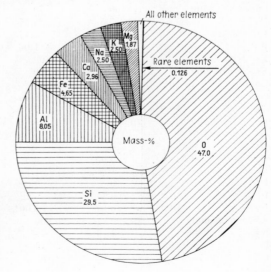

Fig. 53 Distribution of the elements in the earth's crust (after VINOGRADOV, 1962; VLASOV et al., 1964). According to TAYLOR, the concentrations of the following elements are [%]: O = 46.4; Si = 28.15; Al = 8.23; Fe = 5.63; Ca = 4.15; Na = 2.09; Mg = 2.33

logical and exploration problems; in these cases, the knowledge of the average contents of chemical elements in parts of the earth's crust, i.e. in the main types of rock, is of paramount importance. The latest data relative to the magmatic and sedimentary range are given in Table 95. Table 96 shows the geochemically important pairs of elements in meteorites and in the main types of rock.

Table 92. The Clarke Values [ppm] according to Data from Several Authors
(The dashes mean that data are not available)

Symbol	CLARKE and WASHINGTON (1924)	FERSMAN (1933 to 1939)	GOLDSCHMIDT (1937)	VINOGRADOV (1949)	(1962)	TAYLOR (1964)
Ac	—	—	—	$x \cdot 10^{-10}$	—	—
Ag	0.0X	0.1	0.02	0.1	0.07	0.07
Al	75,100	74,500	81,300	88,000	80,500	82,300
Ar	—	4	—	—	—	—
As	X	5	5	5	1.7	1.8
Au	0.00X	0.005	0.001	0.005	0.004,3	0.004
B	10	50	10	3	12	10
Ba	470	500	430	500	650	425
Be	10	4	6	6	3.8	2.8
Bi	0.0X	0.1	0.2	0.2	0.009	0.17
Br	X	10	2.5	1.6	2.1	2.5
C	870	3,500	320	1,000	230	200
Ca	33,900	32,500	36,300	36,000	29,600	41,500
Cd	0.X	5	0.18	5	0.13	0.2
Ce	—	29	41.6	45	70	60
Cl	1,900	2,000	480	450	170	130
Co	100	20	40	30	18	25
Cr	330	300	200	200	83	100
Cs	0.00X	10	3.2	7	3.7	3
Cu	100	100	70	100	47	55
Dy	—	7.5	4.47	4.5	5	3.0
Er	—	6.5	2.47	4	3.3	2.8
Eu	—	0.2	1.06	1.2	1.3	1.2
F	270	800	800	270	660	625
Fe	47,000	42,000	50,000	51,000	46,500	56,300
Ga	$x \cdot 10^{-5}$	1	15	15	19	15
Gd	—	7.5	6.36	10	8	5.4
Ge	$x \cdot 10^{-5}$	4	7	7	1.4	1.5
H	8,800	10,000	—	1,500	—	—
He	—	0.01	—	—	—	—
Hf	30	4	4.5	3.2	1	3
Hg	0.X	0.05	0.5	0.07	0.083	0.08
Ho	—	1	1.15	1.3	1.7	1.2
I	0.X	10	0.3	0.5	0.4	0.5
In	$x \cdot 10^{-5}$	0.1	0.1	0.1	0.25	0.1
Ir	$x \cdot 10^{-4}$	0.01	0.001	0.001	—	—
K	24,000	23,500	25,900	26,000	25,000	20,900
Kr	—	$2 \cdot 10^{-4}$	—	—	—	—
La	—	6.5	18.3	18	29	30
Li	40	50	65	65	32	20

230

Table 92 continued

Symbol	Clarke and Washington (1924)	Fersman (1933 to (1939)	Goldschmidt (1937)	Vinogradov (1949)	Vinogradov (1962)	Taylor (1964)
Lu	—	1.7	0.75	1	0.8	0.50
Mg	19,400	23,500	20,900	21,000	18,700	23,300
Mn	800	1,000	1,000	900	1,000	950
Mo	X	10	2.3	3	1.1	1.5
N	300	400	—	100	19	20
Na	26,400	24,000	28,300	26,400	25,000	23,600
Nb	—	0.32	20	10	20	20
Nd	—	17	23.9	25	37	28
Ne	—	0.005	—	—	—	—
Ni	180	200	100	80	58	75
O	495,200	491,300	466,000	470,000	470,000	464,000
Os	$x \cdot 10^{-4}$	0.05	—	0.05	—	—
P	1,200	1,200	1,200	800	930	1,050
Pa	—	$7 \cdot 10^{-7}$	—	10^{-6}	—	—
Pb	20	16	16	16	16	12.5
Pd	$x \cdot 10^{-5}$	0.05	0.010	0.01	0.013	—
Po	—	0.05	—	$2 \cdot 10^{-10}$	—	—
Pr	—	4.5	5.53	7	9	8.2
Pt	0.00X	0.2	0.005	0.005	—	—
Ra	$x \cdot 10^{-6}$	$2 \cdot 10^{-6}$	—	10^{-6}	—	—
Rb	X	80	280	300	150	90
Re	—	0.001	0.001	0.001	$7 \cdot 10^{-4}$	—
Rh	$x \cdot 10^{-5}$	0.01	0.001	0.001	—	—
Rn	—	?	—	$7 \cdot 10^{-12}$	—	—
Ru	$x \cdot 10^{-5}$	0.05	—	0.005	—	—
S	480	1,000	520	500	470	260
Sb	0.X	0.5	(1)	0.4	0.5	0.2
Sc	0.X	6	5	6	10	22
Se	0.0X	0.8	0.09	0.6	0.05	0.05
Si	257,500	260,000	277,200	276,000	295,000	281,500
Sm	—	7	6.47	7	8	6.0
Sn	X	80	40	40	2.5	2
Sr	170	350	150	400	340	375
Ta	—	0.24	2.1	2	2.5	2
Tb	—	1	0.91	1.5	4.3	0.9
Tc	—	0.001	—	—	—	—
Te	0.00X	0.01	(0.001,8)?	0.01	0.001	—
Th	20	10	11.5	8	13	9.6
Ti	5,800	6,100	4,400	6,000	4,500	5,700
Tl	$x \cdot 10^{-4}$	0.1	0.3	3	1	0.45
Tm	—	1	0.20	0.8	0.27	0.48
U	80	4	4	3	2.5	2.7
V	160	200	150	150	90	135
W	50	70	1	1	1.3	1.5
Xe	—	$3 \cdot 10^{-5}$	—	—	—	—
Y	—	50	28.1	28	29	33
Yb	—	8	2.66	3	0.33	3.0
Zn	40	200	80	50	83	70
Zr	230	250	220	200	170	165

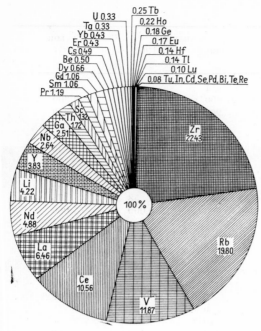

U 0.33
Ta 0.33
Yb 0.43
Er 0.43
Cs 0.49
Be 0.50
Dy 0.66
Gd 1.06
Sm 1.06
Pr 1.19
Sc
Th 1.32
Ga 1.72
2.51
Nb 2.64
Y 3.83
Li 4.22
Nd 4.88
La 6.46
Ce 10.56
V 11.87
Rb 19.80
Zr 22.43
100%

0.25 Tb
0.22 Ho
0.18 Ge
0.17 Eu
0.14 Hf
0.14 Tl
0.10 Lu
0.08 Tu, In, Cd, Se, Pd, Bi, Te, Re

Fig. 54 The relative contents of rare elements in the earth's crust (sum of the rare elements = 100%; absolute content = 0.125 mass-%, after VLASOV et al., 1964

Table 93. Composition of Igneous Rocks to which the Mean Values Given by VINOGRADOV and Shown in Table 95 Refer (from VINOGRADOV, 1956); data in [mass-%]

| | ultrabasic rocks | | basic rocks | | | intermediate rocks | | | acid rocks | | |
	NOCKOLDS (1954)	VINOGRADOV (1955)	NOCKOLDS (1954)	DALY (1933)	VINOGRADOV (1955)	NOCKOLDS (1954)	DALY (1933)	VINOGRADOV (1955)	NOCKOLDS (1954)	DALY (1933)	VINOGRADOV (1955)
SiO_2	43.8	43.6	48.4	49.0	50.0	54.6	58.5	57.0	68.9	70.18	71.00
TiO_2	1.7	0.72	1.8	1.29	1.29	1.5	0.76	0.79	0.5	0.39	0.34
Al_2O_3	6.1	4.72	15.5	16.2	16.48	16.4	16.80	17.5	14.50	14.47	14.30
Fe_2O_3	4.5	4.62	2.8	5.0	4.22	3.3	3.08	3.72	1.70	1.57	1.54
FeO	8.7	8.01	8.1	6.3	6.80	5.2	3.67	3.31	2.20	1.78	1.85
MnO	0.18	0.14	0.17	0.28	0.23	0.15	0.14	0.17	0.07	0.12	0.05
MgO	22.5	24.8	8.6	6.62	6.30	3.8	3.22	3.64	1.10	0.88	0.74
CaO	10.0	12.2	10.7	9.10	9.75	6.5	6.01	6.70	2.60	1.99	1.82
Na_2O	0.8	0.73	2.3	3.02	2.78	4.2	3.56	3.62	3.90	3.48	3.62
K_2O	0.7	0.38	0.7	1.41	1.24	3.2	2.06	2.01	3.80	4.11	4.02
H_2O	0.6	—	0.7	1.58	1.17	0.7	1.26	0.83	0.60	0.84	0.75
P_2O_5	0.3	0.21	0.27	0.44	0.36	0.42	0.26	0.25	0.16	0.19	0.145

ble 94. Composition of Shales and Clays to which the Mean
lues Given by Vinogradov and Shown in Table 95 Refer
m Vinogradov, 1956); data in [mass %]

	Shales (Clarke, 1924)	Clays of all Ages of the Russian Table (Vinogradov and Ronov, 1955)	Clays of all Ages of the Russian Table from Statements in Publications (Vinogradov and Ronov, 1955)
O₂	58.11	50.65	53.62
O₂	0.65	0.77	1.20
₂O₃	15.40	15.10	20.0
₂O₃	6.70	6.47	4.58
₁O	—	0.10	0.10
₉O	2.44	3.32	2.10
O	3.10	7.19	3.5
₁₂O	1.30	0.81	0.67
₂O	3.24	3.50	2.10
O₅	0.17	—	0.46
O₂	2.63	6.10	4.48

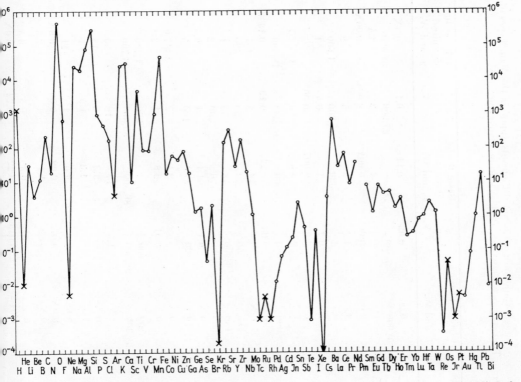

Fig. 55   Abundance of the elements in the lithosphere [ppm]; values after Vinogradov (1962) and other authors (x), cf. Table 92

Table 95. Average Abundance of Chemical Elements [ppm] in the Main Types of Igneous and Sedimentary Rocks of the Lithosphere (T. and W.: from TUREKIAN and WEDEPOHL, 1961; V.: from VINOGRADOV, 1962)

Magmatites

Symbol	Ultrabasic Rocks T. and W.	V.	Basic Rocks (Basalts) T. and W.	V.	Intermediate Rocks (Syenites) T. and W.	(Diorites) V.	Acid Rocks (High Ca) T. and W.	(Low Ca) T. and W.	(Granites etc.) V.
Ag	0.06	0.05	0.11	0.1	0.0X	0.07	0.051	0.037	0.05
Al	20,000	4,500	78,000	87,600	88,000	88,500	82,000	72,000	77,000
As	1	0.5	2	2	1.4	2.4	1.9	1.5	1.5
Au	0.006	0.005	0.004	0.004	0.00X	—	0.004	0.004	0.004,5
B	3	1	5	5	9	15	9	10	15
Ba	0.4	1	330	300	1,600	650	420	840	830
Be	0.X	0.2	1	0.4	1	1.8	2	3	5.5
Bi	?	0.001	0.007	0.007	?	0.01	?	0.01	0.01
Br	1	0.5	3.6	3	2.7	4.5	4.5	1.3	1.7
Ca	25,000	7,000	76,000	67,200	18,000	46,500	25,300	5,100	15,800
Cd	0.X	0.05	0.22	0.19	0.13	—	0.13	0.13	0.1
Ce	0.X	—	48	4.5	161	—	81	92	100
Cl	85	50	60	50	520	100	130	200	240
Co	150	200	48	45	1	10	7	1.0	5
Cr	1,600	2,000	170	200	2	50	22	4.1	25
Cs	0.X	0.1	1.1	1	0.6	50	2	4	5
Cu	10	20	87	100	5	35	30	10	20
Er	0.X	—	2.1	2	7.0	—	3.5	4.0	4
Eu	0.X	0.01	0.8	1	2.8	—	1.4	1.6	1.5
F	100	100	400	370	1,200	500	520	850	800
Fe	94,300	98,500	86,500	85,600	36,700	58,500	29,600	14,200	27,000
Ga	1.5	1.5	17	18	30	20	17	17	20
Gd	0.X	—	5.3	5	18	—	8.8	10	9
Ge	1.5	1	1.3	1.5	1	1.5	1.3	1.3	1.4
Hf	0.6	0.1	2.0	1	11	1	2.3	3.9	1
Hg	0.0X	0.01	0.09	0.09	0.0X	—	0.08	0.08	0.08
Ho	0.X	—	1.1	1	3.5	—	1.8	2.0	2
I	0.5	0.01	0.5	0.5	0.5	0.3	0.5	0.5	0.4
In	0.01	0.013	0.22	0.22	0.0X	—	0.0X	0.26	0.26

Table 95 continued

Magmatites

| Symbol | Ultrabasic Rocks | | Basic Rocks (Basalts) | | Intermediate Rocks | | | Acid Rocks | | (Granites etc.) |
	T. and W.	V.	T. and W.	V.	(Syenites) T. and W.	(Diorites) V.		(High Ca) T. and W.	(Low Ca) T. and W.	V.
K	40	300	8,300	8,300	48,000	23,000		25,200	42,000	33,400
La	0.X	—	15	27	70	—		45	55	60
Li	0.X	0.5	17	15	28	20		24	40	40
Lu	0.X	—	0.6	0.6	2.1	—		1.1	1.2	1
Mg	204,000	259,000	46,000	45,000	5,800	21,800		9,400	1,600	5,600
Mn	1,620	1,500	1,500	2,000	850	1,200		540	390	600
Mo	0.3	0.2	1.5	1.4	0.6	0.9		1.0	1.3	1
N	6	6	20	18	30	22		20	20	20
Na	4,200	5,700	18,000	19,400	40,400	30,000		28,400	25,800	27,700
Nb	16	1.0	19	20	35	20.0		20	21	20
Nd	0.X	—	20	20	65	—		33	37	46
Ni	2,000	2,000	130	160	4	55		15	4.5	8
P	220	170	1,100	1,400	800	1,600		920	600	700
Pb	1	0.1	6	8	12	15		15	19	20
Pd	0.12	0.12	0.02	0.019	?	—		0.00X	0.00X	0.01
Pr	0.X	—	4.6	4	15	—		7.7	8.8	12
Rb	0.2	2	30	45	110	100		110	170	200
S	300	100	300	300	300	200		300	300	400
Sb	0.1	0.1	0.2	1.0	0.X	0.2		0.2	0.2	0.26
Sc	15	5	30	24	3	2.5		14	7	3
Se	0.05	0.05	0.05	0.05	0.05	0.05		0.05	0.05	0.05
Si	205,000	190,000	230,000	240,000	291,000	260,000		314,000	347,000	323,000
Sm	0.X	—	5.3	5	18	—		8.8	10	9
Sn	0.5	0.5	1.5	1.5	X	—		1.5	3	3
Sr	1	10	465	440	200	800		440	100	300
Ta	1.0	0.018	1.1	0.48	2.1	0.7		3.6	4.2	3.5
Tb	0.X	—	8	0.8	2.8	—		1.4	1.6	2.5
Th	0.004	0.005	4	3	13	7		8.5	17	18
Ti	300	300	13,800	9,000	3,500	8,000		3,400	1,200	2,300
Tl	0.06	0.01	0.21	0.2	1.4	0.5		0.72	2.3	1.5

235

Table 95 continued

Magmatites

Symbol	Ultrabasic Rocks		Basic Rocks (Basalts)		Intermediate Rocks (Syenites)	(Diorites)	Acid Rocks (High Ca)	(Low Ca)	(Granites etc.)
	T. and W.	V.	T. and W.	V.	T. and W.	V.	T. and W.	T. and W.	V.
Tm	0.X	—	0.2	0.2	0.6	—	0.3	0.3	0.3
U	0.001	0.003	1	0.5	3.0	1.8	3.0	3.0	3.5
V	40	40	250	200	30	100	88	44	40
W	0.77	0.1	0.7	1	1.3	1	1.3	2.2	1.5
Y	0.X	—	21	20	20	—	35	40	34
Yb	0.X	—	2.1	2	7.0	—	3.5	4.0	4
Zn	50	30	105	130	130	72	60	39	60
Zr	45	30	140	100	500	260	140	175	200

Table 95 continued

Sedimentites

Symbol	Slates	Slates + Clays	Sandstone	Carbonates	Carbonates	Deep-sea Sediments	
	T. and W.	V.	T. and W.	T. and W.	V.	Carbonates T. and W.	Clays T. and W.
Ag	0.07	0.1	0.0X	0.0X	0.0X	0.0X	0.11
Al	80,000	104,500	25,000	4,200		20,000	84,000
As	13	6.6	1	1		1	13
Au	0.00X	0.001	0.00X	0.00X		0.00X	0.00X
B	100	100	35	20		55	230
Ba	580	800	X0	10		190	2,300
Be	3	3	0.X	0.X		0.X	2.6
Bi	?	0.01	?	?		?	?
Br	4	6	1	6.2		70	70

Table 33 continued

Sedimentites

Deep-sea Sediments

Symbol	Slates T. and W.	Slates + Clays V.	Sandstone T. and W.	Carbonates T. and W.	Carbonates T. and W.	Clays T. and W.
Ca	22,100	25,300	39,100	302,300	312,400	29,000
Cd	0.3	0.3	0.0X	0.035	0.0X	0.42
Ce	59	50	92	11.5	35	345
Cl	180	160	10	150	21,000	21,000
Co	19	20	0.3	0.1	0.7	74
Cr	90	100	35	11	11	90
Cs	5	12	0.X	0.X	0.4X	6
Cu	45	57	X	4	30	250
Er	2.5	2.5	4.0	0.5	1.5	15
Eu	1.0	1	1.6	0.2	1.6	6
F	740	500	270	330	540	1,300
Fe	47,200	33,300	9,800	3,800	9,000	65,000
Ga	19	30	12	4	13	20
Gd	6.4	6.5	10	1.3	3.8	38
Ge	1.6	2	0.8	0.2	0.2	2
Hf	2.8	6	3.9	0.3	0.41	4.1
Hg	0.4	0.4	0.03	0.04	0.0X	0.X
Ho	1.2	1	2.0	0.3	0.8	7.5
I	2.2	1	1.7	1.2	0.05	0.05
In	0.1	0.05	0.0X	0.0X	0.0X	0.08
K	26,000	22,800	10,700	2,700	2,900	25,000
La	92	40	30	X	10	115
Li	66	60	15	5	5	57
Lu	0.7	0.7	1.2	0.2	0.5	4.5
Mg	15,000	13,400	7,000	47,000	4,000	21,000
Mn	850	670	X0	1,100	1,000	6,700
Mo	2.6	2	0.2	0.4	3	27

Table 95 continued

Sedimentites

Deep-sea Sediments

Symbol	Slates T. and W.	Slates + Clays V.	Sandstone T. and W.	Carbonates T. and W.	Carbonates T. and W.	Clays T. and W.
N	?	600	?	?	?	?
Na	9,600	6,600	3,300	400	20,000	40,000
Nb	11	20	0.0X	0.3	4.6	14
Nd	24	23	37	4.7	14	140
Ni	68	95	2	20	30	225
P	700	770	170	400	350	1,500
Pb	20	20	7	9	9	80
Pd	?	—	?	?	?	?
Pr	5.6	5	8.8	1.1	3.3	33
Rb	140	200	60	3	10	110
S	2,400	3,000	240	1,200	1,300	1,300
Sb	1.5	2	0.0X	0.2	0.15	1.0
Sc	13	10	1	1	2	19
Se	0.6	0.6	0.05	0.08	0.17	0.17
Si	73,000	238,000	368,000	24,000	32,000	250,000
Sm	6.4	6.5	10	1.3	3.8	38
Sn	6	10	0.X	0.X	0.X	1.5
Sr	300	450	20	610	2,000	180
Ta	0.8	3.5	0.0X	0.0X	0.0X	0.0X
Tb	1.0	0.9	1.6	0.2	0.6	6
Th	12	11	1.7	1.7	X	7
Ti	4,600	4,500	1,500	400	770	4,600
Tl	1.4	1	0.82	0.0X	0.16	0.8
Tm	0.2	0.25	0.3	0.04	0.1	1.2
U	3.7	3.2	0.45	2.2	0.X	1.3
V	130	130	20	20	20	120
W	1.8	2	1.6	0.6	0.X	X
Y	26	30	40	30	42	90
Yb	2.6	3	4.0	0.5	1.5	15
Zn	95	80	15	20	35	165
Zr	160	200	220	19	20	150

Table 96. Ratios of Geochemically Important Pairs of Elements (from VINOGRADOV, 1962)

Ratios of Elements	Stony Meteorites (Chondrites)	Ultrabasic Rocks (Dunites)	Basic Rocks (Basalts, Gabbro)	Intermediate Rocks (Andesites, Diorites)	Acid Rocks (Granites, Granodiorites)	Sedimentary Rocks (Clays, Slates)	Mean Composition of the Earth's Crust (2 parts granite + 1 part basalt)
Al/Ga	4,333	2,250	4,866	4,425	3,850	3,483	4,235
K/Rb	170	150	184	230	167	114	163
Ca/Sr	1,400	700	153	58	53	56	87
Ni/Co	17	70	3.6	5.5	1.6	4.7	3.2
Nb/Ta	15	55	41	26	6	6	8
Mo/W	4	2	1.4	0.9	0.7	1.0	0.8
Zn/Cd	500	600	700	—	600	300	638
Rb/Tl	5,000	200	225	500	133	200	150
Th/U	2.7	1.7	6.0	5.0	5.1	3.4	5.2

Bibliography

AHRENS, L. H.: A Note on the Relationship between the Precision of Classical Methods of Rock Analysis and the Concentration of Each Constituents. Mineralog. Mag. *30* (1954), 30

AHRENS, L. H.: A Survey of the Quality of Some of the Principal Abundance Data of Geochemistry Phys. Chem. of the Earth *2* (1957), 30—45

BERG, G.: Das Vorkommen der chemischen Elemente auf der Erde. J. A. Barth, Leipzig 1932

CAMERON, A. G. W.: Abundance of the Elements. Geol. Soc. Amer. Mem. *97* (1966), 7—10

CHOUBERT, B.: Sur le Comportement Géochimique des Elements de la Lithosphère en Fonction de Temps. Geol. Rdsch. *55* (1965), 2, 239—261

CLARKE, F. W.: The Relative Abundance of the Chemical Elements. Bull. Phil. Soc. Washington *11* (1889), 131

CLARKE, F. W., and H. S. WASHINGTON: The Composition of the Earth's Crust. U.S. Dep. Interior, Geol. Surv., Profess. Paper 127 (1924)

CLARKE, F. W.: The Data of Geochemistry, 5th ed. U.S. Dep. Interior, Geol. Surv. 770 (1924), 518

CORYELL, C. D., I. W. CHASE, and I. W. WINCHESTER: A Procedure for Geochemical Interpretation of Terrestrial Rare Earth Abundance Patterns. J. Geophys. Res. *68* (1962), 559—566

DALY, R. A.: Igneous Rocks and Their Origin. McGraw Hill. New York 1914

FERSMAN, A. E.: Geochemistry, Vol. I—IV. Nature and Technology. ONTI 1933, 1934, 1937 and 1939 (Russian)

FLEISCHER, M.: Recent Estimates of the Abundance of the Elements in the Earth's Crust. U.S. Dep. Interior, Geol. Surv., Circ. 285 (1953)

FLEISCHER, M.: Estimates of the Abundance of Some Chemical Elements and Their Reliability. Geol. Soc. Amer. Spec. Paper 62 (1955)

FLEISCHER, M., and E. C. T. CHAO: Some Problems in the Estimation of Abundance of Elements in the Earth's Crust. Rep. XXI, Sess. Nord. Copenhague 1960, 1

GIBSON, D. T.: The Terrestrial Distribution of the Elements, Quart. Rev. Chem. Soc., London *3* (1949), 263—291

GOLDSCHMIDT, V. M.: Geochemische Verteilungsgesetze der Elemente, IX. Die Mengenverhältnisse der Elemente und der Atomarten. Skrifter Norske Videnskaps-Akad. Oslo, I. Mat.-naturw. C 1. No. 4, 1937 (1938)

GREEN, J.: Abundance and Distribution of Elements. Douglas Adv. Res. Labor. Res. Comm. *91* (1969)

HARKINS, W. D.: The Evolution of the Elements and the Stability of Complex Atoms. I. A New Periodic System which Shows a Relation between the Abundance of the Elements and the Nuclei of Atoms. J. Amer. Chem. Soc. *39* (1917), 856

HEVESY, G. VON: Chemical Analysis by Means of X-rays and Its Application. New York/London 1932

HEVESY, G. VON, et al.: Über die Häufigkeit verschiedener Elemente. Z. anorg. allg. Chem. *194* (1930), 316 (Vanadium group); *212* (1933), 212 (Mo and W); *219* (1934), 192 (Cr and Mn); *216* (1934), 305 (Zr); *216* (1934), *312* (Sr)

LAMBERT, I. B., and K. S. HEIER: Estimates of the Crustal Abundances of Thorium, Uranium and Potassium. Chem. Geol. *3* (1968), 233—248

LYUSTICH, E. N., and A. J. SALTYKOVSKIĬ: On the Problem of the Origin of the Granite Stratum of the Earth's Crust. Geokhimiya *4* (1961), 293—297 (Russian)

NODDACK, I., and W. NODDACK: Die Häufigkeit der chemischen Elemente. Naturwissenschaften *18* (1930), 757

NODDACK, I., and W. NODDACK: Die geochemischen Verteilungskoeffizienten der Elemente. Svensk kem. Tidsskr. *46* (1934), 173

POLDERVAART, A.: Chemistry of the Earth's Crust. Bull. Geol. Soc. America, Spec. Paper *62* (1955)

SEDERHOLM, J. J.: The Average Composition of the Earth's Crust in Finland. Fennia *45* (1925), 18

TOLSTOĬ, M. I., and J. M. OSTAFICHUK: Regularities in the Statistical Distribution of Elements in Rocks and Their Use for Geochemical Purposes. Geokhimiya *10* (1963), 952—956 (Russian)

TONGEREN, W. VAN: Contribution to the Knowledge of the Chemical Composition of the Earth's Crust in the East Indian Archipelago. D. B. Center, Amsterdam 1958

TUREKIAN, K. K., and K. H. WEDEPOHL: Distribution of the Elements in Some Major Units of the Earth's Crust. Bull. Geol. Soc. America *72* (1961), 175

REY, H. C.: On the Concentration of Certain Elements at the Earth's Surface. Proc. Roy. Soc., ondon *219* A (1953), 281—292

INOGRADOV, A. P.: The Laws of the Distribution of Chemical Elements in the Earth's Crust. Geohimiya *1* (1956), 6 (Russian)

INOGRADOV, A. P.: On the Origin of the Substance of the Earth's Crust. Geokhimiya *1* (1961), 3—29 Russian)

INOGRADOV, A. P.: The Average Contents of the Chemical Elements in the Main Types of Eruptive Rocks. Geokhimiya *7* (1962) (Russian)

ISTELIUS, A. B.: The Skew Frequency Distributions and the Fundamental Law of the Geochemical rocesses. J. Geology *68* (1960), 1—22

OGT, J. H. L.: Über die relative Verbreitung der Elemente, besonders der Schwermetalle, und über ie Concentration des ursprünglich fein verteilten Metallgehaltes zu Erzlagerstätten. Z. prakt. Geol. 1898), 225, 314, 377, 413; (1899), 10

OGT, J. H. L.: On the Average Composition of the Earth's Crust, with Particular Reference to the ontents of Phosphoric and Titanic Acid. Skrifter Norske Videnskaps-Akad. Oslo, I. Mat.-naturw. Cl. Jo. 7 (1931)

VASHINGTON, H. S.: The Chemical Composition of the Earth's Crust. Amer. J. Sci. *9* (1925), 53

VASHINGTON, H. S.: The Chemical Analysis of Rocks. John Wiley & Sons, New York 1930

.4.3. Average Abundance of Chemical Elements in the Hydrosphere

ince the average abundance of the chemical elements in the hydrosphere is given practically y the chemical composition of the seawater, Table 97 shows the concentration of the elements measured in seawater.

Table 97. Composition of Seawater (*pH* 8.2 ± 0.2)

Element	Concentration [ppm]	Element	Concentration [ppm]
Cl	18,980	N (as nitrite)	0.000,1—0.0
Na	10,561	N (as ammonia)	>0.005—0.05
Mg	1,272	As (as arsenite)	0.003—0.024
	884	Fe	0.002—0.02
Ca	400	P (as organic phosphorus)	0—0.016
K	380	Zn	0.005—0.014
Br	65	Cu	0.001—0.09
C (inorganic)	28	Mn	0.001—0.01
Sr	13	Pb	0.004—0.005
(SiO$_2$)	0.01—7.0	Se	0.004
B	4.6	Sn	0.003
Si	0.02—4.0	Cs	≈ 0.002
C (organic)	1.2—3.0	U	0.000,15—0.001,6
Al	0.16—1.9	Mo	0.000,3—0.002
F	1.4	Ga	0.000,5
N (as nitrate)	0.001—0.7	Ni	0.000,1—0.000,5
N (as organic nitrogen)	0.03—0.2	Th	0.000,5
Rb	0.2	Ce	0.000,4
Li	0.1	V	0.000,3
P (as phosphate)	>0.001—0.10	La	0.000,3
Ba	0.05	Y	0.000,3
I	0.05	Hg	0.000,3

Table 97 continued

Element	Concentration [ppm]	Element	Concentration [ppm]
Ag	0.000,15—0.000,3	W	$0.01 \cdot 10^{-6}$
Bi	0.000,2	Cd	
Co	0.000,1	Cr	
Sc	0.000,04	Tl	present in
Au	0.000,004—0.000,008	Sb	marine
Fe (in true solution)	10^{-9}	Zr	organisms
Ra	$2 \cdot 10^{-11}$—$3 \cdot 10^{-10}$	Pt	
Ge	present		
Ti	present		

Bibliography

BROECKER, W. S., et al: Geochemistry and Physics of Ocean Circulation. "Oceanography" Amer. Assoc. Adv. Sci., Washington 1961, 301—322

DIETRICH, G., and K. KALLE: Allgemeine Meereskunde. Borntraeger, Berlin 1957

FAUST, S. D., and J. V. HUNTER: Principles and Applications of Water Chemistry. Wiley, New York, London 1967

HARVEY, H. W.: Recent Advances in the Chemistry and Biology of Sea Water. Cambridge University Press, Cambridge 1955

KULP, J. L.: Origin of the Hydrosphere. Bull. Geol. Soc. America 62 (1951), 326—329

RONOV, A. B.: On the Post-precambric Geochemical Development of the Atmosphere and Hydrosphere. Geokhimiya (1959) (Russian)

VERNADSKIĬ, V. I.: Sur la Classification et sur la Composition Chimique des Eaux Naturelles. Bull. Soc. Franc. Minéralog. 53 (1930), 417

WATTENBERG, H.: Zur Chemie des Meerwassers: Über die in Spuren vorkommenden Elemente. Z. anorg. allg. Chem. 236 (1938), 319

WEDEPOHL, K. H.: Einige Überlegungen zur Geschichte des Meerwassers. Fortschr. Geol. Rheinl. u. Westf. 10 (1963), 129—150

WHITE, D. E., J. D., HEM, and G. A. WARING: Chemical Composition of Subsurface Waters (Chap. of the Data of Geochemistry, 6th ed.) – Geol. Surv. Prof. Paper, 440-F, 67 pp. Washington (1963)

7.4.4. Average Abundance of Constituents in the Atmosphere

The average contents of constituents in the atmosphere are given in Table 98. For information about the composition of the early atmosphere reference is made to RUBEY, 1955. The changes in the atmospheric composition depending on the altitude are dealt with in Section 8.6.1. ISRAËL (1967) quotes the following additional contents of trace gases (in vol. %):
$CO = 8 \cdot 10^{-7}$ to $1.6 \cdot 10^{-5}$; CH_2O = maximum $1.2 \cdot 10^{-6}$; N_2O = 2.5 to $6 \cdot 10^{-5}$; NO_2 = maximum $3 \cdot 10^{-7}$; NH_3 = maximum $1.9 \cdot 10^{-6}$; SO_2 = maximum $1.8 \cdot 10^{-6}$; H_2S = 0.2 to $2 \cdot 10^{-6}$; Cl_2 = maximum $1.6 \cdot 10^{-7}$; I_2 = 4 to $40 \cdot 10^{-6}$.

Bibliography

DONAHOE, T. M.: Ionospheric Composition and Reactions. Science 159 (1968), 3814, 489—497

HUTCHINSON, G. E.: The Biogeochemistry of the Terrestrial Atmosphere, in: G. P. KUIPER, The Earth as a Planet. Univ. Chicago Press 1954, 371—433

Table 98. Composition of the Present Atmosphere
(up to an altitude of about 90 km; from UREY, 1959, and others)

Constituents	[vol.-%]	[mass-%]	Total Mass [t]
N_2	78.09	75.53	$3.86 \cdot 10^{15}$
O_2	20.95	23.14	$1.18 \cdot 10^{15}$
Ar	0.93	1.28	$6.55 \cdot 10^{13}$
H_2O-vapor	varying	varying	10^{13}
CO_2	0.03	0.046	$2.33 \cdot 10^{12}$
Ne	$1.8 \cdot 10^{-3}$	$1.3 \cdot 10^{-3}$	$6.4 \cdot 10^{10}$
He	$5.2 \cdot 10^{-4}$	$7.2 \cdot 10^{-5}$	$3.7 \cdot 10^{9}$
Kr	$1.5 \cdot 10^{-4}$	$4.5 \cdot 10^{-4}$	$1.5 \cdot 10^{10}$
CH_4	$1.4 \cdot 10^{-4}$	$7.8 \cdot 10^{-5}$	$3.9 \cdot 10^{9}$
H_2	$5 \cdot 10^{-5}$	$3.5 \cdot 10^{-6}$	$2.3 \cdot 10^{8}$
Xe	$8 \cdot 10^{-6}$	$4 \cdot 10^{-5}$	$2.5 \cdot 10^{9}$
O_3[1])	$\approx 10^{-8}$	$\approx 10^{-8}$	$\approx 10^{9}$
Ra[2])	$\approx 10^{-18}$	$\approx 10^{-17}$	82 kg to 22 t

[1]) increases with increasing altitude; [2]) decreases with increasing altitude

ISRAËL, H.: Spurengase in der Atmosphäre. Naturwiss. Rdsch. *20* (1967), 5114, 329—336
NAKAI, N.: Geochemische Studien zur Genesis natürlicher Gase. J. Earth Sci. Nagoya Univ. *10* (1962), 71—111
RUBEY, W. W.: Development of the Hydrosphere and Atmosphere with Special Reference to Probable Composition of the Early Atmosphere, in: Crust of the Earth. Spec. Pap. Geol. Soc. America *62* (1955), 631—650
SOKOLOV, V. A.: Geochemistry of the Gases in the Earth's Crust and Atmosphere. Izd. "Nedra", Moscow 1966 (Russian)
UREY, H. C.: The Atmosphere of the Planets, in: Handbuch der Physik, Vol. 52. Springer, Berlin/Göttingen/Heidelberg 1959
ZVEREV, V. P.: The Role of Atmospheric Precipitations in the Circulation of Chemical Elements between the Atmosphere and the Lithosphere. Dokl. Akad. Nauk SSSR *181* (1968), 3, 716—719 (Russian)

7.4.5. Average Abundance of Chemical Elements in the Biosphere

The acquisition of data on the average abundance of chemical elements in the biosphere is more complicated and the results are more uncertain than in other spheres. Therefore, the values listed in Table 100 are only given to indicate trends.

Table 99. Mean Chemical Composition of the Most Important
Structural Units of Organic Matter [mass-%]
(from RANKAMA and SAHAMA, 1950)

Element	Carbohydrates	Fats	Albuminous Substances
O	49.38	17.90	22.4
C	44.44	69.05	51.3
H	6.18	10.00	6.9
P		2.13	0.7
N		0.61	17.8
S		0.31	0.8
Fe			0.1

Table 100. Average Abundance of Chemical Elements in Organisms
(arrangement according to VINOGRADOV)

Element Groups	Order of Magnitude [mass-%]	Elements 1st variant	2nd variant
Macroelements	>10	O, H	O, C
	10— 1	C, N, Ca	H
	1—10⁻¹	S, P, K, Si	P, Ca, N, K, S, Cl
	10⁻¹—10⁻²	Mg, Fe, Na, Cl, Al	Mg, Na, Si, Al, Fe,
Microelements	10⁻²—10⁻³	Zn, Br, Mn, Cu	Sr, B
	10⁻³—10⁻⁴	I, As, B, F, Pb, Ti, V, Ni	Mn, Br, Zn, I, Cu, Ti, Ba, F, Li, Rb, Mo
	10⁻⁴—10⁻⁵	Ag, Co, Ba, Th	Rb, Ni, As, Co, V, Cr, U, Se
Ultraelements	10⁻⁵—10⁻⁶	Au, Rb	Cd
	10⁻⁶—10⁻¹¹	Hg	
	10⁻¹¹—10⁻¹²	Ra, Rn	Ra

Note:

The abundance of the following macroelements in organisms
    is higher than in the lithosphere: S, H, N, P, Cl, C, O
    is lower than in the lithosphere: Si, Al, Fe, Na, K, Mg

Bibliography

BAUER, H. M.: Biochemisches Taschenbuch, 2nd ed. Springer, Berlin/Göttingen/Heidelberg 1964
BOWEN, H. T. M.: Trace Elements in Biogeochemistry. Acad. Press, London/New York 1966
SHAW, W. H. R.: Studies in Biochemistry. I. A. Biogeochemical Periodic Table. II. Discussions and References. Geochim. cosmochim. Acta, *19* (1960), 196—207 and 207—217
VERNADSKIĬ, V. I.: The Biosphere and the Noösphere. Amer. Scientist *33* (1945), 1
VINOGRADOV, A.: The Chemical Fundamental Composition of Living Organisms and the Periodic System of the Chemical Elements. C. R. Acad. Sci. *197* (1933), 1673 (French)

3. Geochemistry of Geologic-geochemical Processes

3.1. Geochemistry of Magmatic and Postmagmatic Processes

3.1.1. Geochemistry of Magmatic Processes

STRECKEISEN (1967) gives a summary of the modern systematic classifications of igneous rocks. Further, reference should be made to the petrographic textbooks and monographs by NIGGLI (1923), TRÖGER (1935), BARTH (1962) and RONNER (1963), TURNER and VERHOOGEN. A schematic representation of the succession of rock-forming minerals is given in Fig. 56, whereas Fig. 57 shows the approximate mineral composition of the most important igneous rocks.

The average abundance of elements in igneous and groups of igneous rocks has already been given in Table 93 and 95. As supplements, a simplified graphical representation (Fig. 58) and a tabulated summary of the chemical composition of important prototypes of rock (Table 101) are given below. The general distribution of the main and trace elements in the main groups of rock (ultrabasic to acid) is shown in Figs. 59 to 63, whereas Table 102 demonstrates the range of distribution of trace elements in granitic rocks of local importance.

The distribution of the elements in a magmatic process takes place according to the regularities which hold for fractional differentiation, that is to say, largely under the influence of the gravitational field of the earth (GREEN and POLDERVAART, 1958).

By extending Goldschmidt's rules (Section 3.2.3.2.), RINGWOOD (1955) tried to explain the behaviour of ions during magmatic crystallization:

a) Ions not forming tetrahedral complexes are rarely incorporated in silicate minerals. They are concentrated in the magmatic residual phase (e.g. CO_3^{2-}, BO_3^{2-}).

b) For ions forming tetrahedral complexes the following holds:

1. the higher the charge on the central cation, the less likely is the incorporation of the complex into the silicate structure. This is the reason why complexes whose central

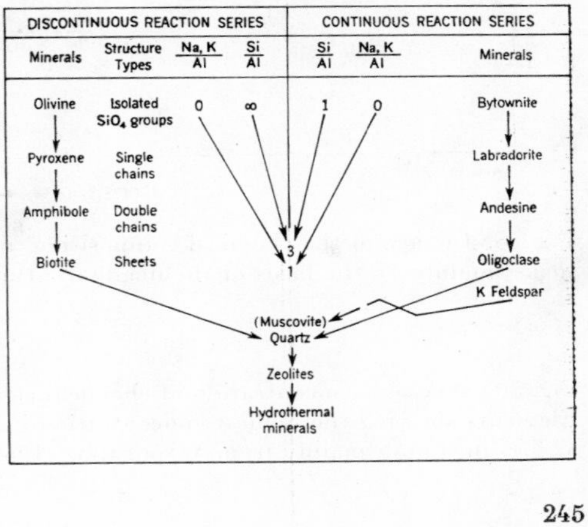

Fig. 56   The crystallization products (Bowen reaction series) of a magma (after ECKERMANN, 1944, from MASON, 1966)

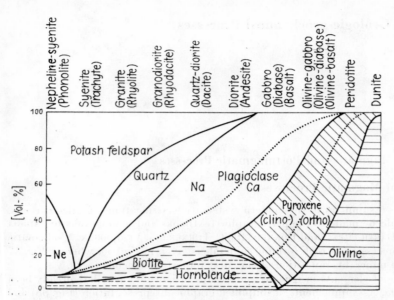

Fig. 57 Approximate mineral composition of the igneous rocks (after ADAMS, 1956; Ne = nepheline)

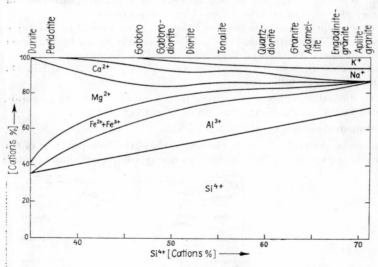

Fig. 58. Change of the chemical composition (major components) from ultrabasic to acid calc-alkalic rocks (mainly on the basis of the magmatic types according to NIGGLI)

Fig. 59 Average concentration of chemical elements in the main types of igneous rocks; Elements showing the highest concentration in ultrabasic igneous rocks.
Values in [ppm], mainly from VINOGRADOV (1962)   →

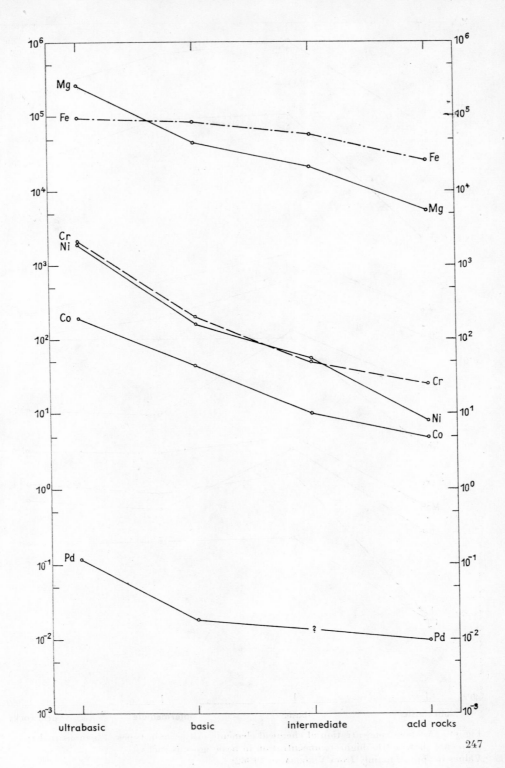

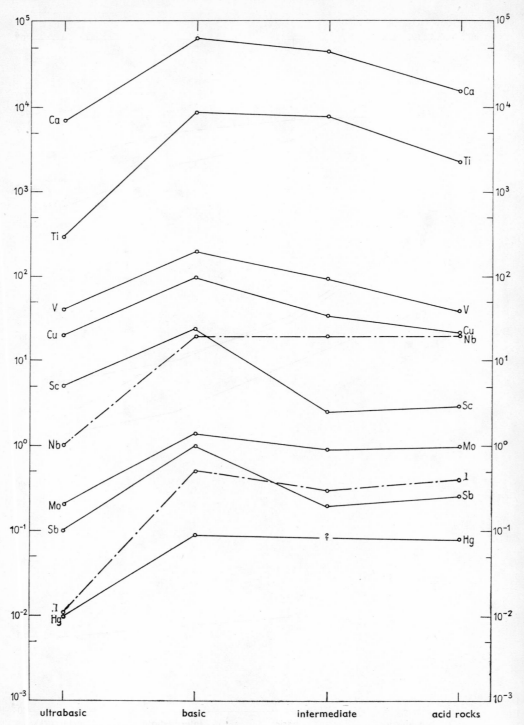

Fig. 60 Average concentration of chemical elements in the main types of igenous rocks;
Elements showing the highest concentration in basic igneous rocks.
Values in [ppm], mainly from VINOGRADOV (1962)

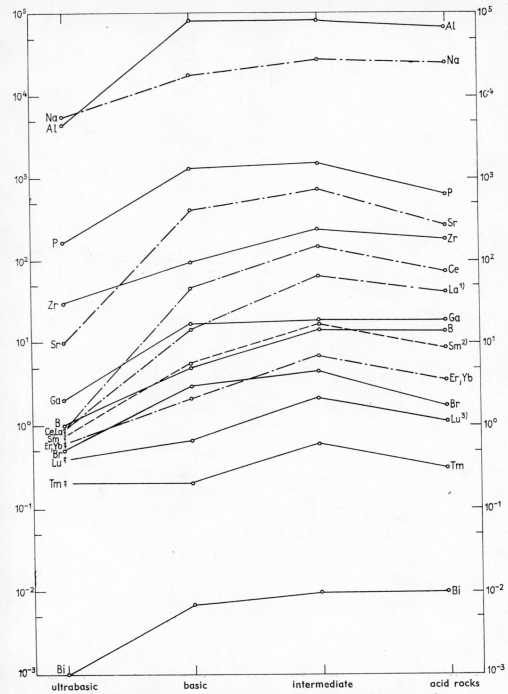

Fig. 61 Average concentration of chemical elements in the main types of igneous rocks;
Elements showing the highest concentration in intermediate igneous rocks.
Values in [ppm], mainly from VINOGRADOV (1962)

1) La: a similar order of magnitude and tendency shows Nd
2) Sm: a similar order of magnitude and tendency show Gd, Pr
3) Lu: a similar order of magnitude and tendency show Eu, Ho, Tb

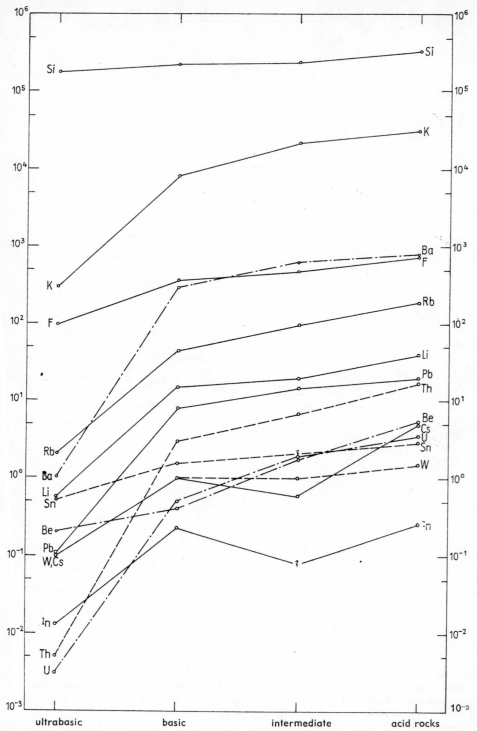

Fig. 62 Average concentration of chemical elements in the main types of igneous rocks;
Elements showing the highest concentration in acid igneous rocks.
Values in [ppm], mainly from VINOGRADOV (1962)

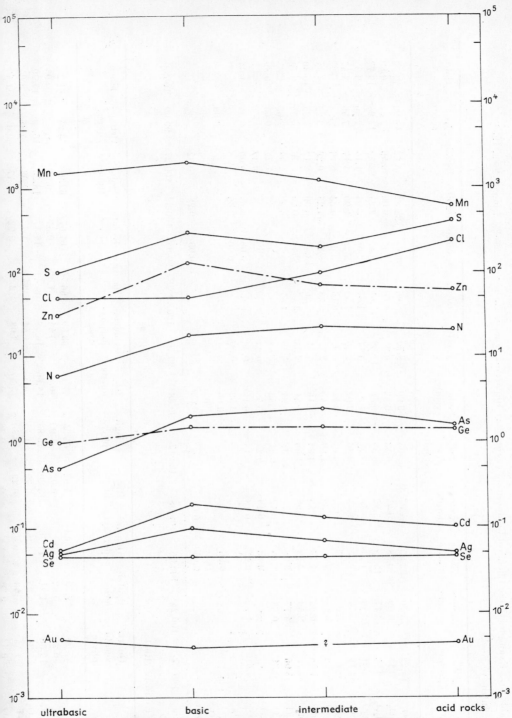

Fig. 63   Average concentration of chemical elements in the main types of igenous rocks;
Elements whose concentrations vary within relatively narrow limits (<1 decimal)
Values in [ppm], mainly from VINOGRADOV (1962)

Table 101. Examples of the Chemical Composition of the Most Important Igneous Rocks (in mass-%, from Tröger, 1930)

A. Intrusive Rocks

	Alkali-granite	Aplite-granite	Grano-diorite	Quartz-diorite	Ortho-clase syenite	Soda-syenite	Syenite	Diorite	Gabbro	Dunite	Foyaite	Es-sexite	Ijolite
SiO_2	75.22	75.70	63.85	64.07	62.03	60.00	58.70	56.06	48.61	38.82	55.22	46.99	43.70
TiO_2	0.13	0.09	0.58	0.45	0.53	0.42	0.95	0.60	0.17	0.00	0.59	2.92	0.89
Al_2O_3	9.93	13.17	15.84	15.82	16.39	16.88	17.09	17.61	17.83	2.24	22.59	17.94	19.77
Fe_2O_3	2.31	0.43	1.91	3.40	0.72	1.83	3.17	1.65	2.08	3.04	1.14	2.56	3.35
FeO	2.19	0.74	2.75	1.44	0.86	3.02	2.29	7.59	5.23	4.90	1.17	7.56	3.47
MnO	0.17	n.d.	0.07	tr.	n.d.	0.12	n.d.	0.16	n.d.	0.28	0.13	tr.	tr.
MgO	0.09	0.15	2.07	3.39	1.60	1.40	2.41	3.38	8.23	44.28	0.28	3.22	3.94
CaO	1.08	0.92	4.76	4.43	3.60	3.16	4.71	7.26	13.72	0.00	2.12	7.85	10.30
Na_2O	4.78	3.59	3.29	4.06	1.08	9.31	4.38	3.47	2.63	0.20	8.76	6.35	9.78
K_2O	4.06	4.77	3.08	2.27	12.38	0.94	4.35	1.67	0.32	n.d.	5.59	2.62	2.87
H_2O^+	0.31	0.68	1.65	0.42	0.61	1.53	0.89	0.95	0.99	5.68	1.77	0.65	0.89
H_2O^-	—	—	0.28	0.10	0.24	0.43	0.23	—	—	—	0.39	—	—
P_2O_5	n.d.	tr.	0.13	0.18	0.13	0.14	0.23	n.d.	0.08	n.d.	0.00	0.94	1.34
CO_2	—	—	—	—	—	0.59	0.00	—	0.00	0.60	—	—	—
BaO	—	—	0.06	—	—	0.06	—	—	—	—	—	0.00	—
SrO	—	—	tr.	—	—	0.02	—	—	—	—	—	—	—
others	—	—	0.04	0.05	—	0.03	—	—	0.05	0.28	0.52	—	—
Sum	100.27	100.24	100.36	100.08	100.17	99.88[1]	99.40	100.40	99.94	100.32	100.27	99.00	100.30

B. Extrusive Rocks

	Quartz-kerato-phyre	Pantel-lerite	Quartz-porphyry	Rhyo-lite	Plagio-phyre	Rhyo-dacite	Quartz-porphy-rite	Dacite	Kerato-phyre	Soda-trachyte	Ortho-phyre	Trachyte
SiO_2	67.90	69.79	76.03	74.02	64.54	67.16	63.39	66.76	61.67	62.91	63.24	61.25
TiO_2	0.24	0.89	n.b.	0.02	1.09	0.87	0.44	1.02	0.34	0.94	tr.	0.05
Al_2O_3	14.36	11.91	11.76	13.20	15.83	14.86	16.58	14.41	17.47	18.25	16.83	17.70
Fe_2O_3	4.36	5.35	1.99	0.75	1.75	0.43	1.41	2.74	1.37	2.08	4.86	2.95
FeO	1.44	1.43	n.d.	0.29	2.80	3.87	3.08	2.23	3.92	1.47	0.07	1.40
MnO	0.32	0.20	n.d.	tr.	0.26	0.07	tr.	n.d	tr.	n.d.	tr.	n.d

Table 101 continued

MgO	0.07	0.57	0.20	2.13	2.01	2.15	1.61	1.01	0.06	0.27	0.25	0.22
CaO	3.10	0.72	0.87	0.18	3.64	4.76	2.84	2.13	0.56	0.45	0.25	1.34
Na$_2$O	3.40	4.02	6.87	8.52	3.02	3.47	2.48	5.25	4.18	3.36	5.66	6.89
K$_2$O	8.08	7.37	5.85	3.38	2.51	2.79	3.77	2.95	4.82	5.61	4.59	1.85
H$_2$O$^+$	1.35	1.13	0.47	0.45	0.62	1.87	0.90	1.39	1.86	0.63	0.17	1.52
H$_2$O$^-$	—	—	—	—	0.77	0.22	0.12	0.42	n.d.	—	0.04	n.d.
P$_2$O$_5$	1.10	0.16	0.19	0.06	0.39	0.14	0.53	0.00	—	n.d.	0.13	—
CO$_2$	—	0.00	—	0.05	—	—	0.20	0.27	—	—	—	—
BaO	—	—	—	—	—	0.11	—	0.06	—	—	—	—
others	—	0.43	—	—	—	—	0.02	0.15	tr.	—	—	tr.
Sum	100.45	99.40	100.10	99.54	100.12	100.41	99.74	99.91[1])	99.76	100.10	100.66	100.44

	Por-phyrite	Ande-site	Diabase	Basalt	Sidero-melane	Pala-gonite	Phono-lite	Leucite-phono-lite	Nephe-linete-phrite	Leucite-tephrite	Nephe-linite	Olivine Leuci-tite	Picrite
SiO$_2$	54.94	57.35	50.20	50.32	51.90	33.00	56.56	55.87	46.26	48.74	40.99	42.20	40.02
TiO$_2$	1.11	0.64	1.21	3.10	1.60	2.30	0.23	0.79	1.69	1.04	2.41	2.44	0.59
Al$_2$O$_3$	18.38	17.54	16.08	12.83	14.70	8.30	21.31	20.85	18.98	16.38	16.50	12.13	8.32
Fe$_2$O$_3$	3.15	3.33	9.30	1.74	1.60	15.20	1.03	2.34	7.39	1.64	10.62	7.27	1.51
FeO	3.02	3.87	3.87	9.93	8.60	—	1.79	1.10	3.27	5.30	n.d.	4.62	11.14
MnO	n.d.	tr.	0.54	0.10	tr.	0.10	0.11	n.d.	n.d.	0.14	0.35	n.d.	0.85
MgO	3.59	4.29	6.82	7.39	8.70	5.00	0.15	0.48	3.09	7.07	3.29	9.24	27.63
CaO	6.29	6.91	7.85	11.06	10.40	7.00	1.24	3.07	10.59	12.19	12.63	14.32	4.04
Na$_2$O	3.97	4.01	2.34	2.38	2.60	0.70	9.47	4.81	5.51	2.01	5.95	2.75	0.65
K$_2$O	2.31	2.54	1.24	0.41	0.40	0.30	5.25	10.49	1.99	4.95	2.36	3.69	0.32
H$_2$O$^+$	2.39	0.79	0.67	0.33	0.20	9.30	0.25	0.34	0.96	0.55	2.63	0.66	4.30
H$_2$O$^-$	—	—	n.d.	0.05	0.16	18.30	1.70	—	—	0.16	—	—	0.70
P$_2$O$_5$	0.27	0.08	—	0.30	—	—	0.06	0.11	0.49	0.18	0.89	0.80	n.d.
CO$_2$	0.69	—	—	—	—	—	0.24	0.00	—	—	—	—	—
BaO	—	—	—	—	—	—	—	0.09	—	—	—	—	—
others	0.12	0.03	—	0.04	0.04	—	0.61	0.21	—	0.07	1.00	0.04	0.51
Sum	100.23	101.38	100.12	99.98	100.86	99.50	100.00	100.55	100.22	100.42	99.62	100.16	100.58

Table 102. Average Concentrations of Elements in Granitoid Rocks in Saxony and Thuringia (Bräuer, 1965) n.d. = not detectable

Granite Massif	Element Concentrations [ppm]																Number of Test Samples
	Sn	Li	Be	Ti	Cr	V	Co	Ni	Mn	Ba	Sr	La	Y	Pb	Cu	U	
A. Saxony																	
Zinnwald (greisen)	62	8,250	41	94	<1	<5	—	—	975	50	8	10	<10	5	1.0	12	4
Markersbach granite	5	300	4.0	90	1	—	—	<1	85	50	5	10	10	30	3.0	6	1
Altenberg granite	205	1,200	1.1	15	<1	—	—	<1	150	10	8	<10	—	14	2.5	32	2
Schellerhau granite	86	830	10	277	2	5	—	<1	83	150	51	10	13	32	3.0	12	4
Bobritzsch granite	10	220	<1	1,950	43	38	7	13	147	1,000	458	77	167	22	21	5	12
Granite from Geyer	37	813	1.6	136	<1	—	—	<1	98	47	10	13	<10	—	2.1	12	6
Granite from Ehrenfriedersdorf	120	1,468	1.3	87	<1	<5	—	<1	78	30	69	<10	<10	5	2.2	7	10
Auerhammer granite	10	300	1.0	2,700	3	18	<3	10	84	1,000	300	100	122	8	1.3	8	4
Schwarzenberg granite	13	100	3.7	362	6	<5	—	<1	61	106	73	20	30	10	1.8	5	4
Eibenstock granite (coarse grained)	27	562	2.0	556	<1	<5	—	<1	69	78	42	25	24	8	2.4	7	20
Eibenstock granite (medium grained)	23	600	3.6	398	1	<5	—	<1	72	43	12	15	16	<5	2.4	6	8
Eibenstock granite (fine grained)	45	603	2.4	338	<1	<5	—	<1	49	28	11	15	12	6	2.7	11	14
Kirchberg granite (medium grained)	10	272	1.1	1,528	7	19	<3	<1	135	400	540	26	91	36	7.3	8	5
Kirchberg granite (small grained)	10	280	1.3	1,300	18	13	<3	<1	101	475	378	25	66	22	2.7	9	4
Kirchberg granite (fine grained)	10	195	3.6	425	<1	<5	<3	<1	52	67	36	11	18	25	4.7	12	6
Bergen granite (coarse grained)	13	373	2.4	756	1	7	<3	<1	138	234	75	53	33	13	3.0	11	3
B. Thuringia																	
Sparnberg granite	26	100	1.7	82	<1	<5	—	<1	92	75	8	10	10	10	1.8	4	2
Granite from Henneberg	24	152	4.0	1,725	11	19	3	<1	66	625	600	20	71	33	1.0	7	4
Granite from Sormitztal	45	100	1	1,600	25	30	7	10	107	300	300	30	120	40	9.5	4	2
Schleusetal granite	<10	<100	—	213	<1	<5	—	<1	20	75	18	<10	<10	7	1.0	1	3
Suhl granite	<10	110	—	2,914	61	120	14	21	75	568	390	192	197	15	12	4	45
Diorite from Brotterode	<10	133	—	5,000	300	233	50	>300	n.d.	692	450	233	757	5	7.5	4	3
Ruhla granite	<10	103	—	2,440	47	56	8	37	78	776	348	55	142	9	6.4	2	25

cation possesses a valence higher than 4 are concentrated in the residual phase (e.g. PO_4^{3-}, VO_4^{3-}, TaO_4^{3-}, AsO_4^{3-}, NbO_4^{3-}, WO_4^{2-}, MoO_4^{2-}, SO_4^{2-}), and

2. the larger the central cation, the less likely the incorporation of the complex into the silicate structure. Complexes whose central cations have a radius larger than that of Si^{4+} (>0.42 Å) are also concentrated in the residual phase (e.g. GeO_4^{4-}, TiO_4^{4-}, SnO_4^{4-}, ZrO_4^{4-}, HfO_4^{4-}).

For this reason, correlations between many elements and isotopes are particularly pronounced in the magmatic processes (e.g. U/K, Th/K, K/Rb, Rb/Sr, Sr isotopes), in contrast with sedimentary and metamorphic formations.

The quantitative geochemical determination of differentiation processes is of great interest. At this point, mention should be made of the following examples of attempts:

a) dependence on the contents of SiO_2

b) the "Larsen Index" $1/3\ SiO_2 + K_2O - (CaO + MgO + FeO)$ (LARSEN, 1938)

c) the "Felsic Index" $F = \dfrac{100\ (Na + K)}{Ca + Na + K}$

the "Mafic Index" $M = \dfrac{100\ Fe}{Mg + Fe}$

both expounded by POLDERVAART and ELSTON (1954)

d) the "Differentiation Index" (D.I.) according to THORNTON and TUTTLE (1960).

The Skaergaard Intrusion (Table 103) can be considered as a good numerical example of the change in contents of trace elements during differentiation.

Table 103. Minor Elements in Differentiates of the Skaergaard Intrusion Compared to the Initial Magma (from WAGER, 1947, values in ppm)

	Initial Magma	Gabbro Picrite and Eucrite (earliest)	Olivine Gabbro	Olivine-free Gabbro	Ferrogabbros		Hedenbergite Granophyre	Granophyre (latest)
Rb	—	—	—	—	—	—	30	200
Ba	40	25	25	45	50	150	450	1,700
Sr	300	200	700	450	700	450	500	300
La	—	—	—	—	—	—	25	150
Y	—	—	—	—	—	125	175	200
Zr	40	40	35	25	20	100	500	700
Sc	15	7	20	15	10	—	—	—
Cu	130	70	80	175	400	200	500	20
Co	50	80	55	40	40	5	10	4
Ni	200	600	135	40	—	—	5	5
Li	3	3	2	3	3	15	25	12
V	150	170	225	400	15	—	—	12
Cr	300	700	175	—	—	—	—	3
Ga	15	12	23	15	20	20	40	30
Proportion of rocks in %		65	14	10	$7^1/_2$	$2^1/_2$	$^1/_2$	$^1/_2$

Starting from different concentrations of elements in the various main magmatic rock types, RECHARSKIĬ (1964) determined so-called concentration coefficients which display close relations with the periodic system of elements.

The contents of volatile constituents, especially of water, is of extraordinary importance to the development and crystallization of silicate magmas. The fundamental knowledge of the solubility of water in granitic melts goes to the credit of GORANSON (1931; Fig. 64); of a more recent date are works by KHITAROV (1963; Fig. 65), LE GRAND (1958), TOLSTICHIN (1961), and others.

SZÁDECZKY-KARDOSS (1952), BARDOSSY and BOD (1961), and KUCHEV (1964, cf. Fig. 66) give full particulars of the degree of oxidation ($2Fe_2O_3/FeO$ or Fe_2O_3/FeO) of rocks of different chemical composition.

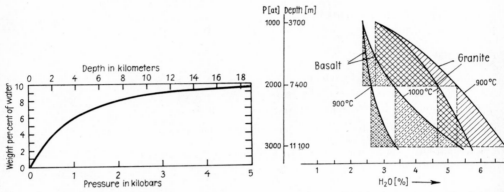

Fig. 64   The solubility of water in molten granite as a function of pressure at the 900 °C isotherm. (After GORANSON, 1931, from BARTH, 1962)

Fig. 65   The solubility of water in molten silicates (granite and basalt) as a function of pressure and temperature (after KHITAROV, 1960)

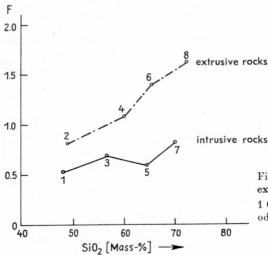

Fig. 66   The F ratio (Fe_2O_3/FeO) in intrusive and extrusive types of rock (from KUCHEV, 1964)

1 Gabbro; 2 Basalt; 3 Diorite; 4 Andesite; 5 Granodiorite; 6 Dacite; 7 Granite; 8 Liparite

Bibliography

AHRENS, L. H.: Geochemical Studies on Some of the Rarer Elements in South African Minerals and Rocks. The Geochemical Relationship between Thallium and Rubidium in Minerals of Igneous Origin. Trans. Geol. Soc. South Africa *XLVIII* (1946)

AHRENS, L. H., W. H. PINSON, and M. M. KEARNS: Association of Rubidium and Potassium and Their Abundance in Common Igneous Rocks and Meteorites. Geochim. Cosmochim. Acta *2* (1952), 4

AHRENS, L. H.: Element Distributions in Specific Igneous Rocks. VIII. Geochim. Cosmochim. Acta *9* (1966), 1, 109—122

ALASHOV, YU. A.: Differentiation of Rare Earths in the Magmatic Process. Khimiya Zemnoĭ Kory Vol. I, Izv. Akad. Nauk SSSR, Moscow (1963), 352—365 (Russian)

ARDOSSY, G., and M. BOD: A New Method to Characterize the State of Oxidation of Rocks. Acta geol. Acad. Sci. Hung. *7* (1961), 29—36

BARTH, T. F. W.: Theoretical Petrology. 2nd ed. John Wiley & Sons, New York/London 1962

BELOV, N. V.: The Basic Magmatic Process in the Light of Crystallochemistry. Khimiya Zemnoĭ Kory Vol. I, Izv. Akad. Nauk SSSR, Moscow (1963), 13 (Russian)

BORISENKO, L. F.: Rare and Trace Elements in Ultrabasites of the Ural. Izd. "Nauka", Moscow 1966, 213 pp. (Russian)

BRÄUER, H.: Spurenelementgehalte in Graniten Thüringens und Sachsens, Dissertation. Bergakademie Freiberg 1969

BROOKS, C. K.: On the Interpretation of Trends in Element Ratios in Differentiated Igneous Rocks, with Particular Reference to Strontium and Calcium. Chem. Geol. *3* (1968), 1, 15—20

BURNS, R. G., and W. S. FYFE: Trace Element Distribution Rules and Their Significance. Chem. Geol. *2* (1967), 2, 89—104

CARMICHAEL, I., and A. McDONALD: The Geochemistry of Some Natural Acid Glasses from the North Atlantic Tertiary Volcanic Province. Geochim. Cosmochim. Acta *25* (1961), 189—222

CLANGEAUD, L., and R. LÉTOLLE: La Théorie des deux Magmas Fondamentaux dans le Volcanisme Intracontinental et l'Evolution Géochimique des Laves du Mont-Dore. Geol. Rdsch. *55* (1965), 2, 316—329

COMPSTON, W., J. McDOUGALL, and K. S. HEIER: Geochemical Comparison of the Mesozoic Basaltic Rocks of Antarctica, South Africa, South America and Tasmania. Geochim. Cosmochim. Acta *32* (1968), 2, 129—149

COPPENS, R., G. DURAND, and G. JURAIN: Etat Actuel de nos Connaissances sur l'Équilibre et le Déséquilibre dans les Roches et ses Conséquences. C. R. 89ᵉ Congr. Soc. Savantes, Lyons (1964), Sect. Sci. (1965), 165—185

COPPENS, R., and G. JURAIN: Considérations Genérales sur la Distribution de l'Uranium dans les Roches Granitoides. Geol. Rdsch. *55* (1965), 2, 427—436

DALY, R. A.: Igneous Rocks and the Depth of the Earth. New York/London 1933

DESBOROUGH, G. A.: Closed System Differentiation of Sulfides in Olivin Diabases, Missouri. Econ. Geology *62* (1967), 5, 595—613

EITEL, W.: The Physical Chemistry of the Silicates. Univ. of Chicago Press, Chicago 1954

ESKOLA, P.: On the Origin of Granitic Magma. Mineral. Petrogr. Mitt. *42* (1932), 155 pp.

FERSMAN, A. J.: Pegmatite I. Leningrad 1931 (Russian)

GERASSIMOVSKIĬ, V. I.: Main Characteristics of the Geochemistry of Nepheline Syenite from the Kola Peninsula. Geokhimiya (1967), 11, 1320—1237 (Russian)

GEVORKIAN, R. G.: The Effect of Alkalinity and Temperature on the Alumina Content of Rocks Crystallizing in Response to the Differentiation of a Common Magmatic Chamber. Dokl. Akad. Nauk *176* (1967), 2, 424—427 (Russian)

GORANSON, R. W.: The Solubility of Water in Granite Magmas. A. Sci. *22* (1931), 481

GREEN, J., and A. POLDERVAART: Petrochemical Fields and Trends. Geochim. Cosmochim. Acta, *13* (1958), 87—122

GREENLAND, L., and I. F. LOVERING: Fractionation of Fluorine, Chlorine and Other Trace Elements during Differentiation of a Tholeiitic Magma. Geochim. Cosmochim. Acta *30* (1966), 963—982

GRUZA, V. V.: Acid-basic Differentiation of Elements in Magmatic Rocks. Dokl. Akad. Nauk SSSR *173* (1967), 5, 1177—1180 (Russian)

HIGAZY, R. A.: Observations on the Distributions of Trace Elements in the Perthite Pegmatites of the Black Hills, South Dakota. Amer. Mineralogist *30* (1963)

HURLEY, P. M., H. W. FAIRBAIRN, and W. H. PINSON: Rb-Sr Isotopic Evidence in the Origin of Potash-rich Lavas of Western Italy. Earth Planet. Sci. Lett. *1* (1966), 5, 301—306

JAHNS, R. H.: The Study of Pegmatites. Econ. Geol. 50th Ann. Vol. (1956), 1025—1130

KADIK, A. A., and N. I. KHITAROV: Possible Ranges of Some Kinds of Mass Exchange with Magmas. Geokhimiya (1966), 9, 1019—1034 (Russian)

KHITAROV, N. I.: Wechselbeziehungen zwischen Wasser und magmatischer Schmelze. Geokhimiya *7* (1960), 586—589 (Russian)

KHITAROV, N. I., A. A. KADIK, and E. B. LEBEDEV: Chief Regularities of Water Separation from Magmatic Melts of Granitic Composition. Geokhimiya (1967), 11, 1274—1284 (Russian)

KHITAROV, N. I., A. A. KADIK, and E. B. LEBEDEV: Solubility of Water in Basalt Melt. Geokhimiya (1968), 7, 763—772 (Russian)

KHOMICHEV, V. L.: Comparison of Magmatic Complexes by the Method of Fractional Geochemical Analysis. Geokhimiya (1967), 3, 296—303 (Russian)

KOLBE, PM. S., and R. TAYLOR: Major and Trace Element Relationships in Granodiorites and Granites from Australia and South Africa. Beitr. Miner. Petrogr. *12* (1966), 139—167, 202—222

KOWALSKI, W.: Geochemistry of K, Na, Ca, Rb, Ba and Cr in Granitoids of the Sudetes and Their Pegmatites. Archiwum Mineralogiczne, *XXVII* Warsaw, 1967 (Polish)

KUCHEV, YU. S.: On a Few Petrogenetic Aspects of the Fe_2O_3:FeO ratio in Magmatic Rocks. Izv. Akad. Nauk SSSR *11* (1964), 63—67 (Russian)

KUSHIRO, J., H. S. YODER, and M. NISHIKAWA: Effect of Water on the Melting of Enstatite. Geol. Soc. Amer. Bull. *79* (1968), 1685—1692

LEBEDEV, V. I.: Magmatic Crystallisation in the Light of Crystallochemistry and Geochemistry. Izv. Akad. Nauk SSSR, Geol. Ser. (1967), 11, 87—97 (Russian)

LE GRAND, H. E.: Chemical Character of Water in the Igneous and Metamorphic Rocks of North Carolina. Econ. Geol. *53* (1958), 178—189

MOENKE, H.: Spurenelemente in variszischen und praevariszischen deutschen Graniten. Chem. d. Erde *20* (1959/60)

MORKOVKINA, V. F.: Chemical Analyses of Igneous Rocks and Rock-forming Minerals. Izd-vo Nauka, Moscow 1964 (Russian)

MUKHERJEE, B.: Genetic Significance of Trace Elements in Certain Rocks of Singhbum, India. Mineral. Mag. *36* (1968), 281, 661—670

NABOKO, S. I. (Edit.): Volcanism and Geochemistry of Acid Products. Izd. "Nauka", Moscow 1967 (Russian)

NIGGLI, P.: Das Magma und seine Produkte. Leipzig 1937

NIEUWENKAMP, W.: Geschichtliche Entwicklung der heutigen petrogenetischen Vorstellungen. Geol. Rdsch. *55* (1965), 2, 460—478

NOCKOLDS, S. R.: Chemical Composition of Igneous Rocks. Bull. Geol. Soc. Amer. *65* (1954), 1007

NOCKOLDS, S. R., and R. ALLEN: The Geochemistry of Some Igneous Rock Series, Part I: Calc-alkalic Rocks. Geochim. Cosmochim. Acta *4* (1953), 105—142

NOCKOLDS, S. R., and R. ALLEN: The Geochemistry of Some Igneous Rock Series, Part II: Alkali Igneous Rock Series. Geochim. Cosmochim. Acta *5* (1954), 245—285

NOCKOLDS, S. R., and R. L. MITCHELL: The Geochemistry of Some Caledonian Plutonic Rocks. Trans. Roy. Soc. Edinburgh *61* (1948), 533—575

PAPEZIK, V. S.: Geochemistry of Some Canadian Anorthosites. Geochim. Cosmochim. Acta *29* (1965), 673—709

PFEIFFER, L.: Beiträge zur Petrologie des Meißner Massivs. Freiberger Forschungsh., C *179* (1964)

RAMBERG, H.: Chemical Bonds and Distribution of Cation in Silicates. J. Geology *60* (1952), 4, 331 to 335

RECHARSKIĬ, V. I.: Particular Aspects of the Distribution of Elements in Eruptive Rocks from the Viewpoint of Periodicity of Concentration Coefficients. Dokl. Akad. Nauk SSSR *156* (1964), 3, 594 to 97 (Russian)

REYNOLDS, D.: The Sequence of Geochemical Changes Leading to Granitization. Quart. J. Geol. Soc., London *102* (1946), 389—446

RICHARDSON, W. A., and G. SNEEBY: The Frequency-distribution of Igneous Rocks; I. Frequency-distribution of the Major Oxides in Analyses of Igneous Rocks. Mineralog. Mag. *19* (1922), 303—313

RINGWOOD, A. E.: The Principles Governing Trace Element Distribution during Magmatic Crystallisation. Geochim. Cosmochim. Acta *7* (1955), 189—202 (Part I), 242—254 (Part II)

RONNER, F.: Systematische Klassifikation der Massengesteine. Springer, Vienna 1963

RONOV, A. B., and A A. YAROSHEVSKIĬ: Chemical Structure of the Earth's Crust. Geokhimiya (1967), 11, 285—1309 (Russian)

RÖSLER, H. J.: Zur Petrographie, Geochemie und Genese der Magmatite und Lagerstätten des Oberdevons und Unterkarbons in Ostthüringen, Freiberger Forschungsh. C *92* (1960)

SCHROLL, E., and H. GROHMANN: Beiträge zur Kenntnis des K-Rb-Verhältnisses in magmatischen Gesteinen. Geol. Rdsch. *55* (1965), 2, 261—274

SHAW, D. M.: Interprétation Géochimique des Eléments en Traces dans les Roches Cristallines. Masson et Cie, Paris 1964, 237 pp.

SHCHERBINA, V. V.: On the Geochemistry of Silicate Melts. Zap. Vsesoy. Min. Obshch. *43* (1964), 5, 537—544 (Russian)

SINHA, R. C., and S. G. KARKARE: Geochemistry of Deccan Basalts: Rep. XXII. Session Intern. Geol. Congr., India 1964, Part VII, 85—103

SNYDER, I. L.: Distribution of Certain Elements in the Duluth Complex. Geochim. Cosmochim. Acta, *6* (1959), 250

STRECKEISEN, A. L.: Classification and Nomenclature of Igneous Rocks. N. Jb. Miner. Abh. *107* (1967) 2, 144—214

SZÁDECZKY-KARDOSS, E. et al.: Complex Experimental Petrologic Investigation on the Interchange of Rocks and Magma. Acta Geol. Acad. Sci. Hung. *8* (1964), 1—4, 71—82

TAUSON, L. V.: The Geochemistry of Rare Elements in Granites. Izd. AN SSSR, Moscow 1961 (Russian)

TAYLOR, S. R., C. H. EMELEUS, and C. S. EXLEY: Some Anomalous K/Rb Ratios in Igneous Rocks and their Petrological Significance. Geochim. Cosmochim. Acta, *10* (1956), 4

THORNTON, C. P., and O. F. TUTTLE: Chemistry of Igneous Rocks. I. Differentiation Index. Amer. J. Sci. *258* (1960), 664—684

TOLSTICHIN, O. N.: On the Quantity of Juvenile Water Separated during the Formation of Effusive Rocks. Geokhimiya *11* (1961), 1005—1008 (Russian)

TRÖGER, W. E.: Spezielle Petrographie der Eruptivgesteine. Berlin 1935

TUROVSKIĬ, S. D., U. U. USMANOV, and A. V. NIKOLAYEVA: Rare-earth Distribution in the Series of Successively Crystallizing Minerals. Dokl. Akad. Nauk SSSR *178* (1968), 5, 1179—1182 (Russian)

TUTTLE, O. F., and I. GITTINS: Carbonatites. Interscience Publ. (Wiley and Sons), New York, London, Sydney 1966

WAGER, L. R., and R. L. MITCHELL: Preliminary Observations on the Distribution of Trace Elements in the Rocks of the Skaergaard Intrusion, Greenland. Mineralog. Mag. *26* (1943), 283

WAGER, L. R., and R. L. MITCHELL: The Distribution of Trace Elements during Strong Fractionation of Basic Magma: A Further Study of the Skaergaard Intrusion. East Greenland. Geochim. Cosmochim. Acta *1* (1951), 129—208

WEDEPOHL, K. H.: Die Untersuchung petrologischer Probleme mit geochemischen Methoden. Fortschr. Miner. *41* (1963) 1

WEDEPOHL, K. H.: Entgasung und Differentiation basaltischer Schmelzen. Fortschr. Miner. *41* (1964) 2, 190

WICKMANN, F. E.: Some Aspects of the Geochemistry of Igneous Rocks and of the Differentiation by Crystallisation. Geol. Foren. Stockholm Förh. *65* (1943), 371—396

Group of Authors: Petrography and Geochemical Peculiarities of the Complex of Ultrabasites, Alkaline Rocks, and Carbonatites. Izd-vo "Nauka", Moscow 1965 (Russian)
Group of Authors: Geology and Geochemistry of Granitic Rocks. Izd-vo "Nauka", Moscow 1965 (Russian)

8.1.2. Geochemistry of Postmagmatic Processes

The postmagmatic processes comprise the pegmatitic, pneumatolytic and hydrothermal stages. The physico-chemical conditions prevailing during magmatic crystallisation and in the postmagmatic processes are represented in the well-known Niggli diagram (Fig. 67).

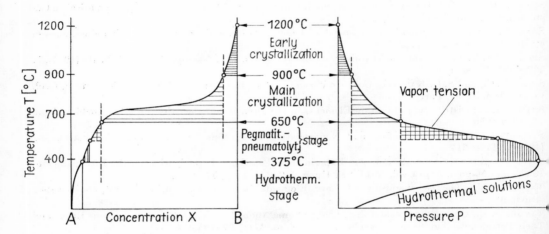

Fig. 67   Physico-chemical conditions during the crystallization of a magma (modified NIGGLI diagram) Diagram at the left: Distribution curve of not volatile (B) and volatile (A) components as a function of temperature T; diagram at the right: Pressure/temperature curve

8.1.2.1. Magmatic Gases

It is quite sure that magmas contain primary gases. It is also sure that magmas take up a certain amount of gases when rising and setting.

The composition of the gases is usually analyzed

a) in the gas inclusions in rocks (see also Section 8.1.2.2.)

b) in the lavas and exhalations of volcanoes.

The chemical composition of the gases varies considerably.
A survey is given in Table 104. Table 105 shows the classification of volcanic exhalations.

Table 104. Chemical Composition of Volcanic Gases (without water, from SOKOLOV, 1966, and SHEPHERD, 1938)

Origin	Klyuchevskaya Sopka, Kamchatka			Ebeko, Kuril Is.	Volcano Oshima (Japan)		
Kind	Fumaroles			Fumaroles	Magmatic Gas	Fumaroles	
Temperature	400 °C	200 °C	100—200 °C	92—110°	400°	400—100°	110—90°
CO₂	2.0	20.0	55	89—98	33.4	50.1	70.7
CO	26.2	25.0	8.0	0.05	9.8	—	—
H₂	29.5	23.0	5.3	0.04	90.2	13.9	—
HCl (Cl₂)	30.0	18.5	13.8	0.6—2.8	20.8	1.0	25.5
HF (F₂)	1.8	1.0	0.2	—	3.1	15.5	0.5
H₂S	} 5.5	} 12.5	} 17.0	} 1—5.6	11.7	17.0	} 0.5
SO₂					22.3		
N₂	5	—	0.7			2.5	3.3
others	—			0.1 (CH₄)	8.7		
H₂O					98%	96%	96—99%

Origin	Mauna Loa Kilauea	Etna	Mt. Pelée	California		
Kind	Lava Lake	Lava (Basaltic)	Lava (Andesitic)	Bomb (Andesitic)	Obsidian	Pumice
Temperature	1,100 °C					
CO₂	56.7	18.2	58.24	21.91	12.09	37.50
Co	1.5	0.3	11.56	5.69	6.08	0.82
H₂	1.6	10.35	1.45	11.36	0.87	4.09
HCl (Cl₂)			2.23 (Cl₂)	26.36 (Cl₂)	25.56 (Cl₂)	11.73 (Cl₂)
HF (F₂)			19.11 (F₂)	27.61 (F₂)	27.27 (F₂)	12.13 (F₂)
H₂S			} 2.43 (S₂)	} 1.44 (S₂)	} 3.16 (S₂)	} 0.33 (S₂)
SO₂	27.8 (+10.9 SO₃)	21.7	4.92	5.66	24.97	33.39
N₂		47.55 (+ noble gas)				
others	1.5	1.90 (O₂ + CH₄)				
H₂O			82.5%	96.23%	88.38%	96.30%

Table 105. Classification of Volcanic Exhalations (from NABOKO, 1959)

Character of Exhalations	Point of Separation	Temperature [°C]	Chemical Composition
I. Exhalations from surface magma	1. from the crater at the instant of eruption		little data available; assumed are H_2O, H_2, CO, H_2S, SO_2, HCl, HF
	2. from lava masses above the volcanic vent	varies	a) fumaroles of the haloid stage: H_2O, H_2, CO, HCl, HF, SO_2, halides and sulphates of Na, K, Fe, Cu b) fumaroles of the sulphidic stage: H_2O, CO, H_2, SO_2, H_2S, sulphates of Na, K, Ca, etc. c) fumaroles of the carbonic stage: H_2O, CO_2, traces of H_2S d) fumaroles of the water vapor stage: water vapors
	3. from lava masses detached from the volcanic vent	1,200—700 500—300	a) fumaroles from liquid lava: H_2O, H_2, CO, HCl, HF, S, SO_2, H_2S, halides and sulphates of Na, K, Fe, oxides of Si, Cr, Fe b) fumaroles from solidified lava; salt fumaroles: NaCl, KCl, $FeCl_3$, NH_4Cl, Na_2SO_4, K_2SO_4, $CaSO_4$, etc. ammonium chloride fumaroles: NH_4Cl, $FeCl_3$ water vapor fumaroles
II. Exhalations from magma located at great depth	calderas, craters, volcanic slopes	above the critical temperature	a) fumaroles of the haloid stage b) fumaroles of the sulphidic stage c) fumaroles of the carbonic stage
		near the critical temperature	a) solfataras: water vapors, CO_2, SO_2, H_2S, H_2SO_4 b) mofettes: water vapor, CO_2, H_2S c) jets of water vapor: water vapor
		under the critical temperature	hydrosolfataras: water vapor, CO_2, CH_4, H_2S

Bibliography

ELLIS, A. J.: Chemical Equilibrium in Magmatic Gases. An Application of Thermodynamic Data to the Problem of the Composition of Magmatic Gases. Amer. J. Sci. *255* (1957), 416—431

NABOKO, S. I.: Volcanic Exhalations and Their Reaction Products. Studies of the Laboratory for Volcanology; Vol. 6. Izd. AN SSSR, Moscow 1959 (Russian)

CHAIGNEAU, M.: Sur les Gaz Volcaniques de l'Etna (Sicile) Compt. Rend. Hebd. Séances Acad. Sci. *254* (1962), No. 23

GALINOV, E. M.: Die Isotopenzusammensetzung des Kohlenstoffs in den Gasen der Erdkruste. Z. Angew. Geol. *15* (1969) 2, 63—70

GREENWOOD, H. J., and H. L. BARNES: Binary Mixtures of Volatile Components. Geol. Soc. Amer., Mem. (1966) 97, 385—400

JAGGER, T.: Magmatic Gases. Amer. J. of Science (1940) May

KRAUSKOPF, K. B.: The Use of Equilibrium Calculations in Finding the Composition of a Magmatic Gas Phase. Researches in Geochemistry. John Wiley and Sons, New York 1959, 260—278

KRAUSKOPF, K. W.: On the Relative Volatility of Fluorides and Chlorides of the Rock-forming Metals in Magmatic Gases. Omnibus Volume: Probl. Geokhim. 147—157. Izd-vo "Nauka", Moscow 1965 (Russian)

MAZOR, E., and G. I. WASSERBURG: Helium, Neon, Argon, Krypton and Xenon in Gas Emanations from Yellowstone and Lassen Volcanic National Parks. Geochim. Cosmochim. Acta 29 (1965), 443—454

PETERSILIE, I. A.: The Gas Constituents and Trace Bitumens of Igneous and Metamorphic Rocks of the Kola Peninsula. Rep. XXII Session Intern. Geol. Congr., India 1964, Part I, 19—37

SHEPHERD, E. S.: The Magnetic Gases. Amer. J. Sci. 35-A (1938)

SHEPHERD, E. S.: The Gases in Rocks and Some Related Problems. Amer. J. Sci. 35-A (1938), 311—351

SOKOLOV, V. A.: Geochemistry of Gases in the Earth's Crust and in the Atmosphere. Izd. Nedra, Moscow 1966 (Russian)

VOITOV, G. I., et al.: Certain Peculiar Features in the Chemical Composition of Plutonic Gases from Data of the Aralsor Abyssal Wall (Pre-Caspian Bassin). Dokl. Akad. Nauk SSSR 172 (1967) 4, 942—945 (Russian)

8.1.2.2. Hydrothermal Solutions

Most hydrothermal solutions are slightly dissociated electrolytic solutions rich in alkali chloride. At supercritical and low temperatures, the *pH* values of these solutions are near the neutral point, whereas they become acid within the mesothermal range. Formation and dissociation of complexes in hydrothermal solutions is largely dependent on temperature, pressure, composition of the solution, thermochemical properties of the substances in the solution, and the solvent. Further information and a comprehensive bibliography are given by HELGESON (1964).

Conclusions regarding the chemical composition of hydrothermal solutions can be drawn from the composition of hot springs (of volcanic origin), liquid inclusions in hydrothermally formed minerals, phase relations and thermochemical parameters (cf. Tables 106 and 108).

A survey of the possibilities of the transport of chemical elements in hydrothermal solutions is given in Table 107.

Table 106. Partial Analyses of Liquid Inclusions; values in mole-% (from HELGESON, 1964)

Constituent	Liquid Inclusions in Quartzes (from quartz veins in the Alps; from NEWHOUSE)		Liquid Inclusions in Quartzes (from auriferous quartz veins of the Grass-Valley District; from ROEDDER)						Liquid Inclusions in Calcites (from veins in Siberia; from KHITAROV)			Mean Composition of Liquid Inclusions in Sphalerite (Creede, Colorado; from ROEDDER)
Na	21.0	34.3	42.4	38.3	36.7	36.9	38.0	43.9	9.3	5.3	19.4	34.3
K	4.4	10.4	26.2	22.5	21.3	24.5	26.9	29.7	—			3.9
Li	7.1	10.4	1.0	1.5	0.7	1.0	1.3	1.8				
Ca	1.7	2.2							27.6	30.6	19.4	6.7
Cl	11.0	13.2	26.7	35.0	39.3	31.8	25.1	16.1	62.1	64.0	61.1	49.5
SO₄	1.2	2.2	3.5	2.6	2.1	5.7	8.5	8.4	—	—	—	<3.6
CO₃	0.7	1.9										
HCO₃									0.1	—	—	
CO₂	52.7	39.8										
Rb			0.1	tr.	tr.	tr.	0.1	0.1				
Cs			tr.	tr.	tr.	tr.	0.2	tr.				
Mg												1.1
B												0.9

263

Table 107. Transport of Elements in Aqueous Solutions (from GUNDLACH, 1964)

Kind of Transport	Acid Solutions (pH <6)	Neutral Solutions (pH 6—8)	Alkaline Solutions (pH >8)
Simple ions, hydratised ions, O-complexes (e.g. SbO^+ or WO_4^-)	Ag, Al, As, Ba, Be, Bi, Ca, Cd, Co, Cr, Cu, Fe, Hg, Mg, Mn, Mo, Ni, Pb, Sb, Sn, Sr, Ti, Tl, U, V, Zr[1]	As, Ba, Ca, Mg, Mo, Sb, Se, Sr, V, W	Al, As, Ba, Ca, Cr, Mo, Nb, Sb, Sn, Sr, Ta, Ti, Tl, U, V, W, Zn, Zr[1]
Isopoly and heteropoly acids	As, Mo, Nb, Sb, Sn, Ta, Te, V, W	—	—
Halide and oxyhalide complexes (e.g. $FeCl_4^-$, VOF_4^-)[2]	Ag, Au, Be, Bi, Cd, Cu, Fe, Ga, Ge, Hg, In, La[1], Mo, Nb, Pb, Pt[1], Re, Sb, Sn, Ta, Th, Ti, Tl, U, V, Zn, Zr[1]	as in acid solution (soluble in brines: $CaCO_3$, $CaSO_4$, CuS, PbS, $SrSO_4$, ZnS)	—
Other anion complexes (above all sulphate, phosphate, carbonate)[3]	Cr, La[1], Th, Ti, Tl, U, V, Zr[1]	Ag, Be, Cr, Cu, La[1], Mn, Mo, Sb, Th, Ti, U, V, Zr[1]	Ag, Be, Cu, La[1], Mo, Th, U
as hydrocarbonate[4]	—	Ba, Ca, Co, Fe, Mg, Mn, Sr	—
Sulphur hydride complex (acid) sulphide complexes (alkaline)[5]	Ag, Ca, Cu, Fe, Hg, Ni, Pb, Zn	Ag, Cu, Fe, Hg, Ni, Pb, Zn	Ag, As, Au, Bi, Cu, Ga, Ge, Hg, In, Mo, Ni, (Pb), Pt[1], Re, Sb, Se, Sn, Te, Tl, W, Zn
Thiosulphate and polythionate complexes	—	Ag, Au, Cu, Hg, In, Pb, Zn	Ag, Au, Cu, Hg, In, Pb, Zn
Colloidal[6]	Au, Fe (oxide sol) Ni (sulphide sol), Sn (oxide sol)	Au, Fe (oxide sol), Ni (sulphide sol), Sn (oxide sol)	Au (SiO$_2$ as protective colloid), Fe (oxide sol), Ni (sulphide sol)

Notes:

[1] La represents all rare earths; Pt all Pt-metals; Zr also represents Hf.

[2] Part of these complexes are stable only at higher halide concentrations.

[3] Part of these complexes are stable only at higher concentrations of the given anion.

[4] Fe and Mn hydrocarbonates are only stable if no oxidation occurs.

[5] The sulphur hydride complexes are only stable under special conditions (H_2S partial pressure, etc.), part of them are only known in small concentrations. In the neutral range, in transition to the sulphide complexes, mostly precipitating of sulphides.

[6] Only a few stable sols are mentioned. Colloidal intermediate states are frequent, however, do not belong to the transport forms.

Table 108. Analyses of Hot Springs, presumably of volcanic origin. (values in mole-%, from HELGESON, 1964)

	Steamboat Springs, Nevada	Morris Basin, Yellowstone	Wairakei, New Zealand	Frying Pan Lake, New Zealand
Temperature [°C]	89.2	84	100	59
pH	7.9	9.45	8.6	3.0
SiO_2	6.80	16.20	5.35	10.63
Fe				0.08
Al				0.22
Ca	0.17	0.26	0.54	0.63
Mg	0.05	0.01	tr.	0.24
Sr	tr.			
Na	39.70	35.14	40.94	39.40
K	2.53	3.48	3.11	2.67
Li	1.52	2.22	1.46	
NH_4	tr.	tr.	0.04	
As	0.05	0.07		
Sb	tr.	tr.		
CO_3	0	0		
HCO_3	6.98	0.81	0.47	—
SO_4	1.45	0.73	0.30	4.07
Cl	34.07	38.61	45.26	40.11
F	0.13	0.47	0.27	
Br	tr.	tr.		
I	tr.	tr.		
B	6.32	1.95	2.00	1.92
H_2S	0.19	0	0.03	—
CO_2			0.21	

Bibliography

BARNES, H. L. (Editor): Geochemistry of Hydrothermal Ore Deposits. Holt, Rinehart and Winston (1967), 670 pp

BARTON, P. B. jr.: Some Limitations on the Possible Composition of the Ore-forming Fluid. Econ. Geol. 52 (1957), 333—357

DAY, A. L., and E. T. ALLEN: The Volcanic Activity and Hot Springs of Lassen Peak. Washington 1925

GARRELS, R. M., and C. I. CHRIST: Solutions Minerals and Equilibria. Harper and Row, New York 1965

GERMANOV, A. I.: Geological Significance of Organic Matter in the Hydrothermal Process. Geokhimiya (1965) 7, 834—843 (Russian)

GRATON, L. C.: Nature of the Ore-forming Fluid. Econ. Geology 35 (1940), 197—358

GUNTER, B. D., and B. C. MUSGRAVE: Gaschromatographic Measurements of Hydrothermal Emanations at Yellowstone National Park. Geochim. Cosmochim. Acta 30 (1966), 1175—1189

HALL, W. E., and I. FRIEDMANN: Composition of Fluid Inclusions, Cave-in-rock Fluorite District. Illinois, and Upper Mississippi Valley Zinc-Lead District. Econ. Geol. 58 (1963), 886—911

HELGESON, H. C.: Complexing and Hydrothermal Ore Deposition. Pergamon Press, Oxford/London/New York/Paris 1964

HELGESON, H. C.: Thermodynamics of Hydrothermal Systems at Elevated Temperatures and Pressures. Am. Journ. 267 (1970), 729—804

INGERSON, E.: The Concept of a Separable Pneumatolytic Stage in Postmagmatic Ore Formation. Symp. Problems of Postmagmatic Ore Deposition, Vol. II, 457—458, 463—471, Prague 1965

KENNEDY, G. C.: The Hydrothermal Solubility of Silica. Econ. Geol. *39* (1944), 1

KERR, P. F.: Hydrothermal Alteration and Weathering. Spec. Paper *62* (Geol. Soc. America) 1955

KHODAKOVSKIĬ, I. L.: On the Hydrosulphide Form of Heavy Metal Transportation in Hydrothermal Solutions. Geokhimiya (1966) 8, 960—971 (Russian)

KÖNIGSBERGER, J., and W. MÜLLER: Über die Flüssigkeitseinschlüsse im Quarz Alpiner Mineralklüfte. Cbl. Mineral. *341* (1906)

KRAUSKOPF, K. B.: Dissolution and Precipitation of Silica at Low Temperatures. Geochim. Cosmochim. Acta *10* (1956), 1—2

KRAUSKOPF, K. B.: The Source of Ore Metals. Geochim. Cosmochim. Acta *35* (1972), 643—659

MARAKUSHEV, A. A.: Dependence of Oxidation Reduction Processes on the pH Conditions in Post-magmatic Media. Geokhimiya (1960), 214—221 (Russian)

MOREY, G. W., and E. INGERSON: The Pneumotolytic and Hydrothermal Alteration and Synthesis of Silicates. Econ. Geol. *32* (1937) Suppl. N 5

MOREY, G. W., and W. CHEM: The Action of Hot Water on Some Feldspars. Amer. Mineralogist *40* (1955), 11—12

NEWHOUSE, W. H.: The Composition of Vein Solutions as Shown by Liquid Inclusions in Minerals. Econ. Geol. *27* (1932), 5

ROEDDER, E.: Technique for the Extraction and Partial Chemical Analysis of Fluid-filled Inclusions from Minerals. Econ. Geology *53* (1958), 235—269

ROEDDER, E.: Fluid Inclusions as Samples of the Ore-forming Fluids. Rept 21 Sess. Internat. Geol. Congr., Copenhague 1960

ROEDDER, E.: Composition of Fluid Inclusions. U.S. Geol. Survey, Prof. Paper (1971), 440—455

ROEDDER, E., B. INGRAM, and W. HALL: Studies of Fluid Inclusions. III Extraction and Quantitative Analysis of Inclusions in the Milligram Range. Econ. Geol. *58* (1963), 3

SENDEROV, E. E.: On the Bulk Coefficient of Component Activity in the Hydrothermal Solution. Geokhimiya (1966) 5, 600—603 (Russian)

TRUFANOV, V. N.: About the Methods of Determining Reaction of Medium of Mineral-forming Solutions. Geokhimiya (1967) 6, 694—702 (Russian)

TUGARINOV, A. I., and G. B. NAUMOV: Physicochemical Parameters of Hydrothermal Ore Formation. Geokhimiya *3* (1972), 259—265 (Russian)

TYURIN, N. G.: On the Problem of the Composition of Hydrothermal Solutions. Geol. rudn. mestorozhd. *4* (1963) (Russian)

UZUMASA, Y.: Chemical Investigations of Hot Springs in Japan. Toukiji Shokan Co., Ltd. Tokyo

WHITE, D. E.: Thermal Waters of Volcanic Origin. Bull. Geol. Soc. Am. *68* (1957), 1637—1658

YERMAKOV, M. P.: The Significance of Gas and Liquid Inclusions in the Search and Exploration of Postmagmatic Deposits. Soviet Geolog. (1966), 9, 77—90 (Russian)

YERMAKOV, M. P., et al.: Research on the Nature of Mineral-forming Solutions with Special Reference to Data from Fluid Inclusions. "Monographs on the Earth Sciences", Pergamon Press, Oxford 1965

8.2. Geochemistry of Metamorphic Processes

Metamorphism means a change in the constitution of a rock effected by changed conditions (of p, T). The new paragenesis caused in this way follows the phase rule (Section 3.3.5.). The metamorphic rocks are classified according to the principle of mineral facies expounded especially by ESKOLA (1951). A survey is given in Fig. 68; new data have been contributed recently by WINKLER (1965).

Metamorphism can take place with and without the supply of substance from outside (allochemic and isochemic). In the majority of metamorphic processes, a chemically active aqueous phase is involved. In contrast with opinions formerly held, it has been established that most of the regional-metamorphic and, to some extent, contact-metamorphic rocks show an isochemic

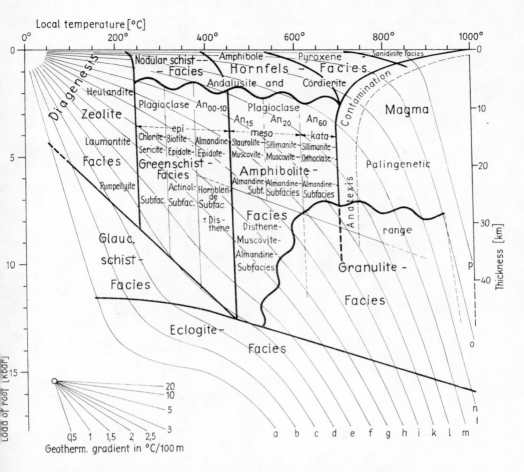

Fig. 68   Relation between the metamorphic facies and the various geothermal gradients in the earth's crust (after TRÖGER, 1963)

conservative) behaviour, that is to say, they mirror the chemical composition of the original paragenetic (sedimentogenetic) or orthogenetic (magmatic) rocks (MEHNERT, 1961; BLACK-BURN, 1968; WIEDEMANN, 1965; cf. Fig. 69). The regularities of the distribution of elements in parametabasites and orthometabasites are shown in Table 109 and Fig. 70. A general survey of the chemical composition of metamorphic rocks is given in Tables 110 and 111.

Allochemic metamorphism is also known as metasomatism or metasomatosis. It occurs especially under the ± direct action of magmas in the upper parts of the earth's crust (e.g. skarns, greisens, spilites, propylites) or deep within the earth (region of fusion; palingenesis, anatexis). Interchange processes of a similar kind are termed transvaporisation by Szádeczky-Kardoss.

267

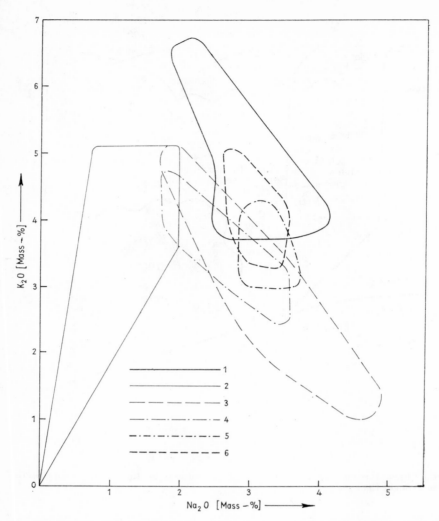

Fig. 69 K_2O/Na_2O ratios of metamorphic rocks from the Saxon Erzgebirge (after WIEDEMANN 1965)
1 Orthogneisses; 2 Mica schist; 3 Marienberg and Annaberg gneisses (paragneisses); 4 Paragneisses of
the mica schist envelope; 5 Upper Freiberg gneisses (paragneisses); 6 Lower Freiberg gneisses

Bibliography

ALTHAUS, E.: Der Einfluß des Wassers auf metamorphe Mineralreaktionen. N. Jb. Miner. Mh. (1968) 9,
289—306
ANGEL, F.: Über Mineralzonen, Tiefenzonen und Mineralfazies, Rückblicke und Ausblicke. Fortschr.
Mineral. *44* (1966) 2, 288—336
BARTH, T. F. W.: Theoretical Petrology. 2nd ed. John Wiley & Sons, New York 1962
BLACKBURN, W. H.: The Spatial Extent of Chemical Equilibrium in Some High-grade Metamorphic
Rocks from the Grenville of Southern Ontario. Contr. Mineral. and Petrol. *19* (1968), 72—92

Table 109. Characteristic Trace Elements for the Identification of Ortho- and Para-metabasites Given Various Publications (Bibliography in LANGE, 1965)

References	Characteristic Trace Elements		
	Ortho-metabasite	Para-metabasite	Unspecific
ENGEL and ENGEL (1951)	higher contents of Co, Cr, Cu, Ni, Sc	corresponding lower values	
ENGEL (1956)	Co, Cr, Ni only locally suited		
DE WIDT (1957)	Cr ($>$170 ppm)	Cr ($<$170 ppm)	
APADU-HARGUES (1958)	higher Ti contents		
WILOX and POLDERVAART (1958)	differentiation according to content of Sr		Ba, Co, Cr, Cu, Ga, Ni, Pb
HAHN-WEINHEIMER (1959)	in eclogites: 0.1—0.5% C, $^{12}C/^{13}C$ = 90.38—91.07		
WALKER et al. (1960)	with increasing metamorphism and metasomatosis, the contents of trace elements change slightly higher contents of Cr, Co, Cu, Ni $TiO_2 > 1.0\%$ (only locally suited)	$TiO_2 < 1.0\%$ (only locally suited)	
LEAKE (1963)	negative correlation of Cr and TiO_2 also of Ni and TiO_2 Cr + Ni $>$ 250 ppm further the relations of certain main elements (or NIGGLI values) and their distribution trends (e.g. C/mg, K/mg, Alk/mg, Ba/mg, Zr/mg, Ti/mg)	positive correlation of Cr and TiO_2 also of Ni and TiO_2	Cr + Ni $<$ 250 ppm
LANGE (1965)	Cr, Ni, Co; negative correlation of Cr/Ti; Ni/Ti; positive correlation of Cr/Ni; the Ni-Co relation; higher Cr and Ni contents; gabbroid macro-chemism	positive and negative correlations, respectively	Ci, V; Ti (as individual value), Cr $<$ 150 ppm, Cr + Ni $<$ 200 ppm

CHINNER, G. A.: Pelitic Gneisses with Varying Ferrous/Ferric Ratio from Glen Clova. J. Petr. *1* (1960)

COOMBS, D. S.: Lower Grade Mineral Facies in New Zealand. XXI Geol. Congr. Norden, Part XIII, 1960

DEVORE, G. W.: The Role of Adsorption in the Fractionation and Distribution of Elements. J. Geology *63* (1955), 159—190

DOMORACKY, N. A.: Definition of the Origin of Metamorphic Rocks on the Basis of Their Content of Inert Matter. Mezd. geol. Kongr., 22 Sess., Dokl. sovet. Geol. Probl. *16* (1964), 166—179 (Russian)

Table 110. Chemical Composition of a Few Metamorphic Rock-types (from CORRENS, 1950; data in mass-%)

	Leptite	Noritic Granulite	Biotite Gneiss	Amphibolite	Eclogite	Andradite Skarn	Disthene Quartzite	Quartz-Muscovite-Chlorite-Phyllite	Hornblende (Glaucophane) Schist	Chlorite Schist with Epidote
SiO_2	78.02	52.03	67.46	49.73	46.26	42.84	89.38	58.32	47.54	49.22
TiO_2	0.11	2.27	0.47	0.56	0.28	0.35	0.16	0.98	1.24	0.18
Al_2O_3	12.22	16.39	15.81	16.05	14.45	4.56	6.57	20.00	19.22	18.56
Fe_2O_3	0.61	0.82	0.17	2.44	4.41	10.60	1.56	2.01	4.58	2.22
FeO	0.86	9.13	3.63	7.96	5.82	8.94	0.50	4.98	2.98	5.35
MnO	0.01	0.17	tr.	0.20	n.d.	0.63	0.01	0.22	0.06	0.12
MgO	tr.	7.04	2.35	7.84	11.99	4.03	0.00	1.85	5.36	8.15
CaO	0.41	8.78	2.97	10.22	11.66	26.24	0.06	0.66	7.90	7.17
Na_2O	4.28	2.14	3.18	2.99	2.45	0.00	0.05	1.26	3.63	4.65
K_2O	3.16	1.21	2.08	0.61	1.51	0.12	0.51	4.49	1.89	0.10
H_2O	0.48	0.35	1.28	1.03	1.10	0.78	1.18	4.05	6.00	3.15
others	—	0.23	0.27	0.12	—	1.00	0.14	1.15	0.05	0.45
Sum	100.16	100.56	99.67	99.75	99.93	100.09	100.12	99.97	100.45	99.32

Table 111. Average Chemical Composition of Metamorphic Rocks in Crystalline Shields
(from EADE, FABRIG and MAXWELL, 1966, data in mass-%)

Element	Regions of the according to REILLY, 1965	Canadian Shield according to EADE et al., 1966	Crystalline Shields (average of earth) according to POLDERVAART, 1955
SiO_2	66.3	64.5	66.4
TiO_2	0.47	0.48	0.6
Al_2O_3	15.4	16.1	15.5
Fe_2O_3	1.3	1.5	1.8
FeO	2.9	2.9	2.8
MnO	0.07	0.08	0.1
MgO	2.1	2.3	2.0
CaO	4.0	3.3	3.8
Na_2O	3.9	4.0	3.5
K_2O	2.3	2.78	3.3
H_2O	0.87	0.8	—
P_2O_5	0.13	0.16	0.2
CO_2	0.15	0.2	—
Sum	99.89	99.10	100.0

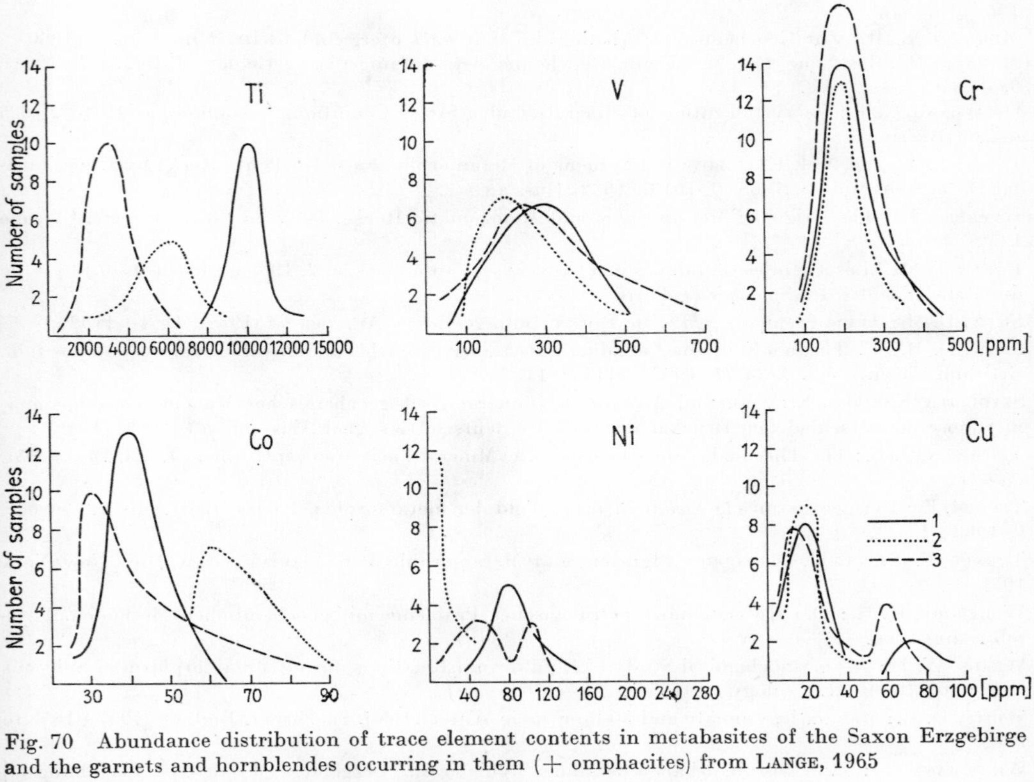

Fig. 70   Abundance distribution of trace element contents in metabasites of the Saxon Erzgebirge and the garnets and hornblendes occurring in them (+ omphacites) from LANGE, 1965

1 Metabasite; 2 Garnet; 3 Omphacite and Hornblende

271

DRESCHER-KADEN, F. K., and J. DITTMANN: Transportvorgänge in Intergranularen. Naturwissenschaften *48* (1961), 217

EADE, K. E., W. F. FABRIG, and J. A. MAXWELL: Composition of Crystalline Shield Rocks and Fractionation Effect of Regional Metamorphism. Nature (1966) 5055, 1245—1249

ESKOLA, P.: Om Mineralfacies. Geol. Fören, Stockholm (1929)

EWERS, W. E.: Physico-Chemical Aspects of Recrystallization. Mineralium Deposita *2* (1967), 221 to 227

FONTEILLES, M.: L'équilibre Chimique dans le Metamorphisme. Bull. Soc. géol. *7* (1965) 1, 136—146

GHOSE, N. C.: Behaviour of Trace Elements during Thermal Metamorphism and/or Granitization of the Metasediments and Basic Igneous Rocks. Geol. Rdsch. *55* (1965) 3, 608—617

GORBATSHEV, R.: Distribution of Elements between Cordierite, Biotite and Garnet. N. Jb. Miner. Abh. *110* (1968) 1, 57—80

GRESENS, R. L.: Composition-volume Relationships of Metasomatism. Chem. Geol. *2* (1967) 1, 47 to 66

HIETANEN, A.: On the Geochemistry of Metamorphism. J. Tennessee Acad. Sci. *29* (1954), No. 4

JAKESH, P.: Studium der koexistierenden Minerale in Gesteinen des Kristallinikums und Probleme der geologischen Thermometrie und Bathymetrie. Geologický průzkum, No. 5, 153—155. Translated by F. Müller, No. 4574

JOVANOVICH, S., and G. W. REED: Hg in Metamorphic Rocks. Geochim. Cosmochim. Acta *32* (1968) 3, 341—346

LANGE, H.: Zur Genese der Metabasite im sächsischen Erzgebirge. Freiberger Forschungsh. C *177* (1965)

MARAKUSHEV, A. A.: Mineral Facies of Metamorphic and Metasomatic Rocks. Izd "Nauka", Moscow 1965 (Russian)

MEHNERT, K. R.: Zur Geochemie der Alkalien im tieferen Untergrund. Beitr. Min. Petr. *7* (1960)

MEHNERT, K. R.: Neue Ergebnisse zur Geochemie der Metamorphose. Geolog. Rdsch. *51* (1961), 384—349

OSTAPENKO, G. T.: Recrystallization of Minerals under Stress Conditions. Geokhimiya (1968) 2, 234 —236 (Russian)

PUGIN, V. A., and N. I. KHITAROV: P-T Scheme of Metamorphic Facies of Pelite Rocks by Experimental Data. Geokhimiya (1968) 9, 1019—1032 (Russian)

RAMBERG, H.: The Origin of Metamorphic and Metasomatic Rocks. Univ. of Chicago Press, Chicago 1952

ROUBAULT, M.: Les Théories Chimiques de l'Orogenèse. Conférence 25. 2. 1965 a l'Institut du Physique du Globe, Faculté des Sciences de Paris

SHAW, D. M.: Trace Elements in Pelitic Rocks. Bull. geol. Soc. America *65* (1954), 1151—1182

SHAW, D. M.: U, Th and K in the Canadian Precambrian Shield and Possible Mantle Composition. Geochim. Cosmochim. Acta *31* (1967), 1111—1113

SZÁDECZKY-KARDOSS, E.: Sinn und Anwendung mineralogisch-geochemischer Modelle. Gesteinsmetamorphose bei verschiedenen Druckarten. Berichte deutsch. Ges. geol. Wiss., B, *13* (1968), 1

THOMPSON, J. B.: The Thermodynamic Basis for the Mineral Facies Concept. Amer. J. Sci. *253* (1955), 65—103

TRÖGER, E.: Der geothermische Gradient im pt-Feld der metamorphen Facies. Beitr. Miner. Petrogr. *9* (1963), 1—12

TURNER, F. J., and J. VERHOOGEN: Igneous and Metamorphic Petrology. McGraw Hill, New York 1951

WEDEPOHL, K. H.: Die Untersuchung petrologischer Probleme mit geochemischen Methoden. Fortschr. Mineralog. *41*

WEEKS, W. F.: A Thermochemical Study of Equilibrium Relations during Metamorphism of Siliceous Carbonate Rocks. J. Geology *64* (1956), 245—270

WHITE, D. E.: Magmatic, Connate and Metamorphic Waters. Geol. Soc. Amer. Bull. *68* (1957), 1659 to 1682

WIEDEMANN, F.: Zum Stoffhaushalt kristalliner Schiefer im Erzgebirge. Freiberger Forschungsh., C *192* (1965)

ꞶINKLER, H. G. F.: Die Genese der metamorphen Gesteine. Springer-Verlag, Berlin/Heidelberg/New York 1965

ꞶODER, H. S.: Role of Water in Metamorphism. Geol. Soc. America, Spec. Paper 62 (1955), 505 ꞏ 524

ꞏroup of authors: Metasomatosis and Other Problems of Physicochemical Petrology. Izd. "Nauka" ꞷeningrad 1968 (Russian)

ꞏ.3. Geochemistry of the Weathering Zone

ꞏ.3.1. Geochemistry of Weathering

Ꞷeathering is one of the most common geochemical processes; it takes place under the ꞷnditions at the earth's surface at low pressure and low temperature. A distinction is made ꞷetween physical weathering (for further information see textbooks of general geology) and ꞷhemical weathering.

ꞷesides the physical destruction of rocks, ores, etc., chemical weathering is of interest within ꞷe scope of this book. Important external conditions for chemical weathering are:

1. the climate (amount of precipitation and temperature),
2. the geological and tectonic state of the earth's crust in the region in question (type of rock, porosity, drainage, tectonic deformation).

ꞷhe main agents involved in the chemical decomposition of rocks are the following:

ꞏ Oxygen

ꞷecrease of the dissolving power of atmospheric oxygen with increasing water temperature. ꞶVater at higher temperature produces a relatively weak oxidation effect.

ꞷor example, the oxidation of pyrite produces free sulphuric acid (besides sulphate and iron ꞷns) and the formation of many new minerals.

ꞷhe zone of oxidation is particularly distinct in the case of a low water table, a pronounced ꞷlief, and a warm or hot climate. In marshy country or regions of permanent frost, it is ꞷractically absent.

ꞏ "Carbonic acid"

CO_2 is produced by the vital processes of organisms, in volcanic processes and by the deꞷomposition of carbonates and organic matter. In surface horizons, carbonic acid is more ꞷctive than silicic acid: decomposition of silicates. Solubility of CO_2 in water decreases with ꞷncrease in temperature.

Ꞷelative contents of CO_2 (related to air = 1; [absolute = 0.03%]):

ꞷ rainwater	300
ꞷ river water and subsurface waters	1,700 to 2,700

ꞏ Water

a) Influence of the hydrogen ion concentration (cf. Section 3.3.7.)
b) Hydration of minerals (silicates, etc.; cf. also Table 115)
c) Water as a solvent, etc.; also in the transport of Fe, Mn and Al as hydroxides.

4. Humic acids and products from the vital processes of organisms

Highly active, decompose silicates, reduce oxidic (Fe-) compounds.
Water from marshy soils is very conducive to weathering.

The resistance to weathering of rocks and their constituent minerals varies considerably (so-called weathering series). A survey of the resistance to weathering offered by heavy minerals and a few other important minerals is given in Table 112.

Table 112. Resistance to Weathering of Some Rock-forming Minerals, Especially Heavy Minerals partly from RUKHIN, 1958)

Non-resistant minerals	Moderately resistant minerals	Resistant minerals	Highly resistant minerals
Olivine	Apatite	Anatase	Limonite
Rhombic pyroxenes	Diopside-	Staurolite	Andalusite
Augite (Fe)	Hedenbergite	Distene	Topaz
Vesuvianite	Allanite	Ilmenite	Brucite
Ordinary amphibole (Na, Fe)	Ca-Fe garnets	Hematite	Leucoxene
Pyrite and many sulphides	Actinolite	Titanite	Chromspinel
Cinnabar	Tremolite	Titanomagnetite	Rutile
Melanite	Epidote	Magnetite	Tourmaline
Glauconite	Zoisite	Monazite	Gold
Biotite	Wolframite	Xenotime	Platinum
Basic plagioclase	Scheelite	Perovskite	Spinel
Feldspathoids	Ottrelite	Columbite	Zircon
Calcite	Axinite	Cassiterite	Corundum
Dolomite	Baryte	Muscovite	Diamond
Gypsum	Sillimanite	Orthoclase-	Quartz
	Interm. plagioclase	Albite	

Geochemically, the degree of weathering can be determined in different ways. A possibility of calculation on the basis of the geochemical stability of Al_2O_3 is given by KRAUSKOPF (Introduction to Geochemistry, 1967, page 100). Another way is the so-called weathering index

$$V_i = \frac{CaO + MgO + Na_2O + K_2O + H_2O}{SiO_2 + Al_2O_3 + Fe_2O_3 + CaO + MgO + Na_2O + K_2O} \quad \text{(in mole)}$$

The relative mobility of the chemical elements in weathering processes is shown in Table 113 and in Table 114. For the solubility of minerals subjected to weathering, the pH value at the boundary surfaces between mineral and water is usually of greater importance than that of the weathering solution in general. This mineral-dependent pH is called abrasion pH and is shown for a few important minerals in Fig. 71.

Table 113. Relative Mobility of Various Elements in Weathering Processes

Type of Rock	Main Elements	Trace Elements
1. Gabbroid Rocks WIEWORA (1967)	$Ca > Na > Mg > Si > K > Al = Fe$	$Mn > Ni > Ti > Cr > V$
2. Porphyrites LISIZYNA (1965)	$Ca > Mg > Na > K > Si > Al = Fe$	$Co > Mn > Ni = Cr = V > Cu > Zr > P = Ti > Ga$

Table 114. Comparison of the Chemical Composition (mass-%) of Crystalline Rocks and of the Dry Residue (mass-%) of River-water, and the Mobility (Schematised) Derived from them of the Chemical Elements or Oxides in Weathering Processes (from RUKHIN, 1958)

Constituents	Average Crystalline Rocks	Average Dry Residue of River-water	Relative Mobility in Weathering Processes	
Al_2O_3	15.35	0.90	0.02 } very	
Fe_2O_3	7.29	0.40	0.04 } insignificant	
SiO_2	59.09	12.80	0.20 insignificant	increasing mobility
K^+	2.57	4.40	1.25	
Mg^{2+}	2.11	4.00	1.31 } moderate	
Na^+	2.97	9.60	2.40	
Ca^{2+}	3.60	14.70	3.00	
SO_4^{2-}	0.15	11.60	60 }	
Cl^-	0.05	6.75	100 } high	

Table 115 shows, in simplified representation, a few important chemical reactions involved in weathering.

Table 115. Chemical Reactions in Weathering of Some Prototype Minerals (simplified representation)

A. Silicates

a) $3\,KAlSi_3O_8 + 2\,H^+$ $\rightleftharpoons KAl_3Si_3O_{10}\,(OH)_2 + 6\,SiO_2 + 2\,K^+$
    $2\,KAlSi_3O_{10}\,(OH)_2 + 2\,H^+ + 3\,H_2O$ $\rightleftharpoons 3\,Al_2Si_2O_5\,(OH)_4 + 2\,K^+$
b) $2\,NaAlSi_3O_8 + 2\,CO_2 + 3\,H_2O$ $\rightarrow 2\,Na^+ + 2\,HCO_3^- + Al_2Si_2O_5\,(OH)_4 + 4\,SiO_2$

a) $Fe_2SiO_4 + 4\,H_2CO_3$ $\rightarrow 2\,Fe^{2+} + 4\,HCO_3^- + H_4SiO_4$
    $2\,Fe^{2+} + 4\,HCO_3^- + {}^1/_2O_2 + 2\,H_2O$ $\rightarrow Fe_2O_3 + 4\,H_2CO_3$
b) $Fe_2SiO_4 + {}^1/_2O_2 + 2\,H_2O$ $\rightarrow Fe_2O_3 + H_4SiO_4$

a) $Mg_2SiO_4 + {}^1/_4H_2O$ $\rightarrow 2\,Mg^{2+} + 4\,OH^- + 4\,H_4SiO_4$
b) $Mg_2SiO_4 + 4\,H_2CO_3$ $\rightarrow 2\,Mg^{2+} + 4\,HCO_3^- + 4\,H_4SiO_4$
 . $SiO_2 + H_2O + OH^-$ $\rightarrow H_3SiO_4^-$

B. Carbonates

 . $CaCO_3 + H_2CO_3$ $\rightleftharpoons Ca^{2+} + 2\,HCO_3^-$
 . $FeCO_3 + {}^1/_2O_2 + 2\,H_2O$ $\rightarrow Fe_2O_3 + 2\,H_2CO_3$
a) $MgCO_3 + H_2O$ $\rightarrow Mg^{2+} + OH^- + HCO_3^-$
b) $MgCO_3 + H_2CO_3$ $\rightarrow Mg^{2+} + 2\,HCO_3^-$

C. Sulphides

 . $2\,FeS_2 + 7^1/_2O_2 + 4\,H_2O$ $\rightarrow Fe_2O_3 + 4\,SO_4^{2-} + 8\,H^+$
a) $PbS + 2\,O_2$ $\rightarrow PbSO_4$
    1st stage $PbS + 2\,H_2CO_3$ $\rightleftharpoons Pb^{2+} + H_2S + 2\,HCO_3^-$
    2nd stage $H_2S + 2\,O_2 + Pb^{2+} + 2\,HCO_3^-$ $\rightarrow PbSO_4 + 2\,H_2CO_3$
 . $ZnS + 2\,O_2$ $\rightarrow Zn^{2+} + SO_4^{2-}$

Table 116. Possibilities of the Formation of Polynuclear Complexes by Hydrolysis (after PACHADZHANOV, 1964)

Cation	Complex $M(A_t M)_n$
Be^{2+}	$Be(OBe)_4^{2+}$; $Be_n(OH)_n^{n+}$
Al^{3+}	within the acid range: $Al_6(OH)_{15}^{2+}$
Al^{3+}	within the alkaline range: $Al(OH)_4^- + Al(OH)_{3t}$
Ga^{3+}	probably analogous to Al^{3+}
Fe^{3+}	$[Fe_4O_3(OH)_5]^+$; $Fe((OH)_2Fe)_n^{n+3}$
Sc^{3+}	$Sc((OH)_2Sc)_{n+1}^{+(3+n)}$
Ti^{4+}	$Ti\left(\dfrac{O}{HTi}\right)_n^{+(n+4)}$
Zr^{4+}, Hf^{4+}	$M(OM)_n^{+(n+4)}$
Th^{4+}	$Th((OH)_3Th)_n^{n+4}$
U^{4+}	$U\left(\dfrac{O}{OHU}\right)_p^{4+n}$
Ce^4	$Ce_6O_4(OH)_4^{12+}$
V^{5+}	$VO_2((OH)_t(VO_2))_n$
Nb^{5+}, Ta^{5+}	$MO_2((OH)_t(MO_2))_n$
U^{6+}	$UO_2\left(\dfrac{(OH)}{O}2\,UO_2\right)_n^{2+}$

M complex forming agent, A bridge group, t constant, n variable

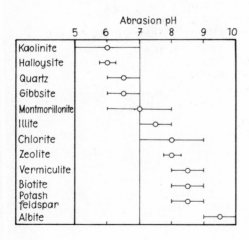

Fig. 71   Abrasion pH values of various mineral (after STEVENS and CARRON, 1948)

Bibliography

ANDERSON, D. H., and H. E. HAWKES: Relative Mobility of the Common Elements in Weathering of some Schist and Granite Areas. Geochim. Cosmochim. Acta *14* (1958), 204—210
BUTLER, I. R.: The Geochemistry and Mineralogy of Rock Weathering, The Lizard Area Cornwall. Geochim. Cosmochim. Acta *4* (1953), 157—178

UTLER, I. R.: The Geochemistry and Mineralogy of Rock Weathering, the Nordmarka Area Oslo.
eochim. Cosmochim. Acta *6* (1954), 268—281

ORRENS, C.: Die experimentelle chemische Verwitterung von Silikaten. Clay min. Bull. *4* (1961),
9—265

ENNEN, W. H., and P. J. ANDERSON: Chemical Changes in the Initial State of Rock Weathering. Geol.
oc. Amer. Bull. *73* (1962), 375—384

OLDICH, S. S.: A Study in Rock Weathering. J. Geology *46* (1938), 17—58

ARRIS, R. C., and I. A. S. ADAMS: Geochemical and Mineralogical Studies on the Weathering of
ranitic Rocks. Amer. J. Sci. *264* (1966), 2

ELLER, W. D.: The Principles of Chemical Weathering. Lucas Bros., Columbia, Miss. 1957, 111 pp.

OUGHNAN, E. C.: Chemical Weathering of Silicate Minerals. Elsevier, Amsterdam/London/New York
969

REICHE, P.: A Survey of Weathering Processes and Products. New Mexico Publ. in Geology (1950)

EUTER, G.: Zur Mineralogie der Verwitterung und Bodenbildung. Ber. deutsch. Ges. geol. Wiss.,
, *13* (1969) 2, 227—236

ONOV, A. B., and A. A. MIGDISOV: Main Features of the Geochemistry of Element-hydrolysates in
rocesses of Weathering and Sedimentation. Geokhimiya (1965) 2, 131—158 (Russian)

TRAKHOV, N. M.: On the Theory of Geochemical Processes in Humid Zones; from: Geochemistry of
edimentary Rocks and Ores. Izd. "Nauka", Moscow 1968 (Russian)

VEYL, P. K.: Environmental Stability of the Earth's Surface—Chemical Consideration. Geochim.
osmochim. Acta *30* (1966), 663—679

VIEWORA, A.: Mineralogical and Geochemical Study of Weathering Processes in Gabbro of Nova Ruda.
rchiwum Mineralogiczne, Warsaw XXVII (1967) (Polish)

.3.2. Geochemistry of Soils and the Oxidation Zone

Soils are the most autochthonic product of rock weathering; the oxidation zone of ore deposits
s an extreme exception. Since the products (minerals) of the soils form the starting material
or the majority of sediments, the knowledge of their development and composition is of great
geochemical importance.

The formation of soil types is above all governed by climate (temperature, precipitation);
a certain minor influence is also exerted by the type of the underlying bedrock, morphology,
vegetation, geological-tectonic situation, and other factors. The soil is divided into several
oil horizons (A, B, C, partly D). For further data, special books of pedology should be con-
ulted (LAATSCH, 1957; FIEDLER, 1968; MILLAR and FURK, 1943 a. o.).

The most important main constituents of soils are, besides rock fragments, inert minerals
(quartz), the silicate clay minerals, and in particular kaolinite, illite, montmorillonite, and
chlorite. Information about their formation and properties are given in the works by GRIM,
1953; JASMUND, 1955; RÖSLER-STARKE, 1969.

The elements occur mostly zonally as they are adsorbed by the clay minerals, and other
constituents (hydroxides, carbonates, SiO_2 minerals, and organic matter). Profiles which are
particularly typical of weathering are exhibited by fossil and recent formations such as podsol
soils, *dark-colored zonal* soils (chernozem), laterite soils, *weather-beaten* kaolin soils, saline soils
(solonchak, solonetz).

In spite of the considerable variance in chemical composition of soils and soil profiles, the
average of all soils does not show a significant difference from the composition of the lithosphere, as is demonstrated by Table 117. As examples of a relatively insignificant and a higly
pronounced differentiation of the minerals and chemical elements. Tables 118 and 119 show
two profiles from humid Central Europe and one profile from tropical South America. Fig. 72
shows a schematic representation of the consistency conditions in soils.

Table 117. Mean Contents of Chemical Elements in Soil Compared
with the Contents in the Lithosphere; compiled from VINOGRADOV (1954, soil)
and (1962, lithosphere)

Element	Lithosphere	Soil
O	47.0 mass-%	49.0 mass-%
Si	29.5	33.0
Al	8.05	7.13
Fe	4.65	3.8
Ca	2.96	1.37
Na	2.5	0.63
K	2.5	1.36
Mg	1.87	0.6
Ti	4,500 ppm	4,600 ppm
Mn	1,000	850
P	930	800
F	660	200
Ba	650	500
S	470	850
Sr	340	300
C	230	20,000
Cl	170	100
Zr	170	300
Rb	150	60
V	90	100
Cr	83	200
Zn	83	50
Ce	70	(50)
Ni	58	40
Cu	47	20
Li	32	30
Ga	19	(x · 10)
N	19	1,000
Co	18	8
Pb	16	10
Th	13	6
B	12	10
Be	3.8	x
Cs	3.7	(5)
Sn	2.5	(x · 10)
U	2.5	1
As	1.7	5
Ge	1.4	x
Mo	1.1	3
I	0.4	5
Cd	0.13	(0.5)
Hg	0.083	0.0x
Se	0.05	0.01
Ra	$1 \cdot 10^{-6}$	$8 \cdot 10^{-7}$

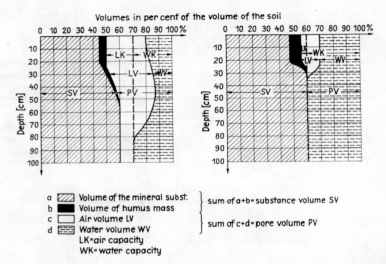

Volumes in per cent of the volume of the soil

Fig. 72   Volume diagrams of two different soil types; on the left: fertile loose soil; on the right: heavy dense soil (after LAATSCH, 1957)

Oxidation zone of ore deposits

The weathering conditions of ore deposits, especially of sulphidic and sulphide-containing bodies, are highly complex and of great importance to the (geochemical) exploration of deposits. At present, the best introduction to these problems is given by S. S. SMIRNOV (1954). This work contains all of the significant data about the solubility of minerals, mineral paragenesis, the chemical behaviour of the individual elements, and information about the possibilities of prospecting.

Table 118. Changes of Chemical Compositions within a Laterite Profile in Venezuela (after SHINTSOVETS 1965, in mass-%)

Depth [m]	Rock	SiO$_2$	TiO$_2$	Al$_2$O$_3$	Fe$_2$O$_3$	FeO	MnO	MgO	CaO	Na$_2$O	K$_2$O	H$_2$O
1—2	iron-rich Bauxite	0.10	3.94	20.70	59.64	—	0.07	—	0.05	—	—	13.53
3—4	iron-rich Bauxite	0.16	3.80	27.09	50.34	—	0.07	—	0.05	—	—	17.20
5—6	iron-rich Bauxite	0.16	3.65	36.56	37.84	—	0.07	—	0.04	—	—	20.81
11—13	iron-rich Bauxite	3.80	3.67	36.82	34.85	—	0.06	—	0.06	—	—	18.55
14—14.3	Laterite	29.79	2.79	28.29	23.62	—	0.05	Tr.	0.10	—	—	11.55
<14.3	Diabase	48.60	1.48	15.83	1.99	11.43	0.12	7.51	9.21	2.45	1.14	0.27

Table 119. Physico-chemical and Chemical Variation in Two Profiles of Soils of the Saxon Erzgeb[i] (Tharandt Forest near Dresden; after LENTSCHIG and FIEDLER, 1967)

Soil Type I: Humus — iron — podsol   Soil Type II: Pseudogley/Brown Earth

Type	Soil Horizon		Depth up to [cm]	pH in KCl	Sorption mval/100 g	Values in % related to dry substa				
						SiO_2	Al_2O_3	Fe_2O_3	TiO_2	Mn
I Podsol on Porphyry	A_0	black fine humus	$\leq$ 3	2.7	10.2					
	A_1	humous loam	— 2	2.9	4.9	66.01	7.20	1.84	0.96	0.0?
	A_2	grey loam	— 7	3.9	0.9	77.96	8.81	2.07	1.02	0.0?
	B_1	red-brown loam	— 18	3.8	3.5	65.41	12.09	3.99	1.02	0.0?
	B_2	yellow-brown loam	— 30	4.1	3.9	68.72	13.73	3.15	0.92	0.0?
	B/C	sandy-loam	—110	4.1	1.4	71.67	14.40	2.43	0.82	0.0?
	C	quartz-porphyry	<110			73.97	14.64	1.55[1]	0.17	0.01
II Brown Earth on Gneiss	A_0	black fine humus	$\leq$ 6	3.1	22.6					
	A_1	humous loam	— 3	3.1	4.0	68.73	8.48	3.15	0.42	0.0?
	B_{1+2}	brown loam 1	— 15	3.3	4.8	75.10	11.25	3.96	0.54	0.0?
		2	— 28	3.7	2.6	74.94	11.125	4.08	0.45	0.0?
	(B)	mottled loam	— 60	4.0	1.9	70.36	14.33	5.52	0.50	0.0?
	B/C$_1$	loamy coarse sand	—105	3.9	5.9	63.22	17.90	7.19	0.61	0.08
	B/C$_2$	coarse-sandy loam	—140	3.8	8.8	65.02	17.90	5.56	0.59	0.0?
	C	Freiberg gneiss	<140			68.31	15.21	1.52[2]	0.48	0.0?

[1]) +0.22% FeO   [3]) GV = Glühverlust (heat loss)
[2]) +2.62 FeO

Bibliography

AUBERT, G.: Les Micro-elements dans les Sols. Comm. Soc. Geochim. Paris XII, 1956, Bull. Pedo ORSTOM 7 (1957)

BERGMANN, W.: Die Bedeutung der Nährstoffuntersuchungen der Böden für die Bodenfruchtbarkeit. Sitz.-Ber. Dt.Akad. Landwirtsch.-Wiss., Berlin 14 (1965) 9

BEAR, F. E. (ed.): Chemistry of the Soil. Reinhold Publ. Corp., New York 1964 (second edition).

CROOKE, W. M.: Chemistry of the Soil. Chem. Industry, G. B. 31 (1965), 1373—1375

FIEDLER, H.-J., and H. REISSIG: Lehrbuch der Bodenkunde. Gustav Fischer Verlag, Jena 1964

GRIM, R. E.: Clay Mineralogy. McGraw Hill, New York/London/Toronto 1953

JASMUND, K.: Die silikatischen Tonminerale. Weinheim 1955

LAATSCH, W.: Dynamik der mitteleuropäischen Mineralböden. Theodor Steinkopf, Dresden/Leipzig 1957

LENTSCHIG, S., and H. J. FIEDLER: Beitrag zur Kenntnis der Böden des Tharandt-Grillenburger Waldes. III. Chemische und mineralogische Untersuchungen usw. Abh. Staatl. Mus. Mineral. Geol., Dresden 12 (1967), 229—258

LENTSCHIG, S., and H. J. FIEDLER: Zum Spurenelementgehalt von Braunerde- und Podsolprofilen des Mittelgebirges. Abh. Staatl. Mus. Miner. Geol., Dresden 11 (1966) 281—306

LUKASHEV, K. I.: Fundamentals of the Lithology and Geochemistry of the Weathering Crust. Acad. Sc. Beloruss. Minsk 1958, 470 pp. (Russian)

MALYUGA, D. P., and A. I. MAKAROVA: On the Contents of Trace Elements in a Few Soils, Developed in the Ore Deposits Pochvaredene (1956) 1, 50—53 (Russian)

MIGDISOV, A. A., and L. A. BORISENKO: On the Geochemistry of Ga in the Humid Lithogenesis. Geokhimiya 12 (1963), 1067—1081 (Russian)

MgO	K$_2$O	Na$_2$O	P$_2$O$_5$	GV[3]	Trace Elements, values in ppm						Pb	Cu	Ga	B
					Ba	Sr	Ni	Co	Cr	V				
					100	100	20	8	20	20	400	40	4	10
0.20	1.84	0.90	0.04	20.95	700	200	20	20	100	60	2,000	10	10	60
0.15	2.29	1.31	0.01	6.75	400	100	10	20	90	30	300	<3	10	30
0.57	2.38	1.00	0.03	12.10	500	60	30	10	90	50	200	<3	<3	30
0.57	2.82	1.07	0.03	9.30	500	70	20	10	80	20	50	10	<3	20
0.07	3.88	2.05	0.02	4.40	500	30	30	20	30	20	100	<3	10	30
0.36	5.30	3.40	0.02	0.62	500	40	10	10	3	10	100	3	10	10
					300	30	20	<1	10	30	200	20	7	10
0.34	2.40	1.14	0.20	15.70	300	80	20	30	50	30	400	70	10	20
0.50	2.28	1.23	0.10	4.42	400	80	20	40	60	30	300	20	10	20
1.00	2.95	1.55	0.10	3.35	300	40	20	40	40	20	40	10	10	<3
1.45	2.46	1.50	0.10	3.16	700	200	40	50	100	30	20	100	10	30
1.82	2.75	2.06	0.14	3.18	600	100	30	10	90	40	30	50	10	20
1.00	4.63	2.00	0.13	3.43	500	50	50	50	80	60	30	40	20	20
1.35	4.33	3.00	0.10	1.27	1,000	200	50	50	40	80	100	3	10	30

MILLAR, C. E., and F. M. FURK: Fundamentals of Soil Science. Wiley, New York 1943

PERELMAN, A. I.: Geochemical Types of Weathering Crusts and Terrestrial Formations on the Territory of the USSR. Kora Vyvetr. 5 (1963), 127—137 (Russian)

PERELMAN, A. I.: Geochemistry of the Weathering Crust (Survey). Priroda (1957) 6, 9—18 (Russian)

PINTA, M., and C. OLLAT: Recherches Physico-chimiques des Eléments-traces dans les Sols Tropicaux. Etudes de Quelques Sols du Dahomey. Geochim. Cosmochim. Acta 25 (1961), 14—23

RUKHIN, L. B.: Grundzüge der Lithologie. Akademie-Verlag, Berlin 1958

SHCHERBINA, V. V.: Geochemistry of Uranium in the Oxidation Zone of Ore Deposits on the Basis of Experimental Study Data. Proc. 2nd U. N. Intern. Conf. Peaceful Uses of Atomic Energy 2 (1958), 211—214

SHCHERBINA, V. V.: The Behaviour of a Few Rare and Disperse Elements in the Zone of Hypergenesis. Sov. Geol. (1962) 6 (Russian)

SCHEFFER, F., E. WELTE, and F. LUDWIEG: Zur Frage der Eisenoxidhydrate im Boden. Chem. d. Erde 19 (1958), 51—64

SMIRNOV, S. S.: Die Oxydationszone sulfidischer Lagerstätten. Akademie-Verlag, Berlin 1954 (Transl. from Russian)

SWAINE, D. J.: The Trace-element Content of Soils. Commonwealth Bur. Soil Sci., Techn. Comm. No. 48, 157

VINE, J. D., V. E. SWANSON, and K. G. BELL: The Role of Humic Acids in the Geochemistry of Uranium. Proc. 2nd U. N. Intern. Conf. Peaceful Uses of Atomic Energy 2 (1958), 187—191

VINOGRADOV, A. P.: Geochemie seltener und nur in Spuren vorhandener chemischer Elemente im Boden. Akademie-Verlag, Berlin 1954 (Transl. from Russian)

WOLFENDEN, E. B.: Geochemical Behaviour of Trace Elements during Bauxite Formation in Sarawak, Malaysia. Geochim. Cosmochim. Acta 29 (1965), 1051—1062

8.4. Geochemistry of Inorganic Sediments and Sedimentites

8.4.1. Transport, Sedimentation and Facies, Diagenesis

The material produced by weathering is transported in ionic, colloidal and clastic forms by various media (air, water, ice, organisms) and deposited in sediments (rate of erosion see Table 120). Further separations into mineral phases and elements take place in this process, leading to an enrichment or a depletion as compared with the pre-existing rock.

Table 120. Periods, Thickness and Rates of Denudation (after MENARD, 1961; from GASKELL, 1967)

Region	Period of Erosion m. y.	Thickness of Denuded Rock [km]	Past Rate of Erosion [mm/yr]	Present Rate of Erosion [mm/yr]
Appalachians	125	7.8	0.062	0.008
Mississippi Basin	150	11.1	0.046	0.042
S. Himalayan Region	40	8.5	0.21	0.100
Rocky Mountains	40	—	0.21	—

1. *Transport*

The best information about the regularities of transport and distribution of coarse clastic components is given in the works on sediment and heavy mineral analyses (MILNER, 1952; PETTIJOHN, 1949; v. ENGELHARDT, 1960; RUKHIN, 1958). Especially interesting geochemically is the fine-clastic, colloidal and ionic transport in river-waters. Fig. 73 and Table 121 demonstrate the paramount importance of the transport in solid form (in suspension). The quantities transported by rivers are given in Tables 122, 123 and 124.

Table 121. Average Content of Some Trace Elements in Rivers of the Soviet Union, in [mg/l]

Element	In Suspension	In Solution
V	50— 165	2— 5
Mn	10—20,000	100—200
Ni	10— 230	5—150
Cu	5— 220	3—247
Co	2— 43	

Table 122. Amount of Substance [t/yr] transported in the (Filtered) Water of a Few Rivers of the Territory of the German Democratic Republic (after LEUTWEIN and WEISE, 1962)

Consti-tuents		Mulde River		Mulde Tributaries	
				Zschopau	Zwickau Mulde
	at km ... of the river	76 (near Nossen)	245 (when joining the Elbe)	115 (mouth)	130 (mouth)
	catchment area [km²]	567.4	7,160	1,845	2,353
SiO_2		1,785	34,400	4,620	7,100
Al		132	592	48	167
Fe		207	2,920	192	604
Mn		176	2,080	183	685
Na		2,898	251,500	1,184	6,910
K		1,572	24,420	3,660	6,980
Mg		3,150	59,200	10,320	16,000
Ca		15,580	256,200	21,400	56,400
Ba		12.3	74.0	38.6	35.5
Sr		12.3	444.0	57.8	106.7
Pb		61.7	74.0	241.0	118.2
Cu		46.3	925.0	106.0	94.5
Zn		179	370	396	331

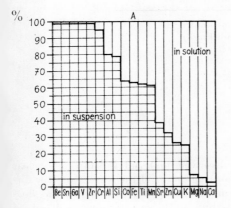

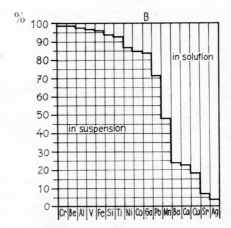

Fig. 73   Type of migration of 19 elements in West Siberian (A) and Central Siberian (B) rivers (after KONTOROVICH, 1968)

283

Table 123. Quantities of SiO_2 Fed into the Mediterranean
(after SCHINK, 1961; values in 10^3 mole/sec.)

	Maximum	Minimum	Optimum Mean
Nile	0.7	0.4	0.4
Other rivers	4.2	1.6	3.2
Black Sea	0.5	0.2	0.3
Sum	5.4	2.2	3.9

Table 124. Mean Absolute Content of Dissolved Substances in Rivers and Lakes

a) Average content (according to MASON, 1966)

b) Filtered water of the Mulde river when discharging into the Elbe (according to LEUTWEIN and WEISE, 1962)

c) according to KRAMER (1961); pH 7.7

	a) Average of Rivers (ppm)	b) Mulde River (mg/l)	c) Lake Erie (mole/l)
Ca	15	69.3	$9.6 \cdot 10^{-4}$
Na	6.3	68.1	$4.3 \cdot 10^{-4}$
Mg	4.1	16.0	$3.7 \cdot 10^{-4}$
K	2.3	6.6	$4.9 \cdot 10^{-5}$
Fe	0.7	0.79	not known
Mn	not known	0.56	not known
Al	not known	0.16	not known
Sr	not known	0.12	not known
Zn	not known	0.10	not known
Cu	not known	0.25	not known
Ba	not known	0.02	not known
Pb	not known	0.02	not known
SiO_2	13.1	9.3	not known
HCO_3	58.5	not known	not known
SO_4	11.2	not known	$2.5 \cdot 10^{-4}$
Cl	7.8	not known	$2.5 \cdot 10^{-4}$
NO_3	1	not known	—
F	not known	not known	$5 \quad \cdot 10^{-6}$

2. Sedimentation and facies

Sedimentation

The sedimentation of coarse clastic, colloidal and dissolved material is subject to clear regularities. In particular it is dependent on the transporting power of the water (and the air) as far as coarse clastic components are concerned, on the possibility of coagulation (flocculation) of colloids, and on the solubility (solubility product) in the case of ionically dissolved substances.

The general laws on the succession of minerals and elements are shown in Fig. 74. As to the meaning of the terms resistates, hydrolysates and evaporates see Table 127, Section 8.4.2.1.

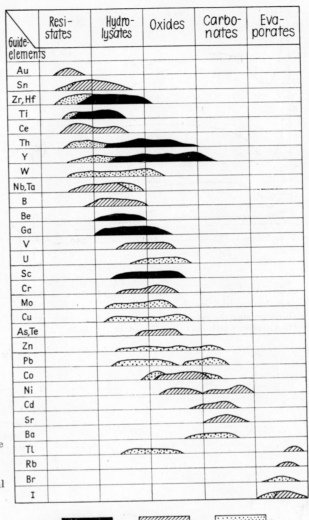

Guide elements	Resi-states	Hydro-lysates	Oxides	Carbo-nates	Eva-porates
Au					
Sn					
Zr, Hf					
Ti					
Ce					
Th					
Y					
W					
Nb, Ta					
B					
Be					
Ga					
V					
U					
Sc					
Cr					
Mo					
Cu					
As, Te					
Zn					
Pb					
Co					
Ni					
Cd					
Sr					
Ba					
Tl					
Rb					
Br					
I					

Fig. 74 Precipitation succession of some chemical elements in sediments (from GINSBURG, 1963)

1 after GOLDSCHMIDT; 2 after several authors; 3 after SMULIKOVSKI

Figs. 75 to 78 give an idea of the distribution of coarse clastic products with respect to distance of transportation, climate and range of sedimentation.
Fig. 79 gives an example of the distribution of substances which are mainly transported in the colloidal state. As to the fundamentals, reference is made to Section 3.3.8.
The usefulness of the ionic potential for the interpretation of the distribution of solute elements in sediments is shown in Fig. 80.
Table 125 gives a survey of the rates of sedimentation.

285

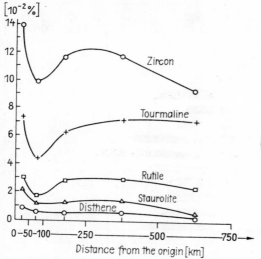

Fig. 75 Variations in the mean concentration of zircon, tourmaline, rutile, staurolite and disthene with increasing distance from their origin, in sands of the Russian platform (after RONOV et al., 1963)

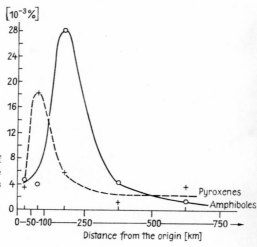

Fig. 76 Variations in the mean concentration of pyroxenes and amphiboles with increasing distance from their origin in sands of the Russian platform (after RONOV et al., 1963)

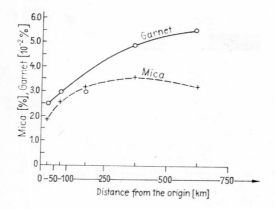

Fig. 77 Variations in the mean concentration of garnet and mica in sands of the Russian platform with increasing distance from their origin (after RONOV et al., 1963)

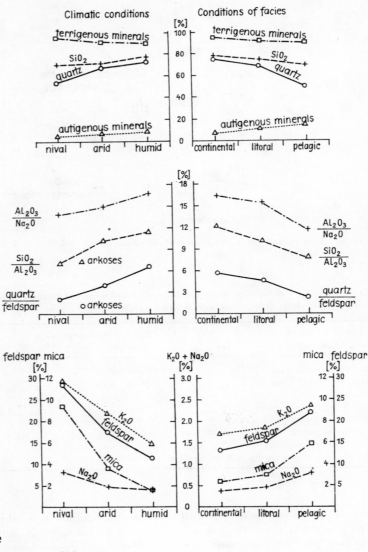

g. 78 Distribution of
rigenous minerals and
mponents in sandy rocks
the Russian platform
a function of climate
d facies (after RONOV et
, 1963)

Climatic conditions

Conditions of facies

terrigenous minerals

terrigenous minerals

SiO_2

SiO_2

quartz

quartz

autigenous minerals

autigenous minerals

nival   arid   humid

continental   litoral   pelagic

$\dfrac{Al_2O_3}{Na_2O}$

$\dfrac{Al_2O_3}{Na_2O}$

$\dfrac{SiO_2}{Al_2O_3}$ arkoses

$\dfrac{SiO_2}{Al_2O_3}$

$\dfrac{quartz}{feldspar}$ arkoses

$\dfrac{quartz}{feldspar}$

nival   arid   humid

continental   litoral   pelagic

feldspar mica [%]

$K_2O + Na_2O$ [%]

mica feldspar [%]

K_2O

K_2O

feldspar

feldspar

mica

mica

Na_2O

Na_2O

nival   arid   humid

continental   litoral   pelagic

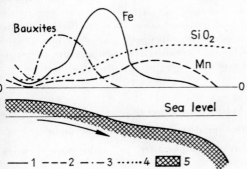

Fe

Bauxites

SiO_2

Mn

Sea level

1 ——— 2 - - - 3 —··— 4 ······ 5

Fig. 79 Schematic representation of the sequence of deposition of clays, iron oxides,
manganese oxides, silicic acid with increasing distance from the place of erosion
(after STRAKHOV; from RUKHIN, 1958)
Distribution curves for the deposition of

1 iron ores, 2 manganese ores, 3 bauxites, 4 siliceous rocks, 5 earth's surface

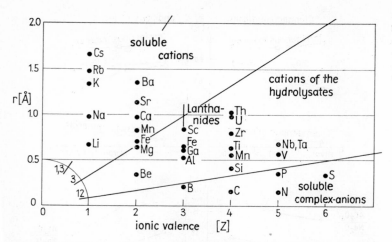

Fig. 80   Distribution of some chemical elements in accordance with their ionic potential in sedimentary formations

Table 125. Annual Rate of Sedimentation of Argillaceous Sediments and Evaporites (according to BORCHERT, 1965, and STRAKHOV, 1965)

Type of Sediment or Area of Sedimentation	Rate of Sedimentation [mm/yr]
Pacific (pelagic part)	0.001—0.000,5
Atlantic (pelagic part)	0.009—0.002
Black Sea (shelf)	0.2 —0.3
(pelagic part)	0.05
Bituminous shale and marl shale	0.05—0.2
Dolomites and bituminous anhydrites	0.4 —1
Non-bituminous anhydrites	2— 20 ($\varnothing$ 5)
Rock salt and potassium salt minerals	20—100 ($\varnothing$ 30)

Facies

The concept of sediment facies is defined as a system of factors of palaeogeographic-climatic kind which determines the lithologic-geochemical character of a sediment. Facies are divided into two main groups, namely, terrestrial-limnal and marine facies; they are further divided into red-bed facies, delta facies, lagoonal, brackish, littoral, sapropelic and other facies types. Besides the geological-palaeontologic methods of facies determination such as grain structure, color, bedding conditions, age, contents of fossil remains, there are geochemical methods.

The most important of the latter are

   a) mineral composition of the sediments (including clay minerals)

   b) macrochemistry and microchemistry and isotope distribution of the sediments

   c) environmental factors such as redox potential, temperature, pH value and chemical composition of the water.

equently used geochemical methods are trace element analysis or relations of elements ich are critical with respect to facies (e.g. B, V, V/Cr), *Eh-pH* measurements, and oxygen tope measurements for the determination of palaeotemperature. Detailed discussions of ese problems (with comprehensive bibliography) are given by DEGENS, 1959; KREJCI-GRAF, 66; RÖSLER et al., 1971, and others. The practical use and the importance of "geochemical rriers" is discussed at some length in Section 10.4. For the use of isotopes in facies analyses Section 10.3.

g. 81 shows the regularities of the formation of minerals in sediments with respect to *pH* d *Eh*. Figs. 82 and 83 show the results of laboratory model tests, while Fig. 84 is an instruc- ze geological example of the dependence on morphology, salinity, *Eh* and *pH*. The role of rption is shown in Table 126.

ble 126. Geochemical Concentration of Elements in Sediments Due to Sorption

ements	Adsorptive Adhesion to
, Rb, Cs; heavy metals ch as Au, Ag, V, Hg; H₄, As	argillaceous sediments and soils
, V, P	iron hydroxides
V, Mo	siliceous sediments
, Ba, Ni, Co, Cu, Zn, Hg, ₁, W and many others	manganese hydroxides

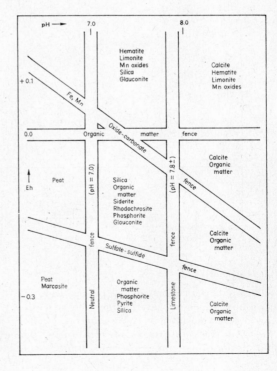

ig. 81 Mineral formation in sediments as a nction of *Eh* and *pH* (after KRUMBEIN and ARRELS, 1952; from MASON, 1966)

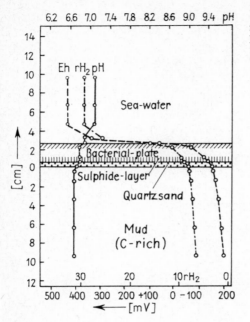

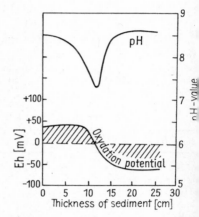

Fig. 82 *Eh, pH* characteristics of a seawater/mud medium with a sulphuretum and precipitation of heavy metals; after a model test, duration 48 days (from SUCKOV, 1963)

Fig. 83 *pH* and *Eh* values in sediment of the gyttja type (after TROFIMOV, 1939)

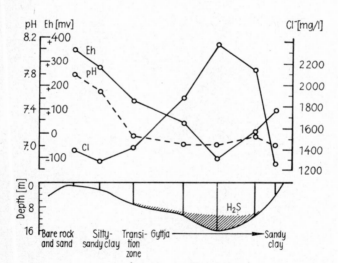

Fig. 84 The distribution of *Eh*, *pH* and Cl⁻ in a fiord near Stockholm (from LANDERGREN and MANNHEIM, 1963)

3. *Diagenesis*

Diagenesis is a reconstructive process in which the products of sedimentation, products which are mostly rich in water (large void spaces, cf. von ENGELHARDT, 1960) are compacted. This process is associated with an emigration of the solvent (water) and eventually leads to considerable recombinations of substances. Among other things, they become manifest by formation of new minerals (concretions, cementation of the clastic supporting structure) and

crystallization of minerals. Frequently, the organic matter content plays an important part
f. GRÜNDEL and RÖSLER, 1963). Full details are given by LARSEN and CHILLINGAR, 1967;
EGENS, 1959; CORRENS, 1950; and STRAKHOV, 1953.

ibliography (See also pp 387 to 389)

GAMIROV, S. S.: The Geochemical Equilibrium of the Radioactive Elements in the Black Sea Basin
eokhimiya (1963) 6, 612—614 (Russian)
HRENS, L. H.: Ionization Potentials and Metal-Amino Acid Complex Formation in the Sedimentary
ycle. Geochim. Cosmochim. Acta 30 (1966), 1111—1119
ORRENS, C. W.: Zur Geochemie der Diagenese. Geochim. Cosmochim. Acta 1 (1950), 49—54
EGENS, E. T.: Die Diagenese und ihre Auswirkungen auf den Chemismus von Sedimenten. Neues Jb.
eol. Paläont., Mh. 2 (1959), 72—83
EGENS, E. T., E. G. WILLIAMS, and M. L. KEITH: Environmental Studies of Carboniferous Sediments.
ull. Amer. Ass. Petrol. Geol. Part I 41 (1957), 2427—2455. Bull. Amer. Ass. Petrol. Geol. Part II
2 (1958), 981—997
NGELHARDT, W. V.: Der Porenraum der Sedimente. Min. Petrogr. Einzeldarst. 2, Berlin 1960
ASKELL, T. F.: The Earth's Mantle. Academic Press, London and New York 1967
INSBURG, I. I.: Grundlagen und Verfahren geochemischer Sucharbeiten auf Lagerstätten der Bunt-
etalle und seltener Metalle. Akademie-Verlag, Berlin 1963 (Transl. from Russian)
LAGOLEVA, M. A.: Migration of the Chemical Elements in Riverwater. Dokl. Akad. Nauk SSSR
21 (1958), 1052—1055 (Russian)
RÜNDEL, J., and H J. RÖSLER: Zur Entstehung der oberdevonischen Kalkknollengesteine Thürin-
ens. Geologie 12 (1963) 9, 1009—1038
ULYAEVA, L. A.: Geochemical Indicators of the Redox Conditions in Terrigenous Sediments.
zd. "Nauka", Moscow 1968 (Russian)
oczy, F. F.: Factors Determining the Element Concentration in Sediments. Geochim. Cosmochim.
cta 1 (1951), 73—85
ONTOROVICH, A. E.: The Forms of Migration of the Elements in Rivers of Humid Zones; from: Geo-
hemistry of Sedimentary Rocks and Ores. Izd. "Nauka", Moscow 1968 (Russian)
RAUSKOPF, K. B.: Factors Controlling the Concentration of Thirteen Rare-metals in Sea Water.
eochim. Cosmochim. Acta 9 (1956), 1
RAUSKOPF, K. B.: Dissolution and Precipitation for Silica at Low Temperatures. Geochim. Cosmo-
him. Acta 10 (1956), 1—26
REJCI-GRAF, K.: Geochemical Diagnosis of Facies. Proc. Yorkshire Geol. Soc. 34 (1964) 4, 23, 469
— 521
RUMBEIN, W. C., and R. M. GARRELS: Origin and Classification of Chemical Sediments in Terms of pH
nd Oxidation-Reduction Potentials. J. Geology 60 (1952), 1—33
ANDERGREN, S., and F. T. MANHEIM: Über die Abhängigkeit der Verteilung von Schwermetallen von
er Fazies. Fortschr. Geol. Rheinl. Westf. 10 (1963), 173—92
ARSEN, G., and G. V. CHILLINGAR (Editors): Diagenesis in Sediments. Elsevier, Amsterdam/London/
ew York 1967, 551 pp.
EUTWEIN, F., and L. WEISE: Hydrogeochemische Untersuchungen an erzgebirgischen Gruben- und
berflächenwässern. Geochim. Cosmochim. Acta 26 (1962), 1333—1348
ARCHENKOV, V. I.: The Importance of the Oxidation Coefficient to the Facies Analysis of Marine
ediments. Lithology and Mineral Resources 4 (1965), 125—137 (Russian)
ILNER, H. B.: Sedimentary Petrography. Murby and Co. 1952
OLKHANOV, V. I.: Sedimentation and Free Hydrogen. Dokl. Akad. Nauk SSSR 182 (1968) 2, 445
— 448 (Russian)
NICHOLLS, G. D.: Environment Studies in Sedimentary Geochemistry. Sci. Progress 51 (1963), 12
— 31

OELSNER, O.: Bemerkungen zur Herkunft der Metalle im Kupferschiefer. Freiberger Forschungsh. C 58 (1959), 106—113

PACHADZHANOV, D. N.: Geochemical Fundamentals of Element Hydrolysates in the Sedimentation Process and Their Explanation. Geokhimiya 12 (1964), 1286—1292 (Russian)

PECK, A. I.: Mass Transport in Porous Rocks. Mineralium Deposits, 2 (1967), 243—247

PERELMAN, A. I.: Geochemistry of the Epigenetic Processes (The Zone of Hypergenesis). 3rd ed. Izd "Nedra", Moscow 1968 (Russian)

PETTIJOHN, F. J.: Sedimentary Rocks. Harper and Br., New York 1949

PORRENGA, D. H.: Clay Mineralogy and Geochemistry of Recent Marine Sediments in Tropical Areas. Diss. Amsterdam 1967

PUSTOVALOV, L. V.: The Geochemical Facies and Their Importance to General and Applied Geology. Problems of Soviet Geology 1 (1933) 1, 57 (Russian)

RAUPACH, F. v.: Die rezente Sedimentation im Schwarzen Meer, im Kaspi und im Aral und ihre Gesetzmäßigkeiten. Geologie 1 (1952), 78—132

RÖSLER, H. J., H. LANGE, et al.: Analyse der geochemischen Ablagerungsbedingungen von Sedimenten. Freiberger Forschungsh. C 272, 9—30, Leipzig 1971

RUKHIN, L. B.: Grundzüge der Lithologie. Akademie-Verlag, Berlin 1958 (Translation from the Russian)

SPENCER, D.: Factors Affecting Element Distributions in a Silurian Graptolite Band. Chem. Geol. 1 (1966), 221—249

STRAKHOV, N. M.: The Problem of the Distribution and Accumulation of the Main Chemical Constituents in the Sediments of the Recent and Old Water Basins. Deliberation on Sedimentary Rocks. Dokl. Akad. Nauk SSSR. 1 (1952), 183 (Russian)

TEODOROVICH, G. I.: Sedimentary Mineralogy-Geochemical Facies; in: Problems of the Mineralogy of Sediments 3—4 (1956), Univ. Lvov (Russian)

TONNDORF, H.: Beiträge zur Geochemie des randnahen Zechsteins in den Mulden von Zeitz-Schmölln und Borna, unter besonderer Berücksichtigung der Stratigraphie, Fazies und Paläogeographie. Freiberger Forschungsh. C 187 (1965), 96 pp.

TROFIMOV, A. V.: Oxidizing Activity and pH of Brown Sediments of the Barentz Sea. Dokl. Akad. Nauk SSSR 23 (1939), 9, 925—928

TSARICHKII, P. V.: On the Intensity of the Diagenetic Double Decomposition of Substances. Lithologiya i poleznye iskopaemye 2 (1968), 119—130 (Russian)

TWENHOFEL, W. H.: Principles of Sedimentation. McGraw Hill, London 1950

VOLKOV, I. I., and V. F. SEVASTYANOV: Double Decomposition of Chemical Elements in the Diagenesis of Sediments of the Black Sea. from: Geochemistry of Sedimentary Rocks and Ores. Izd. "Nauka", Moscow 1968 (Russian)

WATTENBERG, H.: Die Bedeutung anorganischer Faktoren bei der Ablagerung von $CaCO_3$ im Meer. Kieler Meeresforsch. 2 (1937), 81—93

8.4.2. Geochemistry of Argillaceous, Arenaceous, and Carbonate Sediments

8.4.2.1. (Recent) Sediments, especially of the Deep Sea

A geochemical classification which is still widely used was expounded by Goldschmidt; it is shown in Table 128, including brief explanations.

Recent, especially fine-grained sediments are very voluminous and, consequently, show a high porosity which is only reduced in diagenetic compaction and metamorphism. A survey is given in Table 127.

The chemical composition of terrestrial-limnal sediments is largely similar to that of continental sedimentary rocks. Marine sediments of the shelf areas and of the deep-cratonic deep

Table 127. Porosity of Sediments (after v. ENGELHARDT, 1960)

	Grain Size	Porosity (%)	Age	Depth	Place
Lime sludges		87	Recent	surface	NorthernGermany (North Sea)
		78		120 m	
		75		280 m	
		75		340 m	
North Sea sediments	10 μm	80	Recent		North Sea
	240—120 μm	40—44	Recent		
Shallow-sea sediments	10— 15 μm	70—80	Recent		California
Schreibkreide		30—40	Cretaceous	1,500 m	
Basiskalk		22	Mid Triassic		Jena
Dolomite		5—10	Upper Jurassic	3,000 m	Pyrenees
Sandstones		28	Oligocene		Texas
		5	Ordovician		Oklahoma
Poorly consolidated sandstones		28	Lias α		Eddingen near Celle

Table 128. Geochemical Classification of Sediments (after GOLDSCHMIDT)

Type of Sediment	Conditions of Formation	Material	Examples
Resistates	chemically unchanged residues from rocks subject to weathering	from blocks of rock to fine detritus, mainly quartz	sandstones, conglomerates, heavy mineral placers
Hydrolysates	hydrolysed decomposition products, finest particles transported and deposited by water	clay minerals, Al hydroxides	bauxites, soils (partly)
Oxidates	products of oxidation formed in weathering	Fe, Mn oxides	oxidic iron- or manganese-containing sediments or ores
Reductates	deposition under reducing conditions	carbon-rich substances	sapropel with sulphides
Precipitates	precipitation of ionically dissolved elements	Ca, Mg, Fe compounds (carbonates and others)	carbonate, mud and silica sinter and ooze, phosphates (partly), borates (partly)
Evaporates	precipitation products of compounds easily soluble in water due to oversaturation	sulphates, carbonates, chlorides of Na, K, Mg and Ca	rock salt, potassium salts, gypsum
Biolites	deposition of animal and plant remains	liquid and solid hydrocarbons	peat, coal, petroleum, part of natural gases

sea, which recently have been subject to intense investigations, are geochemically of particular interest. Tables 129 and 130 give surveys of the macrochemical and microchemical composition of the most important types of sediments and their averages. A striking feature are the high concentrations of Mn, Ni, Cu, Co, Pb, Mo, Y, Ba and B in the pelagic sediments. The average content of manganese nodules is given in Table 131.

Table 129. Chemical Composition of Typical Marine Shallow-sea and Deep-sea Mud (macro-chemistry in mass-%; trace elements in ppm)

Type	Mississippi Mud	Red Deep-sea Clay	Lime Sludge	Flinty Sludge	Deep-sea Sediments
Origin Author	River Delta NIGGLI, 1952	Average Analyses of 25 Samples EL WAKEEL and RILEY (1961)			Average of all Oceans
SiO_2	69.96	55.34	26.96	63.91	42.72
TiO_2	0.59	0.84	0.38	0.65	0.59
Al_2O_3	10.52	17.84	7.97	13.30	12.29
Fe_2O_3	3.47	7.04	3.00	5.66	4.89
FeO	—	1.13	0.87	0.67	0.94
MnO	0.06	0.48	0.33	0.50	0.41
MgO	1.41	3.42	1.29	1.95	2.18
CaO	2.17	0.93	0.30	0.75	0.60
Na_2O	1.51	1.53	0.80	0.94	1.10
K_2O	2.30	3.26	1.48	1.90	2.10
H_2O^+	1.96	6.54 (total)	3.91 (total)	7.13 (total)	5.35 (total)
Carbonate	1.40 (CO_2)	1.62	52.25	2.13	26.38
SO_3	0.33	—	—	—	—
P_2O_5	0.18	0.14	0.15	0.27	0.16
Others	2.78 (H_2O^-)	0.24 (org. C)	0.31 (org. C)	0.22 (org. C)	
Trace Elements					(Pacific only)
Sr		450	1,110	230	710
Ba		2,000	1,360	1,050	3,900
Cu		400	338	370	740
Pb		175	150	180	150
Co		100	91	200	160
Ni		300	232	330	320
Ga		20	12	18	19
La		90	100	90	140
Sn		20	18	12	—
Zr		126	140	170	180
V		390	300	460	450
Cr		55	51	97	93
Mo		17	20	38	45

ble 130. Some Trace Element Contents [ppm] in Marine Sediments
cording to WEDEPOHL, 1960, EL WAKEEL and RILEY, 1961,
LDBERG and ARRHENIUS, 1958, YOUNG 1954)

ement	Sediments near the Coast	Deep-sea Sediments
	750	2,240
	13	116
	100	77
	48	570
	19	20
	55	293
	20	162
	21	20
	130	330

able 131. Range of Average Contents of Some Chemical Elements in Manganese Concretions of
ll Oceans (from several authors, in mass-%)

n	46.3 —32.04	K	~0.8	Cu	0.06 —0.59
e	9.6 —22.44	Ni	0.42—0.99	Pb	0.09 —0.25
i	4.06—11.0	Ti	0.37—0.81	Mo	0.035—0.2
a	(2.3 — 2.6)	Co	0.17—0.59	V	0.047—0.073
g	1.15— 1.8	Ba	0.17—0.59	B	0.005—0.03
l	0.73— 3.1	P	0.18—0.54	Cr	$\leqq 0.002$
a	0.27— 2.7				

Bibliography

NGINO, E. E.: Geochemistry of Antarctic Pelagic Sediments. Geochim. Cosmochim. Acta 30 (1966),
39—961

TAMAN, G.: Oligoéléments dans les Argiles. Rev. Inst. fr. Pétrole 19 (1964), 958—969

HESTER, R.: Element Geochemistry of Marine Sediments, from: RILEY and SKIRROW: Chemical Oceanog-
aphy, Vol. 2

LOUD, P. E.: Carbonate Precipitation and Dissolution in the Marine Environment. from: RILEY and
KIRROW: Chemical Oceanography, Vol. 2

DEGENS, E. T.: Geochemistry of Sediments — a Brief Survey. Prentice-Hall, Englewood Cliffs, New
ersey 1965, 342 pp.

GLAGOLEVA, M. A.: On the Geochemistry of the Black Sea Sediments. from: Sovrem. Ossadki Morej i
Okeanov, 1961, 448—476 (Russian)

GOLDBERG, E. D., and G. O. S. ARRHENIUS: Chemistry of Pacific Pelagic Sediments. Geochim. Cosmo-
him. Acta 13 (1958), 153—212

HIRST, D. M.: The Geochemistry of Modern Sediments from the Gulf of Paria. Geochim. Cosmochim.
Acta 26 (1962), 309—334, 1147—1187

MILLOT, G.: Géologie des Argiles (Altérations, Sédimentalogie, Géochimie). Edit. Masson et Cie., Paris
964

POTTER, P. E., N. F. SHIMP, and J. WITTERS: Trace Elements in Marine and Fresh-water Argillaceous
Sediments. Geochim. Cosmochim. Acta 27 (1963), 669—694

RONOV, A. B., and Z. V. KHLEBNIKOVA: Chemical Composition of the Main Genetic Clay Types. Geochimiya 6 (1957), 449—469 (Russian)

SEVASTYANOV, V. F.: Redistribution of Arsenic During Formation of Iron-manganese Concretions in
Black-Sea Sediments. Dokl. Akad. Nauk SSSR 176 (1967), 191—193 (Russian)

TUREKIAN, K. K.: Some Aspects of the Geochemistry of Marine Sediments. from: RILEY and SKIRROW: Chemical Oceanography, Vol. 2

VINOGRADOV, A. P.: Introduction into the Geochemistry of the Ocean. Izd. "Nauka", Moscow 1967 (Russian)

VINOGRADOV, A. P.: Iodine in Marine Muds. Trudy Biogeochim. Labor. Akad. Nauk SSSR *5* (1939), 19—46 (Russian)

VOLKOV, I. I.: On Free Hydrogen Sulphide and Sulphureous Iron in the Marine Sullage Sediments of the Black Sea. Dokl. Akad. Nauk SSSR *126* (1959), 163—166 (Russian)

WAKEEL, S. K. E., and J. P. RILEY: Chemical and Mineralogical Studies of Deep-sea Sediments. Geochim. Cosmochim. Acta *25* (1961), 110—146

YOUNG, E. I.: Trace Elements in Recent Marine Sediments. Bull. Geol. Soc. America *65* (1954), 1329

Several authors: Geochemistry, Petrography and Mineralogy of Sedimentary Rocks. Akad. Nauk SSSR 1963 (Russian)

8.4.2.2. Sedimentary Rocks

The distribution of elements in clastic sedimentary rocks is subject to more geological than physico-chemical regularities because it has the character of an open system.

Representations of the geochemical and mineralogical systematics of the carbonate-clay-quartz system are found in RUKHIN (1958), FÜCHTBAUER (1959), PETTIJOHN (1957) and others. The average composition of carbonates and argillaceous rocks is given in Chapter 7. Here, the macrochemistry of a few typical rocks is stated. More comprehensive collections of data are given by GRAF (1960), CRESSMAN (1963), and PETTIJOHN (1963).

In recent years, thorough geochemical investigations into the development of the sediments of the Russian Platform with respect to their age have been carried out. A few results are given in Figs. 85, 86 and 87.

The carbon-rich sediments, the so-called black shales, are of particular scientific and economic interest. A classical ore-bearing type is the Mansfeld copper shale (Kupferschiefer); a few recent data are given in Table 133 and Fig. 88. Table 134 gives another comparative survey of the geochemical composition of black shale.

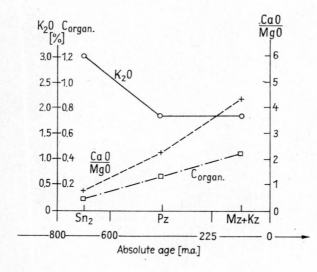

Fig. 85 Changes in the median contents of K_2O, of organic carbon and in the CaO/MgO ratio in the course of geological development in sands of the Russian Platform; after RONOV et al., 1963

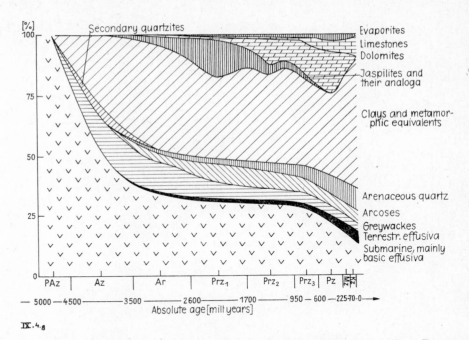

Ⅸ.4.8

Fig. 86 Change in composition of sediments in the course of geological development (from Ronov, 1964)

PAz Pre-azoic; *Az* Azoic; *Ar* Archaic; *Prz₁* to *Prz₃* Proterozoic 1 to 3; *Pz* Palaeozoic; *Mz* Mesozoic; *Kz* Cenozoic

Table 132. Chemical Composition of Sedimentary Rocks from Various Places [mass-%] (from several authors)

	Limestone (Baden)	Flysch Marl (Niesen)	Arkose Sandstone (California)	Greywacke (Hartz)	Glauconite Sandstone (Switzerland)	Lydite (Hartz)
SiO_2	1.65	40.19	61.60	68.85	78.34	90.42
TiO_2	0.04	0.73	tr.	0.74	0.24	—
Al_2O_3	0.46	9.33	12.95	12.05	3.30	3.24
Fe_2O_3	0.38	3.16	2.09	2.72	4.90	1.58
FeO	—	—	3.30	2.03	—	0.63
MnO	tr.	0.01	tr.	0.05	0.04	—
MgO	0.11	1.71	2.33	2.96	2.26	0.28
CaO	54.23	21.48	6.92	0.50	3.35	1.70
Na_2O	0.28	0.71	2.16	4.87	0.25	0.71
K_2O	0.36	3.10	1.41	1.81	1.57	0.39
H_2O^+	—	3.49	3.10	2.30	2.10 ⎫	2.31
H_2O^-	—	0.39	—	0.77	0.12 ⎭	
CO_2	42.76	16.08	5.05	0.08	2.90	—
P_2O_5	tr.	tr.	0.08	0.06	0.09	—
SO_3	0.06	—	0.27	—	0.29	—

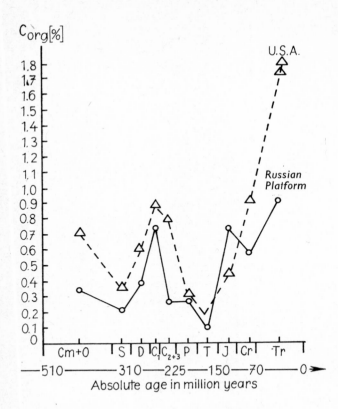

Fig. 87 Development of the average contents of organic carbon in sedimentary rocks of the Russian Platform and in the U.S.A. (after Ronov, 1959)

Table 133. Average Content of the Most Important Major and Trace Elements in the Copper Shale (Kupferschiefer) and Its Direct Bottom and Roof in the Sangerhausen Syncline (from Knitzschke, 1964)

Thickness [cm]			Cu [kg/t]	Pb	Zn	Ag [ppm]	Co	Ni	V	Mo	Ge	Re
	Roof block	Overlying bed	1.4	1.4	1.8	9	16	37	74	43		
12—17	Black dirt		2.3	4.0	5.0	14	28	61	141	73	7.9	21
10—12	Shale top		6.9	5.7	12.5	36	46	78	315	119	8.5	21
25— 4	Crest shell	Copper Shale Bed	17.9	7.6	16.7	107	86	111	751	253	8.1	22
6— 9	Coarse clay		29.0	8.6	18.5	191	144	140	914	308	8.8	21
2— 4	Fine clay		25.7	6.1	9.6	183	159	147	877	251	8.6	21
2	Sand ore I	Underbed	29.5	8.4	10.2	147	102	90	115	79		

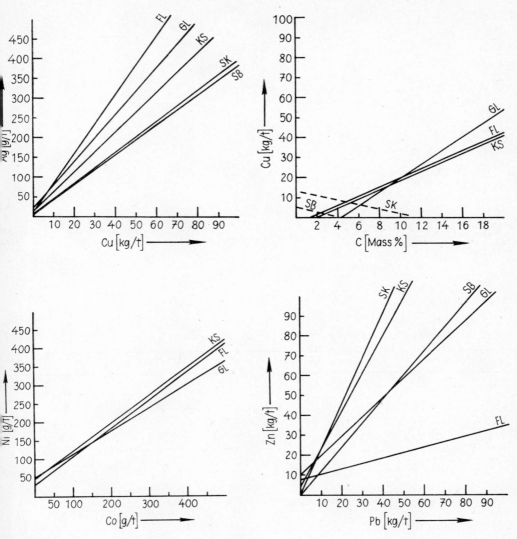

Fig. 88 Correlation diagrams of important elements in the various strata of copper shale (from KNITZSCHKE, 1965)

SB Black dirt (Schwarze Berge); SK Shale top (Schieferkopf); KS Crest shell (Kammschale); GL Coarse clay (Grobe Lette); FL Fine clay (Feine Lette)

For an introduction into the geochemistry of carbonates, special reference is made to INGER-SON (1962), CHILLINGAR et al. (1967), and MÜLLER and FRIEDMAN (1968). Table 135 shows a comparison between sedimentary carbonate rocks and magmatogenic carbonatites. The geo-chemically interesting development of the Ca/Mg ratios are given in Fig. 89. The bibliography included in this Section is only a limited selection.

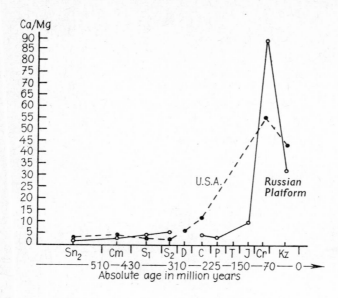

Table 134. Main and Trace Elements in (carbon-rich) Black Shales

Sample No.	1 a	1 b	2	3	4
SiO₂	67.68 %	50.53 %	54.87 %		36.80 %
Al₂O₃	4.89	11.82	8.30	12.00 (Al)	19.98
Fe₂O₃	5.31	6.76	1.09	3.77 (Fe)	—
FeO	—	—	4.23		4.31
CaO	2.13	5.82	6.15		6.22
MgO	0.88	2.01	2.38	0.60 (Mg)	2.49
K₂O	1.48	2.80	2.35	3.49 (K)	—
C_org.	9.07	5.97	13.88 (GV)		12.29
S²⁻	3.05	3.51	1.78		2.82
CO₂	0.79	4.48	4.54		6.87
SO₄²⁻	0.24	0.20	0.17	1.01 (S)	0.63
P₂O₅	0.68	1.07	2.65	0.07 (P)	—
V	1,400 ppm	850 ppm	1,000 ppm	389 ppm	877 ppm
Ti	1,030	3,060	6,600	5,600	n. d.
Ni	353	540	1,300	51	147
Co	77	62	25	49	86
Cu	325	187	300	202	25,700
Mn	162	399	100	700	n. d.
Mo	210	120	130	46	251
Cr	190	208	730	211	n. d.
Ba	573	1,110	n. d.	830	n. d.
Sr	70	226	n. d.	419	n. d.
Pb	23	32	50	101	6,100
Zn	20	38	3,000	n. d.	9,600

nalysis 1 a: Graptolite shale of the silurian system in Thuringia, silicious schist horizon (58% quartz, 17% illite, 3% kaolinite, 5% pyrite, 9% $C_{org.}$; according to Szurowski, 1968)

nalysis 1 b: Graptolite shale of the silurian system in Thuringia, upper alum shale horizon (32% quartz, 34% illite, 3% kaolinite, 7% pyrite, 7% dolomite, 6% $C_{org.}$; Szurovski, 1968)

nalysis 2: Black shale of the silurian-devonian system in the Erzgebirge (Lößnitz, Basin, Nechaev 1968)

_nalysis 3: Black shale (according to Spencer 1966); additional (in ppm): B (100), Be (2), As (25), Ga (18), Li (45), Zr (17a), Rb (283), Cs (12)

_nalysis 4: Mansfeld copper shale (Basin of Sangerhausen; according to Knitzschke, 1965)

able 135. Comparison of Average Sedimentary Carbonate Rock (A) and Carbonatites (B)

	A	B		A	B
	[%]	[%]		[ppm]	[ppm]
SiO_2	5.14	11.99	Co	0.1	17
TiO_2	0.07	0.79	Ni	20	8
Al_2O_3	0.40	3.52	Cu	4	2.5
Fe_2O_3		3.09 (FeO)	Ga	4	1
FeO	0.49	3.74	Zr	19	1,120
MnO	0.14	0.60	Mo	0.4	42
MgO	7.79	5.59	Sn	1	4
CaO	42.30	34.80	Cr	11	48
Na_2O	0.03	0.42	Sc	1	10
K_2O	0.16	1.48	Y	30	96
H_2O	1.63	1.38	Nb	0.3	1,951
P_2O_5	0.05	2.04	La	1	516
F	0.03	0.87	Ce	11.5	1,505
S	0.12	0.61			
BaO	0.01	0.26			
SrO	0.07	0.40			

Bibliography

Ames, L. L.: Metasomatic Replacement of Limestones by Alkaline, Fluoride-bearing Solutions. Econ. Geol. *56* (1961), 730—739

Bashkakov, M. P.: Geochemical Balance and Association of a Few Elements in the Formation of Sediments. Trudy Inst. Geol. Akad. Nauk Usbek. SSR *12* (1956), 33—46 (Russian)

Bernstein, K. H.: Zum Grunddiagramm Sand-Ton-Karbonat. Z. angew. Geol. *7* (1961), H. 9, 469 — 472

Borchert, H., and K. Krejci-Graf: Spurenmetalle in Sedimenten und ihren Derivaten. Bergbauwiss. *6* (1959), 205—215

Chillingar, G. V. et al.: Carbonate Rocks. Elsevier, Amsterdam 1967

Dunham, K. C., and D. M. Hirst: Chemistry and Petrography of the Marl Slate of S. E. Durham, England. Econ. Geol. *58* (1963), 912—940

El Wakeel, S. K., and J. P. Riley: Chemical and Mineralogical Studies of Deep-sea Sediments. Geochim. Cosmochim. Acta *25* (1961), 110—146

FISCHER, R. P., and J. H. STEWART: Copper, Vanadium and Uranium Deposits in Sandstone. Their Distribution and Geochemical Cycles. Econ. Geol. *56* (1961), 509—620

FÜCHTBAUER, H.: Zur Nomenklatur der Sedimentgesteine. Z. Erdöl Kohle *12* (1959) 8, 605—613

GOLDSCHMIDT, V. M., K. KREJCI-GRAF, and H. WITTE: Spurenmetalle in Sedimenten. Nachr. Akad. Wiss. Göttingen *2* (1948), 35—72

GRAF, D. L.: Geochemistry of Carbonate Sediments and Sedimentary Carbonate Rocks. — III. Minor Element Distribution. Illinois Geol. Surv. Circ. 301 (1960), 1—71

HARTMANN, M.: Einige geochemische Untersuchungen an Sandsteinen aus Perm und Trias. Geochim. Cosmochim. Acta *27* (1963), 459—499

HASKIN, L., and M. A. GEHL: The Rare-Earth Distribution in Sediments. J. Geophys. Res. *67* (1962), 2537—2541

HERZBERG, W.: Spurenelemente in den Unterrotliegend-Sedimenten der Saar-Nahe-Senke. Geol. Rdsch. *55* (1965) 1, 48—59

HILL, T. P., et al.: Chemical Composition of Sedimentary Rocks in Colorado, Kansas, Montana, Nebraska, North Dakota, South Dakota and Wyoming. Geol. Surv. Prof. Pap. 561, Washington 1967

HUBER, N. K.: The Environmental Control of Sedimentary Iron Minerals. Econ. Geol. *53* (1958), 123—140

INGERSON, E.: Problems of the Geochemistry of Sedimentary Carbonate Rocks. Geochim. Cosmochim. Acta, *26* (1962), 815—847

IVASHOV, I. V.: Die Geochemie der mitteljurassischen Sande im Nordosten der Russischen Plattform. Z. angew. Geol. *13* (1967) 9, 477—480

JAKUCENI, V. P. et al.: Regularities of Helium Distribution in the Sedimentary Stratum and Formation Conditions of its Commercial Accumulations. Geokhimiya *2* (1969), 192—200 (Russian)

JOST, K.: Über den Vanadiumgehalt der Sedimentgesteine und sedimentären Lagerstätten. Chem. d. Erde *7* (1932), 177—290

KNITZSCHKE, G.: Zur Erzmineralisation, Petrographie, Hauptmetall- und Spurenelementführung des Kupferschiefers im SE-Harzvorland. Freib. Forsch.-H. C 207, Leipzig 1966

KOVALEVSKIĬ, A. L.: The Dependence of the Radioactivity and the Content of Chemical Elements on the Mechanical Composition of Sedimentary Rocks. Geokhimiya (1966) 3, 362—368 (Russian)

LEUTWEIN, F.: Geochemische Untersuchungen an den Alaun- und Kieselschiefern Thüringens. Arch. Lagerstättenforsch., *82*, Berlin (1951)

LUDWIG, G.: Eine Methode zum Nachweis hydrodynamisch-paläogeographisch bedingter Metallverteilungen. Erzmetall *15* (1962 [1962a]) 4, 201—206

LUDWIG, G.: Beziehungen zwischen Metallgehalten und Paläographie des Grauen Hardegsen-Tones im Niedersächsischen Bergland. Geol. Jb. *79* (1962), 537—550

MALYUGA, D. P.: Content of Cobalt, Nickel, Copper, and Other Elements of the Iron Family in Deposits of the Black Sea. Dokl. Akad. Nauk SSSR *67* (1949), 1057 (Russian)

MEMPEL, G.: Verbreitung und Genese der Buntmetallerzspuren in den paläozoischen und mesozoischen Sedimenten Nordwestdeutschlands. Erzmetall *15* (1962), 62—72

MIGDISOV, A. A.: On the Ti/Al Ratio in Sediments. Geokhimiya (1960) *2*, 149—163 (Russian)

MINANI, E.: Gehalte an seltenen Erden in europäischen und japanischen Tonschiefern. Nachr. Ges. Wiss. Göttingen, math.-physik. Kl., Fachgr. IV, *1* (1935), 155

MÜLLER, G., and G. M. FRIEDMAN: Carbonate Sedimentology in Central Europe. Springer-Verlag, Berlin/Heidelberg/New York 1968, 255 pp.

OHRDORF, R.: Ein Beitrag zur Geochemie des Lithiums in Sedimentgesteinen. Geochim. Cosmochim. Acta *32* (1968) 2, 191—208

PETTIJOHN, F. J.: Chemical Composition of Sandstones. Data of Geochemistry, 6th ed., Chapt. S., Washington 1963

RENTZSCH, J.: Der Kenntnisstand über die Metall- und Erzmineralverteilung im Kupferschiefer. Z. angew. Geol. *10* (1964) [1964b] 6, 281—288

RONOV, A. B.: Organic Carbon in Sedimentary Rocks of the Russian Platform. Geokhimiya (1959) (Russian)

RONOV, A. B., et al.: Comparative Geochemistry of Geosyncline and Platform Sedimentary Strata. Geokhimiya (1965) 8, 961—976 (Russian)

ONOV, A. B., YU. A. BALASHOV, and A. A. MIGDISOV: Geochemistry of Rare Elements in the Sedimentary Cycle. Geokhimiya (1967) 1, 3—19 (Russian)

ONOV, A. B., et al.: Sedimentary Differentiation in Platform and Geosyncline Basins. Geokhimiya 1966) 7, 763—776 (Russian)

RUKHIN, L. B.: Grundzüge der Lithologie. Akademie-Verlag, Berlin 1958 (Transl. from Russian)

CHNEIDERHÖHN, H., et al.: Das Vorkommen von Titan, Vanadin, Chrom, Molybdän, Nickel und einigen nderen Spurenmetallen in deutschen Sedimentgesteinen. Neues Jb. Mineralog., Geol. Paläontol. bt. A (1949), 50—72

CHÜLLER, A.: Metallisation und Genese des Kupferschiefers von Mansfeld. Abh. dtsch. Akad. Wiss. Berlin. Kl. Chemie, Geol. Biol., Berlin (1959), 6

HAW, D. M.: Trace Elements in Pelitic Rocks; I. Variation During Metamorphism; II. Geochemical Relations. Bull. Geol. Soc. America 65 (1954), 1151—1182

STRAKHOV, N. M.: On the Geochemistry of Phosphorus, Vanadium, and Copper in Marine Bituminous Rocks. Papers of the Moscow Institute for Geological Exploration 7 (1937), 3 (Russian)

TIKHOMIROVA, YE. S.: On the Geochemistry of Shale Depositions in the Baltic Region. Dokl. Akad. Nauk SSSR 136 (1961), 1209—1212

USDOWSKI, H. E.: The Formation of Dolomite in Sediments, in: Müller-Friedman, Carbonate Sedimentology in Central Europe. Springer-Verlag 1968, 21—32

VINOGRADOV, A. P., and A. B. RONOV: Composition of the Sedimentary Rocks of the Russian Platform. Geokhimiya (1959) 6 (Russian)

ZAV'JALOV, V. A.: Geochemistry and Microelements of the Domanic Sediments of the Southern Pritananya. Izd. "Nauka", Moscow 1966 (Russian)

Several authors: Metals in Sedimentary Rock Strata (Heavy Non-ferrous Metals, Scarcely Occurring and Rare Metals). Izd. "Nauka", Moscow 1965 (Russian)

Several authors: Geochemistry of Silica. Izd. Nauka, Moscow 1966, 420 pp. (Russian)

8.4.3. Geochemistry of Salts

Because not even a more or less complete outline of the geologic-genetic and physico-chemical regularities of salt formation could be given within the scope of this book, we refer readers to the following introductory works.

BORCHERT, H., and R. O. MUIR: Salt Deposits. D. van Nostrand, London 1964

KÜHN, R.: Geochemistry of the German Potash Deposits. Geol. Soc. America Special Paper 88 (1968), 427—504

MATTOX, R. B. (ed.): Saline Deposits. Geol. Soc. America Special Paper 88 (1968), 701

Fig. 90 is included here to give brief information about the general regularities of salt formation. Information about experimental results on the evaporation of seawater is given in Fig. 91; trace elements found in salt minerals are shown in Table 136.

The geochemistry of trace elements in salts is of some importance for geochemical-genetic and stratigraphic reasons. As additional information to Figs. 92, 93 and 94, which deal with the distribution of bromine (the most important element in this connection) the following comments are made:

Br   (see also Figs. 92—94) substitutes for Cl in salt minerals, depending on the salinity (concentration) and, to a lesser degree, on the temperature. It is concentrated in the residual solutions. It has some importance as a stratigraphic indicator within saline complexes, but also shows transformations relatively clearly in comparison

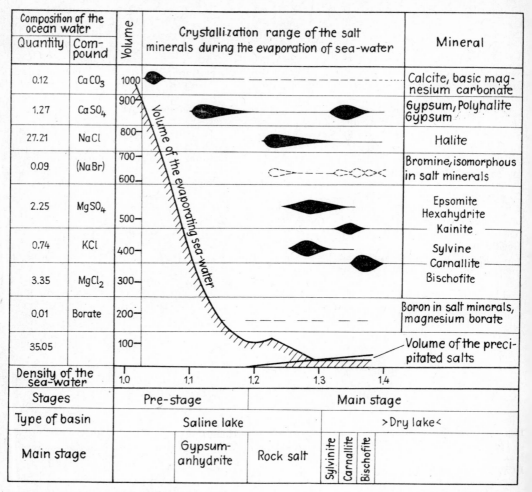

Composition of the ocean water		Volume	Crystallization range of the salt minerals during the evaporation of sea-water	Mineral
Quantity	Compound			
0.12	CaCO₃	1000		Calcite, basic magnesium carbonate
1.27	CaSO₄	900		Gypsum, Polyhalite Gypsum
27.21	NaCl	800		Halite
0.09	(NaBr)	700 / 600		Bromine, isomorphous in salt minerals
2.25	MgSO₄	500		Epsomite Hexahydrite Kainite
0.74	KCl	400		Sylvine Carnallite
3.35	MgCl₂	300		Bischofite
0.01	Borate	200		Boron in salt minerals, magnesium borate
35.05		100		Volume of the precipitated salts

Density of the sea-water	1.0	1.1	1.2	1.3	1.4
Stages	Pre-stage		Main stage		
Type of basin	Saline lake		>Dry lake<		
Main stage		Gypsum-anhydrite	Rock salt	Sylvinite / Carnallite / Bischofite	

Fig. 90   Changes in volume of the seawater on evaporation and the salts separated from it; further, the crystallization ranges of the individual minerals (from VALYASHKO, 1958)

with the absolute contents. In the case of paragenetic crystallization, the ratio of the weight of bromine is as follows (BRAITSCH, 1963):

1 : 10 (±1) : 7 (±1) : 9 (±1)
halite sylvite   carnallite bischofite

Rb   substitutes for K, carnallite showing the highest concentrations. According to KÜHN (1963), the ratio of foot-wall concentrations to roof concentrations (concentration in the first crystallization product) shows a declining trend.

Pb, Cu, and others. The results of studies on the distribution of heavy mineral traces in salt minerals failed to show general regularities. Secondary ɔrmations may show considerᵃ ably higher concentrations, depending partly on the formation conditions.

, P, Sr and B occur both in disperse distribution and in the form of trace minerals in saline rock. Their distribution is dependent on various factors (among other things, also on those of palaeogeography and stratigraphy).

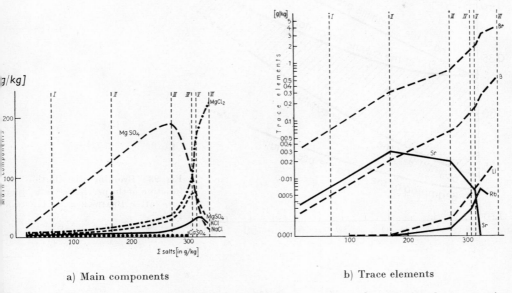

a) Main components

b) Trace elements

Fig. 91 Development and succession of main components and trace elements during the evaporation f Black Sea water (after VALYASHKO, 1964)
Iain precipitation stages:
Carbonates; II Gypsum; III Halite; IV Mg sulphate; V Potassium salts; VI Bischofite

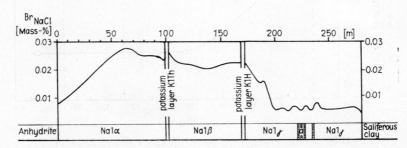

Fig. 92 Bromine distribution in halites of the potassium salt bearing Werra Series of the German Zechstein (Upper Permian) (after DITTRICH, 1961)

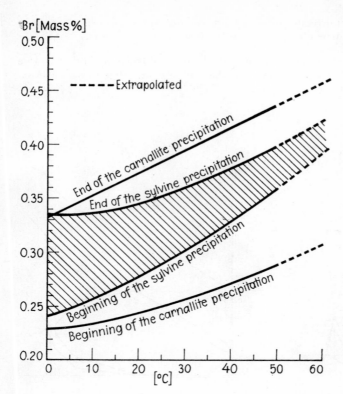

Fig. 93   Bromine distribution as a function of the precipitation temperature (after BRAITSCH and HERRMANN, 1964)

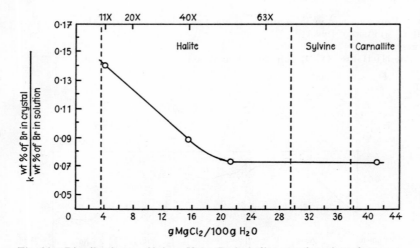

Fig. 94   Distribution coefficient K for Br in halite as a function of

(a) the MgCl₂ concentration (lower scale; BRAITSCH and HERRMANN, 1963)
(b) the appertaining seawater concentration (upper scale; after HOLSER, 1966)

Table 136. Distribution of Some Trace Elements in Important Salt Minerals (from several authors; values in ppm)

Mineral	Element			
	Br	Rb	Cs	Tl
Halite	30—500	—	—	—
Sylvine	1,000—4,000	50—250	Ø 1	} 0.06—2
Carnallite	1,500—3,500	50—1,700	Ø 5	

Mineral	Element			
	Zn	Al	Ag-Sn	F
Halite	<0.1—1 (Ø 0.3)	0.03—6 (Ø 1.4) }	0.01—0.1	2—50
Sylvine	0.2—2 (Ø 0.8)	0.1 —6 (Ø 2.7) }		—
Carnallite	0.1—1 (Ø 0.3)	0.03—6 (Ø 1.3)	—	—
Anhydrite	—	—	—	130—890

Mineral	Element		
	Pb	Cu	Mn
Halite	0.03—4 (Ø 0.2)	0.04—2 (Ø 0.4)	0.03—5 (Ø 0.4)
Sylvine	0.10—5 (Ø 0.5)	0.1 —3 (Ø 1.1)	0.1 —5 (Ø 1.2)
Carnallite	<0.03—0.1 (Ø 0.05)	0.05—4 (Ø 0.4)	0.03—4 (Ø 0.1)

Mineral	Element		
	P	Sr	B
Halite	1—4	0.7—7 (Ø 3)	0.001—0.02
Sylvine	—	Ø 0.5	—
Carnallite	—	1	—
Anhydrite	10—50	700—5,300	1—500

Bibliography

BAAR, A.: Untersuchungen des Bromgehalts im Zechsteinsalz. Bergbautechnik *4* (1954), 281—288

BAAR, A.: Der Bromgehalt als stratigraphischer Indikator in Steinsalzlagerstätten. Neues Jb. Mineralog., Mh. (1963), 145—153

BLANDER, M.: The Topology of Phase Diagrams of Ternary Molten Salt Systems. Chem. Geol. *3* (1968) 1, 33—58

BOIKO, T. F.: Distribution of Trace Elements in Halogenic Deposits. Dokl. Akad. Nauk SSSR *171* (1966) 2, 457—460

BORCHERT, H.: Der Wasserhaushalt bei der Metamorphose der Kalisalze. Ber. Geol. Ges. DDR *7* (1963), Special No. 1, 178—194

BORCHERT, H.: Principles of Oceanic Salt Deposition and Metamorphism, from J. P. RILEY and G. SKIRROW: Chemical Oceanography, Vol. 2 (1967)

BORCHERT, H. E.: Über das Kristallisationsschema der Chloride, Bromide, Jodide usw. Zeitschr. Kristallogr. *45* A (1908), 346—391

BORN, H. J.: Der Bleigehalt der norddeutschen Salzlager und seine Beziehung zu radioaktiven Fragen. Chem. d. Erde *9* (1934/35), 66

BRAITSCH, O.: Zur Geochemie des Broms in salinaren Sedimenten, Part I. Geochim. Cosmochim. Acta, *27* (1963) 4, 361—391

BRAITSCH, O., and A. G. HERRMANN: Zur Bromverteilung in salinaren Salzgesteinen bei 25 °C. Naturwissensch. *49* (1962), 346

BRAITSCH, O., and A. G. HERRMANN: Zur Geochemie des Broms in salinaren Sedimenten. Part II. Die Bildungstemperaturen primärer Sylvin- und Carnallitgesteine. Geochim. Cosmochim. Acta *28* (1964), 1081—1109

BRAITSCH, O.: Bromine and Rubidium as Indicators of Environment during Sylvite and Carnallite Deposition of the Upper Rhine Valley Evaporites, Second Symposium on Salt, Vol. 1, The Northern Ohio Geol. Soc. Inc. Uweland/Ohio, 293—301

CORRENS, C. W.: Über die Geochemie des Fluors und Chlors. Neues Jb. Mineralog., Abh. *91* (1957), 239

D'ANS, J., and R. KÜHN: Über den Bromgehalt von Salzgesteinen der Kalisalzlagerstätten. Kali, verwandte Salze und Erdöl *34* (1940), 42—46, 59—62, 77—83

D'ANS, J.: Die Gewinnung von Rubidium aus Carnalliten und seine Bestimmung neben Kalium. Angew. Chem. *62* (1950), 118

DITTRICH, E.: Zur Gliederung der Werra-Serie (Zechstein 1) im Werra-Kaligebiet. Ber. Geol. Ges. DDR *6* (1961), 296—301

ELERT, K. H.: Mineralogische und geochemische Untersuchungen der 3 Faziesbezirke des Staßfurt-Kalilagers auf der Grube Neusollstedt des Kaliwerks "Karl Marx", Freiberger Forschungsh. C *145* (1963)

FANDRICH, K.: Spurenmetalluntersuchungen im unteren Kalilager der Grube Menzengraben (Werragebiet) und Betrachtungen zur Genese der dortigen Kalisalze. Ber. Geol. Ges. DDR *6* (1962) 1

GENTNER, W., R. PRÄG, and F. SMITH: Argonbestimmung an Kaliummineralen. III. Vergleichende Messungen nach der Kalium-Argon- und Uran-Helium-Methode. Geochim. Cosmochim. Acta *8* (1954), 124

HERRMANN, A. G.: Bedeutung des Verdampfungsgleichgewichtes bei der Anwendung von Kohlerädchen und Stabelektroden für spektrochemische Lösungsanalyse. Chem. Techn. *5* (1955), 132

HERRMANN, A. G.: Geochemische Untersuchungen an Kalisalzlagerstätten im Südharz. Freiberger Forschungsh., C *43* (1958)

HERRMANN, A. G.: Methoden zur Bestimmung von Spurenelementen in salinaren Ablagerungen. Mber. dtsch. Akad. Wiss. Berlin *1* (1959) 11

HERRMANN, A. G.: Zur Geochemie des Strontiums in den salinaren Zechsteinablagerungen der Staßfurt Serie des Südharzbezirks. Chem. d Erde *21* (1961/62), 137—194

HOLSER, W. T.: Bromide Geochemistry of Salt Rocks. The Northern Ohio Geological Society *1* (1966), 248—273

KOCH, K., W. KOCKERT, and V. GRUNEWALD: Geochemische Untersuchungen an Salzen und Laugen von Salzlagerstätten in der DDR, Geologie *17* (1968)

KÜHN, R.: Rubidium als geochemisches Leitelement bei der lagerstättenkundlichen Charakterisierung von Carnalliten und natürlichen Salzlösungen. Neues Jb. Mineralog., Mh. (1963), 107—115

KÜHN, R.: Geochemistry of the German Potash Deposits. Geol. Soc. Amer., Spec. Pap. *88* (1968), 527—594

LADYNINA, I. N., and G. N. ANOSHIN: A Few Regularities of the Distribution of Rb, Tl, and Br in the Process of the Formation of Potash Deposits. Geologiya i geofizika (1962) 3, 64—74 (Russian)

MALIKOVA, Y. N.: Regularities of the Distribution of Rb, Tl and Br in Potash Deposits. Izd. "Nauka", Sibirsk. Otd.; Novosibirsk 1967, 150 pp. (Russian)

MATSUDA: About Rubidium and Cesium in Carnallites. J. Miner. Soc. Japan *3* (1958) 6

McINTIRE, W. L.: Effect of Temperature on the Partition of Rubidium between Sylvine Crystals and Aqueous Solutions, in: Saline Deposits. Geol. Soc. Amer. Spec. Paper *88* (1968), 701

MORACHEVSKIĬ, YU. V.: Compendium of the Geochemistry of Salt Deposits of Verchnekamsk. Paper VNJJG 1939, 17 (Russian)

SCHULZE, G.: Stratigraphische und genetische Deutung der Bromverteilung in den mitteldeutschen Steinsalzlagern des Zechsteins. Freiberger Forschungsh. Ausg. C *83* (1960)

VALYASHKO, M. G., and E. STOLLE: Geochemical Analysis of the Genesis of the Stassfurt Bed Potassium Salts. Geokhimiya (1965) 8, 977—995 (Russian)

VALYASHKO, M. G.: Die wichtigsten geochemischen Parameter für die Bildung der Kalisalzlagerstätten. Freiberger Forschungsh., Ausg. A *123*, 197—236

VALYASHKO, M. G.: Geochemische Gesetzmäßigkeiten der Entstehung der Kalisalz-Lagerstätten. Translation from the Russian, printed as manuscript. TH Leuna-Merseburg: 1964. (Original in Russian: Izd. Moscow University 1962)

ZDANOVSKIĬ, A. B., E. Y. LYAKOVSKAYA, and R. E. SLIMOVIC: Handbook of Solubilities of the Salt Systems. Leningrad 1953/54 (Russian)

Several authors: Mineralsalze ozeanischen Ursprungs. Symposium Berlin 1958, Freiberger Forschungsh., *123* Berlin (1959)

3.5. Geochemistry of the Hydrosphere

3.5.1. Water Distribution on the Earth and Classification of Waters

According to GOLDSCHMIDT (1933), the amount of water per square meter of the earth's surface is 273 litres which comprise:

seawater	268.45	l	or 278.11	kg
continental ice	4.5	l	or 4.5	kg
fresh water	0.1	l	or 0.1	kg
water vapor	0.003	l	or 0.003	kg

According to recent measurements, the water on the earth is distributed as shown in Table 138. This survey shows that about 97 per cent of the total amount of water is salt water and only 3 per cent fresh water (only 20 per cent of the fresh water is useable at present).

The water balance of the earth is largely controlled by a global, complex, "great" cycle which essentially consists of evaporation, precipitation and drainage. This cycle is divided into many subcycles of hemispheric, continental, regional, and local character. The most important characteristic factors resulting from this are shown in Table 137.

From Tables 137 and 138 it follows that the amount of water in the atmosphere must be renewed at least 40 times a year (about every 9 days). In the final analysis, this is the cause of the fact that as a solvent and a means of transportation water is of paramount importance to the exogenic sphere.

Table 137. The Water Balance of the Earth (from DYCK, 1968)

	Area in 10^6 km²	Precipitation cm	km³	Evaporation cm	km³	Drainage cm	km³
Sea	361	112	403,000	125	449,000	13	46,000
Land	149	72	107,000	41	61,000	31	46,000
Earth	510	100	510,000	100	510,000	—	—

Table 138. The Earth's Water Supply (after NACE, from MONSONYT, 1968)

Item	Description	Milliard m³ or km³
1	Oceans (related to a mean depth of the oceans of 3,700 m)	1,320,000,000
2	Saline seas and lakes	104,000
Sum 1—2	Salt water	1,320,104,000
3	Ice caps in the polar regions, glaciers	30,000,000
4	Watercourses	1,250
5	Fresh-water lakes	125,000
Sum 3—5	Surficial fresh waters	30,126,250
Sum 1—5	Surficial water	1,350,230,250
6	Soil moisture	67,000
7	Subsurface waters down to a depth of 800 m	4,200,000
8	Deep-seated water	4,200,000
Sum 6—8	Subsurface water	8,467,000
9	Atmospheric water	12,900
Sum 1—9	Total of the earth's water reserves	1,358,710,150
Sum 3—9	Fresh water reserves of the earth	38,606,150

Classification:

Waters can be classified according to various points of view:

1. according to their origin:

   a) meteoric water (surface and underground waters)
   b) pore or connate waters (water in sediments)
   c) juvenile waters (magmatogenic waters)

2. according to their salinity

   a) fresh waters (no salts or a low salt content, maximum 1%)
   b) salt waters (medium salt content ±3.5%)
   c) brines (high salt content, >5%)

3. according to their relative chemical composition:
   a) sulphatic waters
   b) carbonatic waters
   c) chloride waters

.5.2. Distribution of Elements in Seawater and Salt Lakes

he elements released (absolutely and relatively) as a result of weathering on the continents
re found in seawater in quite different quantities and conditions (see Table 139). Elements
-hich mainly get into the sea are called thalassophile elements (H, O, B, Na, S, Cl, Br, I);
lements which are only found in traces in seawater are called thalassoxene elements (e.g.
1any heavy metals). The main reason for this phenomenon is the selective adsorption of
ertain ions to certain minerals and organic substances (see also Section 3.3.8.), a process
hrough which these ions are withdrawn from seawater. Argillaceous constituents show a
ropensity for preferentially adsorbing alkalis and alkali earths. Fe and Mn hydroxides
referentially adsorb complexing agents (e.g. As, Se, Mo) and heavy metals.
edimentary phosphates mainly adsorb Zn, Cd, In, Bi; V, Mo, Cu, Ni and U are mainly con-
entrated in organic deposits. Generally speaking, the adsorption of elements rapidly increases
-ith increasing atomic weight (cf. DIETRICH and KALLE, 1957). This process results in a decon-
amination of the seawater, removing a great number of elements detrimental to life, and thus
s of great importance to the vital processes taking place in the sea.
'he concentration in seawater of anions such as Cl, Br, S, B is considerably higher than the
upply from weathering solutions. These volatile elements or volatile compound forming
lements, to a great extent pass into the sea via the atmosphere. These elements are added
ontinuously to the atmosphere by volcanic processes.
'alculations of "excess substances" (H_2O, C or CO_2, S, N, Cl, etc.) in the hydrosphere and
tmosphere on the basis of comparisons between the amount due to weathering during the
arth's development and the measurable concentrations in the upper geospheres, were carried
ut by GOLDSCHMIDT (1933) and, more recently, by RUBEY (1951), POLDERVAART (1955), and
VICHOLLS (1965). The so-called residence time of an element in the sea can be derived from
he ratio of the quantity of an element present in solution in seawater to the quantity of this
lement fed into the sea by rivers per year; these values are shown in Table 139 (further details
ıre given by BARTH, 1962).
These regularities are more clearly defined by a number of illustrations and tables given
)elow.
[n addition to the tables showing the distribution of the elements, Table 140 shows the
nolecular composition derived from the latter, whereas Table 141 shows regional averages
of trace element concentration.
The percentage distribution of the three types of combinations formed by CO_2 in seawater,
plotted against the pH of seawater is demonstrated in Fig. 95. The calcium carbonate
content of pelagic sediments and the distribution of silicon and phosphorus with increasing depth of water in the oceans is represented in Figs. 96, 97 and 98.
Information about the concentration and distribution of living and organic matter in seawater, facts which become more and more important, are given in Tables 142, 143 and 144,
whereas particulars of the presence of silicic acid in seawater are shown in Table 145.
The average concentration of radioactive isotopes in seawater is shown in Table 146. Fig. 99
is a graph of a few physico-chemical regularities in the behaviour of gases in minor seas.
A survey of the chemical composition of seas of a high salinity is given in Table 147.

Table 139. General Relations between the Supply of Elements from Weathering Solutions and the Elements in Solution Found in Seawater (from RANKAMA and SAHAMA, 1955; residence time from VINOGRADOV, 1967)

Element	Supply of Elements from Weathering Solutions [ppm]	Portion Dissolved in Sea-water [%]	[ppm]	Residence Time in the Sea [in years]
Li	39	0.3	0.1	$5.8 \cdot 10^6$
B	1.8	256	4.6	$9.1 \cdot 10^6$
C	192	14.6	28	$1.0 \cdot 10^6$
N	27.78	2.5	0.7	
F	540	0.3	1.4	
Na	16,980	62	10,561	$1 \cdot 10^8$
Mg	12,540	10	1,272	$1.55 \cdot 10^7$
Al	48,780	0.004	1.9	$7.8 \cdot 10^3$
Si	166,320	0.002	4	$1.96 \cdot 10^4$
P	708	0.01	0.1	$2.7 \cdot 10^5$
S	312	283	884	$8.8 \cdot 10^6 \ (SO_4)$
Cl	188.4	10,074	18,980	$1.2 \cdot 10^8$
K	15,540	2.4	380	$1.0 \cdot 10^7$
Ca	21,780	1.8	400	$1.2 \cdot 10^6$
Sc	3	0.001	0.000,04	
V	90	0.000,3	0.000,3	$1.2 \cdot 10^6$
Mn	600	0.002	0.01	$7.8 \cdot 10^3$
Fe	30,000	0.000,07	0.02	$5.8 \cdot 10^2$
Co	13.8	0.000,7	0.000,1	$1.96 \cdot 10^4$
Ni	48	0.001	0.000,5	$1.6 \cdot 10^4$
Cu	42	0.03	0.011	$2.3 \cdot 10^4$
Zn	79.2	0.02	0.014	$2.0 \cdot 10^4$
Ga	9	0.006	0.000,5	
As	3	0.8	0.024	$3.9 \cdot 10^4$
Se	0.054	7.4	0.004	
Br	0.972	6,687	65	$1.3 \cdot 10^8$
Rb	186	0.1	0.2	$3.9 \cdot 10^6$
Sr	180	7.2	13	$3.1 \cdot 10^6$
Y	16.86	0.002	0.000,3	
Mo	9	0.008	0.000,7	$4.7 \cdot 10^5$
Ag	0.06	0.5	0.000,3	$5.9 \cdot 10^4$
Sn	24	0.01	0.003	
I	0.18	28	0.05	$9.8 \cdot 10^5$
Cs	4.2	0.05	0.002	$1.5 \cdot 10^5$
Ba	150	0.03	0.05	$2.6 \cdot 10^4$
La	10.98	0.003	0.000,3	
Ce	27.66	0.001	0.000,4	
Au	0.003	0.3	0.000,008	
Hg	0.3	0.01	0.000,03	
Pb	9.6	0.05	0.005	$1.2 \cdot 10^3$
Bi	0.12	0.2	0.000,2	
Ra	$7.8 \cdot 10^{-7}$	0.04	$3 \cdot 10^{-10}$	
Th	6.9	<0.007	<0.000,5	$2.0 \cdot 10^4$
U	0.6	0.3	0.001,6	$1.2 \cdot 10^5$

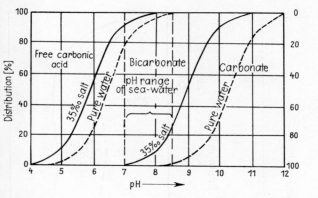

Fig. 95 Percentage distribution of the three types of bond of carbonic acid (free CO_2, HCO_3^-, CO_3^-) in seawater (—) and in pure water (---) as a function of pH (after BUCH, from WATTENBERG, 1943)

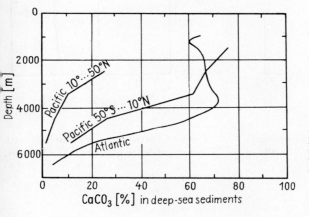

Fig. 96 Mean contents of calcium carbonate of deep-sea sediments as a function of depth (from DIETRICH and KALLE, 1957)

Table 140. Calculated Molecular Composition of Solutes in Seawater (after BRAITSCH, 1962)

Molecule	Mass-%
NaCl	78.03
NaF	0.01
KCl	2.11
$MgCl_2$	9.21
$MgBr_2$	0.25
$MgSO_4$	6.53
$CaSO_4$	3.48
$SrSO_4$	0.05
$CaCO_3$	0.33

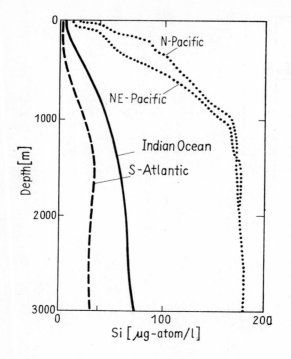

Fig. 97 Distribution of silicon with increasing depth of water in the Atlantic, Indian and Pacific Oceans (after SVERDRUP et al., 1942)

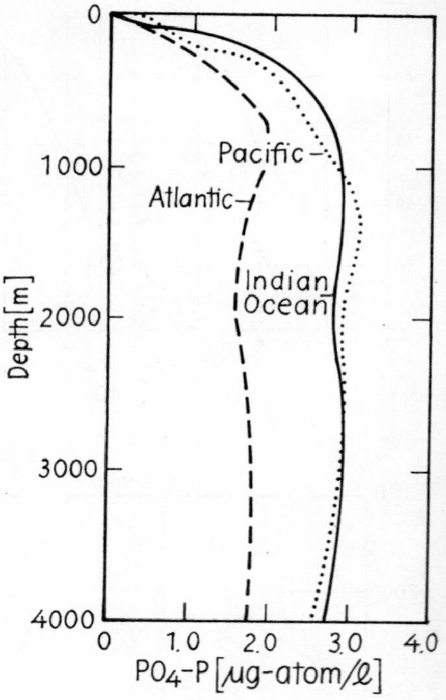

Fig. 98 Distribution of phosphate with increasing depth of water in the Atlantic, Indian and Pacific Oceans (after SVERDRUP et al., 1942)

Table 141. Regional Averages of Trace Element Concentration (after SCHUTZ and TUREKIAN, 1965; in µg/l)

Area	Ag	Co	Ni	Sn	Se	Au
Caribbean	0.25	0.078	2.1	0.26	0.11	
Gulf of Mexico	0.16	0.84	2.0	0.46		
Labrador Sea	0.13	0.16	4.9			
N. W. Atlantic	0.19	0.21	3.5	0.24	0.096	0.021
N. E. Atlantic	0.25	0.13	3.1	0.21	0.088	0.005,6
S. W. Atlantic	0.18	0.22	4.8	0.53	0.075	0.003,6
S. E. Atlantic	0.64	0.25	19.2			
Indian Ocean	0.69	1.4	5.4	0.37		0.010
Central Pacific	0.34	0.75	20	0.51		
East Pacific	0.23	0.18	5.5	0.51	0.087	
Antarctic	0.42	0.088	4.2	0.24	0.052	

Table 142. Content of Organic Substances in Seawater
(after VINOGRADOV, 1967)

Origin	Carbon mg/l
Sea of Azov	4.63—6.02
Black Sea	2.83—3.56
Caspian Sea	5.0 —7.0
Baltic Sea	2.0 —5.2
Arctic Ocean	0.5 —8.0
Atlantic (northern part)	0.2 —1.3
(tropical part)	1.08—1.9
Pacific	0.6 —2.7
Indian Ocean	1.30—1.99

Table 143. Distribution of Bacteria (Biomass,
Amount of Living Matter) in the North-west Pacific
versus Depth (after VINOGRADOV, 1967, abridged)

Depth [m]	Average Concentration of Biomass in mg/m³ of Seawater
0— 10	33.3
25— 50	18.8
75— 100	9.7
200— 250	2.8
500— 600	0.3
1,000—1,500	0.4
3,000—4,000	0.05
6,000—7,000	0.01
8,500—9,000	0.006

Table 144. Contents and Production of Organic Substances (Biomass) in Seawater
(after VINOGRADOV, 1967)

Type of Organisms	Biomass in g		Production in g/yr	
	Raw Mass	C_{org}	Raw Mass	C_{org}
Phytoplankton	$1 \cdot 10^{16}$	$1 \cdot 10^{15}$	$7 \cdot 10^{17}$	$7 \cdot 10^{16}$
Bacteria	$1.4 \cdot 10^{14}$	$1.4 \cdot 10^{13}$	$1.4 \cdot 10^{17}$	$1.4 \cdot 10^{16}$
Zooplankton	$1.5 \cdot 10^{16}$	$1.5 \cdot 10^{15}$	$5 \cdot 10^{16}$	$5 \cdot 10^{15}$

Table 145. Forms of Silicic Acid in Seawater
(after GOLOLOBOV, in BRUEVICH, 1953, values in %)

	Solute	Colloidal	Planktonic
Sea of Azov	60—80	15—20	10—15
Black Sea	70—75	20—23	4— 5
Caspian Sea	60—70	15—20	10—20
Average	70	18	12

Table 146. Average Concentration of Radioactive Isotopes in Seawater
(after VINOGRADOV, 1967)

Isotope	Concentration in %	in Curie/100 g
^3T	$5 \cdot 10^{-19}$	$4.8 \cdot 10^{-15}$
^{10}Be	$1 \cdot 10^{-14}$	$1.4 \cdot 10^{-16}$
^{14}C	$3 \cdot 10^{-15}$	$1.4 \cdot 10^{-14}$
^{40}K	$4.6 \cdot 10^{-6}$	$3.2 \cdot 10^{-11}$
^{87}Rb	$5.6 \cdot 10^{-6}$	$3.7 \cdot 10^{-13}$
^{226}Ra	$1 \cdot 10^{-14}$	$1 \cdot 10^{-14}$
^{227}Ac	$<2 \cdot 10^{-20}$	$1.5 \cdot 10^{-18}$
^{231}Pa	$5 \cdot 10^{-15}$	$2.3 \cdot 10^{-16}$
^{232}Th	$1 \cdot 10^{-9}$	$1.1 \cdot 10^{-16}$
^{235}U	$2.2 \cdot 10^{-9}$	$4.7 \cdot 10^{-15}$
^{238}U	$3 \cdot 10^{-7}$	$1 \cdot 10^{-13}$

Table 147. Composition of the Solute Substances in Salt Lakes (relative content [%] according to RAN-
KAMA and SAHAMA, 1955; absolute content according to BENTOR, 1961)

Component	Dead Sea Surface Water (in %)	at a Depth of 300 m (in %)	Dead Sea Average (in mg/l)	Great Salt Lake (U.S.A) (in %)	Khara Bogaz (USSR; Caspian Sea) (in mg/l)
CO_3	traces	traces	240 (HCO_3^-)	0.09	670 (HCO_3^-)
SO_4	0.31	0.24	540	6.68	67,830
Cl	65.81	67.30	208,020	55.48	132,730
Ca	4.73	6.64	15,800	0.16	traces
Mg	13.28	15.92	41,960	2.76	33,830
Na	11.65	5.50	34,940	33.17	52,250
K	1.85	1.68	7,560	1.66	4,820
SiO_2	traces	traces	60 (Rb)		—
Br	2.37	2.72	5,920		380
Total	100.00	100.00	315,040	100.00	292,510
Salinity	19.215%	25.998%	31.504%	20.349%	29.251%

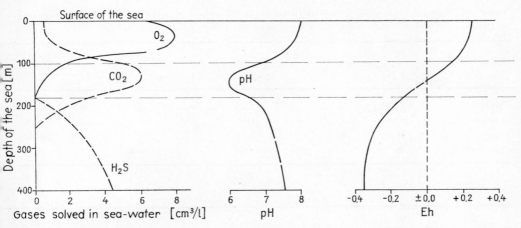

Fig. 99 The behaviour (schematical) of the most important gases in poorly aerated minor seas (from BORCHERT, 1964)

8.5.3. Distribution of Elements in Fresh Waters (River water, Lakes, Rainwater)

The latest and most comprehensive survey of the chemical composition of rivers and lakes is given by LIVINGSTONE (1963).
A number of important data on substances contained in fresh waters and of the amounts of elements transported by them have already been given in Section 8.3. and 8.4. Below follows a survey of the chemical composition of river waters, Tables 148 and 149, and of rainwater, Table 150.
Fig. 100 shows an interesting representation of the oxidation conditions and the concentration of iron depending on it in a lake.

Table 148. Relative Composition of Solute Substances in River water Compared with that of Seawater (in %)

Component	River water (CLARKE, 1924)	River water (MASON, 1966)	River water without cyclic salts (from CONWAY, 1942)	River water average (from CHEBOTAREV, 1955)	Seawater (SVERDRUP, 1942)
CO_3	35.15	48.6 (HCO_3^-)	35.13	36.50	0.41 (HCO_3^-)
SO_4	12.14	9.3	11.35	11.60	7.68
Cl	5.68	6.5	—	6.75	55.04
NO_3	0.90	0.8	0.90	—	—
Ca	20.39	12.5	20.27	14.70	1.15
Mg	3.41	3.4	3.03	4.90	3.69
Na	5.79	5.3	2.63	9.50	30.62
K	2.12	2.0	2.02	4.40	1.10
$Fe_2O_3 + Al_2O_3$	2.75	0.6 (Fe)	2.75	1.30	—
SiO_2	11.67	11.0	11.67	12.80	—
Sr, H_3BO_3, Br	—	—	—	—	0.31
Total	100.00	100.0	89.75	102.45	100.00

Table 149. Migration of Some Chemical Elements in River water in the Soviet Union (percentage in suspension (1) and in solute form (2); after GLAGOLEVA, 1958)

	Fe		Mn		Ca		Cu	
	1	2	1	2	1	2	1	2
Pripet	25.8	74.2	72.7	27.4	n.d.	n.d.	6.3	93.7
Dnieper[1])	80.8	9.2	80.6	19.4	n.d.	n.d.	12.6	87.4
Don[2])	96.0	4.0	70.0	30.0	1.9	98.1	18.0	82.0
Danube[3])	98.3	1.6	98.9	1.1	12.0	88.8	42.2	57.8

	Sr		Ba		P		$C_{org.}$	
	1	2	1	2	1	2	1	2
Pripet	1.7	98.3	6.9	3.1	n.d.	n.d.	n.d.	n.d.
Dnieper[1])	12.5	87.5	20.5	79.5	n.d.	n.d.	17.8	82.2
Don[2])	7.8	92.2	35.4	64.6	61.6	38.4	27.6	72.4
Danube[3])	36.1	63.9	56.5	43.5	75.0	25.0	35.7	64.3

Note: Vanadium, chromium, nickel, gallium and part of beryllium are fully (100 %) transported in suspension

[1]) near Verkhne-Dreprovsk
[2]) near Aksayskoy
[3]) near Izmail

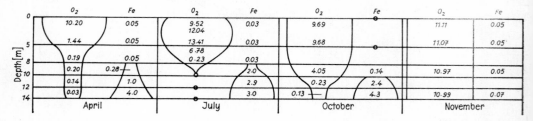

Fig. 100 Distribution of oxygen and iron in the Lunz Upper Lake from spring to autumn (from RUTTNER, 1952; values in mg/l)

318

Table 150. Chemical Composition of Rainwater
(after Gorham, 1955, and Sugawara, 1963; in ppm)

Component	Range of Variations of 42 Samples (soot index: 0—4.3)	Range of Variations of 5 Samples (free from soot)	Mean of 300 Samples (from Wedepohl, 1966)
Na	0.2 — 7.5	1.5 — 7.5	1.1
K	0.05— 0.7	0.05— 0.4	0.26
Ca	0.1 — 2.0	0.1 — 0.3	0.97
Mg	0 — 0.8	0.05— 0.8	0.36
HCO$_3$	0 — 2.8	0.5 — 2.2	1.2
Cl	0.2 —12.6	2.3 —12.6	1.1
SO$_4$	1.1 — 9.6	1.1 — 2.3	1.2
pH	4.0 — 5.8	4.5 — 5.2	

2.5.4. Geochemistry of Deep-seated Waters

In this connection, deep-seated waters are defined as waters occurring below the fresh-water saltwater boundary, especially in sediments. This concept includes terms such as connate water, pore water, oilfield or petroleum water and others.

The deep wells sunk in search for petroleum and natural gas in all parts of the world have shown that huge quantities of water are present in sedimentary basins which sometimes are as thick as 10 km. In general, the degree of mineralization and the temperature of these deep-seated waters increase with increasing depth. The following two main hypotheses on the genesis of deep-seated waters have been propounded.

1. Deep-seated waters are relict (primary) salt-rich seawater.
2. Deep-seated waters have obtained their high salinity by transformation (metamorphosis) of "normal" waters due to reactions with the embedding sediments (ion filtration, reduction, exchange of solutions, decomposition of organic substances, etc.)

The "maceration" of salt deposits should also be taken into consideration.

Table 151 gives information about the quantitative ratios of the important types of water. Fig. 101 shows the so-called Schoeller-Diagram which is one method of representing macro-

Table 151. Representative Waters of Different Genetic Types in Comparison with Lacustrine and Oceanic Waters (from Chebotarev, 1959)

Type of water	Total salinity (ppm)	Reacting value in per cent					
		Na + K	Ca	Mg	Cl	SO$_4$	HCO$_3$ + CO$_3$
Alpine lake water	22.9	15.1	31.2	3.7	9.2	8.2	32.7
Bicarbonate water	387.4	5.3	28.7	16.0	5.4	2.8	41.8
Bicarbonate-chloride water	4,842	47.0	1.1	1.9	13.6	0.4	36.0
Chloride-bicarbonate water	4,710	41.2	5.0	3.8	23.7	9.8	16.5
Chloride-sulphate water	8,953	25.5	13.8	10.7	5.4	37.1	7.5
Chloride water	162,890	32.7	12.9	4.4	50.0	0	0
Oceanic water	35,000	39.3	1.8	8.9	45.2	4.6	0.2

319

Table 152. Chemism of Several Deep-seated Waters in the Thuringian Basin and the North German Plain (after Mü 1965; values in mg/l)

Origin	Geological formation	K	Na	Ca	Mg	NH₄	Fe	Cl	Br
S.W.Mecklenburg	Tertiary	218.1	24,783.4	1,853.7	844.3	tr.	—	43,992.9	5
S.W.Mecklenburg	Upper Cretaceous	n.d.	55,347.4	2,907.4	424.3	—	tr.	91,599.6	15
S.W.Mecklenburg	Hauterivran	309.2	14,099.2	597.6	—	16.2	—	22,371.6	3
N.Brandenburg	Wealden	240.0	47,063.0	1,862	504.4	24.1	—	76,510.0	6
W.Altmark	Kimmeridgran	n.d.	69,143.2	5,450	1,426	—	n.d.	120,850.0	19
S.W.Mecklenburg	Oxfordian	6,139	53,375	5,029.6	1,876.9	76.5	tr.	96,163	28
W.Altmark	Dogger epsilon	382.8	61,972.4	2,685.4	1,002.8	63.7	190.0	103,749	16
W.Altmark	Lias alpha	n.d.	59,434.5	6,772.2	1,584.2	n.d.	n.d.	108,400.0	19
N.W.Mecklenburg	Rhaetian	235.0	40,873	3,756	916.0	n.d.	n.d.	72,205	6
W.Brandenburg	Buntsandstein	2,327	73,530	43,950	4,610	34.2	tr.	206,220	1,40
N.W.Mecklenburg	Buntsandstein	677.3	104.100	17,775	2,160	29.5	49.2	198,730	59
Allmenhausen (Thuringia)	Buntsandstein	562.0	100.331	1,459	463	10.9	24.9	155,285	6
Langensalza (Thuringia)	Buntsandstein	2,447	91,820	21,105	6,225	22.3	28.0	198,700	1,30
N.E.Mecklenburg	Plate dolomite	21,651	21,028	56,112	65,056	260	782	341,223	2,17
Fallstein (Subhercynian)	Main dolomite	10,980	52,980	37,386	18,303	645	354	212,038	1,71
N.E.Mecklenburg	Main dolomite	2,448	71,226	45,917	4,712	966	50	207,466	3,06
Langensalza (Thuringia)	Main dolomite	28,150	59,924	51,052	5,360	695	46	213,126	3,22
Lausitz	Main dolomite	25,646	43,570	10,350	39,245	252	17	221,890	2,56
Mühlhausen (Thuringia)	Rotliegendes	2,250	60,203	41,082	tr.	20.7	14.8	167,147	72
N.E.Mecklenburg	Rotliegendes	1,177	44,487	59,619	912	12.6	125.6	177,128	1,21

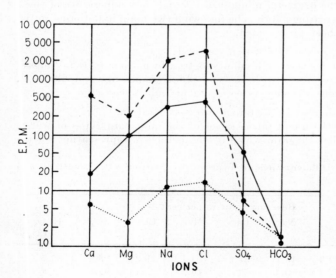

Fig. 101 Schoeller diagram of ion-concentration of water. Equivalents per million (e.p.m.) are plotted on a logarithmic scale. Solid line: sea-water; dashed line: oil-field brine (Pettet Fm., Caddo Parish, La.); dotted line: Red River, Texas (from PARKER, 1969)

chemical composition of waters. Table 152 should be considered as an example of the absolute chemical composition of deep-seated waters. The concentration of elements in so-called petroleum waters with particular reference to the order of magnitude is shown in Table 153.

SO₄	HCO₃	Total Concentration [g/l]	Density at 20 °C	pH	Li	Rb	Sr	B	Mn	Si	Cu	Pb
123.5	89.6	72.29	n.d.	7.5	2.2	—	150	68	95	tr.	—	—
186.7	27.0	150.66	1.102	7.5	n.d.	n.d.	n.d.	n.d.	n.d.	n.d.	n.d.	n.d.
365.2	736.9	38.53		9.3	n.d	n.d.	n.d.	n.d.	n.d.	n.d.	n.d.	n.d.
1,512	230.0	128.0	1.084	7.0	n.d.	n.d.	n.d.	n.d.	n.d.	n.d.	n.d.	n.d.
n.d.	27.45	197.63	1.127	6.0	1	—	490	37	9	tr.	—	—
1,519	95.8	159.48	1.105	6.6	6.4	—	249	68	95	tr.	—	—
tr.	181.0	170.68	1.114	6.0	1	—	200	30	45	tr.	—	—
tr.	43.3	176.71	1.116,3	6.0	2.9	—	240	23	10	tr.	—	—
787.0	30.0	118.88	1.086	6.0	n.d.	n.d.	n.d.	n.d.	n.d.	n.d.	n.d.	n.d.
139.2	—	334.07	1.230	5.8	30.0	4	1,650	30	75	tr.	—	—
233.4	61.6	325.24	1.214	6.0	29.0	1	730	38	24	tr.	—	—
5,448	29.9	258.82	1.172	5.4	3.2	12	21	10	7	—	1	—
368	21.4	322.55	1.210	4.2	56	4	320	—	20	—	tr.	50
—	549	508.84	1.309	3.5	n.d.	n.d.	n.d.	n.d.	n.d.	n.d.	n.d.	n.d.
279	131.8	337.22	1.235	4.3	111	5	1,280	1,000	5	—	—	—
139	170.8	338.41	1.231,1	5.0	140	54	1,550	460	—	—	—	—
200	67.1	357.18	1.243	5.1	150	34	1,800	310	7	—	1	20
536	761	346.38	1.282	6.6	190	2	200	1,000	125	—	1	30
379.2	82.9	272.84	1.187	6.0	88	30	710	13	85	tr.	—	—
209	347.7	285.25	1.205,4	6.2	n.d.	n.d.	n.d.	n.d.	n.d.	tr.	n.d.	n.d.

Table 153. Element Concentration in Mineral Oil Waters from RITTENHOUSE and FULTON, 1969)

%	Na, Cl
% or ppm	Ca, SO₄
≥100 ppm	K, Sr
1—100 ppm	Al, B, Ba, Fe, Li
<1 ppm	Cr, Cu, Mn, Ni, Sn, Ti, Zr (in the majority of waters) Be, Co, Ga, Ge, Pb, V, W (in a few waters)

The works by KREJCI-GRAF (1966), SCHOELLER (1955), VALYASHKO (1965), ANGINO and BIL-LINGS (1969), WHITE and WARING (1963), and CHEBOTAREV (1959) should be recommended as introductory and fundamental publications in this field.

8.5.5. Geochemistry of Mineral Waters

While formerly the ascendant mineral waters (including thermal waters) were considered to be primarily of juvenile origin, today we know that the juvenile (magmatogenic) part is a maximum of 5 per cent, perhaps even less than 1 per cent. This means that mineral waters are also of meteoric origin and, in their mineralization, may behave in a way similar to that shown by deep-seated waters.

Table 154. Chemical Composition of Water from Mineral Springs (values in mg/l = ppm)

Type	Na-Ca-HCO$_3$-SO$_4$ acidic water	Na-SO$_4$-Cl HCO$_3$ acidic water	Na-Cl	Na-Cl Rn-Ra therme	Ca-Fe-SO$_4$ arsenical
Place	Radon spring Bad Brambach	Marienquelle I Bad Elster	Salt Spring Bad Salzungen	Dzhety-Oguz (USSR)	Zuba, Georgian SSR
K$^+$	9.9	23.5	196.2	78.7	21.9
Na$^+$	232.4	938.3	27,450	2,853.8	5.2
Li$^+$	0.9	2.9	2.3		
NH$_4^+$	1.1	0.8	3.4	2.4	3.6
Ca^{++}	171.2	49.0	519.3	1,992.8	485.8
Sr^{++}	2.1[5])	tr.	4.7		20.4
Mg^{++}	34.8	40.1	374.7		165.2
Fe^{++}	11.4	17.4	2.9		499.5
Mn^{++}	0.4	0.5	1.7	1.4	
Fe^{+++}					152.7
Al^{+++}	1.5	2.6	0.5		126.4
F$^-$				0.3	
Cl$^-$	82.9	615.0	43,560	7,574.6	
Br$^-$	0.0^6	0.8	8.8	3.6	
I$^-$	0.0^3	0.0^3	0.0^1		
NO$_3^-$	1.6	0.4	4.4		
HCO$_3^-$	894.7	663.9	301.3	25.0	
HPO$_4^{2-}$	0.0^8	0.6	0.2		6.5[2])
SO$_4^{2-}$	268.9	975.1	1,210	577.2	3,934.3
CO$_2$	2,276	2,259	71.0		
H$_2$SiO$_3$	56.3	37.3	8.2	38.7	122.1
Peculiarity	2,247 ME[1])			containing Ra-Rn	oxid. Gases[3])
Sum	4,348	5,627	73,720	13,185.5	5,574.4
pH	5.9	5.8	6.7		
Temperature	7.6 °C	9.7 °C	11.7 °C	hot spring!	

[1]) = 817.9 Curie   [2]) +4.8 H$_3$PO$_4$ [3]) HAsO$_4^{2-}$ = 10.3; H$_3$AsO$_4$ = 13.1
[4]) HBO$_2$ = 2.1   [5]) 1.3 mg/l Ba

Table 155. Element Ratios in Various (Hot) Springs of Japan (from UZUMASA 1965)

	Ca/Na	Mg/Na	Mg/Ca	Cl/Na	Ca/Cl	Mg/Cl
1. acid hot springs	0.64	0.18	0.28	2.6	0.25	0.068
2. acid cold springs	0.30	0.12	0.40	1.5	0.20	0.078
3. neutral hot springs	0.39	0.03	0.08	1.6	0.24	0.020
4. neutral cold springs	0.50	0.09	0.19	2.3	0.22	0.042
5. alkaline hot springs	0.36	0.04	0.10	1.6	0.23	0.023
6. alkaline cold springs	0.15	0.05	0.36	1.6	0.09	0.033
7. sea-water	0.039	0.12	3.10	1.8	0.02	0.067

a-Ca-SO$_4$ CO$_3$-CO$_2$	Na-Cl-CO$_2$ H$_2$S- containing	Na-Cl-Ca-SO$_4$	Ca-HCO$_3$-SO$_4$ acidic water	Na-SO$_4$	Na-HCO$_3$-SO$_4$
udapest	Sochi-Macesta	Staraya Russa	Kislovodsk	Františ-kovy Lázně	Karlovy Vary
		74.8	9.5	156.7	104.2
19.2	8,094.4	5,154	183.7	6,491.4	1,718.0
			0.6	99.9	3.3
2	24.6	1.5	—		0.1
177.2	1,289.3	1,445	765.9	538.9	102.5
		8.7	9.0		0.3
69.9	357.3	453.0	253.7	146.5	46.5
		4.4	17.7	6.8	0.1
				0.2	
2.2			0.5		2.4
41.5	15,355.9	10,776	25.4	2,541.7	617.0
0.5	58.9	72.8	0.4		1.4
	8.9	0.3	0.2		
88.0	591.4	107.6	2,260.0	3,354.0	2,100.0
					0.2
31.7	10.9	1,403	1,460.0	10,242.0	1,662.0
12.6	202.4	6.2	1,790.0		
19.5	29.6	6.2	50.0		88.4
uc. Gases[4])	H$_2$S 419.4		HBO$_2$ = 6.9		
66.4	26,065	19,514.5	5,094.2		6,447.9
6.8	6.7	7.2	6.4		7.65
spring!	hot spring!			12°	

Ca/SO$_4$	Cl/HCO$_3$	SO$_4$/HCO$_3$	SO$_4$/Cl
0.13	3.1	5.9	1.9
0.05	2.0	8.2	4.2
0.52	3.1	1.5	0.47
0.93	3.7	0.89	0.24
0.55	3.4	1.4	0.41
0.91	2.4	0.25	0.10
0.16	193	26	0.14

Tables 154 and 156 give an insight into the macrochemistry and microchemistry of a few mineral waters, whereas Table 155 gives a summary of the geochemically most important element concentrations in hot springs of Japan.

1*

Table 156. Composition of 3 Hot Springs in Japan [%] (after Uzumasa, 1965)

I: Gongenyu, Yunohanazawa Spring, Hakone, Kan.
II: Motoyu, Yumoto Spring, Nasu, Tch.
III: Tenmangunoyu, Arima Spring, Hyo. (nearly neutral, common salt spring)

Element	Clarke number	I	II	III	Seawater
Si	25.8	7.55	2.52	0.094,2	0.005,8
Al	7.56	4.92	3.12	0.097,5	0.000,3
Fe	4.70	0.89	1.01	0.217	0.000,06
Ca	3.39	4.94	2.33	5.35	1.2
Na	2.63	0.67	0.814	27.0	30.4
K	2.40	0.05	0.26	6.13	1.1
Mg	1.93	1.78	1.12	0.052,6	3.71
Ti	0.46	0.004,7	0.006,2	0.003,4	—
Cl	0.19	0.25	6.19	57.5	55.0
Mn	0.09	0.099	0.012	0.058,2	0.000,01
P	0.08	0.077	0.073	0.000,63	000,1
Rb	0.03	—	0.000,4	0.004,6	0.000,58
F	0.03	—	0.013	0.000,99	0.004,1
Ba	0.023	0.001,2	0.000 n	0.082,0	0.000,03
Sr	0.02	0.003,8	0.000 n	0.002,8	0.038
Cr	0.02	0.000,29	0.000,16	0.000,12	—
V	0.015	0.011,6	0.019	0.007,9	0.000,000,9
Ni	0.01	0.000,0n	0.000,0n	0.000,001	0.000,000,3
Zn	0.008	0.002,9	0.005,1	0.000,252	0.000,01
Cu	0.007	0.000,8	0.001,0	0.000,176	0.000,000,6
Li	0.006	—	0.000,4	0.073,3	0.000,35
Co	0.004	0.000,0n	0.000 n	0.000,001	—
Sn	0.004	—	0.000,01	0.000,000,5	—
Pb	0.001,5	0.003,4	0.002,6	0.000,515	—
Mo	0.001,3	0.000,08	0.000,032	0.000,080	0.001,3
B	0.001	0.005	2.10	0.760	0.014
Ga	0.001	0.000,0n	0.000,0n	0.000,00n	—
Cs	0.000,7	—	0.000,4	0.003,3	0.000,006
Ge	0.000,65	0.000,1	0.000,00n	0.000,000 n	—
Br	0.000,6	—	0.000	0.072,0	0.19
Be	0.000,6	0.000,00n	0.000,00n	0.000,01	—
As	0.000,5	—	0.068	0.000,00	0.000,043
Cd	0.000,05	0.000,0n	0.000,004	0.000,000 n	—
Sb	0.000,05	0.25	0.11	0.060	0.000,03
I	0.000,03	—	0.000,0	0.001,2	0.000,1
Bi	0.000,02	0.000,0n	0.000,004	0.000,000,n	—
Ag	0.000,01	0.000,00n	0.000,04	0.000,000 0	0.000,000,9
In	0.000,01	—	0.000,004	0.000,000 n	—
Ra	140×10^{-12}	71×10^{-12}	59×10^{-12}	278×10^{-12}	0.2×10^{-12}
NH_4^+		0.25	0.11	0.060	—
HSO_4^- } SO_4^{2-} }		61.69	75.06	0.000,0	7.69
CO_2		1.89	0.24	0.486	—
H_2S		6.10	1.23	0.001	—

Bibliography

ANGINO, E. E., and G. K. BILLINGS (Editors): Geochemistry of Subsurface Brines. Special issue of Chem. Geol. *4* (1969), 1—370

ARKHANGELSKIĬ, A., and N. STRAKHOV: Geological Structure and History of the Development of the Black Sea. Akad. Nauk SSSR (1938) (Russian)

BENTOR, Y. K.: Some Geochemical Aspects of the Dead Sea etc. Geochim. Cosmochim. Acta *25* (1961), 239—260

BRUEVICH, S. V.: On the Geochemistry of Marine Silicon. Izv. Akad. Nauk SSSR, Geol. Ser. (1953), 4, 67—78 (Russian)

CARPENTER, A. B., and J. C. MILLER: Geochemistry of Saline Subsurface Water, Saline County (Missouri), in: ANGINO, E. E., and G. K. BILLINGS (editors), Geochemistry of Subsurface Brines. Chem. Geol. *4* (1969), 135—167

CHEBOTAREV, I. I.: Metamorphism of Natural Waters in the Crust of Weathering. Geochim. Cosmochim. Acta *8* (1959), 22—48, 137—170, 198—212

CHOW, T. J., and E. D. GOLDBERG: On the Marine Geochemistry of Barium. Geochim. Cosmochim. Acta *20* (1960), 192—198

CONWAY, E. J.: Mean Geochemical Data in Relation to Oceanic Evolution. Proc. Roy. Irish Acad. *48* (1942), 119—159

DEGENS, E., et al.: Data on the Distribution of Amino-acids and Oxygen Isotopes in Petroleum Brine Waters of Various Geologic Ages. Sedimentology *3* (1964), 199—225

DIETRICH, G., and K. KALLE: Allgemeine Meereskunde. Borntraeger, Berlin 1957

DYCK, S.: Die Wasserhaushaltsbilanz unserer Erde. Wissenschaft und Fortschritt *5* (1968), 208—212

GARRELS, R. M., and M. E. THOMPSON: A Chemical Model for Sea Water at 25 °C and One Atmosphere Total Pressure. Amer. J. Sci. *260* (1962), 57—66

GERMANOV, A. I.: The Oxygen of Underground Waters and Its Geochemical Importance. Izv. Akad. Nauk SSSR, Ser. geol. *6* (1955), 70—82 (Russian)

GERMANOV, A. I., S. G. BATULIN, G. A. VOLKOV, A. K. LISITSIN, and V. S. SEREBRENNIKOV: Some Regularities of Uranium Distribution in Underground Waters. Proc. 2nd. U. N. Intern. Conf. Peaceful Uses of Atomic Energy *2* (1958), 161—177

GLAGOLEVA, M. A.: The Form of Migration of the Elements in River waters; in: On the Knowledge of the Diagenesis of the Sediments. Izd. Akad. Nauk SSSR 1959 (Russian)

GOLDBERG, E. D.: Marine Geochemistry. Amer. Rev. Phys. Chem. *12* (1961), 29—48

GORHAM, E.: On the Acidity and Salinity of Rain. Geochim. Cosmochim. Acta *7* (1955), 231—239

GORHAM, E.: On Some Factors Effecting the Chemical Composition of Swedish Fresh Waters. Geochim. Cosmochim. Acta *7* (1955), 129—150

GULYAEVA, L. A., and J. S. ITKINA: Halogens in Marine and Fresh-water Sediments. Geokhimiya *6* (1962), 524—528 (Russian)

GUNDLACH, H.: Chemische Aspekte des Transportes von Metallen in hydrothermalen Lösungen. Symp. Probl. Postmagm. Ore Depos., Prague 1963, 402—409

HARVEY, H. W.: The Chemistry and Fertility of Sea Waters. University Press Cambridge, 1955

HEIDE, F.: Zur Geochemie der Süßwässer. Chem. d. Erde *16* (1952), 3—21

HEM, J. D.: Geochemistry of Ground Water. Econ. Geol. (1950), 72—81

JORDAN, H.: Bäderbuch der Deutschen Demokratischen Republik. Verlg. G. Thieme, Leipzig 1967

KALLE, K.: Der Stoffhaushalt des Meeres. Akademische Verlagsgesellschaft Geest & Portig, Leipzig 1943

KHARKER, D. P., K. K. TUREKIAN, and K. K. BERTINE: Stream Supply of Dissolved Silver, Antimony, Selenium, Chromium, Cobalt, Rubidium and Cesium to the Oceans. Geochim. Cosmochim. Acta *32* (1968) 3, 285—298

KISSIN, I. G., and S. Y. PAKHOMOV: On the Influence of High Temperatures on the Formation of the Chemical Composition of Underground Waters. Geokhimiya (1967) 3, 341—355 (Russian)

KONTOROVICH, A. E., M. A. SADIKOV, and S. L. SHVARTSEV: The Distribution of a Few Elements in the Surface and Underground Waters of the North-west of the Siberian Plateau. Dokl. Akad. Nauk SSSR *149* (1963) 1, 179—180 (Russian)

KRAMER, J. R.: History of Sea Water, Constant Temperature-pressure Equilibrium Models Compared to Liquid Inclusion Analyses. Geochim. Cosmochim. Acta *29* (1965), 921—945

Krauskopf, K. B.: Factors Controlling the Concentrations of Thirteen Rare Metals in Seawater. Geochim. Cosmochim. Acta 9 (1956), 1—32

Krejci-Graf, K.: Diagnostik der Salinitätsfazies der Ölwässer. Fortschr. Geol. Rheinl. u. Westf. 10 (1963), 367—448

Krejci-Graf, K., et al.: Zur Geochemie des Wiener Beckens. Geol. Mitt. 7 (1966), 49—108

Květ, R.: Die heutigen Ansichten der Genese und Metamorphose von Erdölwässern, in: Geochemische Probleme bei der Erkundung und Förderung von Erdöl und Erdgas, Vol. 1, VEB Deutsch. Verlag f. Grundstoffindustrie, Leipzig 1968, 434—447

Leutwein, F., and L. Weise: Hydrogeochemische Untersuchungen an erzgebirgischen Gruben- und Oberflächenwässern. Geochim. Cosmochim. Acta 26 (1962), 1333—1348

Lisitsin, A. K.: On the Characteristic of the Medium in Hydrogeochemical Investigations. Geokhimiya 2 (1963), 149—157 (Russian)

Livingstone, D. A.: Chemical Composition of Rivers and Lakes. Data of Geochemistry (6th edition). U. S. Geol. Survey Prof. Paper 440-G (1963), 64 pp.

Manheim, F. T.: A Geochemical Profile in the Baltic Sea. Geochim. Cosmochim. Acta 25 (1961), 52—70

Manheim, F. T., and J. L. Bischoff: Geochemistry of Pore Waters etc., in: Geochemistry of Subsurface Brines. Chem. Geol. 4 (1969), 63—82

Michard, G.: Signification du Potential Redox dans les Eaux Naturelles. Conditions d'Utilisation des Diagrammes (Eh, pH). Mineralium Deposita 2 (1967), 34—36

Müller, E. P.: Zur Geochemie der Tiefenwässer und der organischen Substanz im Nordteil der DDR. Z. angew. Geologie 15 (1969) 3, 113—124

Nicholls, G. D.: The Geochemical History of Sea Water, in: Riley and Skirrow, Chemical Oceanography, Vol. 2 (1965)

Ovchinnikov, A. M.: Mineral Waters, 2nd ed. Gosgeoltechizdat, Moscow 1963 (Russian)

Park, K.: Deep-Sea pH. Science 154 (1966) 3756, 1540—1542

Parker, J. W.: Water History of Cretaceous Aquifers, East Texas Basin. Chem. Geol. 4 (1969), 1/2, 111—133

Pavlov, A. N., and V. N. Shemyakin: A Geochemical Classification of Natural Waters. Geokhimiya (1967) 12, 1482—1488 (Russian)

Popov, N. I. (Edit.): Examination of the Radioactive Fall-out in Waters of the World's Oceans. Trudy instit. okeanol., Vol. 82; Izd. "Nauka", Moscow 1966 (Russian)

Posokhov, E. V.: Formation of Chemical Constituents of Subsurface Waters. Leningrad GIMIZ 1966 (Russian)

Reynolds, R. C.: The Concentration of Boron in Precambrian Seas. Geochim. Cosmochim. Acta 29 (1965), 1—16

Riley, J. P.: Analytical Chemistry of Sea Water, in: Riley and Skirrow, Chemical Oceanography, Vol. 2 (1965)

Riley, J. P., and Skirrow: Chemical Oceanography. Vols. 1 and 2; Academic Press, London/New York 1965

Rittenhouse, G., R. B. Tulton et al.: Minor Elements in Oilfield Waters, in: Geochemistry of Subsurface Brines, Chem. Geol. 4 (1969), 189—209

Ronov, A. B.: On the Post-cambric Geochemical History of the Atmosphere and Hydrosphere. Geokhimiya (1959), 397—409 (Russian)

Rubey, W. W.: Geologic History of Sea Water. Bull. Geol. Soc. Amer. 62 (1951), 1111—1148

Schink, D. R.: Budget for Dissolved Silica in the Mediterranean Sea. Geochim. Cosmochim. Acta 31 (1961), 987—999

Schoeller, H.: Géochimie des Eaux Souterrains. Rev. Inst. Franc. Pétrole, Paris (1955)

Schoeller, H.: Les Eaux Souterrains, Masson et Cie. Paris 1962

Schutz, D. F., and K. K. Turekian: The Investigation of the Geographical and Vertical Distribution of Several Trace Elements in Sea Water Using Neutron Activation Analyses. Geochim. Cosmochim. Acta 29 (1965), 259—313

Sigvaldason, G. F.: Chemistry of Thermal Waters and Gases in Iceland. Bull. volcanol. 29 (1966), 589—604

Sillèn, L. G.: The Physical Chemistry of Sea Water, in: Oceanography, American Associated Advisory Sciences Publ. No. 67 (1961), Washington

LLÈN, L. G.: The Ocean as a Chemical System. Science *156* (1967) 3779, 1189—1196

KOPINTSEV, B. A.: On the Equilibrium of Organic Carbon in the Waters of the World Ocean. Dokl. kad. Nauk *174* (1967) 6, 1417—1420 (Russian)

UGAWARA, K., H. NAITO, and S. YANADA: Geochemistry of Vanadium in Natural Waters. J. Earth, agoya-Univ. *4* (1956), 44

VERDRUP, H. U., M. W. JOHNSON, and R. H. FLEMING: The Oceans, their Physics, Chemistry and eneral Biology. X + 1087, New York 1949

UREKIAN, K. K., and D. G. JOHNSON: The Barium Distribution in Sea Water. Geochim. Cosmochim. cta *30* (1966), 1153—1174

DONOV, P. A., and Y. S. PARILOV: On a Few Regularities in the Migration of Metals in Natural Waters. Geokhimiya *8* (1961), 703—707 (Russian)

ALYASHKO, M. G.: Genesis der Salzsolen der Sedimenthülle. Freiberger Forschungsh. C *190* (1965), 1—98 (Russian)

ALYASHKO, M. G.: Principles of the Geochemistry of Natural Waters. Geokhimiya (1967) 11, 1395 to 406 (Russian)

ERNADSKIĬ, V. I.: Ozeanographie und Geochemie. Z. Kristallogr. Mineralog. Petrogr. Abt. B *44* 1933), 168

ERNADSKIĬ, V. I.: Sur la Classification et sur la Composition Chimique des Eaux Naturelles. Bull. Soc. ranc. Minéralog. *53* (1930), 417

INOGRADOV, A. P.: Introduction into the Geochemistry of the Ocean. Izd "Nauka", Moscow 1967 Russian)

INOGRADOV, A. P.: Geochemistry of the Trace Elements in Seawater. Uspekhi chimii *13* (1944) 1, —34 (Russian)

INOGRADOV, A. P.: Recent Tritium Content in Natural Waters. Geokhimiya (1968) 10, 1147—1162 Russian)

VEDEPOHL, K. H.: Die Geochemie der Gewässer. Naturwissenschaften *53* (1966) 14, 352—354

VEDEPOHL, K. H.: Einige Überlegungen zur Geschichte des Meerwassers. Fortschr. Geol. Rheinl. u. Vestf. *10* (1963), 129—150

VHITE, D. E., J. D. HEM, and G. A. WARING: Chemical Composition of Subsurface Waters, Chapter F f the Data of Geochemistry, (6th ed., edited by M. Fleischer). Geol. Surv. Prof. Paper, 440-F, 1963, 7 pp.

VHITE, D. E.: Summary of Chemical Characteristics of Some Waters of Deep Origin. Prof. Paper J. S. Geol. Survey 400 B (1960), 454—456

ZAVODNOV, S. S.: On the Calculation of the CO_2 Content and of the pH-Value in a Few Subsurface Waters. Geokhim. materialy (1964), 121—126 (Russian)

ZYKA, V.: Hydrogeochemische Zonen in Mitteleuropa. Acta Geol. *4* (1957) 3—4, 383—418

Group of authors: Formation and Geochemistry of Underground Water in Siberia and in the Far East. Moscow 1967 (Russian)

Group of authors: The Pacific. Chemistry of the Pacific. Izd. "Nauka", Moscow 1966 (Russian)

8.6. Geochemistry of the Atmosphere and Gases

8.6.1. The Atmosphere

The average composition of the atmosphere has already been stated in Section 7.4.4. In addition to this information, Fig. 102 shows how the composition changes with increasing altitude, while Table 157 gives a detailed classification of gases.

Besides the main components, the atmosphere contains a number of trace elements and impurities. Table 158 shows the concentration of Cl, Br, and I in the atmosphere. According to measurements conducted by WINDOM (1969), the dust deposition in the northern permanent snowfields is of the order of 0.1—1.0 mm per 1000 years on an average.

Table 157. Classification of the Natural Gases (from SOKOLOV, 1966)

Type of Gas	Chemical Composition			Origin of the Gases
	Major Components	Minor Components	Traces	
I. Atmosphere	N_2, O_2, Ar, CO_2	—	O_3, NO_2, N_2O, H_2, noble gases	gas mixtures of chemical, biochemical and radiogenic origin
II. Gases of the surface of the earth and of subaqueous deposits:				
horizons near the surface	CO_2, N_2, O_2 and other gases of the atmosphere	—	CH_4, N_2O, CO, H_2, H_2S, NH_3, volatile organic substances, noble gases (from the air)	gases of primarily biochemical origin with admixtures of other gases. Due to the exchange of gases with the atmosphere, the majority of gases formed in the soil and other gases are admixed with the air.
bog, swamp	CH_4, CO_2, N_2	—	H_2, CO, N_2O, NH_3, H_2S, volatile organic substances, noble gases from the air	Gases mainly of biochemical origin with admixtures of other gases. Sometimes gases from the air are present. Nitrogen mostly is from the air.
marine subaqueous sediments	CO_2, CH_4, N_2	—	H_2, NH_3, H_2S, volatile organic substances, noble gases (from the air)	Gases mainly of biochemical origin. In gases of deep-sea sediments, CO_2, NH_3 and N_2 are the major components.
III. Gases from sedimentary rock:				
petroleum deposits	CH_4, heavy gaseous hydrocarbons	N_2, CO_2, H_2S, He, Ar	H_2, noble gases	Gases mainly of chemical origin, sometimes with admixtures of gases of biochemical origin. In greater depths, where normal activities of microorganisms are impossible because of increased temperatures, biochemical gases are missing.
gas deposits	CH_4	heavy gaseous hydrocarbons, N_2, CO_2, H_2S, He, Ar	H_2, noble gases	
coal	CH_4	CO_2, N_2, H_2	heavy gaseous hydrocarbons, H_2S, NH_3, noble gases	
saline deposits	CH_4, H_2, N_2	H_2S, heavy gaseous hydrocarbons	noble gases	
disperse deposits	CO_2	H_2, heavy gaseous hydrocarbons, H_2S, N_2	noble gases	

Table 157 continued

Types of Gas	Chemical Composition			Origin of the Gases
	Major Components	Minor Components	Traces	
IV. Gases from metamorphic rocks	CO_2, N_2, H_2	CH_4, H_2S, heavy gaseous hydrocarbons	noble gases	Gases of chemical origin with admixtures of gases produced by radioactive decay and radiation.
V. Gases from granitic and basaltic rocks	CO_2, H_2	N_2, H_2S, HCl, HF	CH_4, noble gases	see IV
VI. Volcanic gases: magmatic gases (from lava lakes)	CO_2, H_2, SO_2, HCl HF	CO_2, N_2, NH_3	CH_4, noble gases	Gases of chemical origin with admixtures of gases which are produced by radioactive decay and radiation.
fumaroles	CO_2, H_2, H_2S, SO_2	CO, HCl, HF, N_2	CH_4, noble gases	Volcanic gases are gases of the upper mantle changed to some degree.

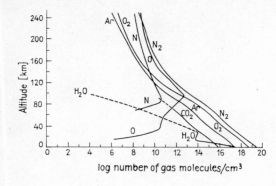

Fig. 102 Composition of the atmosphere, after: Handbook of Geophysics (US Air Force). New York 1960

Table 158. Chlorine, Bromine and Iodine Concentrations in the Atmosphere and Precipitation in Central Europe (from VINOGRADOV, 1954)

Origin	Cl	Br	I	Ratio Cl:Br	Br:I
Air above the Atlantic Coast	$5 \cdot 10^{-5}$	$3 \cdot 10^{-6}$	$2 \cdot 10^{-7}$	17	15
Air above the European Continent	$4 \cdot 10^{-6}$	$2 \cdot 10^{-7}$	$2 \cdot 10^{-8}$	20	10
Continental precipitation (rain, snow)	$9 \cdot 10^{-5}$	$5 \cdot 10^{-7}$	$8 \cdot 10^{-8}$	180	6

8.6.2. Gases in the Lithosphere and Hydrosphere

Figs. 103 and 104 give information about the relative contents of gases in magmatic rocks and volcanogenic products. Table 159 shows the composition of a few gases in seawater. Table 160 gives a few element ratios in gases of the atmosphere and in seawater.

Table 159. Content of a Few Gases in Seawater
(from VINOGRADOV, 1966)

Gas	Content in cm³/l	In Comparison with the Atmospheric Content
N_2	13	781
O_2	2—8	210
Ar	0.32	9.32
CO_2	50.0	0.3 ·
Ne	$1.8 \cdot 10^{-4}$	$1.82 \cdot 10^{-2}$
He	$5 \cdot 10^{-5}$	$5.3 \cdot 10^{-3}$
Kr	$6 \cdot 10^{-5}$	$1 \cdot 10^{-3}$
Xe	$7 \cdot 10^{-6}$	$8 \cdot 10^{-5}$

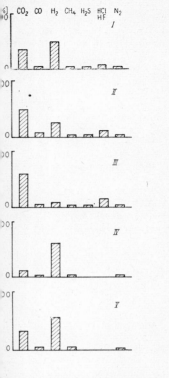

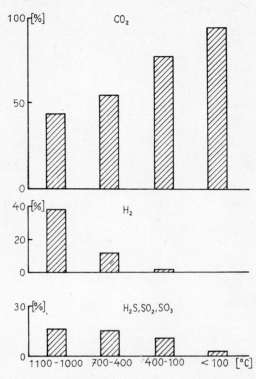

Fig. 103 Average composition of gases from eruptive rocks (released at high temperatures) (after Sokolov, 1967)

I Granites and gneisses; II Basalts and diabases; III Andesites; IV Syenites; V Gabbro, diorite, porphyrite

Fig. 104 Dependence of the composition of volcanic gases on the temperature (after Sokolov, 1967)

Table 160. Ratios of Gases in the Atmosphere and in Seawater (at 20 °C) (after Vinogradov, 1967)

Gases	Ratio in Atmosphere	Ratio in Seawater
He/Ne	0.288	0.246
He/Ar	$5.62 \cdot 10^{-4}$	$1.5 \cdot 10^{-4}$
He/Kr	4.6	0.695
He/Xe	61.0	5.33
N$_2$/Ar	83.8	38.5
Kr/Ne	$6.26 \cdot 10^{-2}$	$35.4 \cdot 10^{-2}$
Xe/Ne	$4.73 \cdot 10^{-3}$	$46.2 \cdot 10^{-3}$
Ne/Ar	$19.5 \cdot 10^{-4}$	$6.08 \cdot 10^{-4}$
Kr/Ar	$1.22 \cdot 10^{-4}$	$2.16 \cdot 10^{-4}$
Xe/Ar	$0.922 \cdot 10^{-5}$	$2.81 \cdot 10^{-5}$
Xe/Kr	$7.54 \cdot 10^{-2}$	$13.0 \cdot 10^{-2}$

Bibliography

BELUSSOV, V. V.: Compendium of the Geochemistry of the Natural Gases. ONTI 1937 (Russian)

BERKNER, L. V., and L. C. MARSHALL: The History of Oxygenic Concentration in the Earth's Atmosphere. Discuss. Faraday Soc. *37* (1964), 122—141

BOTNEVA, V. V., R. G. PANKINA, and V. A. SOKOLOV: Geochemistry of the Gases Accompanying Mineral Oil. Izd. "Nedra" 1960 (Russian)

CHAIGNEAU, M.: Sur les Gaz Volcaniques de l'Etna (Sicile). C. R. Hebd. Séances Acad. Sci *25* (1962), 23

CLOUD, P. E., jr.: Symposium on the Evolution of the Earth's Atmosphere. Proc. Nat. Acad. Sci USA *53* (1965), 6, 1169—1173

COMMONER, B.: Biochemical, Biological and Atmospheric Evolution. Proc. Acad. Sci. USA *53* (1965) 6 1183—1194

DAVIDSON, C. E.: Geochemical Aspects of Atmospheric Evolution. Proc. Nat. Acad. Sci. USA *5* (1965) 6, 1194—1205

DONAHOE, T. M.: Ionospheric Composition and Reactions. Science *159* (1968) 3814, 489—497

EMERY, K. O., and D. HOGGAN: Gases in Marine Sediments. Bull. A. A. P. G. *42* (1958) 9

GALINOV, E. M., N. G. KUSNEZOVA, and V. S. PROCHOROV: On the Problem of the Old Earth Atmosphere Composition in Connection with Results of Isotopic Analysis of Carbon from Precambrian Carbonates Geokhimiya (1968) 11, 1376—1381 (Russian)

HOLLAND, H. D.: The History of Ocean Water and Its Effect on the Chemistry of the Atmosphere Proc. Nat. Akad. Sci. USA *53* (1956) 6, 1173—1183

HUNT, J. M.: Distribution of Hydrocarbons in Sedimentary Rocks. Geochim. Cosmochim. Acta *2* (1961) 1

ISRAEL, H.: Spurengase in der Atmosphäre. Naturw. Rdsch. *20* (1967) 5114, 329—336

JAGGAR, T.: Magmatic Gases. Amer. J. Sci. (1940)

MIRTOV, B. A.: The Gas Content in the Earth's Atmosphere and the Methods of Determination. Izd Akad. Nauk SSSR, 1961 (Russian)

NARIZHNAYA, V. E.: The Geochemistry of the Natural Gases in Central Asia. Izd. "Nedra", 196* (Russian)

PAETZOLD, H. K.: Ozonschicht und Wärmebilanz der Erde. Z. Geomorphol. *8* (1964) 4, 484

PETERSILIE, I. A.: Geology and Geochemistry of Natural Gases and Disperse Bitumina in Various Geological Formations of the Kola Peninsula. Izd. "Nauka" 1964 (Russian)

RUBEY, W. W.: Development of the Hydrosphere and Atmosphere with Special Reference to Probable Composition of the Early Atmosphere. Bull. Geol. Soc. Amer., Spec. Pap. *62* (1953)

SEMENOV, A. D. et al.: On Organic Substances of Atmospheric Precipitates. Dokl. Akad. Nauk SSSR *173* (1967) 5, 1185—1188 (Russian)

SHEPHARD, E. S.: The Magmatic Gases. Amer. J. Sci. *35-A* (1938)

SOKOLOV, V. A.: Geochemistry of Natural Gases. Izd. "Nedra". Moscow 1971, 333 p. (Russian)

TRENDALL, A. F.: Carbon Dioxide in the Precambrian Atmosphere. Geochim. Cosmochim. Acta *3* (1966), 435—437

VINOGRADOV, A. P.: The Regime of Gases on the Earth. Chimiya Zemnoy kory, Vol. 2, Izd. Akad Nauk SSSR, Moscow 1964 (Russian)

WELTE, E.: Kohlenwasserstoffgenese in Sedimentgesteinen usw. Geol. Rdsch. *55* (1965) 1, 131—144

WHITE, D. E., and G. A. WARING: Volcanic Emanations. Data of Geochemistry, Chap. K. U. S. Geol Surv. Prof. Pap. *440-K* (1963)

WINDOM, H. I.: Atmospheric Dust Records in Permanent Snowfields: Implications to Marine Sediments. Geol. Soc. Amer. Bull. *80* (1969), 761—782

8.7. Geochemistry of the Biosphere and Biogenic Products (Organic Geochemistry)

Organic geochemistry can be defined as follows (according to MÜLLER, 1969):

"Organic geochemistry studies the laws of distribution, the forms of appearance and transformation of naturally occurring carbon and its compounds and the regularities of their conditions of formation in space and time, in the course of the earth's history."

From this definition it follows that biogeochemistry as "geochemistry of living matter" is a branch of organic geochemistry and, to a great extent, one of its scientific prerequisites.

8.7.1. Geochemistry of Living Matter (Biogeochemistry)

The paramount importance of living matter to the geochemistry of the upper crust of the earth was expounded for the first time by Vernadskiĭ in the following two works:

VERNADSKIĬ, V. I.: Geochemie. Akademische Verlagsgesellschaft Geest & Portig, Leipzig 1930 ("La Géochimie" in French was published already in 1924).
VERNADSKIĬ, V. I.: Die Biosphäre. Leipzig 1930.

Though living beings occupy only a very small fraction of the upper parts of the earth's crust, their geochemical effect on distribution and redistribution of the chemical elements is rather remarkable.

Besides the main components C, O and H, living organisms take up greatly differing quantities of the other chemical elements. Depending on the physiological importance to the vital functions, a distinction is made between nutritive elements, ballast elements and toxic elements. Though any element could be included in any one of these groups, a certain grouping has been established under the conditions of the surface of the earth. Table 161 shows the physiological effectiveness of a few important biophile elements.

In addition to the data given in Section 7.4.5., Table 162 shows the quantitative distribution of carbon on the earth, whereas Table 163 shows the composition of the major components in the most important parts of the biosphere. Table 164 gives a survey of the total composition of man and plant.

An important, mostly underrated part is played by micro-organisms in the form of viruses, fungi and especially bacteria. Bacteria consist of 75 to 80 percent water, the remaining percentage consists of albuminous substances (amino acids, polypeptides, protein), carbohydrates (monosaccarides, polysaccarides), lipoids (fats, phosphatides), and inorganic substances (K, Ca, Na, Mg, Cl, SO_4, PO_4, CO_3). They decompose ("mineralize") most of the higher organisms and their dead products into simple structural units (degradation), e.g.:

protein $\rightarrow$ polypeptides $\rightarrow$ amino acids $\rightarrow$ ammonia

The autotrophic bacteria are capable of self-nourishment by chemosynthesis or photosynthesis, whereas the heterotrophic bacteria receive the required nutritive substances with the help of other living beings.

Under oxidizing conditions, some aerobic bacteria play an important part in the oxidation of sulphur and iron compounds:

1st scheme $2\,H_2S + O_2 \rightarrow 2\,H_2O + 2\,S$

$$2\,S + 3\,O_2 + 2\,H_2O \rightarrow 2\,H_2SO_4$$

sulphur bacteria: Thiobacteriaceae, e.g.
Thiobacillus thiooxydans
Vibrio desulfuricans
Beggiatoae
Thiotrix

2nd scheme $4FeCO_3 + 6H_2O + O_2 \rightarrow 4Fe(OH)_3 + 4CO_2 (+ 29 \text{ kcal})$

iron bacteria: Gallionella
T. ferrooxydans
Ferrobacillus spp.

3rd scheme $NH_3 + 3O \rightarrow HNO_2 + H_2O (+ 79 \text{ kcal})$

$HNO_2 + O \rightarrow HNO_3 (+ 21 \text{ kcal})$

nitrifying bacteria: Nitrosomonas
Nitrobacter

Under reducing conditions, anaerobic bacteria are active in the reduction of sulphates and nitrates:

1st scheme $SO_4^{2-} + 2C + 2H_2O \rightarrow H_2S + 2HCO_3^-$
desulphurising bacteria: Desulfovibrio
Clostridium nigrificans

2nd scheme $2NO_3^- + 2C + H_2O \rightarrow N_2 + 2CO_2 + H_2O$

denitrifying bacteria: Bacter denitrificans

A few bacteria are highly susceptible to pH, others live in a wide pH interval. Detailed descriptions are given by SILVERMAN and EHRLICH (1964), specific representations and models of oxidations and reductions are, for example, given by HARRISON and THODE (1958), and by KROUSE et al. (1968). A schematic representation of the role of micro-organisms in the sulphur cycle is given in Fig. 105.

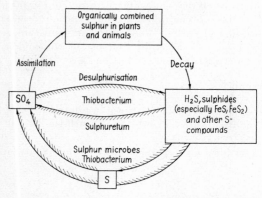

Fig. 105   The role of microorganisms in the sulphur cycle (after SCHWARTZ, 1958; dashed line: biocoenoses of the sulphureticum)

Table 161. Organo-metallic and Organo-metalloid Elements which Play an Important Physiological Role in Organisms (from VINOGRADOV, 1963)

Element	Compounds and Their Occurrence
	In tyrosine — the hormone of the thyroid gland of animals of higher organisation and in the homologue of this gland in lower organisms. I-organic compounds (on the basis of tyrosine and other similarly built molecules) in marine sponges, in alcyonaria, etc., and in seaweed.
	Br-organic compounds in marine sponges, alcyonaria and the like, 6-6 dibromo-indigo pigment in a number of mollusca (Br-tyrosine).
	The boric acid complexes with hydrocarbons in plants (the pollen-tubes of higher floriferous plants will fail to germinate if these complexes are missing), and the like.
	The various sulphur compounds in all organisms — amino acids, essential oils, and the like.
	Replaces S in cystine, in the cysteine of the albumen in astragali and other plants.
n	A great number of oxidation enzymes in plants and animals — proteids with Mn (enzymes and coenzymes) e.g. arginase of the liver and the like.
e	The haemoglobin of the higher animals and other analogous respiratory pigments of the lower animals (proteids). Enzymes — catalase, peroxidase, cytochrome.
u	The respiratory pigments of the lower animals — haemocyanins. Haemocuprin and haemocupreine of the higher animals. Many derivatives of porphyrin (pigments in the feathers of a few birds, egg-shells, etc.). Oxidation enzymes — polyphenolase, tyrosinase, lactase, oxidase; various proteids with copper.
n	Carbonic anhydrase of plants and animals; proteids in fungi, poison of snakes, and many other items.
o	Vitamin B_{12} and other closely allied compounds.
	In the pigment of the blood of Tunicatae Ascididae and others.
o	In enzymes — nitrate reductose, xanthine oxidase.
l	In the form of salts of organic acids in many plants (succinic acid and others).

Table 162. The Quantity and Distribution of Carbon on the Earth (from VINOGRADOV, 1967)

Subject	Carbon C mass[g]	Earth's Surface [g/cm²]
Marine organisms	$1 \cdot 10^{16}$	0.002
Continental organisms	$3 \cdot 10^{17}$	0.06
Atmosphere	$6.3 \cdot 10^{17}$	0.125
Ocean	$3.6 \cdot 10^{19}$	7.5
Coal, crude oil and other caustobiolites	$6.4 \cdot 10^{21}$	663
Slates, clays	$1 \cdot 10^{22}$	2,000
Carbonate rocks	$5.0 \cdot 10^{22}$	2,500

Table 163. Elementary chemical composition of the Organic Substances (from VINOGRADOV, 1967; values in mass-%)

Subject	C	H	O	N
Phytoplankton	45	7	45	3
Zooplankton	50	8	32	10
Pelagic deposits	56	8	30	6
Continental plants	54	6	37	2.8
Humus of the soil	56	4.5	36	3.5

Table 164. Chemical Composition of Man and Plant (in mass-% according to BERTRAND, 1950)

a) Man (60% of water; 35.7% of organic substance; 4.3% of ashes)
b) Plants (Lucerne)

	Man	Plant		Man	Plant
O	62.81	77.90	Br	$2 \cdot 10^{-4}$	$5 \cdot 10^{-5}$
C	19.37	11.34	Sn	$2 \cdot 10^{-4}$	
H	9.31	8.72	Mn	$1 \cdot 10^{-4}$	$3.6 \cdot 10^{-4}$
N	5.14	8.25	I	$1 \cdot 10^{-4}$	$2.5 \cdot 10^{-6}$
Ca	1.38	$5.80 \cdot 10^{-1}$	Al	$5 \cdot 10^{-5}$	$2.5 \cdot 10^{-3}$
S	$6.4 \cdot 10^{-1}$	$1.04 \cdot 10^{-1}$	Pb	$5 \cdot 10^{-5}$	
P	$6.3 \cdot 10^{-1}$	$7.06 \cdot 10^{-1}$	Ba	$3 \cdot 10^{-5}$	
Na	$2.6 \cdot 10^{-1}$		Mo	$2 \cdot 10^{-5}$	$1 \cdot 10^{-4}$
K	$2.2 \cdot 10^{-1}$	$1.70 \cdot 10^{-1}$	B	$2 \cdot 10^{-5}$	$7 \cdot 10^{-4}$
Cl	$1.8 \cdot 10^{-1}$	$7.0 \cdot 10^{-2}$	As	$5 \cdot 10^{-6}$	
Mg	$4 \cdot 10^{-2}$	$8.2 \cdot 10^{-2}$	Co	$4 \cdot 10^{-6}$	$2 \cdot 10^{-6}$
Fe	$5 \cdot 10^{-3}$	$2.7 \cdot 10^{-3}$	Li	$3 \cdot 10^{-6}$	
Si	$4 \cdot 10^{-3}$	$9.3 \cdot 10^{-3}$	V	$2.6 \cdot 10^{-6}$	$1.6 \cdot 10^{-5}$
Zn	$2.5 \cdot 10^{-3}$	$3.5 \cdot 10^{-4}$	Ni	$2.5 \cdot 10^{-6}$	$5 \cdot 10^{-5}$
Rb	$9 \cdot 10^{-4}$	$4.6 \cdot 10^{-4}$	Ti		$4.6 \cdot 10^{-5}$
Cu	$4 \cdot 10^{-4}$	$2.5 \cdot 10^{-4}$	F		$1.5 \cdot 10^{-4}$
Sr	$4 \cdot 10^{-4}$				

8.7.2. Geochemistry of Fossil Organic Matter

In accordance with the varying chemistry of living organic matter, the chemistry of fossils is also marked by a great diversity. The highly reactive organic matter is subject to the action of microorganisms and metamorphism. Fig. 106 shows a classification of fossil organic matter from the genetic angle, whereas Fig. 107 gives a chemically oriented classification of the naturally occurring bituminous substances.

Humic deposits

The average composition of humic sediments and their parent material is shown in Table 165, whereas Fig. 108 shows the proportion of the functional groups in various coals.

Table 165. Composition of Humic Sediments and Their Parent Materials [mass-%], after STUTZER (1923)

	C	H	O	N
Wooden substance	50	6	43	1
Peat	59	6	33	2
Lignite	69	5.5	25	0.8
Bitum. coal	82	5	13	0.8
Anthracite	95	2.5	2.5	traces

Tables 166 and 167 give information about the trace element concentration in German bituminous and brown coals (lignites).

Table 166. Average Trace Element Concentration [ppm] in Coals from GDR (from LEUTWEIN and RÖSLER, 1956)

Geologic age (after HORST, 1952)	Locality	No. of samples	Ge	Cu	Pb	Zn	Ag	As	Sn	Ga	Be	Co	Ni	Mo	V
Mesozoic coals of various places of occurrence		12	54	90	170	85	0.1	35	0.9	25	(6)	8	15	1	12
Lower Rotliegendes — Manebach Stage	Manebach	8	1	14	580	90	0.3	85	2.2	35	3	23	25	15	45
	Freital	53	15	6	130	230	0.2	150	3.5	20	15	6	12	16	30
Gehren Stage	Lauchagrund	5	2	88	180	56	0.5	240	0.6	—	9	17	6	2	5
	Crock	8	2	17	1,600	1,600	2.5	210	2.0	17	—	3	10	7	19
	Ilfeld	7	5	tr.	150	63	0.6	130	4.6	12	—	26	27	6	20
	Meisdorf	3	5	160	670	70	0.4	20	1.3	5	7	65	20	3	48
	Stockheim-Neuhaus	4	1	3	9	17	0.1	20	—	tr.	6	—	2	2	1
Upper Carboniferous — STEFANIAN	Öhrenkammer	4	1	320	8	140	1.2	300	0.5	5	—	26	17	11	24
	Dölau	6	1	37	320	58	0.4	15	9	9	3	1	9	4	22
	Wettin	8	3	17	340	180	0.9	70	1.6	11	16	3	6	5	19
	Löbejün	5	2	190	240	62	0.8	100	0.5	5	—	2	9	4	6
	Plötz	27	1	6	130	41	0.1	130	1.3	37	—	3	17	6	35
WESTFALIAN D	Zwickau	134	29	16	110	130	0.3	76	2.8	22	26	16	25	6	10
	Mülsengrund	24	6	9	31	57	0.1	170	1.1	18	1	8	16	4	6
	Oelsnitz	108	32	25	94	160	0.2	70	1.9	36	8	23	42	4	13
WESTFALIAN B	Flöha	8	3	5	33	13	0.3	5	2.3	77	46	4	15	5	32
	Schönfeld	7	13	23	35	97	0.2	25	1.5	30	38	14	21	4	27
	Zaunhaus	5	1	10	14	38	0.3	60	3.8	8	31	7	5	1	48
Lower Carbon. — VISEAN	Ebersdorf	6	11	<17	17	≦42	0.2	400	<1.7	40	(8)	22	23	11	42
	Berthelsdorf	4	13	27	91	(10)	0.2	310	<0.5	9	(5)	6	8	5	30
	Doberlug	32	1	21	46	125	0.5	15	2.4	85	(tr.)	15	19	5	25

Table 167. Trace Element Concentration [ppm] in Brown Coal (Lignite) and Brown-coal Ashes fr Lusatia and a Few Marginal Basins of the Brown-coal District of the GDR; data from SONTAG (19 KOCH (1962), KRÜGER (1961), MALBERG (1961), ROSETZ (1961), MUCKE (1963)

	Upper Oligocene 4th Lower Lusatian Stratum		Miocene 2nd Lower Lusatian Stratum		Miocene Piskowitz	Miocene Puschwitz	Lower Miocene Olbersd near Zit
	in the ashes	in the coal	in the ashes	in the coal	in the ashes	in the coal	in the as
Pb	4.8	1.3	1.4	0.1	13	<20	230
Ge	0.1	0.01	0.1	0.01	—	—	145
B	105	17	99	6	40	15	165
Cu	73	15	12	0.7	130	55	150
Ni	0.7	0.2	18	0.3	<4	<35	180
Mn	310	43	970	61	300	65	500
Cr	11	2.3	4.7	0.5	60	—	150
V	4.8	1.2	4.5	0.2	$\leqq$125	—	6,220
Ti	3,700	1,100	390	35	2,000	2,100	500
Sr	1,290	250	580	31	5	5	125
Ba	360	50	450	24	530	200	760

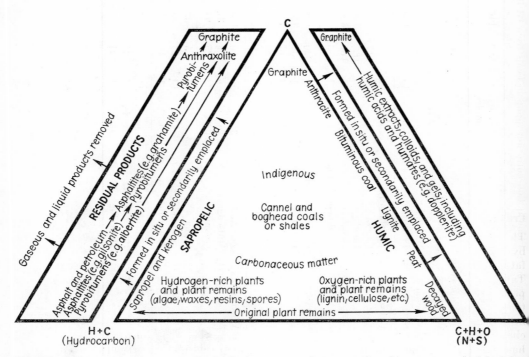

Fig. 106   Diagram showing generalised relation among the types of carbonaceous materials (after VINE et al.: Humic acids in uranium geochemistry)

ocene/Eocene astedt-Holden- t near Eisleben ae ashes	Upper Eocene Opencast Zipsendorf near Zeitz in the ashes	Upper Eocene/ Lower Oligocene Opencast at Etzdorf near Eisleben in the ashes	Upper Oligocene Opencast at Holzweißig near Bitterfeld in the ashes	Lower Miocene Mittweida near Karl-Marx-Stadt in the ashes
—	—	—	—	—
—	—	—	—	250
0	190	1,000	610	330
7	45	55	680	570
—	—	—	40	150
0	2,150	740	780	1,725
—	—	—	400	—
0	—	11	305	450
0	1,950	4,600	5,620	8,810
0	—	6,920	3,030	250
0	2,000	970	1,615	2,800

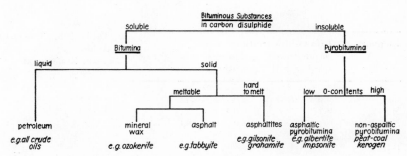

Fig. 107   Terminology and classification of the naturally occurring bituminous substances (after ABRAHAM, 1960, and HUNT, 1963)

Crude oil and organic pigments

The average composition of hydrocarbons in crude oil and natural gases has been expounded by SOKOLOV (1967) as follows: 32% CH_4; 6% C_2—C_4; 28% petrol, kerosene, gas oil; 19% fats, and others totalling 15%.

Tables 168, 169 and 170 survey the trace element concentrations in petroleum ashes and petroleum fractions.

Certain organic pigments which can build organo-metallic compounds and thus can be instrumental in fixing a number of chemical elements in crude oil are of particular biogeochemical importance. The most important pigments are the chlorophyll in plants and the haemoglobin in animals. Their common basic structure is the porphyrin core. Comprehensive information about the importance of the porphyrins and relevant bibliographies are given by TREIBS (1936), VANNOTTI (1954), DUNNING (1963), and ZUL'FUGARLY (1964), EGLINTON (1965).

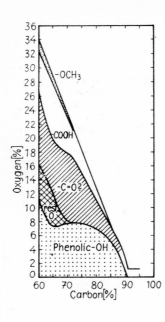

Regarding the combination of trace elements with organic matter of coals, the following statements can be made (after Leutwein and Rösler, 1956):

1. elements which are almost exclusively combined with the coal substance are:
   Ge, Be, Zr, Cr,
2. elements which are preferentially combined with the coal substance are:
   Cu, Pb, Zn, Ag, As, Ni, Mo, Sb,
3. elements which are primarily combined with foreign ashes are:
   Ga, Co, V,
4. elements which are almost exclusively combined with foreign ashes are:
   Mn, Sr, Ba.

Fig. 108   Functional groups with oxygen in various coals (after van Krevelen, 1961; Blom et al., 1957)

Table 168. Trace Elements in Petroleum Ashes
(after Zul'fugarly, 1964)

Element	Limits of Variations [mass-%] in oils from the U.S.S.R.		in oils from other countries
Ca	0.5	—>3	0.49—18.59
Mg	0.1	—3	0.11— 2.57
Na	0	—>3	0 —22.83
Sr	0	—1	—
Ba	0	—1	—
Al	0.3	—3	2.17—20.55
Ni	0	—2	0 —15.8
Mn	0	—0.1	0.03— 0.36
Cu	0.001	—0.5	—
Ti	0	—0.1	—
Cr	0	—0.3	—
Co	0	—1	—
Sn	0	—0.05	—
Pb	0	—0.5	—
Mo	0	—0.03	—
K	0	—0.01	0— 1.51
V	0	—3	0—36.6

A survey of the geochemistry of amino acids (together with a bibliography) is given by Abelson (1963), a survey of the geochemistry of carbohydrates by Vallentyne (1963), and a survey of the geochemistry of lipides (and other fats and waxes) by Bergmann (1963).

Table 169. Vanadium and Nickel Concentration in
Crude Oil Ashes of Various Deposits
(from ZUL'FUGARLY, 1964)

	V_2O_5 (mass-%)	NiO (mass-%)
Iran	5.03	2.70
Oklahoma	22.14	5.89
Kansas	0.44	0.57
Texas	1.43	1.52
Mid-Continent	traces	0.52
Wyoming	traces	0.33

Table 170. Distribution of Volatile Vanadium
Combinations in Mineral Oil Distillates
(from LIEBETRAU and FAULHABER, 1965)

Fraction		°C	V in ppm
Petroleum ether		35— 88	—
Petrol		88—176	—
Jet fuel		176—240	—
Diesel fuel	1	240—292	—
	2	292—340	0.025
Spindle oil	1	340—374	0.14
	2	374—410	0.40
Neutral-oil	1	410—448	0.57
	2	448—494	0.75
	3	494—510	0.99
Sodium residue		>510	235.0

Bibliography

Organic Geochemistry and Biogeochemistry

BALDWIN, E.: Biochemie. Verlag Chemie, Weinheim 1957

BERTRAND, D.: The Biogeochemistry of Vanadium. Bull. Am. Museum Nat. Hist. *94* (1950), 403 to 456

BOWEN, H. I. M.: Trace Elements in Biochemistry. Acad. Press, London/New York 1966

BREGER, I. A.: Organic Geochemistry. Vol. 16, Elsevier 1963, 658 pp.

BREGER, I. A., and M. DEUL: The Organic Geochemistry of Uranium. U. S. Dep. Interior. Geol. Surv., Prof. Pap. *300*, 505—510

COLOMBO, U., and G. D. HOBSON (editor): Advances in Organic Geochemistry, Elsevier 1964, 498 pp.

EGLINTON, G.: Recent Advances in Organic Geochemistry. Geol. Rdsch. *55* (1965) 3, 551—567

EGLINTON, G., and M. T. I. MURPHY (ed.): Organic Geochemistry-Methods and Results. Springer, Berlin/Heidelberg/New York 1969

GOLDBERG, E. D.: Biogeochemistry of Trace Elements. Bull. Geol. Soc. America, Mem. *67* (1957), 345

HOBSON, G. D., and M. C. LOUIS: Advances in Organic Geochemistry 1964 (Symposium). Internat. Series of Monographs in Earth Sciences, Vol. 24, Pergamon Press, Oxford/London 1966

HOBSON, G. D., and G. C. SPEERS (ed.): Advances in Organic Geochemistry. Proc. 3rd Intern. Congr., Pergamon Press, Braunschweig/Oxford 1970

341

HUTCHINSON, G. E.: The Biogeochemistry of Aluminium and of Certain Related Elements. Quart. Rev. Biol. *18* (1943), 1 a, 2

KARAMATA, S.: Contribution à la Biogéochimie du Chrome. Bull. Sci., Cons. Acad. RSF Yougoslavie, Sect. A *12* (1967) 9—10, 244—247

MANSKAYA, S. M., and T. V. DROZDOVA: Geochemistry of Organic Substances. Pergamon Press, Oxford 1968, 342 pp.

RAUEN, H. M.: Biochemisches Taschenbuch, 2nd ed. Springer-Verlag, Berlin/Göttingen/Heidelberg/ New York 1964

SCHARRER, K.: Biochemie der Spurenelemente. Verlag P. Parey, Berlin/Hamburg 1965

SIEVER, R., and R. A. SCOTT: Organic Geochemistry of Silica, in: Organic Geochemistry. Pergamon Press 1963

ZAJIC, I. E.: Microbial Biogeochemistry. Acad. Pres. New York 1969

Origin and Development of the Living Matter

ABELSON, P. H.: Some Aspects of Paleobiochemistry. Ann. N. Y. Acad. Sci. *69* (1957), 276, 285

ESKOLA, P.: Perioden geochemischer Beschleunigung der Lebensentwicklung. Acta geol. Acad. Sci. Hung. *7* (1961), 115—122

FOX, S. W.: Prebiological Formation of Biochemical Substances, in: E. INGERSON (Edit.), Organic Geochemistry. Pergamon Press 1963

MILLER, S. L.: Production of Some Organic Compounds under Possible Primitive Earth Conditions. J. Amer. Chem. Soc. *77* (1955), 2351—2360

OPARIN, A. I.: The Origin of Life on the Earth. Acad. Sci. USSR (Russian)

PASCHKE, R., R. W. H. CHANG, and D. YOUNG: Importance of X-rays for the Origin of Life. Science, Washington *125* (1957), 881

UREY, H. C.: On the Early Chemical History of the Earth and the Origin of Life. Proc. Nat. Acad. Sci. USA *38* (1952), 351—363

VINOGRADOV, A. P.: Biogeochemical Provinces and Their Role in the Organic Evolution. Geokhimiya (1963) *3*, 199—213 (Russian)

Classification of Organic Matter

BREGER, I. A.: Origin and Classification of Naturally Occurring Carbonaceous Substances, in: E. INGERSON (editor), Organic Geochemistry. Pergamon Press 1963

GOTHAN, W.: Über Klassifikation des Bitumens. Erdöl und Kohle 2 (1949), 173—175

KREULEN, D. J. W.: Elements of Coal Chemistry. Nijgh and van Ditmar, Rotterdam 1948

KREVELEN, D. W. VAN: Classification of coal. Het Gas *73* (1953), 51—57

PACLT, J.: A system of caustolites. Tschermaks mineralog. petrogr. Mitt. *3* (1953), 332—347

POTONIÉ, H.: Eine Klassifikation der Kaustobiolithe. S.-B. preuß. Akad. Wiss., physik.-math. Kl. *2* (1908), 48—57

VAROSSIEAU, W. W., and I. A. BREGER: Chemical Studies on Ancient Buried Wood and the Origin of Humus. Proc. 3rd Cong. on Stratigraphy and Geol. of the Carboniferous, Heerlen 1951, 637 to 646

VINOGRADOV, A. P.: Fundamentals of the Genetic Classification of Bitumen. Jzd. "Nedra", Leningrad 1964, 266 pp (Russian)

Chemistry of Plants and Animals

GERICKE, S.: Die Versorgung von Pflanze und Tier mit Mikronährstoffen. Phosphorsäure *17* (1957), 203—217

HAYEK, E.: Die Mineralsubstanz der Knochen. Klin. Wschr. *45* (1967), 857—863

LEUTWEIN, F.: Contribution sur le Connaissance de la Distribution des Oligoéléments dans les Coquilles Actuelles. Sciences de la Terre, Nancy 1960

NODDAK, I., and W. NODDAK: Häufigkeit der Schwermetalle in Meerestieren. Ark. Zool. *32-A* (1939), 1—35

ROBINSON, W. O., and G. EDGINGTON: Minor Elements in Plants and Some Accumulator Plants. Soil Sci. *60* (1945), 15

TYLES, V.: The Trace Elements in Plant and Animal Life. Publishing-house for Foreign Literature, Moscow 1949 (Russian)

VINOGRADOV, A. P.: La Composition Chimique Élémentaire des Organismes Vivants et le Système Periodique des Éléments Chimiques. Compt. Rend. *197* (1933)

VINOGRADOV, A. P.: The Elementary Chemical Composition of Marine Organisms. Sears Foundation for Marine Res., Yale-Univ., N. H. 1953

WARREN, H. V., and R. E. DELAVAULT: Observations on the Biogeochemistry of Lead in Canada. Trans. Roy. Soc. Can., Ser. III, Sect. IV, *54* (1960) 11

WASKOWIAK, R.: Geochemische Untersuchung an rezenten Molluskenschalen mariner Herkunft. Freiberger Forschungsh. C *136* (1962), 155 pp.

WEBB, D. A.: Observations on the Blood of Certain Ascidians with Special Reference to the Biochemistry of Vanadium. J. Exp. Biology *16* (1939), 499—523

Geochemical Effectiveness of Living Beings (Including Microorganisms)

BERGER, W.: Die geochemische Rolle der Organismen. Tschermak's Mineralog. Petrogr. Mitt. *2* (1948), 136—140

BOĬCHENKO, E. A.: On Oxidizing Functions in the Evolution of the Biosphere. Geokhimiya (1967) *8*, 971—976 (Russian)

BOĬCHENKO, E. A., G. N. SAENKO, and T. M. UDEL'NOVA: Evolution of the Concentration Function of Plants in the Biosphere. Geokhimiya (1968) *10*, 1260—1264 (Russian)

DAVIS, I. B.: Petroleum Microbiology, Elsevier, Amsterdam 1967

DAVIS, I. B., and H. F. YARBROUGH: Anaerobic Oxidation of Hydrocarbons by Desulfovibro Desulfuricans. Chem. Geol. *1* (1966), 137—144

DOSTÁLEK, M.: Die Assimilation von Kohlenwasserstoffen durch Mikroorganismen, in: Geochemische Probleme bei der Erkundung und Förderung von Erdöl und Erdgas. Vol. 1, VEB Deutscher Verlag für Grundstoffindustrie, Leipzig 1968, 546—549

FEELY, H. W., and J. L. KULP: Origin of Gulf-coast Salt-dome Sulfur Deposits. Bull. Amer. Assoc. Petrol. Geologists *41* (1957), 1802—1853

HARRISON, A. G., and H. G. THODE: Mechanism of the Bacterial Reduction of Sulphate from Isotope Fractionation Studies. Trans. Faraday Soc. *4* (1958), 84—92

KARASKIEWICZ, J.: Eine mikrobiologische Methode zur Lokalisierung von Erdgasexhalationen, in: Geochemische Probleme bei der Erkundung und Förderung von Erdöl und Erdgas. Vol. 1, VEB Deutscher Verlag für Grundstoffindustrie, Leipzig 1968, 540—545

KIMATA, M., H. KADOTA, Y. HATA, and H. MIYOSHI: Growth of Sulfate Reducing Bacteria in Relation to the Oxidation Reduction Potential of Culture Media. Records Oceanogr. Works Japan (1959), Special No. 3, 167—174

LINDBLOM, G. P., and M. D. LUPTON: Microbiological Aspects of Organic Geochemistry. Development in Industrial Microbiology, Vol. 2, Plenum Press 1961

NICHOLAS, D. I. D.: Biological Sulphate Reduction. Mineralium Deposita *2* (1967), 169—180

PAULI, F. W.: Some Recent Developments in Biogeochemical Research. Proc. Geol. Ass. Canada *19* (1968), 45—49

PEJVE, YA. V.: The Role of Trace Elements in the Nutrition of Plants and Animals. Biochemistry *20* (1955) 3, 265 (Russian)

SADLER, W. R., and P. A. TRUDINGER: The Inhibition of Microorganisms by Heavy Metals. Mineralium Deposita *2* (1967), 158—168

SCHWARTZ, W., and A. MÜLLER: Methoden der Geomikrobiologie. Freiberger Forschungsh. C 48, Berlin 1958

SCHWARTZ, W.: Die Bakterien des Schwefelkreislaufes und ihre Lebensbedingungen. Freiberger Forschungsh. C *44* (1958), 5—13

TRUDINGER, P. A., and B. BUBELA: Microorganisms and the Natural Environment. Mineralium Deposita *2* (1967), 147—157

WALLHÄUSSER, K. H., and H. PUCHELT: Sulfatreduzierende Bakterien in Schwefel- und Gruben-wässern Deutschlands und Österreichs, 12—30

ZOBELL, C. E.: Studies on the Bacterial Flora of Marine Bottom Sediments. J. Sediment. Petrol. 8 (1938), 10—18

Humus, Peat, Coal, Graphite

ALEKSANDROV, I. V.: The Process of Humus Formation in a Few Primitive Mountainous Soils. Poch-vovedenie 67 (1951), 604—616 (Russian)

APUSHKIN, K. K.: Coagulation and Peptication of Humic Acids by Phosphates. I. Colloid Z. (USSR) 1 (1935), 495—507 (Russian)

BREGER, I. A.: Chemical and Structural Relationship of Lignin to Humic Substances. Fuel 9 (1951), 204—208

BREGER, I. A.: Geochemistry of Coal. Econ. Geol. 53 (1958), 823—841

BREGER, D., and MEYROWITZ: Geochemistry and Mineralogy of a Uraniferous Subbituminous Coal. Econ. Geol. 50 (1955), 600—624

CARNOT, A.: Sur l'Origine et la Distribution du Phosphore dans la Houille et la Cannel-Coal. C. R. hebd. séances Acad. sci. 99 (1884), 154—160

DRAGUNOV, S. S.: Chemical Nature of Humic Acids. J. Petrology USSR (1950), 151—157 (Russian)

DUCHESNE, J., J. DEPIREUX, and J. M. VAN DER KEA: Origin of Free Radical Carbonaceous Rocks. Geochim. Cosmochim. Acta 23 (1961), 209—218

ENDERS, C.: Origin of Humus in Nature. Die Chemie 56 (1943), 281—285

ERASMUS, P.: Über die Bildung und den chemischen Bau der Kohlen. Verlag F. Enke, Stuttgart 1938

ERSHOV, V. M.: Trace Elements in Coal from the Kishelovsk Coal Basin. Geokhimiya (1961) 3, 274—275 (Russian)

FÖLDVARY, A.: The Geochemistry of Radioactive Substances in the Mecsek Mountains. Acta Geologica Akad. Sci. Hung. 1 (1952), 37—48

FOTIEV, A. V.: On the Nature of the Humic Substances in Water. Dokl. Akad. Nauk SSSR 179 (1968) 2, 443—446 (Russian)

FUCHS, W.: Die Chemie der Kohle, Springer-Verlag Berlin 1931

FUCHS, W.: Fortschritte in der Kenntnis der Kohle. Brennstoff-Chem. 32 (1951), 12—19

GANGULI, N. C., and D. R. DUTTA: Spectrographic Determination of Germanium in Coals. Journ. Sci. Indust. Research. (India) 15 B (1956), 327—328

GULYAEVA, L. A., and J. S. ITKINA: Halogens, V, Ni and Cu in Coal. Geokhimiya (1962) 4, 345—355 (Russian)

JANDA, I., and E. SCHROLL: Geochemische Untersuchungen an Graphit-Gesteinen. Rep. 21st. Int. Geol. Congr., Norden 1 (1960), 40—53

JULIEN, A. A.: On the Geological Action of the Humic Acids. Proc. Amer. Assoc. Adv. Sci. 28 (1880), 311—410

KREULEN, D. J. W., and F. J. KREULEN-VAN SELMS: Humic Acids and Their Role in the Formation of Coal. Brennstoff-Chem. 37 (1956), 14—19

KREULEN, D. W. VAN: Geochemistry of Coal, in: E. INGERSON (Edit.), Organic Geochemistry. Perga-mon Press 1963

LEUTWEIN, F., and H. J. RÖSLER: Geochemische Untersuchungen an paläozoischen und mesozoischen Kohlen Mittel- und Ostdeutschlands. Freiberger Forschungsh., C 19. Akademie-Verlag, Berlin 1956

MANSKAYA, S. M., L. A. KODINA, and V. N. GENERALOVA: On the Geochemical Role of Lignite. Geo-khimiya (1968) 8, 980—988 (Russian)

OTTE, M. U.: Spurenelemente in einigen deutschen Steinkohlen. Chem. d. Erde 16 (1953), 239

POTONIÉ, H.: Die Entstehung der Kohle und der Kaustobiolithe. Borntraeger, Berlin 1910

RAKOVSKIĬ, V. E.: Theories and Facts on the Origin of Peat. Khimiya Tverd. Gornykh Iskopaemykh, Trudy I. Vsesoyuz. Soveshaniya. Moscow 1950, 44—54 (Russian)

RATYNSKIĬ, V. M., and S. V. GLUSNEV: The Regularities in the Distribution of a Number of Metals in Fossil Coals. Dokl. Akad. Nauk SSSR 177 (1967) 5, 1193—1196 (Russian)

STACH,. E.: Lehrbuch der Kohlenpetrographie. Borntraeger, Berlin 1935

STADNIKOV, G., and P. KORSHEV: Zur Kenntnis der Humussäuren. Kolloid-Z. 47 (1929), 136—141

VAIN, F. M.: Geochemistry of Humus, in: E. INGERSON (Edit.), Organic Geochemistry. Pergamon ess 1963

ÁDECZKY-KARDOSS, E., and M. FÖLDVÁRINÉ-VOGL: Geochemische Untersuchungen an den Aschen garischer Steinkohlen. Földtani Közlöny 85 (1955) 1, 7

ALAY, A.: The Significance of Humus in the Geochemical Enrichment of Uranium. Proc. 2nd U. N. tern. Conf. Peaceful Uses Atom. Energy 2 (1958), 187—191

AYLOR, E., and MACKENZIE: Base Exchange and Its Bearing on the Origin of Coal. Fuel Sci. Pract. (1926), 195—202

EICHMÜLLER, R.: Zur Metamorphose der Kohle. C. R. 3. Congr. Strat. Geol. Carbonifère, Heerlen 51, 615—623

LADYCHENSKII, S. S.: Colloid-chemical Properties of Soil Humus. Colloid-Z., Voronezh 6 (1940), 683 - 694 (Russian)

HURAVLEVA, E. G.: The Content of Trace Elements in Soil Organic Matter. Sov. Soil Sci. 12 (1965), 387—1392 (Russian)

UBOVIC, P.: Physicochemical Properties of Certain Minor Elements as Controlling Factors in Their istribution in Coal. Coal Science, Sec. III (1966) 13

isperse Organic Substances (Including Organopelitic Products)

ODENHEIMER, W., L. HELLER, and SH. YARIF: Organo-metallic Clay Complexes, Part I, IV, V: Israel . Chem. 1 (1963), 69 and 391, 2 (1964) 201. Part II, VII: Clay Mineral. Bull. 5 (1962), 145. Part III, I: Proc. Intern. Clay Conf., Stockholm 2 (1963), 351

ORDOVSKII, O. K., et al.: Organic Matter in Marine Sediments. Special Issue of "Marine Geology" 3, /2, Elsevier, Amsterdam 1965

EMOLON, A., and G. BARBIER: Conditions de Formation et Constitution du Complexe Argilo-Humique es Sols. Compt. Rend. 188 (1929), 654—656

ISMA, E., and J. W. JURG: Basic Aspects of the Diagenesis of Organic Matter and the Formation of etroleum. Proc. 1967 World Petroleum Congress, 112—138 PD/D.C. 1

EDOSEEV, A. D., and E. V. KUCEARSKAYA: Organic Derivates of Kaoline. Proc. Intern. Clay Conf., tockholm 2 (1963), 365—371

LOROVSKAYA, V. N., T. A. TEPLITSKAYA, and R. I. PERSONOV: Results of Investigating the Quantitative ontent of Polynuclear Aromatic Hydrocarbons in the Organic Matter of Rocks of Different Origin. eokhimiya (1966) 5, 538—545 (Russian)

ORSMANN, I. P.: Geochemistry of Kerogen, in: E. INGERSON (Edit.), Organic Geochemistry, Pergamon ress 1963

GRIM, R. E. W., H. ALLAWAY, and F. L. CUTHBERT: Reactions of Different Clay Minerals with Organic ations. J. Amer. Ceram. Soc. 30 (1947), 137—242

GROBLER, H. J., and F. W. PAULI: Adsorption Equilibria of Humic Substances. S. Afr. J. Agr. Sci. (1964), 187—192

GULYAEVA, L. A.: Geochemische Indikatoren und die Bitumenführung mariner und kontinentaler alinarer Serien, in: Geochemische Probleme bei der Erkundung und Förderung von Erdöl und Erdgas, Vol. 1, VEB Deutscher Verlag für Grundstoffindustrie, Leipzig 1968, 77—90

HUNT, J. M.: Distribution of Hydrocarbons in Sedimentary Rocks. Geochim. Cosmochim. Acta 22 (1961) 1

HUNT, J. M.: Geochemical Data on Organic Matter in Sediments. Proc. 3rd International Scientific Geochemical Conference, Budapest 1963

JOHNS, R. B., et al.: The Organic Geochemistry of Ancient Sediments–Part II. Geochim. Cosmochim. Acta 30 (1966), 1191—1222

KASATOCHKIN, V. I., and O. I. ZILBERBRAND: The Chemical Structure of Shale Kerogen. Dokl. Akad. Nauk SSSR 111 (1956), 1031—1034 (Russian)

MACKENZIE, R. C.: Complexes of Clays with Organic Compounds. Trans. Farad. Soc. 47 (1948), 368 to 375

OLPHEN, H. V.: Clay-organic Complexes and the Retention of Hydrocarbons by Source-rocks. Proc. Intern. Clay Conf., Jerusalem 1 (1966), 307—317

PETERSILIE, J. A., and E. M. GALINOV: Über die Natur der Kohlenwasserstoffgase und der dispersen Bitumen in Eruptivgesteinen im Zusammenhang mit der Isotopenzusammensetzung des Kohlenstoffs, in: Geochemische Probleme bei der Erkundung und Förderung von Erdöl und Erdgas, Vol. 1, VEB Deutscher Verlag für Grundstoffindustrie, Leipzig 1968, 367—378

PRASHNOWSKY, A. A., and M. SCHIDLOWSKI: Investigation of Pre-Cambrian Thucholite. Nature *216* (1967), 560—563

RODIONOVA, K. F.: Geochemistry of Organic Matter and Mineral Oil Bedrocks. Izd. "Nedra", Moscow 1967 (Russian)

RODIONOVA, K. F.: Die Gesetzmäßigkeiten in der Verteilung und Veränderung der Zusammensetzung der organischen Substanz in paläozoischen und meso-känozoischen Gesteinen in den wichtigsten Erdöl-Erdgasprovinzen der UdSSR, in: Geochemische Probleme bei der Erkundung und Förderung von Erdöl und Erdgas, Vol. 1, VEB Deutscher Verlag für Grundstoffindustrie, Leipzig 1968, 131 — 136

VEBER, V. V., and L. I. GORBUNOVA: On the Role of the Clay Content in the Formation of Bitumen. Sovetsk. Geologiya (1961) 11, 73—85 (Russian)

WEISS, A.: Organische Derivate der glimmerartigen Schichtsilikate. Angew. Chem. *75* (1933), 113—122

Anonymous: Mikroelemente in Kaustobilithen und Sedimentgesteinen. Izd. "Nauka", Moscow 1965 (Russ.)

Sapropelites, Oil Shale, Petroleum, Natural Gas, etc.

ABELSON, P. H.: Geochemistry of Amino Acids, in: E. INGERSON (Edit.), Organic Geochemistry. Pergamon Press 1963, 431—455

ABRAHAM, H.: Asphalts and Allied Substances, Vol. 1. Van Nostrand, New York 1945

ANDREEV, P. F.: Geochemical Transformation of Crude Petroleum in the Lithosphere. Geokhimiya (1962), *10* 880—889 (Russian)

BAKER, E. G.: Origin and Migration of Oil. Science, Washington *129* (1959) 3353, 871—874

BELL, K. G., and J. M. HUNT: Native Bitumens Associated With Oil Shales, in: E. INGERSON (Edit.), Organic Geochemistry. Pergamon Press 1963

BERGMANN, W.: Geochemistry of Lipids, in: E. INGERSON (Edit.), Organic Geochemistry. Pergamon Press 1963, 503—542

BONHAM, L. C.: Geochemical Investigation of Crude Oils. Bull. Amer. Assoc. Petroleum Geologists *40* (1956), 897—908

BOTNEVA, T. A.: Geochemistry of Gases Accompanying Crude Petroleum. Izd. "Nedra", Moscow 1966, 204 pp. (Russian)

BOTNEVA, T. A., and SHULOVA: Die Anwendung mathematischer Methoden bei geochemischen Erdöluntersuchungen, in: Geochemische Probleme bei der Erkundung und Förderung von Erdöl und Erdgas, Vol. 1, VEB Deutscher Verlag für Grundstoffindustrie, Leipzig 1968, 275—300

CARISON, M. T., and E. I. GUNN: Determination of Trace Metallic Components in Petroleum Oils by Means of the Spectrograph. Analyt. Chem. *22* (1950), 1188—1121

COLOMBO, U., E. DENTI, and G. SIRONI: Radiation Effects of Hydrocarbons — a Geochemical Study. Proc. VI World Petroleum Congress 1963

DOBRYANSKIĬ, A. F.: Geochemistry of Crude Petroleum. Gostoptekhnizdat 1958 (Russian)

DUNNING, H. N.: Geochemistry of Organic Pigments, in: E. INGERSON (Edit.), Organic Geochemistry. Pergamon Press 1963, 367 —430

DUNNING, H. N., J. W. MOORE, and A. T. MYERS: Properties of Porphyrins in Petroleum. Ind. Engng. Chem. *46* (1954), 2000—2007

ELDRIDGE, G. H.: The Asphalt and Bituminous Rock Deposits of the United States. U. S. Dep. Interior., Geol. Surv., *23*. Ann. Rep. 1901, 209—464

GLOGOCZOWSKI, J. J.: Erwägungen über Faktoren, die die Veränderung der Zusammensetzung des Erdöls hervorrufen und die Frage deren Modellierung, in: Geochemische Probleme bei der Erkundung und Förderung von Erdöl und Erdgas, Vol. 1, VEB Deutscher Verlag für Grundstoffindustrie, Leipzig 1968, 173—183

HEIDE, F.: Die Chemie von Rohölaschen. Naturwissenschaften *26* (1938), 693

HILL, R.: Hemoglobin in Relation to Other Metallo-hematoporphyrins. Biochem. J. *19* (1925), 341 to 349

HIMUS, G. W.: Oil Shales, Torbanites, Their Allies, and Some of Their Problems. Petroleum, London *4* (1941), 9—13

346

ODGSON, G. W., and B. L. BAKER: Porphyrin Abiogenesis from Pyrrole and Formaldehyde under mulated Geochemical Conditions. Nature 216 (1967), 29—32

UNT, J. M., F. STEWART, and P. A. DICKEY: Origin of Hydrocarbons of Uinta Basin, Utah. Bull. mer. Assoc. Petroleum Geologists 38 (1954), 1671—1698

ALINKO, M. K.: Non-organic Origin of Crude Petroleum in the Light of Modern Data (A Critical nalysis). Izd. "Nedra", Moscow (Russian)

ALINKO, M. K., and T. A. BOTNEVA: Rationelle Komplexe geochemischer Untersuchungen in verhiedenen Stadien der Such- und Erkundungsarbeiten auf Erdöl und Erdgas, in: Geochemische Probleme bei der Erkundung und Förderung von Erdöl und Erdgas, VEB Deutscher Verlag für Grundoffindustrie, Leipzig 1968, 42—56

AZAKOV, E. J.: Origin and Chemism of Fresh Water Sapropels. Trudy Inst. Gornykh Iskopaemykh kad. Nauk SSSR 2 (1950), 278—284 (Russian)

OGERMAN, P. K.: The Chemical Constitution of the Estonian Oil Shale "Kukersite". Proc. Oil Shale oal, 1. Conf. 1938, 114—123

REJCI-GRAF, K.: Zur Geochemie der Erdölentstehung. Erdöl Kohle 8, (1955), 393—401

ÜLLER, E. P. (Edit.): Geochemische Probleme bei der Erkundung und Förderung von Erdöl und rdgas. Vol. 1, VEB Deutscher Verlag für Grundstoffindustrie, Leipzig 1968

AGY, B., and U. COLOMBO (Edit.): Fundamental Aspects of Petroleum Geochemistry. Elsevier, msterdam/London/New York 1967, 388 pp.

RIEN, C. H.: Oil Shales and Shale Oil. Proc. Oil Shale Cannel Coal, 2. Conf. 1950, 76—111

ODIONOVA, K. F.: Organic Substances and Petroleum Bedrocks of the Devonian in the Volga-Ural etroleum District. Izd. "Nedra", Moscow 1967, 243 pp. (Russian)

USSEL, W. L.: The Composition of Petroleum Ashes. Principles of Petroleum Geology 24 (1951)

EMENOV, S. S., YU. I. KORNILOV, and YU. E. GUREVICH: Discovery and Determination of Functional roups in the Organic Substance of the Baltic Shale. Trudy Vses. Nauch.-Issl. Inst. Pererab. Slan, No. 3, 11—15 [Fuel Abstr. 20 (1956), 47—48] (Russian)

TAROBINETS, I. S.: On the Connection between the Distribution of the Geochemical Types of Oils and he Tectonic Conditions. Dokl. Akad. Nauk. SSSR 180 (1968) 6, 1464—1466 (Russian)

TEGENA, L.: Über die prinzipiellen Grundlagen der geochemischen Erdölforschung. Papers read at he 3rd International Scientific Conference for Geochemistry, Microbiology and Petroleum Chemistry. Budapest 1963

ISSOT, B.: Problèmes Géochimiques de la Génèse et de la Migration du Pétrole. Rev. Inst. Franc. Pétrole 21 (1966) 11, 1621—1671

REIBS, A.: Chlorophyll und Häminderivate in organischen Mineralsubstanzen. Angew. Chem. 49 1936), 682—686

USPENSKIĬ, V. P.: Introduction in the Geochemistry of Crude Oil. Izd. "Nedra", Leningrad 1970 Russian)

VALLENTYNE, J. R.: Geochemistry of Carbohydrates, in: E. INGERSON (Edit.), Organic Geochemistry. Pergamon Press 1963, 456—502

VANNOTTI, A.: Porphyrins, Their Biological and Chemical Importance. Hilger&Watts, London 1954, 258 pp

WHITEHEAD, W. L.: Geochemistry of Petroleum, in: E. INGERSON (Edit.), Organic Geochemistry. Pergamon Press 1963

WILLIAMS, R. J. P.: The Properties of Metalloporphyrins. Chem. Reviews 56 (1956), 299—328

WOODLE, R. A., and W. B. CHANDLER, jr.: Mechanism of Occurrence of Metals in Petroleum Distillates. Ind. Engng. Chem. 44 (1952), 2591—2596

ZUL'FUGARLY, D. J.: Verbreitung der Spurenelemente in Kaustobiolithen, Organismen, Sedimentgesteinen und Schichtwässern. VEB Deutscher Grundstoffverlag, Leipzig 1964, 290 pp (Transl. Russian)

ZUL'FUGARLY, D. J.: Geochemische Untersuchungen an Kaustobiolithen, insbesondere an Erdöl, Ber. deutsch. Ges. geol. Wiss. B, 13 (1969) 2, 251—262

Biogenic Inorganic Products

BAAS-BECKING, L. G. M., and D. MOORE: Biogenic Sulfides. Econ. Geol. 56 (1961), 259—272

KOCHENOV, A. V., and V. V. SINOVEV: Distribution of the Elements of the Rare Earths in the Phosphates of Fish Remains etc. Geokhimiya (1960), 714—725 (Russian)

KREJCI-GRAF, K., and H. DITTMAR: Zur Chemie von Gesteinen mit Korallen und Organismen ähnlicher Umweltbedingungen. Chem. Erde *27* (1968), 121—142

MÜLLER, E. P.: Geochemische Gesetzmäßigkeiten und Isotopengeochemie der organischen Substanz im Oberperm des Nordteils der DDR, in: Geochemische Probleme bei der Erkundung und Förderung von Erdöl und Erdgas. Vol. 1. VEB Deutscher Verlag für Grundstoffindustrie, Leipzig 1968, 150 — 162

SCHNEIDER, A.: Der Schwefelgehalt in Knochen und Zähnen rezenter und fossiler Organismen. Contr. Miner. and Petrol. *18* (1968), 310—325

WEDEPOHL, K. H.: Schwermetallgehalt der Kalkgerüste einiger mariner Organismen. Nachr. Akad. Wiss. Göttingen *5* (1955), 79—86

WYCKOFF, R. W. G., and A. R. DOBERENZ: The Strontium Content of Fossil Teeth and Bones. Geochim. Cosmochim. Acta *32* (1968) 1, 109—115

The representation of the geochemistry of individual elements in a form which is easy to survey is practically possible only with the help of geochemical cycles.

.1. Geochemical Cycles

The use of geochemical cycles as a form of representation is based on the assumption that the development of the earth, in particular of the earth's crust, is cyclic. Thus, it is no small wonder that well-grounded statements on the cyclicity of elements were first made in the biogeochemical field (cycles of K and P by Dumas, Boussingault, Liebig, Bischof, Mohr). The first graphical representations were made at the beginning of this century (e.g. BLACK-WELDER, 1916, with respect to P; FELLENBERG, 1926, with respect to I). The decisive achievement in this regard goes to the credit of Vernadskiĭ, a consistent champion of the migrational and cyclic nature of the chemical elements. His schemes in three parts (see Fig. 109) formed the basis of all further studies; improvements were made by Goldschmidt and, usually starting from the forms developed by him, by the following geochemists who increased the vividness of representation and, in the recent two decades, added quantitative statements to the qualitative ones (e.g. Figs. 110 and 111).

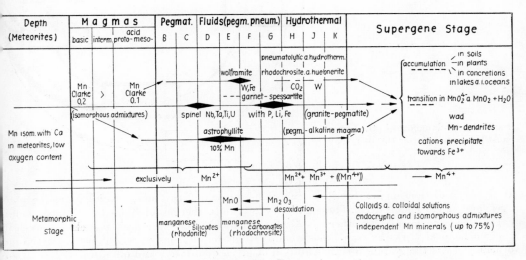

Fig. 109   Geochemical diagram (after VERNADSKIĬ by FERSMAN)

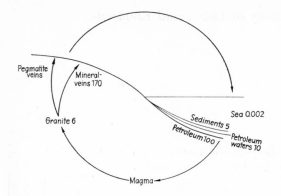

Fig. 110 Geochemical cycle of uranium (after TOMKEIEFF, 1946; values in ppm, partly obsolete)

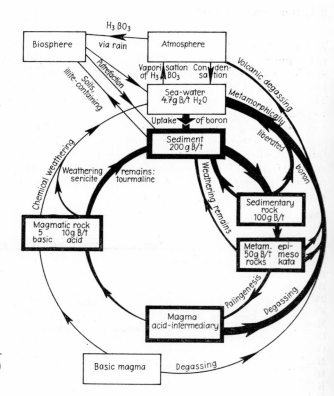

Fig. 111 Cycle of boron in the uppermost lithosphere (after HARDER, 1959)

Basically, a distinction is made between a major cycle and two minor cycles, an endogenous cycle and an exogenous cycle. These minor cycles can further be subdivided into biogeochemical, hydrogeochemical, and other cycles. A great number of other cycle models are given in the geochemical works by Rankama-Sahama, Goldschmidt, Mason, Smulikovski, Barth, and others. Besides the graphical representation of given qualitative and quantitative facts, geochemical cycles are an excellent means to mirror the geological reality, the structure of the

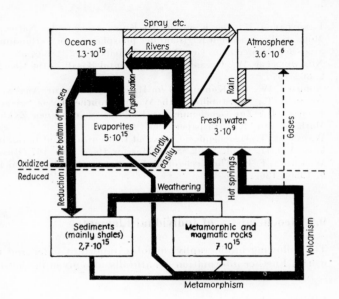

Fig. 112   Sulphur cycle after
Holser and Kaplan (1966)

earth's crust. The graphs from Borchert, which have already been published several times, are an attempt to combine geological conditions and accepted geochemical laws and regularities into one common representation.

Bibliography

Barth, T. F. W.: The Geochemical Evolution of Continental Rocks — a Model
n: Ahrens, L. H. (Edit.): Origin and Distribution of the Elements. Pergamon Press 1968, 587—597
Barth, T. F. W.: Abundance of Elements, Areal Averages and Geochemical Cycles. Geochim. Cosmo-him. Acta 23 (1961), 1—8
Barth, T. F. W.: On the Geochemical Cycle of Fluorine. J. Geol. 55 (1947), 420
Borisenko, L. A., and A. A. Saukov: Geochemical Cycle of Gallium. Internat. Geol. Congr.-Rep. 21 Sess. North (1960), Copenhague P. 1. Geochem. Cycles, 96—105
Fellenberg, T. V.: Das Vorkommen, der Kreislauf und der Stoffwechsel des Jods. Verlag Bergmann, Munich 1926
Galimov, Z. M.: On the Evolution of Carbon on the Earth. Geokhimiya (1967) 5, 530—536 (Russian)
Goldschmidt, V. M.: Kreislauf der Metalle in der Natur. Metallwirtschaft 10 (1931), 265 ff.
Harder, H.: Das Bor im Kreislauf der Gesteine. Rept. 21. Intern. G. Geol. Congr. Norden 1 (1960), 10—13
Kalle, K.: Der Stoffhaushalt des Meeres. Akad. Verl.-Ges. Becker u. Erler, Leipzig 1943
Kubach, I.: Darstellung zur Geochemie des Siliziums. Heidelberger Beitr. Mineralog. Petrogr. 3 (1952), 53—59
La Rivière, J. W. M.: The Microbial Sulphur Cycle and Some of Its Implications for the Geochemistry of Sulphur Isotopes. Geol. Rdsch. 55 (1966) 3, 568—581
Linck, G.: Kreislauf der Stoffe in der anorganischen Natur. Handwörterbuch der Naturwiss. Vol. 5, Fischer, Jena 1912
Livingstone, D. A.: The Sodium Cycle and the Age of the Ocean. Geochim. Cosmochim. Acta 27 (1963), 1055—1069

LUNDEGARDH, H.: Der Kreislauf der Kohlensäure in der Natur. G. Fischer, Jena 1924
MOHR, C. F.: Über den Kreislauf der phosphorsauren Verbindungen und der Fluorsäure auf der Erde. Sitz.-Ber. Niederrhein. Ges. Natur- u. Heilkde. *22* (1865), 88—91
NIEUWENKAMP, W.: Oceanic and Continental Basalt in the Geochemical Cycle. Geol. Rdsch. *57* (1968) 2, 362—372
NODDACK, W.: Der Kohlenstoff im Haushalt der Natur. Angew. Chemie *50* (1937), 505 ff.
ODUM, H. T.: The Stability of the World Strontium Cycle. Science Washington *114* (1951), 407—411
PUCHELT, H.: Zur Geochemie des Bariums im exogenen Zyklus. Sitz. Ber. Heidelb. Akad. Wiss., Math.-naturw. Kl., 4th paper (1967) 4
WATANABE, T.: Geochemical Cycle and Concentration of Boron in the Earth's Crust. Khimiya Zemnoi kory, Vol. II, Akad. Nauk SSSR, Moscow 1964, 156—167 (Russian)
WICKMANN, F. E.: The Cycle of Carbon and the Stable Carbon Isotopes. Geochim. Cosmochim. Acta, *9* (1956), 136—153

9.2. Geochemistry of Individual Elements

The following bibliography includes important older and classical publications and modern works. Further information is naturally also given in the standard works on geochemistry.

Bibliography

Antimony (Sb)

ONISHI, H., and E. B. SANDELL: Notes on the Geochemistry of Antimony. Geochim. Cosmochim. Acta, *8* (1955)

Arsenic (As)

ONISHI, H., and E. B. SANDELL: Geochemistry of Arsenic. Geochim. Cosmochim. Acta *7* (1955), 1—33

Barium (Ba)

CHOW, T. J., and E. D. GOLDBERG: On the Marine Geochemistry of Barium. Geochim. Cosmochim. Acta *20* (1960), 192—198
ENGELHARDT, W. v.: Geochemie des Bariums, Chem. d. Erde *10* (1936), 187
PUCHELT, H.: Zur Geochemie des Bariums im exogenen Zyklus. Sitz. Ber. Heidelbg. Akad. Wiss., Math.-naturw. Kl. (1967) 4th paper

Beryllium (Be)

BEUS, A. A.: Geochemistry of Beryllium. Geokhimiya (1956) *5*, 75—92 (Russian)
FLEISCHER, M., and E. N. CAMERON: Geochemistry of Beryllium. U. S. Atomic Energy Comm., TID-5212 (1955), 80—92
GIUSCA, D., and F. POPEA: Contribution to the Geochemistry of Beryllium in a Few Rocks from the People's Republic of Romania. Comm. Acad. Rep. Pop. Rom. *IX* (1959) 8, 853—858 (Romanian)
GOLDSCHMIDT, V. M.: Zur Geochemie des Berylliums. Nachw. Ges. Wiss. Göttingen, Math.-naturw. Kl. *3* (1932) 23; *4*, No. 25, 360
MERILL, J. R., M. HONDA, and J. R. ARNOLD: Beryllium Geochemistry and Beryllium-10 Age Determination. Proc. 2nd. U. N. Intern. Conf. Peaceful Uses of Atomic Energy *2* (1958), 251—254

Bismuth (Bi)

BROOKS, R. R.: Apparent Geochemical Association of Bismuth and Thallium, Nature, London *189* (1961), 910—911

Boron (B)

BARSUKOV, V. L.: Die Geochemie des hypogenen Bors. Volume: Geochemical Cycles. Intern. Geol. Congr. XXI Session (Norden), Copenhagen 1960
GOLDSCHMIDT, V. M., and CL. PETERS: Zur Geochemie des Bors. Sitz. Ber. Göttingen, Math.-phys. Klasse (1932)
HARDER, H.: Beitrag zur Geochemie des Bors. II. Nachr. Akad. Wiss. Göttingen, math.-phys. Kl. 5 (1959), 123
HEIDE, F., and A. THIELE: Zur Geochemie des Bors. Chem. d. Erde *19* (1958), 329
LANDERGREN, S.: Contribution to the Geochemistry of Boron, 2. Ark. Kem. Miner. Geol. *19*-A (1945), 31 pp.

Bromine (Br)

BENNE, W.: Untersuchungen zur Geochemie des Cl und Br. Geochim. Cosmochim. Acta *3* (1953), 186—215

Cadmium (Cd)

BURYANOVA, E. Z.: On the Mineralogy and Geochemistry of Cd in Sedimentary Rock of Tuva. Geochimiya *2* (1960), 177—182 (Russian)
IVANOV, V. V.: Geochemistry of Cadmium, in: Geochemistry of Rare Elements. Izd. "Nauka", Moscow 1964 (Russian)

Carbon (C)

BORCHERT, H.: Zur Geochemie des Kohlenstoffs. Geochim. Cosmochim. Acta *2* (1951), 62—75
HOEFS, I.: Ein Beitrag zur Geochemie des Kohlenstoffs in magmatischen und metamorphen Gesteinen. Geochim. Cosmochim. Acta *29* (1965), 399—428

Cesium (Cs)

SLEPNOV, YU. S., N. A. SOLODOV, and T. F. BOIKO: Geochemistry of Cesium, in: Geochemistry of Rare Elements. Izd. "Nauka", Moscow 1964 (Russian)

Cerium (Ce) see: Rare Earths

Chlorine (Cl)

BENNE, W.: Untersuchungen zur Geochemie des Cl und Br. Geochim. Cosmochim. Acta *3* (1953), 186—215
CORRENS, C. W.: Über die Geochemie des Fluors und Chlors. Neues Jb. Miner. *91* (1957), 239 to 256
MASURENKOV, YU. P., and S. Y. PAKHAMOV: On the Geochemistry of Chlorine. Dokl. Akad. Nauk SSSR *139* (1961) 2, 453 (Russian)

Chromium (Cr)

FRÖHLICH, F.: Beiträge zur Geochemie des Chroms. Geochim. Cosmochim. Acta *20* (1960), 215—240
LUNDEGARDH, P.: Aspects of the Geochemistry of Chromium, Cobalt, Nickel and Zinc. Sveriges Geol. Unders., Ser. C. *513* (1949), 1—56

Cobalt (Co)

CARR, M. H., and K. K. TUREKIAN: The Geochemistry of Cobalt. Geochim. Cosmochim. Acta *23* (1961), 9
JUNG, R. S.: Geochemistry of Cobalt, in: Geochemistry of Rare Elements. Izd. Nauka Moscow 1959, 511 (Russian)
MALYUGA, D. P.: Distribution of Cobalt in the Earth's Crust, in: Trace Elements and Their Importance to Plant and Animal Life, edited by A. P. VINOGRADOV. Moscow 1952, 417—435 (Russian)

Copper (Cu)

HEIDE, F., and E. SINGER: Zur Geochemie des Cu und Zn. Naturwissenschaften *37* (1950), 541

MORITA, Y.: Distribution of Copper and Zinc in Various Phases of the Earth Materials. J. Earth Sci. Nagoya Univ. *3* (1955), 33—57

SUGAWARA, K., and M. JOSHIMI: On the Revision of the Clarke Numbers of Copper and Zinc. Mikrochemie (1951), 36—37, 1093

WEDEPOHL, K. H.: Beiträge zur Geochemie des Kupfers. Geol. Rdsch. *52* (1962), 492—504

Fluorine (F) and Halogens

BORCHERT, H.: Zur Geochemie des Fluors. Heidelberger Beitr. Mineralog. Petrogr. *3* (1952), 36—43

BUDZINSKIĬ, YU. A.: On the Geochemistry of Halogenes. Geokhimiya (1965) 6, 707—723 (Russian)

CORRENS, C. W.: The Geochemistry of the Halogens. Phys. Chem. of the Earth *1* (1956), 181—233

CORRENS, C. W.: Über die Geochemie des Fluors und Chlors. Neues Jb. Mineralog. *91* (1957), 239 to 256

KORITNIG, S.: Ein Beitrag zur Geochemie des Fluors. Geochim. Cosmochim. Acta *1* (1951), 89—116

STUEBER, A. M., W. H. HUANG, and W. D. JOHNS: Chlorine and Fluorine Abundance in Ultramafic Rocks. Geochim. Cosmochim. Acta *32* (1968) 3, 353—358

TAGEEVA, N. V.: Geochemistry of Boron and Fluorine. Priroda *6* (1943), 25—35 (Russian)

Gallium (Ga)

BORISENOK, C. A.: Distribution of Gallium in Rocks of the Soviet Union. Geokhimiya (1959), 52—59 (Russian)

BURTON, S. D., F. CULKIN, and J. P. RILEY: The Abundance of Gallium and Germanium in Terrestrial Materials. Geochim. Cosmochim. Acta *16* (1959), 166

MARKUSEV, A. A., and YU. K. POLIN: On the Distribution of Ga in Minerals of Archaic Metamorphites of the Aldan Shield. Geokhimiya (1961) 2, 181—183 (Russian)

SANDELL, E. B.: Gallium in Rocks. Am. J. Sci. *247* (1949), 40

VERSOVSKAYA, O. V.: Geochemistry of Gallium, in: Geochemistry of Rare Elements. Izd. "Nauka", Moscow 1964 (Russian)

Germanium (Ge)

BURKSER, E. S., B. F. MITSKEVICH, and K. J. LAZEBNIK: Ge in Granitoids of the Ukrainian Crystalline Shield. Geokhimiya *6* (1961), 515—520 (Russian)

FLEISCHER, M., and J. O. HARDER: Geochemistry of Germanium. U. S. Atomic Energy Comm. TJD *5212* (1955), 93—105

HEIDE, F., and D. KÖRNER: Zur Geochemie des Germaniums. Chem. d. Erde *23* (1963), 104—115

MANSKAYA, S. M., et al.: On the Biogeochemistry of Ge. Dokl. Akad. Nauk SSSR *143/6* (1962), 1435—1437 (Russian)

RÖSLER, H. J., W. SCHRÖN, and B. VOLAND: Germanium und Indium: ein Vergleich zwischen den regionalen und absoluten Clarke-Werten. in: Origin and Distribution of the Elements (ed. L. H. Ahrens) Pergamon Press (1968)

SCHRÖN, W.: Beitrag zur Geochemie des Germaniums. I. Petrogenetische Probleme. Chem. d. Erde *27* (1968) 3, 193—251. II. Lagerstättengenetische Probleme. Freiberger Forschungsh. C *246* (1968)

WARDINI, S. A.: On the Geochemistry of Germanium. Geochim. Cosmochim. Acta *13* (1957), 5—19

ZAGYANSKIĬ, A. L.: On the Geochemistry of Ge. Dokl. Akad. Nauk SSSR *143/6* (1962), 1435—1437 (Russian)

ZUKOVA, A. S.: On the Geochemistry of Germanium. Trudy Miner. Geokh. i Kristallokhim. redkikh Elementov, Akad. Nauk SSSR *3* (1959), 26—43 (Russian)

Gold (Au)

DE GRAZIA, A. R., and E. HASKIN: On the Gold Content of Rocks. Geochim. Cosmochim. Acta *28* (1964), 559—564

GONI, J., C. GUILLEMIN, and C. SARCIA: Géochimie de l'Or Exogène. Mineralium Deposita *1* (1967), 259—268

SHCHERBAKOV, Y. G., and G. A. PEREZOGIN: Geochemistry of Gold. Geokhimiya (1964) *6*, 489—496 (Russ.)

VINCENT, E. A., and J. H. CROCKET: Studies in the Geochemistry of Gold. I + II Geochim. Cosmochim. Acta *18* (1960), 130 and 143

Hafnium (Hf): see: Zirconium

Helium (He)

MAYNE, K. J.: Terrestrial Helium. Geochim. Cosmochim. Acta *9* (1956), 174—182

Indium (In)

FLEISCHER, M., and J. O. HARDER: Geochemistry of Indium. Geochemistry of Niobium and Tantalium. U. S. Atomic Energy Comm. TJD *5212*, 106—117, 118—131

IVANOV, V. V.: *In* in a Few Magmatic Rocks of the USSR. Geokhimiya *12* (1963), 1101—1110 (Russ.)

IVANOV, V. V.: Geochemistry of Indium, in: Geochemistry of Rare Elements. Izd. "Nauka", Moscow 1964 (Russian)

SHAW, D. M.: The Geochemistry of Indium. Geochim. Cosmochim. Acta *2* (1952), 185—206

VOLAND, B.: Zur Geochemie des Indiums. Freiberger Forschungsh. C *246* (1968)

Iodine (I)

GULYAEVA, L. A.: Iodine in Sedimentary Rock of the Devonian Formation. Dokl. Akad. Nauk SSSR, *80*. Moscow (1951) 5, 787—789 (Russian)

Lanthanum (La) see: Rare Earths

Lead (Pb)

HEIDE, F., and H. LERZ: Zur Geochemie des Bleis. Chem. d. Erde *17* (1955), 217—222

WEDEPOHL, K. H.: Untersuchungen zur Geochemie des Bleis. Geochim. Cosmochim. Acta *10* (1956), 69

Lithium (Li)

GINZBURG, A. I.: A Few Peculiarities of the Geochemistry of Lithium. Trudy miner. Mus. Akad. Nauk SSSR (1957), 29—41 (Russian)

SLEPNOV, YU. S., N. A. SOLODOV, and T. F. BOIKO: Geochemistry of Lithium, in: Geochemistry of Rare Elements. Izd. "Nauka", Moscow 1964 (Russian)

STROCK, L. W.: Geochemie des Lithiums. Nachr. Ges. Wiss. Göttingen, math.-phys. Kl. IV (1936), N. F., No. 15, p. 171

Mercury (Hg)

HEIDE, F., and G. BÖHM: Geochemie des Quecksilbers. Chem. d. Erde *19* (1957), 198

MELNIKOV, S. M.: Mercury. Metallurgizdat, Moscow 1951 (Russian)

SAUKOV, A. A.: Geochemistry of Mercury. Izd. Akad. Nauk SSSR (1946) (Russian)

Molybdenum (Mo)

KURODA, P. K., and E. B. SANDELL: Geochemistry of Molybdenum. Geochim. Cosmochim. Acta, *6* (1954), 35—63

RECHENBERG, H. P.: Die metallischen Rohstoffe: Vol. 12 — Molybdän. F. Enke, Stuttgart 1960

STUDENIKOVA, Z. V., M. I. KLINKINA, and L. Y. PAVLENKO: On the Distribution of Molybdenum in Intrusive Rocks. Geokhimiya (1957), 111—119 (Russian)

Nickel (Ni)

EDELSTEĬN, Y. Y.: On the Geochemistry of Nickel. Geokhimiya (1960) *7*, 601—609 (Russian)

KÜHN, K.: Zur Geochemie von Nickel und Kobalt. Thesis TU Clausthal 1968

INKSOV, V. A., and N. V. LONDOCNIKOVA: The Distribution of Co and Ni in Eruptive Rock in the Earth's Crust. Geokhimiya (1961) *1*, 732—741 (Russian)

Niobium (Nb)

ESKOVA, E. M.: Geochemistry of Niobium; in: Geochemistry of Rare Elements. Izd. "Nauka", Moscow 1964 (Russian)
FLEISCHER, M., and J. O. HARDER: Geochemistry of Niobium and Tantalum. U. S. Atomic Energy Comm. TJD *5212*, 118—131
KUZMENKO, M. V.: On the Geochemistry of Ta and Nb. Trudy Inst. Miner. Geokh. i Kristallokh. redk. elem. (IMGRE). Akad. Nauk SSSR *3* (1959), 3—25 (Russian)
ZNAMENSKIĬ, E. B.: On the Problem of the Average Concentration of Niobium and Tantalum in Rocks in the Earth's Crust. Geokhimiya (1967) *8*, 730—736 (Russian)

Nitrogen (N)

SOKOLOV, V. A., and V. F. VOLYNEZH: Zur geochemischen Bedeutung der verschiedenen Formen des Stickstoffs, die bei der Metamorphose organischer Stoffe entstehen; in: Geochemische Probleme bei der Erkundung und Förderung von Erdöl und Erdgas, Vol. 1, VEB Deutscher Verlag für Grundstoffindustrie, Leipzig 1968, 357—366
WLOTZKA, F.: Untersuchungen zur Geochemie des Stickstoffs. Geochim. Cosmochim. Acta *24* (1961), 106—154

Phosphorus (P)

GEIJER, P.: Some Aspects of Phosphorus in Precambrian Sedimentation. Ark. Miner. Geol. *3* (1962), 165—186
RONOV, A. B., and G. A. KORZINA: P in Sedimentary Rocks. Geokhimiya (1960) *8*, 667—687 (Russian)

Platinum (Pt)

YUSHKO-ZAKHAROVA, O. E., et al.: On the Geochemistry of Elements of the Platinum Group. Geokhimiya *11* (1967), 1381—1394 (Russian)

Potassium (K) and Alkaline Metals

GOLDSCHMIDT, V. M. et al.: Zur Geochemie der Alkalimetalle. Nachr. Ges. Wiss. Göttingen, math.-phys. Klasse IV, N. F. 1, 1st Part (1933), 2nd Part (1934)
MOROZOV, N. P.: Zur Geochemie von Alkalielementen in marinen Sedimenten. Litologiya i poleznye iskopaemye *6* (1968), 4—16
SHCHERBINA, V. V.: The Differences between Geochemical Processes in which K and Na Take Part. Geokhimiya (1963) *3*, 229—236 (Russian)

Rare Earths

BALASHOV, YU. A.: Regularities in the Distribution of Rare Earths in the Earth's Crust. Geokhimiya (1963) *2*, 99—114 (Russian)
BANDURKIN, G. A.: On the Behaviour of Rare Earths in a Surrounding Containing F. Geokhimiya (1961) *2*, 143—149 (Russian)
FLEISCHER, M.: Some Aspects of the Geochemistry of Yttrium and the Lanthanides. Geochim. Cosmochim. Acta *29* (1965), 755—772
GERASSIMOVSKIĬ, V. I. (Edit.): Rare Earth Elements (Translation from the Russian). Israel Program for Scientific Translations, Jerusalem 1960
HASKIN, L. A., F. A. FREY, R. A. SCHMITT, and R. H. SMITH: Meteoritic, Solar and Terrestrial Rare-earth Distributions. Pergamon Press 1966. Russian Translation in Izd. "Mir", Moscow 1968, 186 pp.
KOSLOVA, C. G.: Rare Earths in Fluorspars. Izvestiya vyshikh uchebnykh zavedeniĭ. Geologiya i razvedka (1961) 7 (Russian)
MINEEV, D. A.: Geochemical Differentiation of Rare Earths. Geokhimiya (1963) *12*, 1082—1100 (Russ.)
SCHMITT, R. A., et al.: Abundances of the Fourteen Rare-earth Elements, Scandium and Yttrium in Meteoritic and Terrestrial Rocks. Geochim. Cosmochim. Acta *27* (1963), 577—622

SCHOFIELD, A., and L. HASKIN: Rare-earth Distribution Patterns of Eight Terrestrial Materials. Geo-him. Cosmochim. Acta *28* (1964), 437—446
SEMENOV, E. I.: Geochemistry of Lanthanides (Rare Earths); in: Geochemistry of Rare Elements. Izd. "Nauka", Moscow 1964 (Russian)

Rhenium (Re)

BROWN, H. S., and E. D. GOLDBERG: A New Determination of the Relative Abundance of Rhenium in Nature. Physic. Rev. *76* (1949), 1260—1261
GOROKHOVA, V. N.: Geochemistry of Rhenium, in: Geochemistry of Rare Elements. Izd. "Nauka", Moscow 1964 (Russian)
MORRIS, D. C. F., and F. W. FIFIELD: Rhenium Contents of Rocks. Geochim. Cosmochim. Acta *25* 1961), 232—233
NODDACK, I., and W. NODDACK: Die Geochemie des Rheniums. Z. physik. Chem., Abt. A *154* (1931), 207—244
TISCHENDORF, G., W. LIMPACK, and H. UNGETHÜM: Beitrag zur Geochemie des Rheniums. Geologie *17* 1968) 10, 1208—1218

Rubidium (Rb)

HURLEY, P. M.: Absolute Abundance and Distribution of Rb, K and Sr in the Earth. Geochim. Cosmo-him. Acta *32* (1968) 3, 273—283
SLEPNOV, YU. S., N. A. SOLODOV, and T. F. BOIKO: Geochemistry of Rubidium; in: Geochemistry of Rare Elements. Izd. "Nauka", Moscow 1964 (Russian)

Ruthenium (Ru)

SVYAGINTSEV, O. E.: Geochemistry of Ruthenium. Izv. Sektora Platiny i drugikh blagorodnykh Metallov *25* (1950), 129—139 (Russian)

Scandium (Sc)

BORISENKO, L. A.: On the Distribution of Sc in Intrusive Rock with Particular Reference to a Few Massifs in the USSR. Geokhimiya (1959), 623—627 (Russian)
BORISENKO, L. A.: Geochemistry of Scandium; in: Geochemistry of Rare Elements. Izd. "Nauka", Moscow 1964 (Russian)
FRYKLUND, V., M. FLEISCHER, and FLEISCHER, JR.: The Abundance of Scandium in Volcanic Rocks, a Preliminary Estimate. Geochim. Cosmochim. Acta *27* (1963), 643—664
KALENOV, A. D.: A Few Peculiarities of the Sc Concentration. Geokhimiya (1961) *3*, 243—251 (Russian)
KUKHARENKO, A. A., and L. P. MURAV'EVA: On the Geochemistry of Scandium in Alkali Gabbros of the Karelo-Finnish A.S.S.R. Sborn. Miner. Geokhim., Leningr. Univ. (1964), 181—191 (Russian)
NERMAN, J. C., and L. A. HASKIN: The Geochemistry of Sc. Geochim. Cosmochim. Acta *32* (1968) 1, 93—108
SCHMITT, R. A., et al.: Abundances of the Fourteen Rare-earth Elements, Scandium and Yttrium in Meteoritic and Terrestrial Rocks. Geochim. Cosmochim. Acta *27* (1963), 577—622
SHCHERBINA, V. V.: On the Geochemical Distribution of Scandium. Geokhimiya (1959) 8, 711—715 (Russian)

Selenium (Se)

ROCKENBAUER, W.: Zur Geochemie des Selens in österreichischen Erzen. Tscherm. min.-petr. Mitt. *7* (1960) 3, 149—185
SINDEEVA, N. D.: Geochemistry of Selenium; in: Geochemistry of Rare Elements. Izd. "Nauka", Moscow 1964 (Russian)
STROCK, L. W.: The Distribution of Selenium in Nature. Amer. J. Pharm. *107* (1935), 144—157
TISCHENDORF, G.: Zur Verteilung des Selens in Sulfiden. Freiberger Forschungsh. C *208*, Leipzig 1966, 162 pp.

Silver (Ag)

KERSCHAGL, R.: Die metallischen Rohstoffe: Vol. 13, Silber. F. Enke, Stuttgart 1961

Silicon (Si)

KRAUSKOPF, K. B.: Die Geochemie der Kieselsäure unter sedimentären Bedingungen. Soc. econ. Paleontologist Mineralogists, Spec. Publ. 7 (1959), 4—19

Sodium (Na) see: Potassium (K)

Strontium (Sr)

BURKOV, V. V.: Geochemistry of Strontium; in: Geochemistry of Rare Elements. Izd. "Nauka", Moscow 1964 (Russian)
FORNASERI, M., and L. GRANDI: Contenuto in Stronzio di Serie Calcaree Italiane. Giornale di Geologia, Annali del Museo Geologico di Bologna, Ser. 2, *XXXI* (1963)
NOLL, W.: Geochemie des Strontiums. Chem. d. Erde *8* (1933/34), 507
TUREKIAN, K. K., and J. L. KULP: The Geochemistry of Strontium. Geochim. Cosmochim. Acta, *10* (1956), 245—296

Sulphur (S)

RICKE, W.: Ein Beitrag zur Geochemie des Schwefels. Geochim. Cosmochim. Acta *21* (1960), 35
ZOBELL, C. E.: Organic Geochemistry of Sulfur. Chapt. 13 in "Organic Geochemistry of Sulfur". Internat. Ser. of Monographs on Earth Sciences, Vol. 16, 543—578. Pergamon Press. Oxford/London/New York/Paris 1963

Tantalum (Ta)

FLEISCHER, M., and J. O. HARDER: Geochemistry of Niobium and Tantalum. U. S. Atomic Energy Comm., TJD *5212*, 118—131
KUZMENKO, M. V.: Geochemistry of Tantalum; in: Geochemistry of Rare Elements. Izd. "Nauka", Moscow 1964 (Russian)
RANKAMA, K.: On the Geochemistry of Tantalum. Bull. Comm. Geol. Finlande *133* (1944), 1

Tellurium (Te)

SÍNDEEVA, N. D.: Geochemistry of Tellurium; in: Geochemistry of Rare Elements. Izd. "Nauka", Moscow 1964 (Russian)
SÍNDEEVA, N. D.: A Few Geochemical Features of Selenium and Tellurium. Intern. Geol. Congr., XXI Sess. (1960), Vol. Geochem. Cycles, Copenhagen (1960)

Thallium (Tl)

BUTLER, J. R.: Tl in a Few Eruptive Rocks. Geokhimiya (1962) *6*, 514—523 (Russian)
IVANOV, V. V.: Geochemistry of Thallium; in: Geochemistry of Rare Elements. Izd. "Nauka", Moscow 1964 (Russian)
SHAW, D. M.: The Geochemistry of Thallium. Geochim. Cosmochim. Acta *2* (1952), 118—154
VOSKRESENSKAYA, N. T.: On the Geochemistry of Thallium and Rubidium in Eruptive Rocks. Geokhimiya (1959) *6*, 495—504 (Russian)

Thorium (Th) see: Uranium (U)

Tin (Sn)

BARSUKOV, V. L.: On the Geochemistry of Tin. Geokhimiya (1957) *1*, 36—46 (Russian)
BORCHERT, H., and I. DYBECK: Zur Geochemie des Zinns. Chem. d. Erde *20* (1960), 137—154

DE LAETER, I. R., and P. JEFFERY: Tin: Its Isotopic and Elemental Abundance. Geochim. Cosmochim. Acta *31* (1967), 969—985

HAMABUCHI, H., et al.: The Geochemistry of Tin. Geochim. Cosmochim. Acta *28* (1964), 1039—1053

ONISHI, H., and E. B. SANDELL: Meteoritic and Terrestrial Abundance of Tin. Geochim. Cosmochim. Acta *12* (1957), 262—270

Titanium (Ti)

BARDOSSY, G., and S. L. BARDOSSY: Contribution to the Geochemistry of Titanium. Acta Geol. Hung. *?* (1954), 191—203

Uranium (U) and Thorium (Th)

ADAMS, J. S. A., J. K. OSMOND, and J. ROGERS: The Geochemistry of Thorium and Uranium. Phys. and Chem. of the Earth. *3* (1959)

BARANOV, V. I.: On the Geochemistry of Thorium and Uranium. Geokhimiya (1956), 3—8 (Russian)

HEIDE, F., and G. PROFT: Zur Geochemie des U. Chem. d. Erde *20* (1960), 169—182

KOCZY, F., E. TOMIC, and F. HECHT: Zur Geochemie des Urans im Ostseebecken. Geochim. Cosmochim. Acta *11* (1957), 86—102

TOMKEIEFF, S. J.: The Geochemistry of Uranium. Sci. Progr. *34* (1946) 136, 696—712

VINOGRADOV, A. P. (Edit.): Outline of the Geochemistry of Uranium. Izd. Akad. Nauk SSSR, Moscow 1963, 366 pp. (Russian)

Vanadium (V)

LEUTWEIN, F.: Geochemie und Vorkommen des Vanadiums. Ber. Freiberger Geol. Ges. (1941), 73 to 83

Wolfram (Tungsten) (W)

JEFFERY, P. G.: The Geochemistry of Tungsten, with Special Reference to the Rocks of the Uganda Protectorate. Geochim. Cosmochim. Acta *16* (1959)

VINOGRADOV, A. P., E. E. VAINSTEIN, and L. I. PAVLENKO: Tungsten and Molybdenum in Eruptive Rock (with Geochemistry of Tungsten). Geokhimiya (1958) *5*, 399—408 (Russian)

Yttrium (Y)

SCHMITT, R. A., et al.: Abundances of the Fourteen Rare-earth Elements, Scandium and Yttrium in Meteoritic and Terrestrial Rocks. Geochim. Cosmochim. Acta *27* (1963), 577—622

SEMENOV, E. I.: Geochemistry of Yttrium; in: Geochemistry of Rare Elements. Izd. "Nauka", Moscow 1964 (Russian)

Zinc (Zn)

HEIDE, F., and E. SINGER: Zur Geochemie des Cu und Zn. Naturwissenschaften *37* (1950), 541

MORITA, Y.: Distribution of Copper and Zinc in Various Phases of the Earth Materials. J. Earth Sci. Nagoya Univ. *3* (1955), 33—57

WEDEPOHL, K. H.: Untersuchungen zur Geochemie des Zinks. Geochim. Cosmochim. Acta *3* (1953), 93

Zirconium (Zr)

DEGENHARDT, H.: Untersuchungen zur geochemischen Verteilung des Zr in der Lithosphäre. Geochim. Cosmochim. Acta *11* (1957), 279—309

KUKHARENKO, A. A., and YA. M. KRAVTSOV: On the Geochemistry of Zirconium and Beryllium in Ultrabasic Rocks. Dokl. Akad. Nauk SSSR *134* (1960) 4, 931—934 (Russian)

TIKHONENKOVA, R. P.: Geochemistry of Zirconium and Hafnium; in: Geochemistry of Rare Elements. Izd. "Nauka", Moscow 1964

10. Important Fields of Investigation of Applied Geochemistry

The essence and purpose of any science is the practical use of the knowledge. This knowledge may be of a general theoretical nature; this means an extension of man's knowledge which will become effective indirectly and at a future time (general geochemistry). This applies to a great part of the knowledge, data, laws and regularities discussed in the previous chapters. On the other hand, there are possibilities of using the geochemical knowledge practically, that is, to turn it to good account. This "applied geochemistry" can be subdivided according to the ends pursued in the following way:

1. Prospecting for deposits (geochemical prospecting)

   a) by means of elementary geochemical methods (geochemical prospecting in a narrow sense)

   b) by means of isotopic geochemical and radiometric methods

   c) by means of indirect geochemical methods, such as heavy mineral analysis, geobotanic prospecting, radiation and luminescence measurements

2. Exploration and genesis of deposits (geochemistry of deposits)

   a) Determination of the mineralogic-geochemical composition (geochemistry of deposits in a narrow sense)

   b) Determination of the origin of the deposited matter (metallogenesis, geochemical provinces, etc)

   c) Determination of the conditions of concentration (e.g. geochemical facies analysis)

   d) Determination of the age of deposits (age determination by means of isotopes)

3. Origin and development of magmas and rocks (geochemistry of rocks = lithogeochemistry)

   a) Origin, composition and differentiation of magmas and rocks (petrochemistry, age determination, isotopic geochemistry, trace elements, mineral chemistry)

   b) Lithology of sediments (many methods are employed, especially geochemical facies analysis)

   c) Weathering, transport, sedimentation (geochemical cycles and balances, pedogeochemistry)

4. Elaboration of fundamentals for adjacent fields of study

   a) Supply of nutritive substances for agriculture and forestry (biogeochemical provinces, fertiliser requirements, diseases of plants and animals)

   b) Injurious substances in the soil, air and water (e.g. waste gas from the metallurgical industry, sewage, fall-out from atomic explosions, atomic power stations).

9.1. Geochemical Prospecting

9.1.1. General and Subdivisions

Geochemical prospecting is one of the most important direct applications of geochemistry. Geochemical prospecting aims at the discovery or location of element concentrations, especially of heavy metals, in the upper parts of the earth's crust. A precondition of geochemical prospecting usually is the dissolution and migration of the chemical elements towards the sampling horizon and the formation of a geochemical dispersion aureole. A dispersion aureole may already be formed during the original development of a deposit (primary dispersion aureole) or it may be formed during weathering of the deposit at the surface of the earth (secondary dispersion aureole).

A primary dispersion aureole is rarely found on the surface of the earth, but more usually in geochemical prospecting for deep-seated deposits by sinking wells (geochemistry of wells) or in the development of mines. It is a spatial extension of the "blind" deposit sought for, and the elements constituting the aureole must not necessarily be identical with those contained in the deposit proper (so-called "pathfinder" or "indicator" elements such as Hg, As and Sb; cf. WARREN and DELAVAULT, 1959); the discovery of an aureole enables a planned further exploration (Figs. 113 and 114). To verify the nature of an aureole, plugs and other rock materials are subjected to routine examinations, usually on the basis of spectrographic analyses.

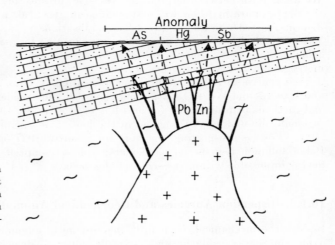

Fig. 113 Schematic representation of a magmatogenic Pb-Zn deposit whose outermost primary dispersion aureole crops out together with the indicator (pathfinder) elements As, Hg and Sb

In surface prospecting for secondary dispersion aureoles, the following geochemical methods are the most commonly used:

1. lithogeochemical and pedogeochemical prospecting (rock and soil samples)
2. biogeochemical prospecting (analysis of plant samples and their ashes)
3. hydrogeochemical prospecting (analysis of waters and their residues)
4. atmogeochemical or gas prospecting (analysis of gas)
5. indirect geochemical prospecting (e.g. by radiometry, mineralogic phase analysis, bacteriological methods).

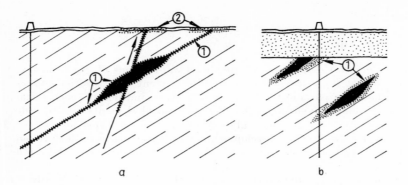

Fig. 114 Schematic representation of ore deposits concordantly embedded in crystalline schist with indicated primary dispersion aureoles. The possibility of exploring these dispersion aureoles by means of depth and surface prospecting is demonstrated
1 = primary aureoles; 2 = secondary aureoles

We distinguish between large-scale (strategic) prospecting, in regions where the geology and the deposits are little known, and small-scale (tactical) prospecting, in regions with known deposits, in order to find additional mineralisation.

The particularly successful geochemical prospecting for metallic materials (ores) is sometimes called metallometry and its sub-divisions are known as cuprometry, plumbometry, radiometry, stannometry, mercurometry, etc.

A good concise introduction into geochemical prospecting has been given by HAWKES (1957) and HAWKES and WEBB (1962); a more detailed and excellent description of the fundamentals and methods of prospecting for ores has been given by GINSBURG (1963). A modern, though regionally limited survey is given in the book "Geochemical Prospecting in Fennoscandia" (1967). For gas prospecting, the work by SOKOLOV (1947) is recommended.

Below follow a few special particulars; a more comprehensive representation of the whole field is impossible within the scope of this book.

10.1.2. Dispersion Aureoles and Geochemical Anomalies

Besides the mechanical, frequently macroscopical anomalies (glacial aureoles, debris-slides, oxidation zones, pedologic and fluviatile (alluvial) dispersions, placers, etc.), the invisible geochemical dispersion aureoles, which owe their element concentration to diffusion and rearrangement by water transport, come especially under this heading.

The extent of a secondary dispersion aureole depends largely on the solubility or migration capability of an element; these factors are governed by the solubility product of the parent mineral, its paragenesis, its country rock, the tectonic situation, the climate, the morphology, and the precipitation and sorption conditions. The absolute extent mostly varies within a range of from a few decametres to several hundred metres, maximum a few kilometres; hydrogeochemical "aureoles" in water courses may have a length of many kilometres (2 to 6 km on an average). The shape of the aureole depends, among other things, on the morphology of the ground and the corresponding hydrologic conditions (formation of so-called pispersion streams or brooks). Primary dispersion aureoles (gas aureoles excluded) have an tent of a few metres up to a maximum of a few hundred metres. For the identification and interpretation of dispersion aureoles, the following concepts are of importance:

a) geochemical background: The range of variation of the local or regional Clarke values of the country rock and of the waters, the air, etc.

b) geochemical anomaly: The regional and numerical range of concentrations which exceed those of the geochemical background. The limit between them is sometimes termed as „Störpegel" or threshold concentration (Fig. 115). For the calculation of the threshold value, statistical methods are used.

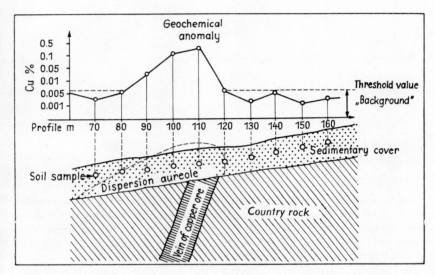

Fig. 115   Schematic representation of pedogeochemical prospecting for a vein of copper ore

c) Kurtosis (contrast, slope, relief) of a geochemical anomaly expressed by the quotient

$$\frac{\text{maximum}}{\text{threshold value}} \quad \text{or} \quad \frac{\text{maximum}}{\text{background}}$$

d) Homogeneity of a geochemical anomaly: (uniform rise in values; variations at short distances will bear on the density of sampling).

e) Autochthonous anomaly: Geochemical anomaly about a spring usually produced by diffusion; directly indicates the deposit. Allochthonous anomaly: Anomaly mostly produced by the mechanical transport of decomposition (weathering) products; is not found above a deposit or in direct connection with it.

f) Variance or coefficient of variation of the contents in the samples, according to GRANIER (1962) dependent on:

1. the homogeneity of the distribution of the elements in the samples,
2. the reproducibility of the chemical method,
3. the type of dispersion of the element.

g) Productivity of the anomaly or the dispersion aureole.

h) Coefficient of correlation between two and more elements as well as between elements and geological parameters.

i) "False" anomaly: Accumulation of an element without any connection to a deposit (brought about by special sorption conditions, human actions such as waste matter from smelting plants, waste dumps, etc.).

For a more detailed mathematical, genetic and economic study of dispersion aureoles and geochemical anomalies reference is made to the bibliography; the works by SAFRONOV, 1959, and KRASNIKOV, 1959, are of particular importance.

10.1.3. Sampling

A. Sampling for pedogeochemical, lithogeochemical and, to some extent, biogeochemical prospecting:

  a) random sampling to obtain a general survey along roads, river,s and in special development areas (granite areas, oxidation and bleaching zones, etc.)

  b) regular sampling in the form of profiles perpendicular to the strike of more or less linear bodies
  distance between profiles: usually 50—200 m
  distance between sampling sites on profiles: usually 10—25 m

  c) regular sampling in the form of networks; distance between lines 10—250 m.

For soil sampling, it is advisable to use the so-called three-point method (three samples taken at a distance of 1—2 metres from the sampling point and equally spaced about it provide the desired collective sample). Most favourable depth of sampling is 10—60 cm, largely depending on the type of soil and the purpose. The amount of one sample is usually 50—250 g. Transport in paper or plastic bags or plastic containers.

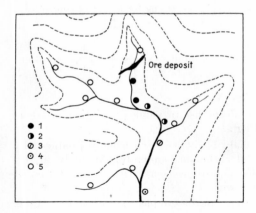

Fig. 116 Schematic representation of hydrogeochemical prospecting for an ore deposit.
(The circles represent the levels of element concentration in riverwater: 1 high concentration; 2 medium concentration; 3 low concentration; 4 traces; 5 element cannot be detected; dashed lines are contour lines)

B. Sampling for hydrogeochemical prospecting takes place at springs, wells, in rivers, especially in the mouth area of tributary rivers (see Fig. 116). Amount of the sample 0.5 to 2 litres. Transport in plastic bottles.

.1.4. Analysis and Evaluation

or the determination of elements transported in dissolved form in water, usually only the est fraction (<100 μm) of the soil or sediment sample is analysed. In the case of weather sistant minerals (heavy minerals, native elements, also baryte, etc.) an analysis of the action of the order of 0.1—1 mm is recommendable. Plant samples are slowly reduced to hes at a temperature $\leq 500\,°C$ and the ashes are analysed.

ne most important methods of analysis are the emission spectrographic analysis (see able 171) and the colorimetric determination of elements, especially in the form of the thizone method (cf. Section 4.4.). In addition, all the other geochemical analyses such as larography, chromatography, flame spectroscopy, X-ray spectroscopy, radiometry, and en neutron activation (e.g. beryllium with a "beryllometer") are used. The desired accuracy the analytical results is usually low and is of the order of 10—50%. Detailed specifications analyses are contained in the standard works and in a great number of special publications ee bibliography summarized under "Methods of Analysis").

he results of analyses may be evaluated in the form of relative values (e.g. density value in ectrographic analysis) or absolute values after calibration. If the distance of the samples is nall enough, the drawing up of isogram charts (anomaly maps) is advisable. The methods quired for this purpose are those employed in applied geophysics.

able 171. Limits of Detection of a Few Elements which were btained by Colorimetry under Field Conditions and by pectrographic Analysis in a Laboratory rom GINSBURG, 1963); data in [mass-%]

lement	Colorimetry	Spectral Analysis
n	0.001	0.01 and 0.03
i	0.001,5	0.001
u	0.001,0	0.000,5 and 0.001
ɔ	0.001,0	0.001 and 0.003
b	0.001	0.002 and 0.001
g	0.000,02	0.000,1
ʃ	0.001	0.01—0.03
o	0.000,1	0.000,5 and 0.003

0.1.5. Notes on Other Methods of Prospecting

. *Biogeochemical Prospecting*

iogeochemical methods in a broad sense can be grouped as follows:

1. Plant geochemical methods (cf. Fig. 117)
2. Zoogeochemical methods (of minor importance)
3. Peat-geochemical methods
4. Bituminological methods
5. Biological methods such as geobotanic and microbiological methods

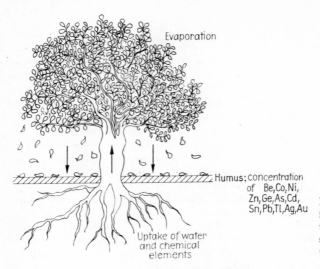

Evaporation

Humus: concentration
of Be,Co,Ni,
Zn,Ge,As,Cd,
Sn,Pb,Tl,Ag,Au

Uptake of water
and chemical
elements

Fig. 117 Uptake of chemical elements from the soil by plants and concentration of the elements in the uppermost soil horizon (classical representation by GOLDSCHMIDT, 1937)

In various countries, prospecting with the help of plants has been successful for the following elements: U, Cu, Zn, Mn, Au, Ni, Mo, Cr, V, Ba, B, Sn, W.

Geochemical methods based on peat are of great importance to the colder climatic zones of the earth, e.g. in Siberia, Scandinavia, and Canada (cf. KVALHEIM, 1967). Geobotanic methods are based on the fact that selection of species and plant growth are influenced by the supply of elements (e.g. halophytes, calamine violet, cf. Tables 172 and 173). Geobotanic methods have been particularly successful in deserts and semi-deserts (cf. TKALICH, 1954). Microbiological methods can be used in surface prospecting for petroleum and natural gas deposits. They are based on the fact that aerobic bacteria develop particularly well in soil and sedimentary horizons where a supply of gaseous hydrocarbon is present (MOGILEVSKI, 1938; STEGENA 1961; and others). The "bituminological method" is based on the fact that hydrocarbons which are capable of migrating, oxidize or condense and are fixed in the sediment or soil. The concentration of bitumen is determined by means of ultraviolet rays (luminescence method) or analysed by extraction (extraction method).

Table 172. Metal Concentrations in Ashes of Normal Plants and Plants Grown above Deposits (after A. P. VINOGRADOV)

Metal	Metal Concentration in %		Degree of Concentration
	in normal plants	in plants grown above deposits	
W	$\sim 5 \cdot 10^{-4}$	$\sim n \cdot 10^{-2}$	100
Cr	$\sim 5 \cdot 10^{-4}$	$\sim 1 \cdot 10^{-2}$	200
Mn	$\sim 1 \cdot 10^{-2}$	~ 10	1,000
Co	$\sim 4 \cdot 10^{-4}$	$5 \cdot 10^{-3}$	10
Ni	$\sim 1 \cdot 10^{-3}$	$n \cdot 10^{-2}$	100
Cu	$\sim 5 \cdot 10^{-3}$	$n \cdot 10^{-1}$	100
Zn	$\sim 1 \cdot 10^{-2}$	$1 \cdot 1.0^{0}$	100
Mo	$\sim 5 \cdot 10^{-4}$	$n \cdot 10^{-2}$	100
Pb	$\sim 1 \cdot 10^{-4}$	$1 \cdot 10^{-2}$	100

Element	Effect
Aluminium	Stubby roots, leaf scorch, mottling
Boron	Dark foliage; marginal scorch of older leaves at high concentrations; stunted, deformed, shortened, internodes; creeping forms; heavy pubescence; increased gall production
Chromium	Yellow leaves with green veins
Cobalt	White dead patches on leaves
Copper	Dead patches on lower leaves from tips; purple stems, chlorotic leaves with green veins, stunted roots, creeping sterile forms in some species
Iron	Stunted tops, thickened roots; cell division disturbed in algae, resulting cells greatly enlarged
Manganese	Chlorotic leaves, stem and petiole lesions, curling and dead areas on leaf margins, distortion of laminae
Molybdenum	Stunting, yellow-orange coloration
Nickel	White dead patches on leaves, apetalous sterile forms
Uranium	Abnormal number of chromosomes in nuclei; unusually shaped fruits; sterile apetalous forms, stalked leaf rosette
Zinc	Chlorotic leaves with green veins, white dwarfed forms; dead areas on leaf tips; roots stunted

B. *Atmogeochemical or Gas Prospecting*

These methods are mainly used in prospecting for deposits of hydrocarbons and for deposits of radioactive substances.

In prospecting for hydrocarbons, air is removed from the soil at a depth of 1.5—2 m (method of free gas) or extracted from deep-seated sedimentary rocks (degassing; method of sorbed gases) and subjected to a microanalysis in a special apparatus (determination of methane and heavy hydrocarbons down to $10^{-5}\%$). In the method of the so-called gas-carottage, the borehole, the plug and the mud are tested for hydrocarbons in a similar manner (cf. SOKOLOV, 1947; KUSNETSOV, 1966).

Radioactive substances, upon decaying, release gaseous products such as helium, radon, thoron, argon, etc. Since they frequently escape through faults, a wide range of possibilities for prospecting is given which complement the direct radiometric close-range methods and sometimes are superior to them.

C. *Heavy Mineral Analysis*

The method of detritus analysis has been perfected in an excellent manner in the Soviet Union. Autochthonous and allochthonous sediments (weathering horizons, river deposits, etc.) are examined by means of mineralogic (phase-analytical) and geochemical methods and used in the exploration for soluble and inert minerals. This analysis combines screen analyses, microscopic examinations and chemical analyses. It is used for large-scale prospecting for deposits.

Bibliography

General Works and Synopses (Including Terminology and Organisation)

BOBROVA, M. M., and N. D. BOBROVA: The Organisation of Laboratory Work in the Central Asian Geophysical Trust. All-union Deliberation on Geochemical Methods for the Exploration of Ore Deposits. Gosgeoltekhizdat, Moscow 1956 (Russian)

Burek, J.: Grundlagen der geochemischen Prospektion. Z. angew. Geol. *4* (1958) 9, 415—422
Granier, C.: La Terminologie des Méthodes de Prospection Géochimique. Bull. Soc. franc. Minér Crist. *LXXXV* (1962), 11—14
Hawkes, H. E.: Geochemical Prospecting; in Abelson, P. H.: Researches in Geochemistry. Wiley New York 1959
Hawkes, H. E., and S. Webb: Geochemistry in Mineral Exploration. Harper and Son, New York 1962, 415 pp.
Heltzen, A. M.: Geokjemisk Prospektering. Tidsskr. Kjemi, Bergvesen Met. *16* (1956), 19
Rösler, H. J.: Die jüngste Entwicklung und der Stand der geochemischen Prospektionsarbeiten au Buntmetalle. Freiberger Forschungsh. C *162* (1963), 61—71
Rosenquist, A. M., and T. Vogt: Geokjemisk ok Geobotanisk Malmleting, Part V, Kgl. Norske Videns kab. Selskabs Forh. *15* (1942) 22
Shatrov, V. V.: Geochemical Methods of Prospecting. Guide to the Complex of Methods for the Exploration of Non-ferrous and Rare Metals in the Eastern Regions of the USSR. WITR, Gosgeoltekhizdat, Moscow 1961 (Russian)
Smirnov, V. I.: Symposium de Exploración Geoquímica. Tomo 1. Mexiko. 1958, XV, 234 pp. (Congrès Geologique International XX, Mexico 1956)
Solovov, A. P.: Instructions for the Metallometric Survey. Gosgeoltekhizdat, Moscow 1957 (Russian)
Solovov, A. P.: Fundamentals of the Theory and Practice of Metallometry. Alma-Ata 1950 (Russian)
Stevens, N. P.: Geochemical Prospecting in the United States. 3rd International Scientific Conference for Geochemistry, Microbiology and Petrochemistry, Budapest 1963
Webb, J. S.: Review of American Progressive Geochemical Prospecting and Recommendation for Future British Work in this Field. Bull. Inst. Min. and Met. (1953) 55—57
Williams: Organisation of Geochemical Prospecting in the British Empire Bull. Inst. Min. Metall, London (1954) 576

Regional Prospecting

Bernstein, K.-H.: Geochemische Prospektion auf Schwerspatgängen im Raum Warmbad Wolkenstein (Erzgeb.). Z. angew. Geol. *6* (1960), 277—279
Boyle, R. V.: Geochemical Prospecting in the Yukon. Canadian Mining Journal *76* (1955), No. 6
Chew, R. T. III, and H. T. Mitten: Geochemical Prospecting in northwest Montana. Geol. Soc. Amer., Spec. Pap. *68* (1961), 85—86
Dvornikov, A. G.: The Types of Dispersion Aureoles of Mercury in the Southeastern Part of the Donbass. Dokl. Akad. Nauk SSSR *154* (1964) 5, 1110—1112 (Russian)
Giusca, D.: La Prospection Géochimique des Gisements de Molybdenite des Monts de Drocea. Rev. Geol. Geogr., Bucuresti *3* (1959) 2, 201—207
Glasovskaya, M. A.: The Geochemistry of the Landscapes and the Exploration of Mineral Raw Materials in the South Ural. Izd. MGU, Moscow-Ufa 1960 (Russian)
Ivantishin, M. N., and B. F. Mitskevich: Geochemical Methods of Prospecting Useful Deposits and their Future Use in the Ukraine. Geol. Jurnal Akad. nauk Ukraine, Kiev *22* (1962) 2, 19—28
Kvalheim, A. (Editor): Geochemical Prospecting in Fennoscandia. John Wiley and Sons, Interscience Publishers, New York/London/Sydney 1967
Leutwein, F., and L. Pfeiffer: Ergebnisse und Anwendungsmöglichkeiten geochemischer Prospektionsmethoden auf hydrosilikatische Nickelerze. Geologie *3* (1954) 6/7, 950—1008
Martinelli, J. A., and J. V. Nogueira Filho: Geochemical Prospecting for Copper in Rio Grande do Sul. Bol. Soc. Brasil. Geol. 8, São Paulo *1* (1959), 5—15 (Portuguese)
Mazzucchelli, R. H., and C. H. James: Arsenic as a Guide to Gold Mineralization in Laterite-covered Areas of Western Australia. Inst. Mining Met. Trans. *76* (1967) Sect. B. Bull. No. 732, 218
Michael, J., and W. Schrön: Pedogeochemische Prospektion auf Kupfer in hydrothermal umgewandelten Rotliegendporphyriten Thüringens. Bergakademie (1964) 1, 2—9
Müller-Kahle, E.: Geochemische Lagerstättenprospektion in der Provinz Mendoza, Argentinien. Z. Erzbergbau Metallhüttenwesen, *19* (1966) 7, 330—339

OZEROVA, N. A.: On the Use of Primary Hg Dispersion Aureoles for the Exploration of Pb-Zn Deposits. Geokhimiya (1959) 7, 638—643 (Russian)

PETRASCHEK, W. E.: Die Anwendung neuerer Prospektionsmethoden im österreichischen Erzbergbau. Z. Erzbergbau Metallhüttenwesen 20 (1967) 7, 295—300

POSKOTIN, D. L.: Geochemical Prospecting for Non-ferrous Metals with Special Geologic Mapping in the District of Pyshminsk (Central Ural). Trudy Sverdlovsk. gorn. Inst. 38 (1961), 94—105 (Russian)

WILLIAMS, X. K.: Geochemical Prospekting for Copper, Nickel and Zinc in the Longwood Range, Southland, New Zealand, N. Z. J. Geol. Geophys. 10 (1967) 3, 742—758

WILLIAMS, X. K.: Chemical Variations in Stream Waters and Sediments in the Make Creek Area, West Otago. N. Z. J. Geol. Geophys. 10 (1967) 3, 759—770

WILLIAMS, X. K., and B. L. WOOD: Geochemical Prospecting for Copper, West Otago. N. Z. J. Geol. Geophys. 10 (1967) 3, 798—830

Migration, Dispersion, Aureole Formation, Geochemical Maps

GINSBURG, I. I., and K. M. MUKANOV: Main Principles Governing the Preparation of Geochemical Maps of Ore Districts for Metallogenetic Investigations. Omnibus volume "Metallogenetic and Prognostic Maps". Alma-Ata 1959 (Russian)

GRIGORYAN, V. V., and E. M. YANISHEVSKIĬ: Über die Hauptkriterien zur Bewertung geochemischer Anomalien beim Aufsuchen verborgener Vererzung. Z. angew. Geol. 12 (1966) 12, 630—633

KADYROV, G. F.: Contribution to the Problem of Geochemical Mapping. Geokhimiya (1965) 11, 1373—1376 (Russian)

KALITA, E. D.: On the Question of the Dispersion Aureoles of Li, Rb, and Be. Mat. geol. rud. mest., petr., miner., geoch., Akad. Nauk SSSR (1959), 205—211 (Russian)

KAPKOV, YU. N., et al.: The Main Principles Governing the Drawing up of Geochemical Maps to the Scale of 1:50,000. Zap. gorn. inst., Leningrad 51 (1966) 2, 3—10 (Russian)

KAZMIN, V. N., and I. V. ORLOV: Basic Principles of the Preparation of Geochemical Maps in Geologic Mapping. Soviet geologiya 9 (1966) 6, 90—100 (Russian)

KONSTANTINOV, V. M., and E. M. YANISHEVSKIĬ: On the Use of Primary Dispersion Aureoles for the Evaluation of the Presene of Ore in Disjunctive Faults. Geologiya rudnykh mestorozhdenii, Moscow 5 (1963) 2, 126—127 (Russian)

KRASNIKOV, V. I.: Auffinden tiefliegender Erzlagerstätten durch Untersuchung der primären Dispersionshöfe. Z. angew. Geol. 5 (1959) 11, 533—535

KUKAROV, K. M.: On a Few Peculiarities of the Formation of Primary Dispersion Aureoles of Ore Deposits. Omnibus volume: "Materials Regarding the Ore Deposit Geology, Petrography, Mineralogy, and Geochemistry". Izd. Akad. nauk SSSR, Moscow 1959 (Russian)

RONOV, A. B., and A. I. YERMISHKINA: The Method of Drawing Up a Quantitative Lithological-geochemical Map. Akad. nauk SSSR 91 (1953) 5 (Russian)

SAFRONOV, N. I.: On the Theory of the Primary Dispersion Aureoles. Informative omnibus volume of the WITR, No. 27 (1959) (Russian)

SAFRONOV, N. I.: On the Problem of the "Dispersion Aureoles of Deposits of Mineral Raw Materials" and Their Use in Exploration and Prospecting. Probl. Soviet geol. (1934) 4 (Russian)

SAUKOV, A. A.: The Migration of the Chemical Elements as a Basis of Geochemical Prospecting. Omnibus volume "Geological Results of Applied Geochemistry and Geophysics". Gosgeolizdat, Moscow 1960 (Russian)

SHCHERBINA, V. V.: Concentration and Dispersion of the Chemical Elements Due to the Redox Potentials. Dokl. Akad. nauk SSSR 27 (1940) 7 (Russian)

WARREN, H. V., and R. E. DELAVAULT: Pathfinding Elements in Geochemical Prospecting. Congr. Geol. Intern. (XX. Sess.), Vol. 2. Mexico 1959, 255—260

Analytical Methods and Sampling

BLOOM, H.: Field Method for the Determination of Ammonium Citrate-soluble Metals in Soils and Alluvium. Econ. Geol. 50 (1955), 533

HUFF, L. C.: A Sensitive Field Test for Heavy Metals in Water. Econ. Geol. 43 (1948), 675

HUFF, L. C., et al.: A Comparison of Analytical Methods Used in Geochemical Prospecting for Copper. Econ. Geol. *56* (1961) 5, 855—872

HUNT, E. C., R. E. STANTON, and R. A. WELLS: Field Determination of Beryllium in Soils for Geochemical Prospecting. Bull. Inst. Mining and Metallurgy *69* (1960), 361—369

LAKIN, H. U., H. ALMOND, and F. N. WARD: Compilation of Field Methods Used in Geochemical Prospecting by the U. S. Geological Survey. U. S. Geol. Survey Circular *161* (1952)

MARSHALL, N. J.: Rapid Determination of Molybdenum in Geochemical Samples Using Dithiol. Econ. Geol. *59* (1964), 142—148

RODIONOV, D. A., Y. V. PROKHOROV, and V. M. ZOLOTAREV: Average Sample Use in Geochemical Prospecting Methods. Geokhimiya (1965) *6*, 747—756 (Russian)

SOKOLOV, I. YU.: Das halbstationäre Feldlaboratorium GXL-1 (Typ VSEGINGEO, 1960) für hydrochemisches Aufsuchen von Erzlagerstätten. Razv. i ochrana nedr. *28* (1962), 44—45

STANTON, R. E., and M. A. GILBERT: Analytical Procedures Employed at the Geochemical Prospecting Research Center at Imperial College, London. Imper. Coll. Sci. Technol., Geoch. Prosp. Res. Center, Techn. Commun. 1—5, 10 (1957)

WARD, F. N., H. U. LAKIN, F. C. CANNEY et al.: Analytical Methods Used in Geochemical Exploration by the U. S. Geological Survey. U. S. Geol. Surv. Bull. 1152-A (1963)

Group of authors: Methods of Chemical Analysis of Mineral Raw Materials (Vanadium, Tungsten, Copper, Mercury, Lead, Sulphur, Titanium, Zinc, Zirconium, and others). Gosgeoltekhizdat, Moscow 1955 (Russian)

Geochemical Prospecting for Ore Deposits
General

ARISTOV, V. V.: Mineral Dispersion Aureoles as Criteria in Prospecting (with particular reference to the deposits in the Transbaykal region). Trudy moskovskogo geolorazvedochnogo instituta. *33* (1958)

BEUS, A. A., et al.: Geochemical Prospecting for Endogenous Deposits of Rare Elements (Ta). Izd. "Nedra", Moscow 1968, 263 pp.

FERSMAN, A. E.: Geochemical and Mineralogic Methods of Exploring for Minerals; in: Selected Works, Vol. II, Publ. Acad. Sc. USSR, Moscow (Russian)

FUJIHARA, T.: On Geochemical Prospecting of Metallic Ore Deposits. J. Geol. Soc. Japan *57*, 331

GINSBURG, I. I.: The Geochemical Exploration for Hidden Ore Bodies. Omnibus volume "Geological Results of Applied Geochemistry and Geophysics". Gosgeolizdat, Moscow 1960 (Russian)

GINSBURG, I. I.: Wege der geochemischen Sucharbeiten auf verborgene Erzkörper polymetallischer Lagerstätten und Lagerstätten seltener Metalle. Z. angew. Geol. Berlin *6* (1960) 1, 13—14

GINSBURG, I. I.: Geochemische Erkundung von Metallen. Z. angew. Geol. Berlin *6* (1960) 12, 59 to 597

GINSBURG, I. I., et al.: The Use of Geochemical Methods in the Metallogenic Exploration of Ore Regions Izd. "Nedra", Moscow 1966 (Russian)

GINSBURG, I. I.: a) Grundlagen und Verfahren geochemischer Sucharbeiten auf Lagerstätten der Buntmetalle und seltenen Metalle. Akademie-Verlag, Berlin 1963. b) Principles of Geochemical Prospecting. Pergamon Press, London/Oxford 1960

GRANIER, C.: Mise en Evidence de Structures Minéralisées par Prospection Microchimique Tactique. Chron. Mines d'Outre-mer et Rech. Min. *26* (1958) 270

HAWKES, H. E.: Annotated Bibliography of Papers on Geochemical Prospecting for Ores. US Geol. Survey Circ. *28* (1948)

HAWKES, H. E.: Principles of Geochemical Prospecting. United States Government Printing Office, Washington 1957. Geological Survey Bulletin 1000-F

KRASNIKOV, V. I.: Exploration of Deep-seated Ore Deposits by Investigations into the Primary Dispersion Aureoles. Z. angew. Geol., Berlin *5* (1959) 11, 533—535

KRASNIKOV, V. I. (Edit.): Omnibus volume "Geochemical Prospecting for Ore Deposits in the USSR". Gosgeoltekhizdat, Moscow 1957 (Russian)

LOVERING, T. S., V. P. SOKOLOFF, and H. T. MORRIS: Heavy Metals in Altered Rocks over Blind Ore Bodies. East Tintic District, Utah. Econ. Geol. *43* (1948) 5, 384—399

UKASHEV, K. I.: Geochemical Prospecting for Elements in the Zone of Hypergenesis, Vol. 1: Theoreti-
al Fundamentals of Geochemical Prospecting. Vol. 2: Geochemical Prospecting for Individual
lements. Izd. "Nauka i tekhnika", Minsk 1967 (Russian)

AZOR, E.: The Rare Gases as Geochemical Tracers. Israel J. Earth-Sci. *13* (1965) 187, 3—4

ERELMAN, A. I., and A. A. SAUKOV: The Geochemical Fundamentals of Prospecting for Ore Deposits.
: "Geochemical Ore Deposit Prospecting". (edited by V. I. KRASNIKOV), Gosgeoltekhizdat, Moscow
)57 (Russian)

OLIKARPOCHKIN, V. V.: The Geochemical Prospecting for Ore Deposits on the Basis of Dispersion
ureoles. Soviet Geol. (1962) 4 (Russian)

ÖSLER, H. J.: Möglichkeiten und Perspektiven geochemischer Erkundungsarbeiten in Mitteleuropa.
er. Geol. Ges. DDR *10* (1961) 4, 400—407

ATOLOKINA, V. A.: The Method of Prospecting and Exploring Polymetallic Ore Bodies in Carbonatic
eries which Do not Appear at the Surface (with particular reference to the Achisay deposit). Soviet
eol. *53* (1956) (Russian)

AUKOV, A. A.: Geochemical Indicators for Deep-seated Deposits. Z. angew. Geol. *5* (1959) 8, 355
) 356

ERGEEV, Y. A.: The Method of Mercurometric Investigations. In: "Geochemical Ore Deposit Pro-
pecting in the USSR" (edited by W. I. KRASNIKOV). Gosgeoltekhizdat, Moscow 1957 (Russian)

HACKLETTE, H. T.: Copper Mosses as Indicators of Metal Concentrations. Contribution to Geochemical
rospecting for Minerals. G.; Geol. Surv. Bull. 1198-G.; 18 p., Washington 1967

OLOVOV, A. P.: Theory and Practice of Metallometric Prospecting. In: "Geochemical Ore Deposit
rospecting in the USSR" (edited by W. I. KRASNIKOV). Gosgeoltekhizdat, Moscow 1957 (Russian)

OLOVOV, A. P., and N. YA. KUNIN: Metallometric Mapping According to the Dispersion Flows in
Iountainous Regions. Soviet. Geol. *5* (1960), 32—46 (Russian)

UNDBERG, R.: Moderna malmletningsmetoder. Jernkontorets Ann. *121* (1937)

ooms, J. S., and J. S. WEBB: Geochemical Prospecting Investigation in the Northern Rhodesian
opperbelt. Econ. Geol. *56* (1961) 4, 815—846

USHKO, S. A.: The Use of Chemical Reactions in the Exploration for Mineral Raw Materials. Gosgeol-
ekhizdat, Moscow 1953 (Russian)

roup of authors: Geochemical Methods of Exploration for Ore Deposits (Bibliography from 1928
o 1963); Vol. 1: General Problems of Lithogeochemical Exploration. Vol. 2: Hydrogeochemistry,
Biogeochemistry, Complex Methods, etc. Izd. "Nedra", Moscow 1966 and 1967 (Russian)

Particular Geochemical Prospecting

ALBOV, M. N., and P. P. ZELOBOV: On the Method of Prospecting for Hidden Gravel Deposits in the
Jral. Razvedka i okhrana nedr. *2* (1960), 43—45 (Russian)

OUKHOVSKIĬ, A. A., and M. G. ILYAEV: Flows of Dispersion Close by a Few Tin Deposits in the Sichote-
Alin Mountains and Their Use in Prospecting. Inform. Sbornik wsesoyuzn. nauchno-issled. geol.
nst. *34* (1960), 93—102 (Russian)

UTT, G. N.: Geochemical Prospecting for Nickel Ores at Kwatha and Nampesha-Humine Areas,
Manipur State. Indian Minerals, *14* (1960), 246—258

OVORNIKOV, A. G.: Mercury Aureoles in Limestone of Polymetallic Deposits in the Graskaya Ravine
Nagolny Mountains). Geokhimiya (1962) *5*, 464—470 (Russian)

LEROV, B. L.: The Use of the Stannometric Analysis for the Prospecting of Primary Tin Deposits.
n: "The Rare Metals", No. 1 (1935) (Russian)

RUTH, I.: Anwendungsmöglichkeiten geochemischer Prospektion im Bereich der Grube "Bayerland".
hemie der Erde *XXI* 48—96

GINSBURG, I. I., K. M. MUKANOV, and N. P. POLUSEROV: Copper and Lead in the Soils of the Uspenskiy
Tin Deposits (Central Kazakh S.S.R.). Geokhimiya (1960), 339—344 (Russian)

GORSHEVSKIĬ, D. I., and G. I. ROSSMAN: Primary Dispersion Aureoles of Polymetallic Ores in Rudny
Altai. In: Omnibus volume: Probl. Geokhim. *1* (1959), 184—189, Izd. Univ. Lwow (Russian)

GRANIER, C.: Aperçu sur la Prospection Géochimique du Plomb et du Zinc. Bull. Soc. franç. Miner
t Crist *78* (1960), 209—215

GRIGOREV, A. M.: On Geochemical Methods of Prospecting for Gold Deposits. Sbornik geol -proizwod.
Inform. burjat. geol. Uprawlen. *1* (1959), 15—21 (Russian)

HILL, V. G.: Geochemical Prospecting for Nickel in the Blue Mountain Area, Jamaica, W. I. Econ. Geol. *56* (1961) 6, 1025—1032

IVASHOV, P. V.: On Mercury Prospecting in Mountainous Taiga Regions. Rasvedka i okhrana Nedr. *6* (1961), 38—39 (Russian)

KAŠPAR, J., and R. KRÁL: Second Contribution to the Mercury Geochemistry of the Prešov Mountains. Sbornik Vys. školy chem. technol., Prague 1961, 19—29 (Czech)

KOROLEV, A. V., and P. A. SHEKHTMAN: Slimes and Plumbometric Analysis in the Prospecting and Evaluation of Lead Deposits. Razvedka nedr. (1952) 1 (Russian)

KRAFT, M.: Ein Beitrag zum Nachweis primärer Dispersionshöfe am Beispiel einer Zinnerkundungsbohrung im Raum Mühlleiten (Erzgebirge). Z. angew. Geologie, Berlin *8* (1962) 12, 619—623

KVYATKOVSKIĬ, YE. M., and I. N. KRIZUK: On Metallometric Mapping in Prospecting for Tin Deposits. Sapiski Leningrad, Gorn. Inst. *39* (1961) 2, 129—135 (Russian)

LITINSKIĬ, V. A.: On the Contents of Ni, Cr, Ti, Nb and a Few Other Elements in Kimberlites and on the Possibility of Using Geochemical Methods in Prospecting for Kimberlitic Bodies. Izd. Akad. Nauk Kazakh., SSR, Alma-Ata 1961, 104 pp.

MILOŠAKOVIĆ, R.: Cuprometric Investigations into the Hydrothermally Transformed Andesites of the Timok Eruptive Region. Vesnik Zav. Geol. i Geofiz. Istraživ. Srbije, Beograd *17* (1959) 1962, 175—188 (Serbo-Croatian)

NIKIFOROV, N. A., and V. P. FEDORCHUK: On the Study and Evaluation of Ore Indicators in the Exploration for Ore Deposits with Rock Superimposed. Trudy sredneaziyat. politekhn. Inst. (1959) 6, 178—190 (Russian)

OSEROV, I. M.: Stannometry as a Method of Prospecting. Razvedka nedr. (1937) 24 (Russian)

OSEROVA, N. A.: On the Use of Primary Dispersion Aureoles of Mercury in the Prospecting for Lead-Tin Deposits. Geokhimiya (1959) 7 (Russian)

OSEROVA, N. A., and A. A. SAUKOV: Mercury as an Indicator in Ore Exploration. In: Omnibus volume: Geol. Resul'taty prikladn. Geokhim. i Geofiz. 1, Gosgeoltekhizdat, Moscow 1960 (Russian)

PÁCAL, Z.: Geochemical Prospecting in Deep-seated Greisens. In: Omnibus volume of the Geol. Central Institute. Sbornik *26* (1959), 373—396

SHILO, N. A., and I. P. KARTASHOV: Use of Geochemical Methods in Prospecting for Gold Deposits in the North-East of the U.S.S.R. Izd. Akad. Nauk SSSR, Moscow 1959 (Russian)

VALYASKO, M. G., et al.: Geochemical Methods in the Exploration for Potash Deposits. Izd. Moskovsk. Univ. 1966, 22, 62—70 (Russian)

WAMBEKE, L. VAN: Geochemical Prospecting and Appraisal of Niobium-bearing Carbonatites by X-ray Methods. Econ. Geol. *55* (1960), 732—758

WERSHKOVSKAYA, I. V.: Primary Mercury Dispersion Aureoles as Prospecting Guide for Hg-Sb Deposits. Razvedka i Odrana nedr. (1956) 4 (Russian)

Hydrogeochemical Prospecting

ANTONOV, A. A.: On the Hydrochemical Method of Prospecting for Copper-Nickel Sulphide Deposits in the Kola Peninsula. Razvedka i okhrana nedr., Moscow *28* (1962) 3, 15—19 (Russian)

BALYAKOVA, J. J., A. A. RESNIKOV, L. J. KRAMARENKO, A. A. NECHAEVA, and T. F. KRONIDOVA: Hydrogeochemical Methods for Prospecting Ore Deposits. WSEGEI, Leningrad 1962 (Russian)

BELYAKOVA, E. E.: Principles and Methods of Preparing Prognostic Hydrochemical Maps for Ores. Sovetskaya Geologyia *5* (1962) 1, 104—130 (Russian)

BRODSKI, A. A.: Hydrochemical Methods of Prospecting for Ore Deposits. In omnibus volume "Geochemical Prospecting for Ore Deposits in the U.S.S.R. (edited by V. I. KRASNIKOV). Gosgeoltekhizdat, Moscow 1957 (Russian)

BUGELSKIĬ, YU. YU., and V. I. VINOGRADOV: Hydrochemical Methods of Prospecting for Molybdenum and Polymetallic Ores; in omnibus volume: Geol. Resultat v priklad. Geokh. i. Geofiz. Izd. Gosgeol tekhizdat *1* (1960), 64—70, Moscow (Russian)

CHERNYAEV, A. M.: The Assessment of the Hydrogeochemical Conditions of a Region for the Correc Interpretation of Anomalies. Razv. i okhrana nedr., Moscow *28* (1962) 3, 36—41 (Russian)

DALL'AGLIO, M., and F. TONANI: Recording and Assessment of Hydrogeochemical Data. Com. Naz Ric. Nucliari: Studi e Ricerche Della Division Geomineraria III, 353—386, Rome 1960 (Italian)

DALL'AGLIO, M., and F. TONANI: Methods of Hydrogeochemical Prospecting. Com. Naz. Ric. Nucliari: Studi e Ricerche Della Division Geomineraria III, 299—327, Rome 1960 (Italian)

DE GEOFFROY, J., S. M. WU, and R. W. HEINS: Geochemical Coverage by Spring Sampling Method in the South West Wisconsin Zinc Area. Econ. Geol 62 (1967) 5, 671—697

GLUSKOV, V. G.: Questions of Theory and Methods of Hydrogeologic Investigations. Izd. Akad. Nauk SSSR, Moscow 1961, 415 pp. (Russian)

KRAĬNOV, S. R., and M. CH. KROL'KOVA: Main Principles Governing the Use of Ore-forming Elements in Carbon Dioxide Waters as an Aid in the Exploration for Ore Deposits, with Particular Reference to the Central Part of the Minor Caucasus. Geokhimiya, (1962) 5, 453—463 (Russian)

OVCHINNIKOV, A. M.: The Basic Questions Regarding the Use of Hydrologic Criteria in the Prospecting of Mineral Deposits. Voprosy teoreticheskoĭ i prikladnoĭ Geologii. Sbornik MGRI 2 (1947) (Russian)

SOKOLOV, I. YU. (Edit.): Methodic Guide to the Determination of Micro-components in Natural Waters for Prospecting Ore Deposits. Izd. Gosgeoltekhizdat, Moscow 1961, 287 pp. (Russian)

Biogeochemical and Geobotanic Prospecting

BOTOVA, M. M.: Attempt to Use a Biogeochemical Method of Prospecting for Uranium under Desert Conditions. Geokhimiya (1963) 4, 361—369 (Russian)

KHAMRABAYEV, I. CH., and R. M. TALIPOV: A Few Results of Biogeochemical (Geobotanic) Investigations in the West Uzbek S.S.R. Uzbek. Geol. Zhurn. (1960) 5, 3—11 (Russian)

EPSTEIN, J. F.: The Flora-geochemical Method (Florae Metallography) for the Prospecting of Mineral Deposits. Inf. of the "Artyom" Mining Institute in Dnepropetrovsk, 20 (1948) (Russian)

GAWEL, A.: Memorial by M. Borch, the First (1780) Trial in Poland of Geo-biochemical Methods for Ore Research. Praze Muzeum Ziemi, Warsaw (1966) 8

KOVALEVSKIĬ, A. L.: A Few Problems of Theory and Practice of Biogeochemical Methods of Prospecting. Geol. Geof. Novosibirsk 6 (1963), 68—77 (Russian)

LINSTOV, O.: Bodenanzeigende Pflanzen. Abh. preuß. geol. Landesanstalt, N. F., No. 114, Berlin 1929

LUKICHEVA, A. N.: The Vegetation Cover as an Indicator of Kimberlite Channels. Geologiya i geofizika 11 (1960), 35—48 (Russian)

MAKAROVA, A. I.: Biogeochemical Investigations on a Polymetallic Deposit. Geokhimyia (1960), 624 to 633 (Russian)

MALYUGA, D. P., et al.: The Use of the Biogeochemical Method of Prospecting under Alpine Conditions. Geokhimiya (1960) 4, 330—338 (Russian)

MARMO, V.: Biogeochemical Prospecting of Ores in Finland. Econ. Geol. 48 (1953), 215

McGAVOCK, E. H.: An Evaluation of Biogeochemical Prospecting for Zinc in the Shenandoah Valley, Virginia. Virginia Journ. Sci. 13 (1962), 284—285

MOSKALYOV, P.: Luminescent Bitumenological Mapping in Solid Rock. Vestnik MGU (State Univ. Moscow), 3 (1953) (Russian)

OSBERGER, R.: Die geobiochemische Prospektionsmethode. Berg- und Hüttenmänn. Mh. Montan. Hochschule Leoben 98 (1953) 9, 197—202

POSKOTIN, D. L., and M. V. LYUBIMOVA: The Use of the Biogeochemical Method of Prospecting for Copper Pyrite Deposits. Geokhimiya (1963) 6, 603—611 (Russian)

SALMI, M.: Prospecting for Bog-covered Ore by Means of Peat Investigation. Bull. Comm. Geol. Finlande 169 (1955)

TKALICH, S. M.: Practical Guide to the Biogeochemical Method of Prospecting for Ore Deposits. Gosgeoltekhizdat, Moscow 1959 (Russian)

TKALICH, S. M.: The Botanic Method of Prospecting for Ore Deposits, 1954 (Russian)

VOGT, T., and H. BERGH: Geokjemisk og geobotanisk malmleting Kgl. Norske Videnskab. Selskabs Forh. 19 (1947) 21; 20 (1948) 20; 20 (1948) 27

VINOGRADOV, A. P.: Geochemical Prospecting for Ores on the Basis of the Metal Concentrations in Plants and Soils. Trudy Biogeokhim. Labor, Akad. Nauk SSSR 10 (1954), 3—27 (Russian)

VOSTOKOVA, E. A.: Geobotanic Methods of Prospecting for Subsurface Waters in the Arid Regions of the Soviet Union. Gosgeoltekhizdat, Moscow 1961, 87 pp. (Russian)

WORTHINGTON, J. E.: Biogeochemical Prospecting at the Shawangunk Mine. Econ. Geol. *50* (1955). 420—429

Prospecting for Petroleum and Natural Gas

GODARD, J. M., O. ISSENMAN, and REBILLY: Développements Nouveaux de la Prospection Géochimique de Surface. Méthode Directe de Recherche Petrolière. Pétrole Informations, 5 Juillet, 1965

KARTSEV, A. A., et al.: Geochemical Methods of Prospecting and Exploration for Oil and Gas Deposits. Gosgeoltekhizdat, Moscow 1954 (Russian)

KARTSEV, A. A., Z. A. TABASARASUKIĬ, M. S. SUBBOTA, and G. A. MOGILEVSKIĬ: Geochemical Methods of Prospecting and Exploration for Petroleum and Natural Gas. Univ. Calif. Press., Berkeley and Los Angeles 1959

MOGILEVSKIĬ, G. A.: Serviceability of Gas Measurements in Prospecting. Internat. Geol. Rev. *3* (1961) 11, 1068—1075

MÜLLER, E. P., and W. FREUND: Die Gaschromatographie bei der Erdgas- und Erdölprospektion und -förderung. Z. angew. Geol. *8* (1962) 6, 304—307

NEGLIA, S., I. SALVADORI, and P. ZUFFARDI: The Geochemical Prospecting for Hydrocarbons. "La scuola in azione". Varese 1961, 1—136 (Italian)

PIRSON, S. G.: New Concepts in Geochemical Prospecting for Petroleum Oil in Canada. *13* (1961), 41

POMEYEROL, R., F. BIENNER, and M. LOUIS: Examples of Geochemical Prospecting by Means of Gas Analyses from Surface Samples in the Basin of Fort Polignac. Revue Inst. France. Petrole *16* (1961) 7/8, 868—874 (French)

SOKOLOV, V. A.: On the Theory and Method of Gas Surveying. In: Geochemical Methods of Prospecting for Petroleum. Moscow/Leningrad 1950 (Russian)

STEGENA, L.: On the Principles of Geochemical Oil Prospecting. Geophysics Tulsa *26* (1961) 4, 447 —451

Group of authors: Geochemical Methods of Prospecting for Oil and Gas. No. 1 and 2, Gostoptekhizdat (1954) (Russian)

Group of authors: Geochemical Methods of Prospecting for Oil and Natural Gas. Geochem. News *31* (1962), 7

Radiometry, Well Geochemistry

MOGILEVSKIĬ, G. A.: Gas Surveying after Sinking Wells. Razv. i ochrana nedr., Moscow *28* (1962) (Russian)

ROTHE, K.: Radiometrische Bestimmungen an Mineralien und Gesteinen. Freiberger Forschungsh. C *68* (1958)

SPANDERASHVILLI, G. J., and F. I. RUBAKHA: Experiences in the Use of Lithochemical and Radiometric Methods of Prospecting for Phosphorites in Gorny Shory. Razv. i okhrana nedr. *28* (1962), 20—24 (Russian)

Group of authors: Field and Well Geochemistry; omnibus volume No. 1, Gostoptekhikdat, Moscow/ Leningrad 1953 (Russian)

Group of authors: Geochemical and Radioactive Methods of Prospecting and Exploration. Zap. Gorn. Inst. *51*, No. 2, Izd. "Nedra", Leningrad 1966 (Russian)

10.2. Geochemistry of Mineral Deposits

Geochemical investigations into deposits of every description are so numerous and versatile that only a brief information of the main trends can be given here.

In pursuit of their main aim, the forecasting of the locations and extent of deposits, geoscientists avail themselves of geologic, geophysical and above all geochemical experiences and methods. Deposits are not brought about by chance, but are the result of a geologic-geo-chemical development. Therefore, special attention must be paid to the relation between geological factors such as orogenesis, magmatism, erosion and sedimentation, and to the substantial geochemical differentiations in question. These efforts are characterized by such concepts as metallogenesis (minerogenesis), metallogenetic and geochemical provinces, magmatic and sedimentary geochemical specialisation, ore-controlling factors, regional geology and geochemistry, etc. The knowledge gathered in these fields enables us to forecast the occurrence of deposits.

Fig. 118 gives a survey of the most important elements in various types of deposits.

10.2.1. Magmatogenic Ore and Mineral Deposits

The following tasks and problems have to be approached from the geochemical point of view:

1. Geochemical relation between magmatic rock and deposit, especially on the basis of trace elements and isotopes (S, O, C) in the rock and in rock-forming minerals.
2. Age relations between magmatic rock, deposit and country rock (isotopic methods).
3. Distribution of the mineral- and element-paragenetic conditions with respect to space and time (succession of minerals).
4. Distribution of the trace elements, in particular in the minerals of the deposit
   a) for the purpose of determining isothermal zones

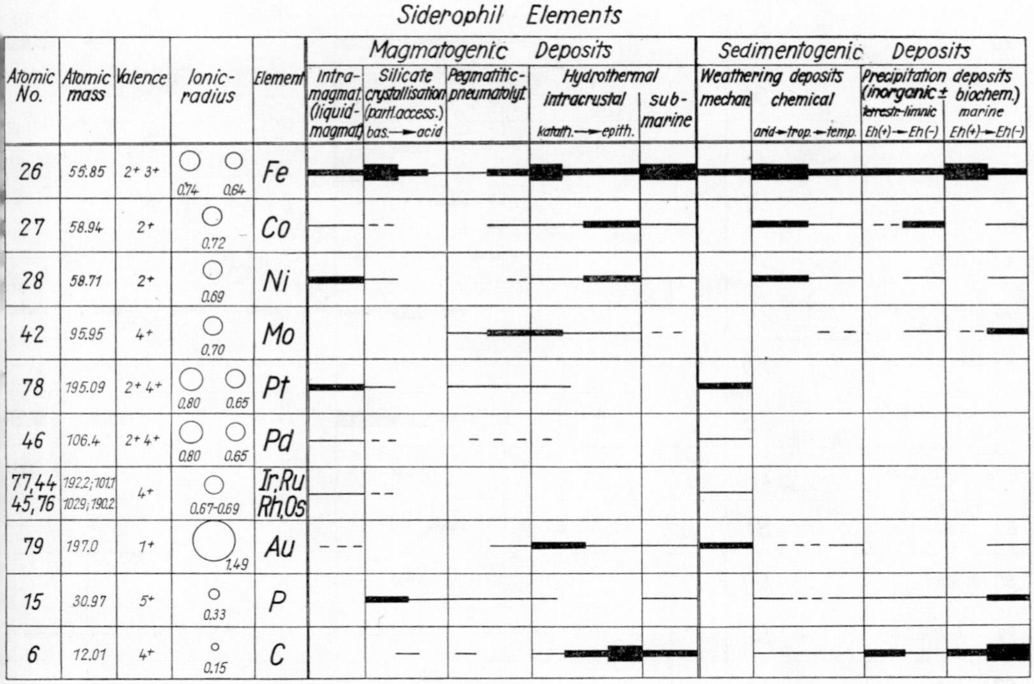

Fig. 118 Geochemical distribution of the most important elements among the main genetic types of deposits (after CISSARZ, compiled by BAUMANN)

b) for the purpose of determining magmatic and postmagmatic displacements of elements

c) to judge their utilization in dressing and in the metallurgical process

5. Investigation of the transformation of country rock and of mineral inclusions (liquid and gaseous) to determine the character of the ore-forming solution.

Tables 174 and 175 show two examples of the application of geochemical parameters to postmagmatic ore formation.

Chalcophil Elements

Atomic No.	Atomic mass	Valence	Ionic radius	Element	Magmatogenic Deposits					Sedimentogenic Deposits		
					Intra-magmat (Liquid-magmat)	Silicate cristallisation (part.access.) bas.→ acid	Pegmatitic-pneumatolyt.	Hydrothermal intracrustal katath.→ epith.	sub-marine	Weathering deposits mechan.	chemical arid→trop.→temp.	Precipitation deposits (inorganic± biochem.) terrestr-limnic Eh(+)→Eh(−) / marine Eh(+)→Eh(−)
16	32.07	2− 6+	1.85 / 0.29	S								
29	63.54	1+	0.96	Cu								
50	118.7	4+	0.71	Sn								
33	74.91	3+	0.58	As								
30	65.38	2+	0.74	Zn								
31	69.72	3+	0.62	Ga								
32	72.6	4+	0.50	Ge								
48	112.41	2+	1.01	Cd								
49	114.82	3+	0.81	In								
47	107.88	1+	1.31	Ag								
51	121.74	3+	0.76	Sb								
82	207.21	2+ 4+	1.28 / 0.84	Pb								
81	204.4	1+	1.57	Ti								
83	209.0	3+	0.96	Bi								
80	200.61	2+	1.14	Hg								
34	78.96	2−	1.91	Se								
52	127.61	2−	2.11	Te								

Fig. 118   continued

Table 174. Geological and Geochemical Features for Assessing the Tin-bearing State of Granitic Rock (after Tischendorf, 1968)

Geologic Phenomenon	Perspective Formation	Non-perspective Formation
Geologic-tectonic position of the granitoid rock	palingenetic granites with intrusive contacts; post-tectonic, post-orogenetic-subsequent; multi-phase, intrusive bodies of complicated structures	migmatites, granitisation phenomena, syntectonic, syn-orogenetic; frequently single-phase, monotonous intrusive bodies
Relative age in orogen	belonging to younger intrusive complexes	belonging to older intrusive complexes
Granite morphology	distinctly ragged, steep stocks and crests, protuberances, granite tongues, flat to medium steep dips of the contact	from insignificantly rugged to non-rugged, smooth course and relatively steep dip of the contact
Type of granitoid rock	biotite-granite, partly bearing muscovite, alaskite-granite, adamellite, tourmaline-granite; acid — ultraacid	granodiorite, granite, partly bearing hornblende, inter-mediately acid
Feldspar Dark mica	Or > Pl; Ab-rich plagioclase protolithionite, Li-bearing sidero-phyllite, Li-bearing lepidomelane	Pl > Or; plagioclase poor in Ab Fe-biotite, Mg-biotite
Accessories	cassiterite, topaz, tourmaline (fluorite)	titanite, allanite, apatite
Chemistry of the granitoid rock	relatively rich in Si, K, Na, F and Li; K > Na	relatively rich in Ca, Mg, Mn, Ti
Sn concentration in the granitoid rock (ppm)	15—50	3—10
Sn concentration in dark mica (ppm)	(100) 150—500	10—100 (150)
Trace element distribution in dark mica	rich in Sn, Ga, F, Li	rich in Pb, Ni, Cu
Trend of tin distribution in the rock	concentration in the most recent granite phases of the complex	partly concentration in the older granite phases of the complex
Autometasomatism	albitisation, microclinisation, muscovitisation, sericitisation	insignificant or none
Dispersion of trace elements, especially of Sn, Li, F	relatively great (more than double that of the non-perspective formation)	relatively small

10.2.2. Sedimentary and Sedimentogenic Ore and Mineral Deposits

Emphasis in geochemical investigations into these deposits is on the geochemical facies analysis (cf. Section 10.4.). This also applies to sedimentogenic (paragenic) deposits in metamorphic rocks. Particular attention is paid to certain ore bodies of parallel strata because of their mostly indistinct genetic situation (skarn deposits, iron deposits, sulphide deposits). Besides microscopic examinations of the ores, investigations into trace elements and isotopes usually are very useful.

Table 175. Chart of Postmagmatic-metasomatic Processes after BEUS & SOBOLEV (1964); stages 3 and 4 can be subsumed under "sequence" (from TISCHENDORF, 1968)

Stage	Precipitation of	Mobilisation or maintenance of mobile state	pH range of the solution about
1. Early microclinisation (early muscovitisation)	K, Rb	Ca, (Na), (Li), (F)	$9 \rightarrow 8$, due to precipitation of K—Rb and concentration of Ca and presence of small amount of Na and F
2. Early albitisation	Na	Ca, Si, (Li), (F)	$8 \rightarrow 6$, due to precipitation of Na and concentration of Ca, Si and F
3. Silicification	Si	Na, K, Li, Ca, Al, F	$6 \rightarrow 4$, due to precipitation of Si and higher concentration of F
4. Topazisation muscovitisation	Si, Al, Ca, Li, F	K, Na	$4 \rightarrow 6$, due to precipitation of F and concentration of Na and K
5. Late albitisation	Na	K	$6 \rightarrow 8$, due to precipitation of Na and concentration of K
6. Late microclinisation	K	—	—

10.2.3. Deposits of Caustobiolites

Geochemical investigations into coal serve the analysis of facies, because several quality features are dependent on it, and the exploration of inorganic major and trace elements (Fe, As, Ge, and others). The most important data have already been given in Section 8.7.
In the exploration of deposits of petroleum and natural gas, geochemical methods are used on an increasing scale. In this field, the geochemical facies analysis is also the most important method used. Genetic problems can be solved by macrochemical and microchemical examinations and by isotopic analyses of the organic matter (C, O, H, S). Table 176 gives information about several hydrogeochemical indicators of the presence of petroleum.

Table 176. Important Hydrochemical Indicators of the Presence of Petroleum (after GOMULKA, 1968)

1. Total mineralisation of waters (in g/l)

2. Geochemical characteristic of the waters according to Shulin (genetic type and group of water on the basis of the predominant anions and cations)

3. Chemical characteristic of waters according to Palmer (salinity, alkalinity)

4. Equivalent coefficients

   a) Na_x/Cl_x = metamorphism coefficient

   b) $SO_4 \cdot 100/Cl_x$ = sulphate coefficient
      (Cl_x = sum of the halogens Cl, Br, I)
      (Na_x = sum of the alkaline earth metals Na, K, Li)

   c) Mg/Ca = dolomitization coefficient

   d) $\dfrac{\text{mineralisation of water in mg/l}}{\text{depth of the water-bearing horizon in } m}$ = coefficient of the "hydrogeologic density" (after Gatalski)

e) $$\frac{\text{Br in mg} \cdot 100}{\text{depth of the water-bearing horizon in } m} = \text{bromine coefficient}$$

. Concentration of bromide, iodide; K, Li and NH_4 salts, and "chloride" milligram-coefficients
Cl/K, Cl/Li, Cl/Br, CL/NH_4 or equivalent coefficients of Na/K and NH_4/SO_4

. Trace elements

Bibliography

Ore Deposits

AFANAS'EV, G. D.: Magmatism and Mineralization. Izv. Akad. Nauk SSSR, Ser. Geol. *10* (1961), 44
—52 (Russian)
ANANTAIYer, G. V., and A. R. VASUDEVA MURTY: Geology and Geochemistry of Kolar Gold Field and
the Origin of Gold-Quartz Veins in Dharwars of Mysore; Proceedings of the Symposium on Upper
Mantle Project, January 1967, Hyderabad (India)
ANGER, G., and H. BORCHERT: Die Bedeutung der Schwefel-Isotope für die Beurteilung lagerstätten-
kundlich-genetischer Fragen. N. Jb. Miner. Abh. *98* (1962) 3, 295—348
BADALOV, S. T.: On the Role of Predominant Components in the Geochemistry of Minor and Rare
Elements of Ore Deposits. Geokhimiya (1965) 9, 1168—1170 (Russian)
BARABANOV (Edit.): Mineralogy and Geochemistry of Wolframite Deposits. Izd. Leningrad. Univ.
967, 239 pp. (Russian)
BAUMANN, L., and H. J. RÖSLER: Zur genetischen Einstufung varistischer und postvaristischer Miner-
alisationen in Mitteleuropa. Bergakademie *19* (1967) 11, 660—664
BEUS, A. A., and Y. P. DIKOV: Geochemistry of Beryllium in the Process of the Endogenous Formation
of Mineral (On the Basis of the Hydrothermal Experiment). Izd. "Nedra", Moscow 1967, 159 pp.
Russian)
BEUS, A. A., and A. A. SITNIN: Geochemical Specialization of Magmatic Complexes and Criteria for the
Exploration of Hidden Deposits. Report 23. Sess. Int. Geol. Congr. Section 6 (Geochem.), Prague
1968
BEUS, A. A.: Geochemistry of the Greisen Deposits and the Regularities in the Distribution of Trace
Elements. Geokhim. tsykly, Moscow (1960), 52—64 (Russian)
BILIBIN, Y. A.: Metallogenetic Provinces and Metallogenetic Epochs. Gosgeoltekhizdat, Moscow 1955
(Russian. There is now an English translation)
BODECHTEL, J., and D. D. KLEMM: Über die lagerstättenkundliche Stellung und chemische Zusammen-
setzung der Bleiwismutspießglanze: ein Beitrag zur Geochemie des Wismuts. Geol. Rdsch. *55* (1965) 2,
418—427
BORCHERT, H.: Der Wert gesteins- und lagerstättengenetischer Forschung für die Geologie und Roh-
stoffnutzung. Ber. Deusch. Ges. Geol. Wiss. B, *13* (1968) 1, 65—116
BORODIN, L. S.: Die Seltenen Erden als geochemischer Indikator bei der Lösung von Fragen der endo-
genen Mineralbildung. Z. angew. Geol. *13* (1967) 1, 9—16
BOYLE, R. V.: The Geochemistry of Silver and Its Deposits. Geol. Surv. Canada, Bull. *160* (1968),
264
BRAUN, H.: Zur Entstehung der marin-sedimentären Eisenerze. Clausth. Hefte Lagerstättenkunde
Geoch. miner. Rohstoffe *2* (1964), 130 pp.
ČADEK, J., M. MALKOVSKY, and ŠULCEK: Geochemical Significance of Subsurface Waters for the Accu-
mulation of Ore Components. Report 23. Sess. Int. Geol. Congr. Section 6 (Geochem.), Prague 1968
CAMBEL, B., and J. JARKOVSKY: Geochemistry of Nickel and Cobalt in Pyrrhotines of Different Genetic
Types. Report 23. Sess. Int. Geol. Congr. Section 6 (Geochem.), Prague 1968
CHAKRABARTI, A. K.: On the Trace Element Geochemistry of Zawar Sulphides and Its Relation to
Metallogenesis. Can. Mineral. *9* (1967) 2, 258—262
CISSARZ, A.: Die Metallverteilung in einem Profil des Mansfelder Kupferschiefers. Zbl. Mineralog.,
Abt. A (1929), 425—427

Cissarz, A.: Die durchschnittliche Zusammensetzung des Mansfelder Kupferschiefers. Metall und Erz 27 (1930 [1930b]), 316—319

Davidson, C. E.: On the Origin of some Stratabound Sulfide Ore Deposits. Econ. Geol. 57 (1962), 265—273

Deans, T.: The Kupferschiefer and Associated Mineralization in the Permian of Silesia, Germany and England. Rpt. 18th Int. Geol. Congress 7 (1950), 340—351

Feiser, J.: Die metallischen Rohstoffe (Vol. 17: Nebenmetalle — Cadmium, Gallium, Germanium, Indium, Selen, Tellur, Thallium, Wismut). Ferd. Enke, Stuttgart 1967

Ganeev, I. G., Pachadzhanov, and L. A. Borisenko: On the Geochemistry of Gallium, Tin and a Few Other Elements During the Greisen Formation. Geokhimiya (1961) 9, 757—764 (Russian)

Gehlen, K. v.: Neuere Methoden der Erforschung der Genese von Erzlagerstätten. Erzmetall 18 (1965) 10, 536—543

Gehlen, K. v.: Schwefel-Isotope und die Genese von Erzlagerstätten. Geol. Rdsch. 55 (1965) 1, 178 to 197

Govorov, I. N.: The Geochemical Stages of the Process of Ore-formation of Granitoid Intrusions. Prob. Genez. Rud., Dokl. Soviet Geol. Mezhdun. Geol. Congr. XXII, Probl. 5, Moscow 1964, 50—66 (Russian)

Gundlach, H.: Untersuchungen zur Geochemie des Strontiums auf hydrothermalen Lagerstätten, Geol. Jb. 76 (1959), 637—712

Gundlach, H., and D. Weisser: Zur Geochemie der Barytgänge von Dreisler (östl. Sauerland). Geol. Rdsch. 55 (1965) 2, 375—385

Hawkes, H. E., and J.S. Webb: Geochemistry in Mineral Exploration. Harper and Row, New York 1962

Hawkes, H. E., and J. S. Webb: Geochemistry in Mineral Exploration. Elsevier 1962, 401 pp.

Hegemann, F., and F. Albrecht: Zur Geochemie oxydischer Eisenerze. Chemie d. Erde XVII, 81—103

Hertel, L.: Die Fremdelementführung des Bleiglanzes als Hilfe zur Bestimmung der Bildungstemperatur. Z. Erzbergbau-Metallhüttenwesen 19 (1966) 12, 632—635

Hügi, Th.: Zur Geochemie der Uranvererzungen in den Schweizer Alpen. Geol. Rdsch. 55 (1965) 2, 437—445

Ivanov, V. V.: The Geochemistry of the Disperse Elements Ga, Ge, Cd, In, and Tl in Hydrothermal Deposits. Izd. "Nedra", Moscow 1966, 390 pp. (Russian)

Ivanov, V. V.: The Migration of Thallium in the Endogenous Ore Formation; In: "Geokhim. Sykly". Izd. Gosgeoltekhizdat, Moscow 1960, 121—128 (Russian)

Jackisch, W.: Zur Lagerstättenkunde und Geochemie des Titans, Thesis TH Clausthal 1967

Kaemmel, T.: Die Rolle der Metasomatose im Rahmen der Lagerstättengenese. Ber. Deutsch. Ges. geol. Wiss., B, 13 (1969) 2, 183—192

Kalliokoski, J.: Economic Geology and Geochemistry. Econ. Geol. 63 (1968), 567—571

Kautzsch, E.: Die sedimentären Erzlagerstätten des unteren Zechsteins. Freiberger Forschungsh. C 44 (1958), 14—21

Khambrabaev, I. K.: Petrological and Geochemical Criteria of Metal Presence in Magmatic Complexes. Internat. Geol. Congr., 22nd Sess., India 1964, Vol. Abstracts, 234

Khitarov, N. I., and Y. P. Mulikovskaya: On the Geochemistry of Mine Waters of Sulphidic Deposits. Probl. Sov. Geol. (1935) 8 (Russian)

Kholodov, V. N., Yu. E. Baranov et al.: Genetic Types of Sedimentary Deposits of Rare Elements and of the Climatic Zones; in: Geochemistry of Sedimentary Rocks and Ores. Izd. "Nauka", Moscow 1968 (Russian)

Klemm, D. D.: Bemerkungen zur quantitativen Messung von Diffusionsvorgängen in Erzlagerstätten mit Hilfe der Elektronenmikrosonde. Geol. Rdsch. 55 (1965) 2, 359—365

Knitzschke, G.: Vererzung, Hauptmetalle und Spurenelemente des Kupferschiefers in der Sangerhäuser und Mansfelder Mulde. Z. angew. Geol. 7 (1961 [1961b]) 7, 349—356

Korzhinskiĭ, D. S.: The Acidity Conditions Governing Postmagmatic Products. Izv. Akad. Nauk SSSR, Geol. Ser. 12 (1957), 3—12 (Russian)

Korzhinskiĭ, D. S.: Hydrothermal Acidic-alkaline Differentiation. Dokl. Akad. Nauk SSSR 122 (1958) 2, 267—270 (Russian)

Krauskopf, K. B.: Sedimentary Deposits of Rare Metals. Econ. Geol. 59 (1955), 411—463

Kutina, J. (Edit.): Symposium: Problems of Postmagmatic Ore Deposits (with special references to the geochemistry of ore veins). Geol. Surv. Czechoslov., Prague 1963

Lafitte, P., F. Permingeat, and P. Routhier: Cartographie Métallogénique, Métallotecte et Géochimie Régionale. Bull. Soc. franç. Miner. et Crist. 88 (1965) 1, 3—6

Landergren, S.: On the Geochemistry of Swedish Iron Ores. Arsb. Sver. Geol. Unders., C 496 (1948b) 42, 182 pp.

Leutwein, F.: Die Zusammensetzung der Wolframite und ihre lagerstättenkundliche Bedeutung. Acta Geol., Akad. Sci. Hung., Budapest 1952

Leutwein, F., C. Levy, and H. G. Oehlschlegel: Etude Géochimique du Gisement d'Étain et de Tungstène de Montbelleux. Bull. B.R.G.M. 6 (1965), 97—110

Polge, B.: Contribution à l'Etude Métallogénique et Géochimique de la Région de Melle (Deux Sèvres). Sci. Terre 11 (1966) 2, 127—162

Preuss, E., and H. Ziehr: Zur Verbreitung des Quecksilbers in ostbayerischen Flußspatlagerstätten. Geol. Rdsch. 55 (1965) 2, 400—414

Roedder, E.: Report on S.E.G. Symposium on the Chemistry of the Ore-forming Fluids. Econ. Geol. 60 (1965), 1380—1403

Rösler, H. J.: Kriterien und Methoden zur genetischen Untersuchung varistischer und postvaristischer Lagerstätten. Freib. Forsch.-H. C 209 (1967), 7—13

Rouhunkoski, P.: On the Geology and Geochemistry of the Vihanti Zinc Ore Deposit, Finland. Bull. Comm. Geol. Finlande N. 236, 1968

Routhier, P.: Les Gisements Métallifères. Vol. 1 et 2; Masson et Cie, Paris 1963

Savul, M.: Le Vanadium dans les Minérais de Fer. Commun. Acad. Rep. Pop. Rom. 1 (1951), 77 to 83

Schreiter, W.: Seltene Metalle. VEB Deutscher Verlag für Grundstoffindustrie, Leipzig Vol. I (1963), Vol. II (1961), Vol. III (1962)

Schröcke, H.: ZurGeochemie erzgebirgischer Zinnerzlagerstätten. Neues Jb. Mineralog., Abh. 87 (1955) 3, 416—456

Schroll, E.: Die ostalpine Vererzung im Lichte der geochemischen Forschung. Mitt. Österr. Mineralog. Ges., No. 117, in: Schweiz. mineralog. petrogr. Mitt. 6 (1958) 4, 409—411

Semenov, E. I.: Lanthanides. Geochemistry, Mineralogy and Genetic Types of Deposits of Rare Elements. Geokhimiya redkikh elementov, Vol. 1, Izd. Akad. Nauk SSSR 1964 (Russian)

Semenov, E. I., and A. A. Smyslov: Geochemical Criteria of the Relation between Industrial Mineralization and Rocks, and the Problem of the Origin of Mineralization. Zapiski Vsesoj. min. obsc. Leningrad, Series II, 96, Ed. 5 (Russian)

Shcherbakov, D. I., and F. I. Volfson (Editor): Geology of Hydrothermal Uranium Deposits. Izd. "Nauka", Moscow 1966 (Russian)

Shterenberg, L. E., I. V. Belova, et al.: Ga, Ge, Pb, Zn, W and Mo in Manganese Ore of Čiatura and Other Deposits in the Georgian S.S.R. Dokl. Akad. Nauk SSSR 173 (1967), 187—189 (Russian)

Stumpfel, E. F.: Einige neue Ergebnisse zur Geochemie der Zinnlagerstätten. N. Jb. Miner. 1963, 88—95

Smirnov, V. I.: The Source of Ore-forming Material. Econ. Geol. 63 (1968), 380—389

Svensenikov, G. B.: Electrochemical Processes in Sulphidic Deposits. Izd. Leningrad Univ., Leningrad 1967 (Russian)

Szádeczky-Kardoss, E.: Zur Verteilung der Elemente in den sedimentären und magmatischen Sulfiderzen. Freiberger Forschungsh., C 79 (1960), 106—125

Tauson, L. V., V. D. Kozlov, and M. I. Kuz'min: Geochemical Criteria of Potential Ore-bearing in Granite Intrusions. Report 23. Sess. Int. Geol. Congr. Section 6 (Geochem.), Prague 1968

Tischendorf, G.: Das System der metallogenetischen Faktoren und Indikatoren bei der Prognose und Suche endogener Zinnlagerstätten. Z. angew. Geologie 14 (1968) 8, 393—405

Tooms, J. S.: Applied Geochemistry in Marine Mineral Exploration. Report 23rd Sess. Int. Geol. Congr. Section 6 (Geochem.), Prague 1968

Vendel, M.: Zusammenhänge zwischen Gesteinsprovinzen und Metallprovinzen. Mitt. Berg-Hüttenm. Abt., Sopron 17 (1948), 206—324

VLASSOV, K. A. (Chief Editor): Geochemistry, Mineralogy and Genetic Types of Deposits of Rare Elements; Vol. I, Geochemistry of the Rare Elements. Izd. "Nauka", Moscow 1964 (Russian)

WEISS, A., and G. C. AMSTUTZ: Ion-Exchange Reactions on Clay Minerals and Cation Selective Membrane Properties as Possible Mechanism of Economic Metal Concentration. Mineralium Deposita *1* (1966), 60—66

YERMAKOV, N. P.: The Importance of Gas-Liquid Inclusions to the Prospecting and Exploration of Postmagmatic Deposits and Hidden Ore Deposits. Sovyetskaya Geologiya, No. 9, 77—90, Moscow 1966. (Russian) Ref.: Z. angew. Geol. *18* (1967) 8, 419—422 (German)

ZHARIKOV, V. A.: Experimental Investigation on the Acid-base Filtration Effect. Symp. Probl. Postmagm. Ore Depos., Prague 1963, 466—474 (Russian)

Group of authors: Metallogenetic Specialization of Magmatic Complexes. Omnibus volume, Izd. "Nedra", Moscow 1964 (Russian)

Group of authors: Metallogenesis of Sedimentary and Sedimentary metamorphic Rocks. Izd. "Nauka", Moscow 1966 (Russian)

Group of authors: Mineralogy and Geochemistry of Tungsten Deposits. Izd. Leningr. Univ., Leningrad 1967, 240 pp. (Russian)

Group of authors: Problems of Theory and Experimentation in Ore Formation. Akad. Nauk Ukrainsk. SSR, Kiev 1966, 217 pp. (Russian)

Deposits of Salts, Coal, Petroleum and Others

BRAITSCH, O.: Anwendung der Bromverteilungsgleichgewichte auf genetische Fragen der Salzlagerstätten. Fortschr. Miner. *40* (1962) 69, 1963

GULYAEVA, L. A.: Geochemistry of Caustobiolites and Their Deposits (Omnibus Volume of 11 Papers on Petroleum Geochemistry). Izd. Akad. Nauk SSSR, Moscow (1962) (Russian)

KREJCI-GRAF, K., W. ERNST, et al.: Zur Geochemie des Wiener Beckens II. Bor und Jod. Chem. Erde *27* (1968), 143—150

KREJCI-GRAF, K.: Diagnostik der Herkunft des Erdöls. Erdöl und Kohle *12* (1959), 706—712, 805 — 815

KÜHN, R.: Der Einfluß mineralogisch-geochemischer Untersuchungen auf die Vorstellungen zur Bildung von Kalisalzlagerstätten. Ber. deutsch. Ges. geol. Wiss., B, *13* (1969) 2, 193—218

KÜHN, R.: Rubidium als geochemisches Leitelement bei der lagerstättenkundlichen Charakterisierung von Carnalliten und natürlichen Salzlösungen. Fortschr. Miner. *40* (1962) 73, 1963

LEUTWEIN, F., and H. J. RÖSLER: Geochemische Untersuchungen an paläozoischen und mesozoischen Kohlen Mittel- und Ostdeutschlands. Freiberger Forschungsh. C *19* (1956), 196 pp.

MEINHOLD, R.: Die geochemischen Verfahren der Erdölgeologie (Fortschrittsbericht). Z. angew. Geologie *7* (1963), 380—386

MÜLLER, P.: Hydrogeochemische Methoden in der Geologie. Ber. deutsch. Ges. geol. Wiss., B, *13* (1969) 2, 263—274

NAKAI, N.: Geochemical Studies on the Genesis of Natural Gases. J. Earth Sci. Nagoya Univ. *10* (1962), 71—111

PUCHELT, H.: Zur Geochemie des Grubenwassers im Ruhrgebiet. Z. deut. geol. Ges. Part I, *116* (1964), 167—203

Geology and Geochemistry of Petroleum and Natural Gas Deposits (Vol. 2). Naukova dumka, Kiev 1965 (Russian)

10.3. Petrochemical Calculations

The geochemical and petrochemical interpretation of rock analyses, which mostly consist of 10 to 12 components, is difficult. Therefore, these analyses are converted, condensing geochemically similar components into a few groups. Since a detailed discussion of the subject-matter within the scope of this book is impossible, only the most important methods are mentioned here and reference is made to relevant publications.

ood surveys of the methods of petrochemical calculations are given by JOHANNSEN (1931),
URRI (1959), SAVARITSKIĬ (1954), and others.

1. Conversion into cation percentages (ESKOLA, 1954)
2. "Standard cell" according to BARTH (BARTH, 1948)
3. "Equivalent standard" according to NIGGLI (BURRI, 1959)
4. "Basis" according to NIGGLI (BURRI, 1959)
5. NIGGLI's values (NIGGLI, 1920)
6. System according to OSANN (OSANN, 1898—1903)
7. System according to v. WOLFF (v. WOLFF, 1922, 1951)
8. CIPW-system (CROSS et al., 1902, 1912)
9. System according to SAVARITSKIĬ (SAVARITSKIĬ, 1954)
10. Other methods
    a) Family index according to RITTMANN (RITTMANN, 1960)
    b) ACF-diagram according to ESKOLA (BARTH/CORRENS/ESKOLA, 1939)

ibliography

ARTH, T. F. W., C. W. CORRENS, and P. ESKOLA: Die Entstehung der Gesteine. Springer-Verlag,
erlin 1939, Reprint 1970
ARTH, T. F. W.: Oxygen in Rocks. A Basis for Petrographic Calculations. J. Geology 56 (1948),
0—60
URRI, C.: Petrochemische Berechnungsmethoden auf äquivalenter Grundlage (Methoden von Paul
IGGLI). Birkhäuser-Verlag, Basle and Stuttgart 1959
ROSS, W.: Use of Symbols in Expressing the Quantitative Classification of Igneous Rocks. J. Geology
0 (1912), 758—762
ROSS, W., I. P. IDDINGS, L. V. PIRSSON, and H. S. WASHINGTON: A Quantitative Chemico-mineralogical
lassification and Nomenclature of Igneous Rocks. J. Geology 10 (1902), 555—690
ROSS, W., I. P. IDDINGS, L. V. PIRSSON, and H. S. WASHINGTON: Modification of the Quantitative Sys-
em of Classification of Igneous Rocks. J. Geology 20 (1912), 550—561
SKOLA, P.: A Proposal for the Presentation of Rock Analyses in Ionic Percentage. Ann. Acad. Fen-
icae, A, III, 38 (1954)
OHANNSEN, A.: A Descriptive Petrography of the Igneous Rock. Vl. I: Introduction, Textures,
lassifications and Glossary. The University of Chicago Press, Chicago and Illinois 1931
IGGLI, P.: Lehrbuch der Mineralogie. Bornträger-Verlag, Berlin 1920
SANN, A.: Versuch einer chemischen Klassifikation der Eruptivgesteine. Tschermak's Min. Petr.
Mitt. 14 (1899), 351—469; 20 (1901), 399—558; 21 (1902), 365—488; 22 (1903), 322—356, 403—436
RITTMANN, A.: Vulkane und ihre Tätigkeit. 2nd ed., Ferdinand Enke, Stuttgart 1960
SAVARITSKIĬ, A. N.: Einführung in die Petrochemie der Eruptivgesteine. Akademie-Verlag, Berlin
954 (Translation from the Russian)
. WOLFF, F.: Die Prinzipien einer quantitativen Klassifikation der Eruptivgesteine, insbesondere der
ungen Eruptivgesteine. Geol. Rundschau 13 (1922), 9—14
. WOLFF, F.: Gesteinskunde — Die Eruptivgesteine. Rudolf A. Lang, Pößneck 1951

10.4. Geochemical Facies Analysis and Geochemical Barriers

The fundamentals of geochemical facies analysis have already been dealt with in Section 3.3.
and 8.4.1. It is of practical importance especially to recent and fossil sediments because of
the possibilities of

a) distinguishing between marine, brackish, and continental depositions, and
b) determining the conditions of deposition such as redox conditions, acidity, temperature.

383

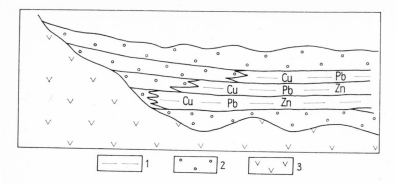

Fig. 119 Vertical facies differences (schematised) in metalliferous depositions near the submerged border, showing how deposition is influenced by the marine environment due to a displacement of the shoreline (Tonndorf 1965)

1 marine deposits; 2 continental deposits; 3 continental rock

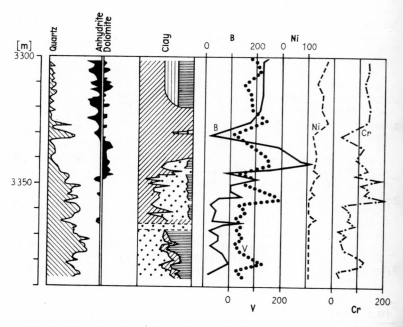

Fig. 120 Example of a combined mineralogical-geochemical examination of a boring core for facies analysis in prospecting for petroleum (after KÜBLER, 1965; slightly modified)

Explanations as to "clay": oblique hatching = mica
dotted = kaolinite
horizontal hatching = chlorite
vertical hatching = corrensite

384

Table 177. Classification of Geochemical Barriers and of the Elements Fixed in Them (affter PERELMAN)

I. Biogeochemical barrier
O, Si, Fe, Ca, Na, K, Mg, H, Mn, P, F, Ba, S, Sr, Cl, Rb, V, Zn, Cu, Ni, C, N, Co, **B, As, Ge, Mo, I, Se, Ra**

II. Physico-chemical barrier
1. Oxidation barrier (oxygen barrier)
Fe, Mn, S, Co

2. Reduction hydrogen sulphide barrier
Fe, V, Zn, Ni, Cu, Co, Pb, U, As, Cd, Hg, Ag, Se

3. Sulphate and carbonate barrier
Ca, Ba, Sr

4. Alkaline barrier
Fe, Ga, Mn, Sr, V, Cr, Zn, Cu, Ni, Co, Pb, Cd, Ca, Mg

5. Acid barrier
SiO_2

6. Vaporization (evaporation) barrier
Ca, Na, K, Mg, F, S, Sr, Cl, (Pb), Zn, U, (Ni), Li, Mo, N, V, Rb, I

7. Adsorption barrier
Ca, K, Mg, P, S, Rb, V, Cr, Zn, Ni, Cu, Co, Pb, U, As, Mo, Hg, Ra

III. Mechanical barrier
Zr, Cr, Nb, Th, Sn, W, Hg, Pt, Pd, Os, Au, Ru, Ir, Rh, and others

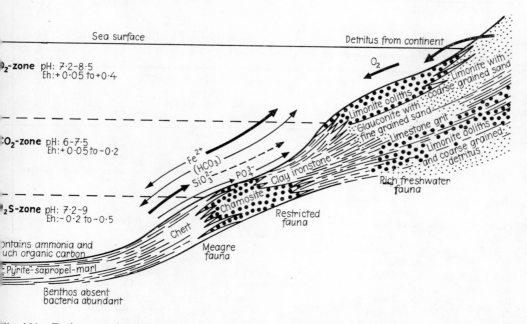

Fig. 121   Facies types in marine environment with corresponding Eh and pH values (after BORCHERT, 1967)

5 Geochem. Tables

Methods of facies analysis comprise, besides geologic-palaeontological methods and mineralogical phase analysis, the determination of the macrochemism and microchemism of the sediments (including the entrapped pore waters), the measurement of *Eh* and *pH*, and of palaeotemperature with the help of oxygen isotopes, for instance.

The formation of sedimentary element concentrations (deposits) is largely dependent on the primary facies conditions. Examples are given in Figs. 119 and 120 and in the Tables 178 and 179. The primary facies indicators can be changed to different degrees by diagenesis and metamorphism, nevertheless they can sometimes be detected and used even in highly metamorphed rocks.

Geochemical barriers are (according to Perelman) sections of supergene zones which occur in syngenesis, epigenesis and diagenesis and which are characterised by changes in the conditions of migration within short distances and, eventually, lead to a concentration of elements.

Table 178. Geochemical Data for the Identification of Gyttja and Sapropel (after SZUROWSKI)

Feature	in Gyttja	in Sapropel
C:N	70—350 50—100 (fossil coal)	3 (recent) 20—30 (fossil)
V:Cr	~ 1	$\geqq 10$
V:Mo	< 5	$\geqq$ 5—10
Phosphorite	present	missing

Table 179. Differences in the Geochemistry of Deep-waters as a Function of Genesis and Facies in Thuringia and in the North German Plain (after MÜLLER, 1968)

	1. Saline relict solutions (stratum waters in the Staßfurt carbonate and sheet dolomite)	2. Diagenetically changed seawater; infiltration solutions (stratum waters of the Mesozoic)	3. Ocean water
Mineralisation:	300 g/l (independent of depth)	330 g/l (dependent on depth)	35 g/l
Type	earth alkaline chloride waters	alkaline chloride waters	alkaline chloride waters
^{2}D	$\geqq 0.015,4$ at % ^{2}D concentrated	$\leqq 0.015,4$ at % ^{2}D depleted	0.015,4 at % average
^{18}O	—0.199,5 at % ^{18}O concentrated	$\leqq 0.199,5$ at % ^{18}O depleted	0.199,5 at % average
δ^{34}S	$\geqq$ evaporites $\pm 5^0/_{00}$ of Zechstein	$+40^0/_{00}$ to $+10^0/_{00}$	—
Br/Cl	$\geqq 0.003,4$	$\leqq 0.003,4$	Φ 0.003,4 average
Sr	> 200 mg/l	< 200 mg/l	13 mg/l
Rb	$>$ 5 mg/l	$<$ 5 mg/l	0.2 mg/l
Li	$>$ 10 mg/l	$<$ 10 mg/l	0.2 mg/l
K	$>$ 2 g/l	$<$ 2 g/l	380.0 mg/l
B	>100 mg/l	<100 mg/l	4.6 mg/l

Causes of the formation of barriers may be of biogeochemical, physico-chemical and mechanical kind. Within the range of geochemical barriers, the fixation of the elements mentioned in Table 177 may occur.

Fig. 121 may serve to illustrate a primary (syngenetic) barrier and Fig. 122 to illustrate a secondary (epigenetic) barrier.

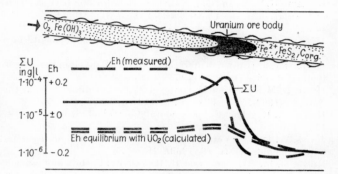

Fig. 122 Schematic representation of the most important geochemical parameters in the epigenetic "strata oxidation" of a sediment horizon. Example of a well developed geochemical barrier (after PERELMAN et al., 1968)

Bibliography

BORCHERT, H.: Formation of Marine Sedimentary Iron Ores, in: RILEY-SKINNER, Chemical Oceanography 1967

BORCHERT, H.: Über Faziestypen von marinen Eisenerz-Lagerstätten. Ber. Geol. Ges. DDR 9 (1964) 2, 163—193

BRAITSCH, O.: Rubidium und Brom als Indikatoren der Bildungsbedingungen der Oberrhein-Kalisalze in: III. Intern. Kalisymp. 1965, Part II, 65—95, Verlag Grundstoffind., Leipzig 1967

CAMERON, E. N.: Studies on the Geochemical Correlation of Sediments. Geol. Surv. Canada, Paper 64/2 (1964), 55—56

CHESTER, R.: Elemental Geochemistry of Marine Sediments, in: Chem. Oceanogr. 2 (1965), 23—80

CHILLINGAR, G. V.: Relationship between Ca/Mg Ratio and Geologic Age. Bull. Amer. Petrol. Geol. 40 (1956), 2256—2266

CLAYTON, R. N., and E. T. DEGENS: Use of Carbon-isotope Analyses for Differentiating Fresh-water and Marine Sediments. Bull. Amer. Assoc. Petrol. Geol. 43 (1959), 80

DANTSEV, V. I., and T. A. LAPINSKAYA: Deposits of Radioactive Substances. Izd. "Nedra", Moscow 1965 (Russian)

DEGENS, E. T., and M. BAJOR: Die Verteilung der Aminosäuren in limnischen und marinen Schieferone ndes Ruhrkarbons. Fortschr. Geol. Rheinld.-Westf. 3 (1962) 2, 429—440

DEGENS, E. T.: Die Chemofazies der Sedimente. Thesis, Würzburg 1959

DEGENS, H.: Geochemische Untersuchungen zur Faziesbestimmung im Ruhrkarbon und im Saarkarbon. Glückauf 94 (1958), 513—520

DODD, B.: Environmental Control of Strontium and Magnesium in Mytilus. Geochim. Cosmochim. Acta 29 (1965), 385—398

EPSTEIN, S., and BUCHSBAUM et al.: Revised Carbonate-water Isotopic Temperature Scale. Bull. Geol. Soc. Amer. 64 (1953), 1315—1326

ERNST, W.: Stratigraphisch-fazielle Identifizierung von Sedimenten auf chemisch-geologischem Wege. Geol. Rdsch. 55 (1966) 1, 21—28

ERNST, W.: Geochemical Facies Analysis. Elsevier, Amsterdam/London/New York 1971

ERNST, W., K. KREJCI-GRAF, and H. WERNER: Parallelisierung von Leithorizonten im Ruhrkarbon mit Hilfe des Bor-Gehaltes. Geochim. Cosmochim. Acta *14* (1958), 211

FRANKE, D.: Zu Fragen der geologischen Terminologie und Klassifikation; II. Der Begriff Fazies (Part 1—3). Zt. angew. Geol. *9* (1963), 39—45, 97—102, 153—157

FREDERICKSON, A. F., and R. C. REYNOLDS: Geochemical Method for Determining Paleosalinity. Proc. 8. Nat. Conf. Clays and Clay-minerals *8* (1960), 203—213

GORLICHKIĬ, V. A., and G. I. KALYUEV: Relationship between Trace Element Distribution and Conditions of Formation in the Upper Part of the Krivoy Rog Series. Gèokhimiya (1962) *12*, 1245—1250 (Russian)

HARDER, J. O.: Inwieweit ist Bor ein marines Leitelement? Fortschr. Geol. Rheinl.-Westf. *10* (1963), 239—352

HESEMANN, J. (Edit.): Unterscheidungsmöglichkeiten mariner und nichtmariner Sedimente. Fortschr. Geol. Rheinld.-Westf. *10* (1963), 482 pp.

HIRST, D. M.: The Geochemistry of Modern Sediments from the Gulf of Paria. Geochim. Cosmochim. Acta *26* (1962), 303—334, 1147—1187

HUBER, N. K.: The Environmental Control of Sedimentary Iron Minerals. Econ. Geol. *53* (1958), 123—140

JOHNS, W. D.: Die Verteilung von Chlor in rezenten marinen und nichtmarinen Sedimenten. Fortschr. Geol. Rheinld.-Westf. *10* (1963), 213—230

KEITH, M. L., REICHLER, and R. H. PARKER: Carbon and Oxygen Isotope Ratios in Marine and Fresh-water Mollusk Shells. Prog. 1960 Am. Meetings. Geol. Soc. Amer. (1959), 133—134

KEITH, M. L., and J. N. WEBER: Carbon and Oxygen Isotopic Composition of Selected Limestones and Fossils. Geochim. Cosmochim. Acta *28* (1964), 1787—1816

KEITH, M. L., and E. T. DEGENS: Geochemical Indicators of Marine and Fresh-water Sediments. In: Researches in Geochemistry, Wiley and Sons, New York 1959

KOCH, K., W. KOCKERT, and V. GRUNEWALD: Geochemische Untersuchungen an Salzen und Salzlösungen (Laugen) von Salzlagerstätten der DDR. Geologie *17* (1968) 6/7, 792—803

KOLBE, H.: Fazies und Geochemie der Kreideerze des nördl. Harzvorlandes. Ztsch. deutsch. Geol. Ges. *109* (1957), 36—40

KOSSOVSKAYA, A. G., V. D. SHUTOV and V. A. ALEKSANDROVA: Dependence of the Mineral Composition of the Clays in the Coal-bearing Formations on the Sedimentation Conditions. C.R.V. Congr. Intern. Strat. Geol. Carbonif., Paris 1963, 519—529

KREJCI-GRAF, K.: Geochemische Faziesdiagnostik. Freiberger Forschungsh. C 224, Leipzig 1966

KREJCI-GRAF, K.: Geochemische Fazieskunde. Freiberger Forschungsh. C *210*, Leipzig 1966, 143—153

KREJCI-GRAF, K.: Geochemical Diagnosis of Facies. Proc. Yorkshire Geol. Soc. *34* (1964) 4, 23, 469—521

KREJCI-GRAF, K., et al.: Versuche zur geochemischen Faziesdiagnostik. Chemie der Erde *24* (1965), 115—146

KRUMBEIN, W. C., and R. M. GARRELS: Origin and Classification of Chemical Sediments in Terms of pH and Oxidation-reduction potentials. J. Geol. *60* (1952), 1—33

KÜBLER, B.: Untersuchungen über die Tonfraktion der Trias der Sahara. Ein Beispiel gegenseitiger Abhängigkeit der Salinität und der Tonminerale. Fortschr. Geol. Rheinl. u. Westf. *10* (1963), 319—324

KÜHN, R.: Rb als geochemisches Leitelement bei der lagerstättenkundlichen Charakterisierung von Carnalliten und natürlichen Salzlösungen. N. Jb. Miner. Mh. (1963), 167—215

LANDERGREN, S., and F. T. MANHEIM: Über die Abhängigkeit der Verteilung von Schwermetallen von der Fazies. Fortschr. Geol. Rhein.-Westf. *10* (1963), 173—192

LONKA, A.: Trace-elements in the Finnish Precambrian Phyllites as Indicators of Salinity at the Time of Sedimentation. Bull. Comm. Geol. Finlande *228* (1967)

LUBOMIROV, B. M.: Postsedimentary Rock Transformation as a Factor in the Determination of the Salinity of Water. Soviet Geol. *5* (1965), 83—89 (Russian)

MARCHENKOV, V. I.: The Importance of the Oxidation Coefficient to the Facies Analysis of Marine Sediments. Lithologie u. Bodenschätze *4* (1965), 125—137 (Russian)

MIGDISOV, A. A.: On the Ti/Al Ratio in Sedimentary Rocks. Geokhimiya (1960), 149—163 (Russian)

Motojima, K., A. Ando, and M. Kawano: Chemical Composition and Environment of Sedimentation. J. Jap. Ass. Petrol. Techn. *25* (1960), 298—303

Nicholls, G. D.: Techniques in Sedimentary Geochemistry (2) Determination of the Ferrous Iron Contents of Carbonaceous Shales. J. Sedim. Petrol. *30* (1960), 603

Parham, W. E.: Lateral Variations of Clay Mineral Assemblages in Modern and Ancient Sediments. Proc. Int. Clay Conf., Jerusalem *1* (1966), 135—145

Perelman, A. I., E. A. Golobin et al.: Supergene Epigenetic Changes in Sedimentary Rocks and Their Role in the Formation of Ores; in: Geochemistry of Sedimentary Rocks and Ores. Izd. "Nauka", Moscow 1968, 410—422 (Russian)

Pilkey, O. H., and J. Power: The Effect of Environment on the Concentration of Skeletal Magnesium and Strontium in Dendrester. J. Geol. *68* (1960), 203—216

Porrenga, D.: Bor in Sedimenten als Indiz für den Salinitätsgrad. Fortschr. Geol. Rheinl.-Westf. *10* (1963), 267—270

Potter, P. E., N. F. Shimp, and J. Witters: Trace Elements in Marine and Fresh-water Argillaceous Sediments. Geochim. Cosmochim. Acta *27* (1963), 669—694

Pustovalov, L. V.: Geochemical Facies and Their Importance to General and Applied Geology. Probl. Sovetskoĭ Geol. *1* (1933) (Russian)

Ricke, W.: Geochemie des Schwefels und ihre Anwendung auf Faziesprobleme. Fortschr. Geol. Rheinl.-Westf. *10* (1963), 271—278

Ronov, A. B.: Comparative Geochemistry of Geosynclinal and Platform Sedimentites. Geokhimiya (1965) 8, 961—976 (Russian)

Roche, de la, H.: Sur l'Existence de Plusieurs Faciès géochimiques dans les Schistes Paléozoiques des Pyrénées Luchonnaises. Geol. Rdsch. *55* (1965), 274—301

Rücker, J. B., and J. W. Valentine: Paleosalinity Prediction Using Trace Element Concentration in Oyster Shells. Geol. Soc. Abstracts for 1961, New York 1962, 257

Stadler, G.: Zusammenhänge zwischen Mineralfazies und Borgehalten. Fortschr. Geol. Rheinl.-Westf. *10* (1963), 325—326

Stadnikoff, G.: Ein chemisches Verfahren zur Feststellung der Ablagerungsbedingungen von Tonen und tonigen Gesteinen. Glückauf *94* (1956), 58—62

Starke, R.: Mineralogisch-geochemische Methoden der Faziesdiagnostik. Ber. Deutsch. Ges. Geol. Wiss., B, *13* (1968) 2, 219—226

Teodorovich, G. I., Sedimentary Geochemical Facies. Byull. Moskovsk. obshchestva ispyt. prirody, otd. geol. *22* (1947) 1 (Russian)

Theilig, F., E. P. Müller, and D. Harzer: Die isotope Zusammensetzung des Sauerstoffs in Karbonatgesteinen der Staßfurt-Serie (Perm) und ihre Bedeutung für die Fazies und Paläothermometrie. Z. angew. Geologie *14* (1968), 506—508

Tischendorf, G., and H. Ungethüm: Über die Bedeutung des Reduktions-Oxydationspotentials (Eh) und der Wasserstoffkonzentration (pH) für Geochemie und Lagerstättenkunde. Geologie *13* (1964), 125—158

Tonndorf, H.: Beiträge zur Geochemie des randnahen Zechsteins etc. Freiberger Forschungsh. C *187* (1965)

Turekian, K.: Palaeoecological Significance of the Strontium/Calcium Ratio in Fossils and Sediments. Geol. Soc. Amer. Bull. *66* (1955), 155

Turekian, K.: The Use of Trace Element Geochemistry in Solving Geological Problems. Roy. Soc. Canada, Spec. Pap. No. 6, Univ. Toronto Press 1963

Walter, C. T., and N. B. Price: Departure Curves for Computing Paleosalinity from Boron in Illites and Shales. Bull. Amer. Ass. Petrol. Geol. *47* (1963), 833—841

Weber, J. N.: Paleoenvironmental Significance of Carbon Isotopic Composition of Siderite Nodules. J. Sed. Petrol. *34* (1964), 814—818

Wedepohl, K. H.: Spurenanalytische Untersuchungen an Tiefseetonen aus dem Atlantik. Ein Beitrag zur Deutung der geochemischen Sonderstellung von pelagischen Tonen. Geochim. Cosmochim. Acta *18* (1960), 200—281

Werner, H.: Über das Ca/Mg-Verhältnis in Torf und Kohle. Fortschr. Geol. Rheinl.-Westf. *10* (1963), 279—282

10.5. Mineral Chemistry (including Trace Elements)

Whereas the study of the macrochemical composition of minerals is carried out by mineralogists and geochemists in like manner, microchemistry (trace elements) remains one of the most important subjects of geochemistry.

10.5.1. Macrochemistry of Minerals (mineral chemistry in a narrow sense)

The best representation of the macrochemistry of minerals is the mineral formula from which the ideal chemical composition can be derived quickly. The derivation of mineral formulae from chemical analyses follows certain rules; they can be found in works by MACHATSCHKI (1953), BULACH (1967), HEY (1954), and others. Two simple examples are given below. Absolute values of mineral compositions are given in mineralogical works by SCHÜLLER (1950—1954), v. PHILIPSBORN (1936), DOELTER (1912—1931). Since the number of special publications on the chemistry of individual minerals is very great, a list is not included here.

Examples of calculating mineral formulas:

In any case, the individual components are divided by their molecular weights; conclusions can already be drawn from the molecular quotients of minerals of simple composition.

First example (emplectite from Sadisdorf; after SCHÜLLER 1954)

Element	Mass-%	Molecular quotient × 1000	Ratio
Cu	19.20	3,020	1
Bi	62.27	2,979	1
S	18.53	5,780	2

Result: $CuBiS_2$

The analysis of an amphibole or a pyroxene resembles a rock analysis as far as the number of components is concerned. For evaluation, the numerical proportions of O, OH and F (Cl etc.) ions are derived from the molecular quotients. Their sum must correspond to the given structural facts; it has been found to be 24 for amphiboles. The ratio of this basis to the sum of the O, OH and F proportions results in a factor with the help of which the number of cations, related to the basis, can be determined.

According to Strunz, the best way to determine the analytical formula of complex oxides is to start from the 0 ions per elementary lattice cell; it is found with the help of the following formula:

$$Z_0 = V \cdot D \cdot N \cdot S/\Sigma$$

(where V is the volume of the elementary lattice cell in cm^3; D is the mean density of the mineral; N is Loschmidt's number (also known as Avogadro's number); S is the sum of the atomic quotients of 0; Σ is the analytical sum ~ 100)

..en, the quantity of any individual cation is equal to $Z_0 \cdot K/S$, where K is the atomic quotient
..any one type of cation.

..cond example (triphyline from Hagendorf, after STRUNZ „Mineralogische Tabellen" 1967

..lement	Mass-%	Molecular quotient $\times 100$	Atomic quotient of 0 $\times 100$	Atomic quotient of the cation $\times 100$
..$_2$O	9.01	30.15	30.15	60.30
..eO	33.04	45.99	45.99	45.99
..nO	12.12	17.09	17.09	17.09
..O$_5$	45.60	32.10	160.50	64.20
..$_2$O	0.70	3.89	3.89	7.78
..otal	100.47		257.62	

$..0 = 292.4 \cdot 10^{-24} \cdot 3.541 \cdot 0.6023 \cdot 10^{24} \cdot 2.5762/100.47$

$\quad = 15.99 \sim 16$

..ontent of the elementary cell (related to $Z_0 = 16.0$):

$..i = 16 \cdot 60.30/257.62 = 3.74$

$..e = 16 \cdot 45.99/257.62 = 2.85$ etc.

..esult: $Li_{3.74}Fe^{II}_{2.85}Mn^{II}_{1.06}H_{0.48}P_{3.99}O_{16}$

..deal formula: $Li(Fe, Mn) PO_4$

..0.5.2. Trace Elements

..irst, the following definitions should be given:

..race elements (also known as microelements, oligoelements) are chemical elements whose
..oncentration in minerals and rocks $<0.1\%$ (down to the so-called omnipresent concentration
..n the sense propounded by W. and I. NODDACK, 1937). The notion trace element is relative
..ecause the concentration of an element in a given mineral or rock can be very low so that it
..s considered a trace element, whereas the same element can constitute a main element in
..nother compound (e.g. Fe in quartz is a trace element, Fe in magnetite or pyroxene is a
..ain element).

..isperse elements are those elements in the earth's crust which never occur in significant con-
..entrations and which do not form their own minerals in significant quantities. They are,
..requently due to diadochy, scattered in minerals of frequently occurring elements. Accord-
..ng to Vernadskiĭ, disperse elements comprise Li, Se, Ga, Br, Y, In, I and Cs.

..are elements are disperse elements whose economical importance is on the increase. The
..ollowing elements come under this heading: Be, B, Cs, Ga, Ge, Hf, In, Li, Mo, Nb, Pt, Ra,
..b, Re, Sc, Se, Rare Earth's, Ta, Te, Th, Tl, U, V, W, Y, Zr.

..he distribution of trace elements in minerals can be grouped as follows according to GON[I]
..nd GUILLEMIN, 1968:

A. Trace elements which can be substituted for other ones (diadochy) in a mineral lattice

Table 180. Survey of a Few Trace Elements in the Most Important Rock-forming Minerals (from

Mineral	Rock	Origin	Ba	Co	Sr	Cu	Mn	Ni
Plagioclase	Granite	Saxony	35	1.5	10			
	Gabbro	Minnesota		19		29	90	8
	Diabase	Hartz	1,500	22	1,900	220	1,800	35
Potash Feldspar	Granite	Minas Gerais	90		30			
	Granodiorite	California	3,000		200			
	Tonalite	South Africa	2,000		200			
Biotite	Granite	Saxony	590	69	84			42
	Granodiorite	California	25	25				5
	Pyroxenite	Colorado		60	130	150	700	425
Muscovite	Granite	South Africa			5—16	7—16	195—700	
	Eclogite	Bavaria		15		150	100	83
	Cryst. Schist	Wyoming		11	tr.	18	100	20
Pyroxene	Granite	California		12				10
	Trapp	Siberia	≦200	2—130	20—300	1—68	600—1,700	2—450
	Eclogite	Fichtelgebirge		22	150	14	80	60
Amphibole	Granite	California	60	10	10			5
	Norite	California	30	28	20			15
	Dioritegneiss	Wyoming		55	60	24	2,850	135
Olivine	Gabbro	Minnesota		284		169	3,400	1,050
	Trapp	Siberia		≦75	≦450	4—10	30—1,550	≦200
	Wehrlite	Norway					1,290	2,670
Garnet	Granite-pegmatite	Saxony		25		30		12
	Eclogite	Bavaria		19		10	2,700	20
	Schist	Amph. Fazies	5	43		160	8,000	4
Magnetite	Granite	California		40				20
	Diabase	Hartz		200		22	2,100	250

tr. = traces

B. Trace elements outside the lattice

    a) intercrystalline location (grain boundaries)

    b) intracrystalline location (cleavage, crack, etc.)

Overemphasis on the diadochy of trace elements in minerals (Goldschmidt) was rejected by Goni (1966), Tauson (1966), and Devore (1955).

As has already been mentioned in previous Chapters (especially Chapter 3), the incorporation of elements into mineral lattices is dependent on the conditions of concentration, the possibilities of diadochy due to Goldschmidt's rules and other factors, the temperature of formation, as well as the Eh and pH conditions of the environment of formation. That is the reason why trace elements frequently allow drawing conclusions as to the mineral formation conditions and thus making genetic statements regarding rocks and deposits (according to W. and

i	Ga	Pb	Li	Rb	Cr	Sn	V	Sc	Zr	Ag	Others
770											B 40 U 5.2
	46	60									
			13	720							Cs 0.8
	10		3	300							Fe 800
	10			450							
,530		71	1,560		50	220	320				
	25		175	1,000	10		150	25	50		
.2%	30	4			11	3	170	11	48	0.3	Zn 34
00—	37—	<6—	50—	420—	4—8	4.8--	<8—14	7—45			Cs
,100	65	33	70	1,400		7.0					22—82
,700	37	10			250	2.6	260	tr.	65	0.24	Zn 24
,500	45	12			90	8	540	38	350	0.12	
	5		7		4			30			
,900—					40—		40—	40	37		
,700					1,000		1,600				La 2—20
,200	6	4			6	tr.	380	8	tr.	0.16	Zn tr.
	20		15		30		175	200	250		Y 2,000
	10		1		370		240	80	30		Y 15
7,800	16	16			620	24	610	85	90	0.8	Zn 170 Y 100
2,400					40		25	11	104		
25—							≦20				
1,200											
1,050					230						
2,400		12			180		80				
1,020	2.1	tr.			250		100	72	tr.	0.28	Y tr.
480		1			125	tr.	27	33	150	4.0	Zn 50
	100				250		1,100				Mo 100
	7				700		500		200		

I. NODDACK, (1937), so-called "origin studies"). BRÄUER (1968) stated that, even if a distinct macrochemical differentiation is missing in a granitic rock, a distinct microchemical differentiation can be superimposed.

JEDWAB (1953) classified the trace elements as follows:

a) Crystallochemical elements whose incorporation is only governed by crystallochemical factors.
b) Typochemical elements which are susceptible to the supply and the paragenetic conditions.

Trace elements (and main elements) which are indicative of the genesis of minerals, rocks and ore deposits are also called *guide elements* (Leitelemente, GOLDSCHMIDT, 1937). Several elements have the character of guide elements whose interdependence can be detected by regression and correlation analyses. Usually, general statements cannot be derived from guide elements; they may differ both in quality and quantity in certain geological units, petrogenetic pro-

Table 181. Survey of the Distribution of Some Trace Elements in the Most Important Sulphidic Ore-forming Minerals (mainly from FLEISCHER, 1955; contents in ppm)
(The values in brackets are the "Mineral Clarke" according to Ivanov, 1968)

Element	Galena		Chalcopyrite		Sphalerite		Pyrite	
	Range of Variation	Main Range of Distribution	Range of Variation	Main Range of Distribution	Range of Variation	Main Range of Distribution	Range of Variation	Main Range of Distribution
As	10—10,000	100—5,000			50—10,000	10— 500	10—50,000	10—5,000
Sb	10—30,000	100—5,000			10—30,000	10— 100	50— 200	50
Bi	1—50,000	100—5,000			1— 100	1— 50	10— 100	100
Cd	10— 1,000	100			10—44,000	200—5,000 (2,680)		
Cu	1— 3,000	10— 100			1—50,000	1—60,000	1—60,000	50—5,000
Mn	1— 2,000	1— 50	10—20,000	10— 50	1—54,000	10—5,000	5—10,000	10—5,000
Ni	1— 100	1— 50	10— 2,000	10— 50	1— 300	5— 50	2—25,000	5—5,000
Ag	1—30,000	100—5,000	1— 2,300	10—1,000	1—10,000	5— 500	5— 200	5— 50
Tl	1— 1,000	1— 50 (8)			1— 5,000	1— 100 (19)	10 (34)	10
Sn	3— 1,300	3— 50	10— 770	10— 200	3—10,000	10— 500	10— 400	10
Co			10— 2,000	10— 50	3— 3,000	10— 500	10—25,000	10—5,000
In			1— 1,000	1— 10 (14)	1—10,000	5— 500 (40)		
Se			1— 2,100	10— 50	1— 900	1— 50	1— 300	5— 50
Pb					5—10%	5—5,000	10— 5,000	200—2,000
others	Fe 1— 5,000	1— 50			Ga 1— 3,000 Ge 1— 1,000 Hg 10—10,000	Ga 10— 500 (41) Ge 1— 500 (43) Hg 5— 200	Ti 10—600 V max. 1,000 Zn 100—45,000	Ti 200— 500 V 10— 50 Zn 100—10,000

Table 182. Percentage Distribution of Trace Elements in Rock-forming Minerals of Granites and Granodiorites from the Saxon Erzgebirge and Thuringia (rough calculations according to data from BRÄUER, 1969)

Element	Rock	Place of Origin	Concentration in the Rock [ppm]	Percentage of an Element in the Rock-forming Mineral								
				Plag.	Orthoc.	Quartz	Biotite	Amphibole	HM²	Titanite Allanite	Topaz	Apatite
U	Granite	Bobritzsch	4.4	11	7	16	25		30¹			11
	Granite	Schellerhau	33.5	9	9	16	2		62¹		2	
	Granite	Kirchberg	11.0	14	9	10	8		56¹	1		3
	Granite	Ruhla	3.8	8	10	13	16		34¹	18		1
	Diorite	Brotterode	1.3	43		21	6		30¹			
Sn	Granite	Schellerhau	130	7	6	7	48		10	9	22	
	Granite	Eibenstock	55	11	12	8	45		22		2	
V	Granite	Bobritzsch	37			2	92		6			
	Granite	Eibenstock	5		27		69		1	1	3	1
	Granodiorite	Suhl	95	5		2	56	18	18			1
Co	Granite	Bobritzsch	8.5	6	3	4	68		10	9		
	Granite	Ruhla	12	3	4	3	54		34			2
Ni	Granite	Bobritzsch	12				51		49			
	Granite	Ruhla	40				77		23			
Cr	Granite	Schellerhau	2			35	58		3		4	
	Granite	Kirchberg	7	3		3	58		37³			1
	Granodiorite	Suhl	45	4	1		25	22	30³	1		1
Ti	Granite	Kirchberg	1,530	1	1	1	87		13			
	Granodiorite	Suhl	3,090	1	1	1	54	16	3	27		

Notes:
¹) Mainly zircon and uranium minerals in the heavy mineral concentrate
²) Heavy mineral fractions as sum if concentrations are not given in the following columns
³) Magnetic fractions

vinces or deposits (question of the abundance of elements). Elements which are considered guide elements must comply with the following requirements (HAHN-WEINHEIMER and JOHANNING, 1963):

a) The differences in concentration of these elements must be greater between the minerals, mineral parageneses and mineral associations to be compared than the dispersion of the concentrations within a given mineral etc. of a deposit (of the same genetic type).

b) Their concentrations must be high enough to be clearly above the limit of detection of the selected analytical method.

Examples of the occurrence and behaviour of trace elements are given in Tables 180, 181, 182, and 183, and in Figs. 123 and 124.

Table 183. Schematic Survey of Characteristic Trace Elements in Various Types of Garnet (arranged in the order of their decreasing importance), from LANGE, 1964

Type of Garnet	Trace Element Association	Host Rock
Alm > Spess (Fe) (Mn)	Y, Zr, Sc, Ti	granite
Spess > Alm (Mn) (Fe)	Y, Ge, Ti, (Ga) . . . . + Sn, + Sc	granite-pegmatite in general . . . . young (Mesozoic) granite-pegmatites
Spess > Alm (Mn) (Fe)	Y, Zr, (Sc, Cu)	acid types of gneiss
Alm > Gross > Pyr (Fe) (Ca) (Mg) . . . . Alm > Pyr (Fe) (Mg)	Ti, V, Cr, (Co, Ni, Cu)	more basic
Alm > Gross > Pyr (Fe) (Ca) (Mg) . . . . Alm > Pyr > Gross (Fe) (Mg) (Ca)	Ti, V, Cr, Co, (Cu)	amphibolites and eclogites in gneisses
Pyr > Alm > Gross (Mg) (Fe) (Ca)	Ti, Cr, V, Co, Ni, (Cu)	eclogites in serpentinite, light-brownish garnets of griquaites
Pyr > Alm (Mg) (Fe)	Cr, Ti, V, Co, Ni	serpentinite, kimberlite, violet garnets of griquaites

Meaning of abbreviations: Alm = almandite, Spess = spessartite, Gross = grossularite, Pyr = pyrope

Bibliography

Macrochemistry

BORNEMAN-STARYNKEVICH, I. D.: Guide to the Computation of Mineral Formulas. Izd. "Nedra", Moscow 1964 (Russian)

BULACH, A. G.: Guides and Tables for the Computation of Mineral Formulas. Izd. "Nedra", Moscow 1967 (Russian); German Translation: Berechnung von Mineralformeln. Leipzig 1970

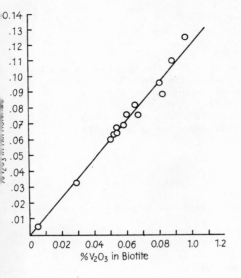

Fig. 123   Example of a good correlation of an element in coexisting minerals (after KRETZ, 1959)

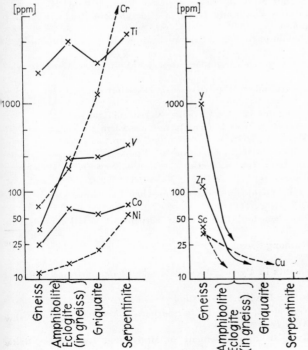

Fig. 124   Behaviour of some trace elements in garnets from acid to ultrabasic metamorphic rocks (from LANGE, 1964)

DEER, W. A., R. A. HOWIE, I. ZUSSMAN: Rock-Forming Minerals. Vol. 1—5, Longmans London 1963
COCKBAIN, A. G.: The Crystal Chemistry of the Apatites. Mineral. Mag. *36* (1968), 281, 654—660
HEY, M. H.: Chemical Index of Minerals. London 1950
KLEBER, W.: Kristallchemie. B. G. Teubner, Leipzig 1963, 128 pp.
LARSEN, E. S., and W. M. DRAISIN: Composition of the Minerals in the Rocks of the Southern Californian Batholith. Rep. 18 Intern. Geol. Congr. 1948, Part. II, Proc. Sect. A, 66—79, 1950
MACHATSCHKI, F.: Spezielle Mineralogie auf geochemischer Grundlage. Springer, Vienna 1953
MICHEL, A., and J. BENARD: Chimie Minérale. Masson et Cie., Paris 1964
RAMBERG, H.: Chemical Bonds and the Distribution of Cations in Silicates. J. Geology *60* (1952), 331
SCHÜLLER, A.: Die Eigenschaften der Minerale. 2 Vols. Akademie-Verlag, Berlin 1950—1954

Trace Elements

BELOV, N. V.: Crystal Chemical Aspects of the Problems of the Diadochism Entry of Boron in Silicates. Geokhimiya (1960) 6, 551 (Russian)
CHAKRABARTI, A. K.: On the Trace Element Geochemistry of Zawar Sulphides and Its Relation to Metallogenesis. Am. Mineral. *9* (1967) 2, 258—262
CRUFT, E. F.: Minor Elements in Igneous and Metamorphic Apatite. Geochim. Cosmochim. Acta *30* (1966), 375—398
DEVORE, G. W.: Crystal Growth and the Distribution of Elements. J. Geology *63* (1955), 471—494
DUDYKINA, A. S.: Paragenetic Associations of Admixed Elements in Cassiterite of Various Genetic Types. Trudy Inst. Geol. rudn. Mesterozhden. Petr. Miner. Geokh., Akad. Nauk SSSR *28* (1959), 111—121 (Russian)
ERNST, W.: Geochemical Facies Analysis Elsevier, Amsterdam/London/New York 1971
FAUST, G. T.: Minor Elements in Serpentine-additional Data. Geochim. Cosmochim. Acta *6* (1963) *27*, 665—668
FLEISCHER, M.: Minor Elements in some Sulfide Minerals. Econ. Geol. *50* (1955), 970—1024
FLEISCHER, M.: Hafnium Content and Hafnium-Zirconium Ratio in Minerals and Rocks. U.S. Dep. Interior. geol. Surv., Bull. 1021 A (1955), 135
GADOMSKI, M.: Rubidium, Caesium and Thallium in Pegmatitic Mica Minerals. Arch. Mineralog. *22* (1958), 207—225
GANEEV, I. G., and N. P. SECHINA: Über geochemische Besonderheiten der Wolframite. Geokhimiya *6* (1960), 518—523
GOLDSCHMIDT, V. M.: The Geochemical Background of Minor Elements Distribution. Soil. Sci., U.S.A. *60* (1945), 1—7
GOLDSCHMIDT, V. M.: Geochemische Verteilungsgesetze der Elemente. Kristiania 1924
GONI, J.: Contribution à l'Étude de la Localisation et de la Distribution des Éléments en Traces dans les Minéraux et les Roches Granitiques. Memoires BRGM No. 45 (1966), 68 pp.
GONI, J., and C. GUILLEMIN: Nouvelles Données sur la Localisation des Éléments en Traces dans les Minéraux et dans les Roches. XXIIᵉ Congr. Géol. Intern. Section Géochimie, New Delhi (1964b)
GOUDOT, A., and D. BERTRAND: Les Oligoéléments. Presses Univ., No. 1010. 1962
HABERLANDT, H.: Die Bedeutung der Spurenelemente in der geochemischen Forschung. Monatshefte für Chemie *77* (1947), 293
HAHN-WEINHEIMER, P.: Akzessorische Mineralien und Elemente in Serpentinit von Leupoldsgrün. Geochim. Cosmochim. Acta *21* (1961) 3/4, 165
HARRIS, P. G.: Distribution of Germanium among Coexisting Phases of Partly Glassy Rocks. Geochim. Cosmochim. Acta *5* (1954), 185
HAWLEY, J. E., and I. NICHOL: Trace Elements in Pyrite, Pyrrhotite and Chalcopyrite of Different Ores. Econ. Geol. *56* (1961) 3, 467—487
HEIER, K. S.: Trace Elements in Feldspars — a Review. Norsk Geol. Tidsskr. *42* (1962) 2, 415—454
HEIER, K. S., and S. R. TAYLOR: Distribution of Ca, Sr and Ba in Southern Norwegian pre-Cambrian Alkali Feldspars. Geochim. Cosmochim. Acta *17* (1959)
HELLWEGE, H.: Zum Vorkommen des Zinns als Spurenelement in Mineralien. Hamburger Beitr. Angew. Mineralog. *1* (1956), 73—136

398

HERTEL, L.: Die Fremdelementführung der Bleiglanze als Hilfe zur Bestimmung der Bildungstemperatur. Z. Erzbergbau Metallhüttenwesen *19* (1966) 12, 632—635

HERZ, N., and C. V. DUTRA: Trace Elements in Alkali Feldspars, Brazil. Amer. Mineralogist *51* (1966)
1593—1607

IVANOV, V. V., et al.: On the Character of the Distribution and the Mean Contents of In in a Few
Minerals from Deposits of Different Genetic Types. Geokhimiya (1963) *11*, 1016—1026 (Russian)

IVANOV, V. V.: Die Geochemie der dispersen Elemente Ga, Ge, Ce, In und Tl in hydrothermalen Lagerstätten. Z. angew. Geol. *14* (1968) 2, 80

JEDWAB, J.: Sur la Définition des Éléments Typochimiques. Bull. Soc. Belge Géol. *62* (1953), 173
to 179

KACHENKOV, S. M.: The Distribution of Chemical Elements in Clay Minerals. Dokl. Akad. Nauk SSSR
134 (1960) 3, 680—683 (Russian)

KHOMYAKOV, A. P.: On the Interrelations between Concentration and Composition of Trace Elements
in Minerals. Geokhimiya (1963) *2*, 115—121 (Russian)

KOSLOVA, O. G.: Seltene Erden in Flußspaten. Z. angew. Geol. (1962), 361—362

KOSTOV, I., and G. PANAYOTOV: Rare Earths in Epidote and Orthite. Comptes Rendus Acad. Sci. Bulg.
20 (1967) 10, 1057—1059

KRETZ, R.: The Distribution of Certain Elements among Coexisting Calcic Pyroxenes, Calcic Amphiboles and Biotites in Skarns. Geochim. Cosmochim. Acta *20* (1960), 161—191

LANGE, H.: Die chemische Zusammensetzung von Granaten aus Metabasiten des Erzgebirges. Geologie *13* (1964) 3, 325—352

LEEDER, O.: Geochemie der Seltenen Erden in natürlichen Fluoriten und Kalziten. Freiberger Forschungsh. C *206*, Leipzig (1966), 137 pp.

MERCY, E., and M. J. O'TTARA: Distribution of Mn, Cr, Ti and Ni in Co-existing Minerals of Ultramafic
Rocks. Geochim. Cosmochim. Acta *31* (1967), 2331—2341

MINEEV, D. A.: The Lanthanide in Minerals. Izd. "Nedra", Moscow 1968 (Russian)

NEMEC, D.: Fluorine in Tourmalines. Contr. Miner. Petrol. *20* (1969), 235—243

NOCKOLDS, S. R., and R. L. MITCHELL: The Geochemistry of some Caledonian Plutonic Rocks: A Study
on the Relationship between the Major and Trace Elements of Igneous Rocks and Their Minerals.
Trans. Roy. Soc., Edinburgh *61* (1948) 2, 20, 533

PARRY, W. T., and M. P. NACKOWSKI: Copper, Lead and Zinc in Biotites from Basin and Range Quartz
Monzonites. Econ. Geology *58* (1963) 7

PEARSON, G. R.: Trace Elements in Aluminium Silicates. Canad. Mining J. Abstr. *77* (1956) 1, 69

PREUSS, E., and H. ZIEHR: Zur Verbreitung des Quecksilbers in ostbayerischen Flußspatlagerstätten.
Geol. Rdsch. *55* (1965) F. Enke, Stuttgart

RANKAMA, K.: On the Use of Trace Elements in some Problems of Practical Geology. Bull. Comm. Geol.
Finlande *126* (1940), 90

ROST, F., and E. GRIGEL: Über accessorische Elemente in mitteleuropäischen Eklogiten und ihren
Mineralien. Geochim. Cosmochim. Acta *28* (1964), 1989—1998

SCHULZE, E. G.: Zur Spurenelementverteilung und Kluftparagenese in initialen Vulkaniten des nördl.
Oberharzes. Chem. d. Erde *26* (1967) 4, 279—303

SEMENOV, E. I.: Relationship between the Rare Earth Composition and the Composition and Structure of Minerals. Geckhimiya (1958) *5*, 452 (Russian)

SEMENOV, E. I.: Über die Zr-Gehalte in Ti-Mineralen. Trudy Inst. Miner. Geokhim. Kristallokhim. redkikh element. Akad. Nauk SSSR *2* (1959), 87—92 (Russian)

SEN, N., R. NOCKOLDS, and R. ALLEN: Trace Elements in Minerals from Rocks of the S. Californian
Batholith. Geochim. Cosmochim. Acta *16* (1959) 1/3, 58

SHAW, D. M.: The Camouflage Principle and Trace Element Distribution in Magmatic Minerals. J.
Geology *61* (1953) 2, 142

SHAW, D. M.: The Geochemistry of Scapolite. J. Petrology *1* (1960), 218—285

SHCHERBINA, V.: Die Verteilungsarten einiger seltener Elemente in Mineralien einer gemeinsamen Para-
genese. N. Jb. Miner., Abh. *94* (1960), 1093—1100

SIMS, P. K., and P. B. BARTON, jr.: Some Aspects of the Geochemistry of Sphalerite, Central City
District. Econ. Geology *56* (1961), 1211—1237

SNYDER, I. L.: Distribution of Certain Elements in the Duluth Complex. Geochim. Cosmochim. Acta *16* (1959), 243—277

SZÁDECZKY-KARDOSS, E.: Seltene Elemente und Geochemie. Freiberger Forschungsh. C *58* (1959) 5—19

SZÁDECZKY-KARDOSS, E.: Zur Verteilung der Elemente in den sedimentären und magmatischen Sulfiden. Freiberger Forschungsh., C *79* (1960), 106—125

TAUSON, L. V.: Factors in the Distribution of the Trace Elements during the Crystallisation of Magmas. Phys. Chem. Earth *6* (1966), 219—249

TISCHENDORF, G.: Zur Verteilung des Selens in Sulfiden. Freiberger Forschungsh. Leipzig C *208* (1966) 162 pp.

WRIGHT, J. B., and J. F. LOVERING: Electron Probe Microanalysis and Geothermometry of Sphalerite etc. Amer. Mineralogist *52* (1967), 524—529

10.6. Isotope Geochemistry

"The geochemistry of isotopes studies the regularities of the formation and reformation of isotopes of chemical elements in nature, the regularities of their abundance and their changes in space and time in the course of the earth's history." (KRÜGER, 1968)

In nature, a number of elements exhibit variations in their isotopic abundance. Changes in such isotopic abundances can occur primarily

a) due to radioactive decay or (more generally) due to nuclear processes (e.g. ^{206}Pb, ^{207}Pb, ^{208}Pb, ^{87}Sr, ^{40}Ar, ^{10}Be).

b) due to effects of physical (diffusion, thermodiffusion), physicochemical (evaporation, condensation, solidification, melting), chemical (equilibrium of isotopic exchanges, see Table 184, kinetic effects associated with reduction and oxidation), and biologic (reduction due to anaerobic bacteria, photosynthesis) characters. The isotopic exchange effects are most distinct at low mass numbers, depend on temperature and disappear at very high temperatures. In many cases, the heavy isotope is concentrated in the stronger combined state.

Geochemical-geologic processes (such as volcanic activity, magmatic intrusions, differentiation, metamorphism, sedimentary processes) provide different conditions for the action of such effects and, in this way, exert a secondary characteristic influence on the variation of the isotopic abundance. The distribution of isotopic abundances characterizes in this way geochemical-geologic processes, geospheres, deposits and other geologic objects.

For instance, the sulphate now present in seawater shows today a very constant composition of S isotopes, a feature which was subject to changes in the course of the earth's history (Fig. 125). Similar constant conditions are found in meteoric sulphur (Fig. 126) whose isotopic composition seems to resemble the original isotopic composition of sulphur and that of the sulphur in the earth's mantle and, thus, is of particular importance to the geochemistry of sulphur isotopes. In contrast with this, all sedimentary sulphides, which were brought about by bacterial sulphate reduction, exhibit an enormously wide range of dispersion (Fig. 126 and Table 185).

The general task of isotope geochemistry is to investigate and interpret the isotopic abundances in geologic objects (i.e. to trace them back to the processes mentioned above in paragraphs a) and b) by a geochemical-geologic interpretation).

The determination of the distribution of isotopic abundances of different geochemical-geologic objects enables geochemists to draw conclusions regarding genetic similarities or differences, e.g. in the case of hydrothermal deposits, differences between variscan (= hercynian)

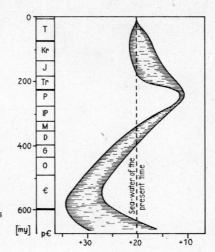

Fig. 125   S isotopes in sea water sulphate throughout earth's history (after NIELSEN, 1965)

Table 184. Chemical Exchange Reactions

				$K_{25°C}$	
HD (Gas)	+ H_2O (Gas)	⇌ HDO (Gas)	+ H_2 (Gas)	3.70	^2D
$H^{12}CN^-$ (Gas)	+ ^{13}CN (Solution)	⇌ $H^{13}CN^-$ (Gas)	+ $^{12}CN^-$ (Solution)	1.030	^{13}C
$^{13}CO_2$ (Gas)	+ $H^{12}CO_3^-$ (Solution)	⇌ $^{12}CO_2$ (Gas)	+ $H^{13}CO_3$ (Solution)	1.014	^{13}C
$^{15}NH_3$ (Gas)	+ $^{14}NH_4^+$ (Solution)	⇌ $^{14}NH_3$ (Gas)	+ $^{15}NH_4^+$ (Solution)	1.035	^{15}N
$^1/_2C^{16}O_2$ (Gas)	+ $H_2^{18}O$ (Liquid)	⇌ $^{18}/_2C^{18}O_2$ (Gas)	+ $H_2^{16}O$ (Liquid)	1.039	^{18}O
$^{32}SO_2$ (Gas)	+ $H^{32}SO_3^-$ (Solution)	⇌ $^{32}SO_2$ (Gas)	+ $H^{34}SO_3^-$ (Solution)	1.019	^{34}S
$H_2^{34}S$ (Gas)	+ $^{32}SO_4^{2-}$ (Solution)	⇌ $H_2^{32}S$ (Gas)	+ $^{34}SO_4^{2-}$ (Solution)	1.074	^{34}S

and postvariscan mineralizations with respect to S isotopes (RÖSLER, et al., (1966) and to D isotopes (HARZER, 1967), conclusions regarding the type of source, e.g. by means of ^{87}Sr FAURE, and HURLEY, 1963; HEDGE and WALTHALL, 1963), Pb and S isotopes; or conclusions regarding the organic and inorganic carbon in different compounds. In this connection reference should be made to the studies in carbon and sulphur isotopes of crude petroleum SILVERMAN, 1964; THODE, 1963; MÜLLER et al. 1968; STAHL and others, 1971).
The interrelations between isotopic abundance and other parameters are of particular importance. On the basis of the radioactive decay law, a correlation with time is formulated which forms the basis of the field of physical (radiometric) dating (Section 10.7.).

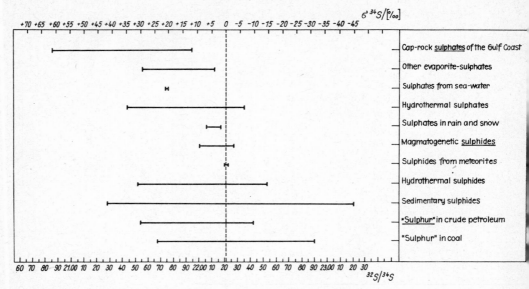

Fig. 126   The distribution of ^{32}S and ^{34}S sulphur isotopes in nature

Table 185. Maximum Fractional Amounts in the Metabolism of Sulphur Combinations Due to Micro organisms (after KROUSE et al., 1966; HARRISON and THODE, 1958; KAPLAN and RITTENBERG, 1962 from PILOT, 1969; fractionation factor $= K_{32}/K_{34}$)

Primary Process	Organism	Original Substance	End Product	Fraction ation Factor
Sulphate reduction	Desulfovibrio desulfuricans	SO_4	H_2S	— 46.0
Sulphite reduction	Desulfovibrio desulfuricans	SO_3	H_2S	— 14.3
Sulphite reduction	S. cerevisiae	SO_3	H_2S	— 41.0
Sulphite reduction	Salmonella Heidelberg	SO_3	H_2S	— 42.0
Sulphate assimilation	Escherichia coli Saccaromyces cerevisiae (yeast)	SO_4	organic S	— 2.8
Hydrolytic Cysteine	Protus vulgaria	cysteine	H_2S	— 5.1
Chemosynthetic oxidation	Thiobacillus concretivorus	H_2S	SO_4	— 18.0
		H_2S	S_xO_y	+ 19.0
Photosynthetic oxidation	Chromatium	H_2S	S^o	—100
		H_2S	SO_4	0.0
		H_2S	S_xO_y	+ 11.2

If isotopic exchanges which lead to equilibrium take place between two chemical compounds a correlation with temperature is given (see Table 184). This is widely known, as far as the determination of palaeotemperatures are concerned; the isotopes of the oxygen in water and those of the carbonate dissolved in it reach the exchange equilibrium within a relatively short time:

$$H_2{}^{18}O + 1/3\ C^{16}O_2{}^{2-} \rightleftharpoons H_2{}^{16}O + 1/3\ C^{18}O_2{}^{2-}$$

If the carbonate is precipitated, the oxygen-isotope ratio will normally remain fixed (e.g. in the calcareous shell of belemnites). Inasmuch as the equilibrium constant

$$K = \frac{[C^{18}O_3^{2-}]^{1/3} : [C^{16}O_3^{2-}]^{1/3}}{[H_2{}^{18}O] \quad : [H_2{}^{16}O]}$$

is dependent on temperature, the temperature of formation of the carbonate can be derived from the O isotope ratio of carbonate and water. The temperature of such marine formations can be calculated according to the empirical formula

$$T = 16.5 - 4.3\,\delta + 0.14\,\delta^2 \,,$$

where

$$\delta \text{ in } \%_0 = \left[\frac{R_{Pr}}{R_{St}} - 1\right] \cdot 1000 \left(\text{with } R = \frac{M\,(46)}{M\,(44)}, \text{ CO}_2 \text{ as test gas for the mass spectrometer}\right).$$

R_{St} is related to the PDB standard (the mean value for ocean water has been taken into account in the constants and is assumed to be constant in this case; BOWEN, 1966). Such studies are of great importance to climatology, especially when they are combined with ^{14}C datings. The O-isotope thermometry is not only suitable for such low temperatures, but also for high temperatures, for instance in the hydrothermal field (O'NEIL and CLAYTON, 1964). If, on the other hand, the temperature of formation is known, conclusions as to the type of water (surface or magmatic water) from which the minerals were settled can be drawn (cf. Fig. 130). The determination of temperatures is generally possible for any isotopic exchange equilibrium (see Fig. 128) if the correlation between the equilibrium constant and the temperature has

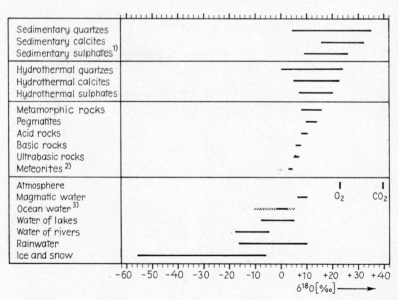

Fig. 127 The distribution of oxygen isotopes in nature (from several authors, reproduced after data from HARZER, 1969)

[1] only marine sulphates
[2] only chondrites
[3] hatched: seawater influenced by melted snow and ice

26*

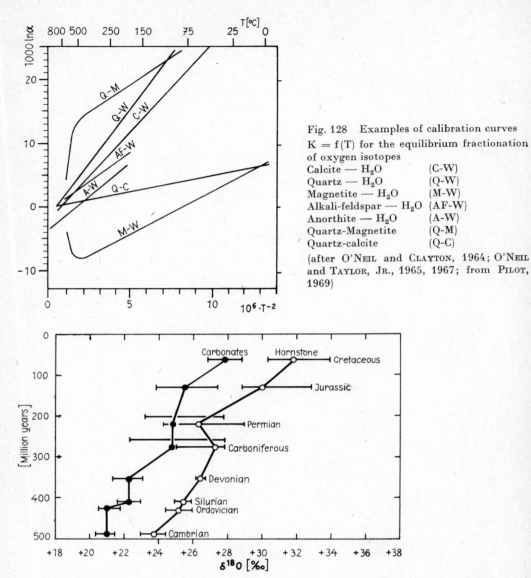

Fig. 128   Examples of calibration curves
K = f(T) for the equilibrium fractionation
of oxygen isotopes
Calcite — H₂O   (C-W)
Quartz — H₂O   (Q-W)
Magnetite — H₂O   (M-W)
Alkali-feldspar — H₂O   (AF-W)
Anorthite — H₂O   (A-W)
Quartz-Magnetite   (Q-M)
Quartz-calcite   (Q-C)
(after O'NEIL and CLAYTON, 1964; O'NEIL
and TAYLOR, JR., 1965, 1967; from PILOT,
1969)

Fig. 129   The changes in the ¹⁸O/¹⁶O ratios as a function of geological ages (after DEGENS and EPSTEIN, 1962; from DEGENS, 1968)

been established and the fractionation is not too small. In several cases, equilibrium has not been established in nature (e.g. frequently in sulphide-sulphate systems with respect to the S isotopes) so that misleading temperatures would then be obtained.

Sometimes, important information and correlations are only obtained if the isotopic composition of several elements is studied. Thus, according to CRAIG (1961), for meteoric waters which

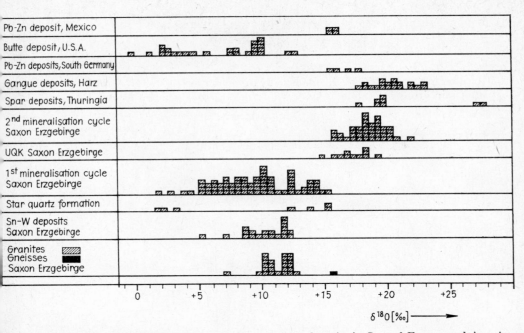

Fig. 130   The distribution of ¹⁸O isotopes in quartzes from deposits in Central Europe and America (after HARZER, 1969)

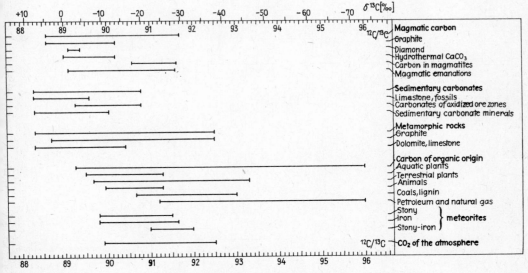

Fig. 131   Distribution of ¹²C and ¹³C carbon isotopes in nature (from several sources)

were subject to equilibration, condensation and evaporation processes, the "world-wide" relation

$$\delta D = 8\delta^{18}O + 10$$

(both values in $^0/_{00}$) holds true. In abnormally dry climates (e.g. South Africa, too rapid evaporation of rain drops), the slope is smaller (5); for large independent lakes (evaporation also a

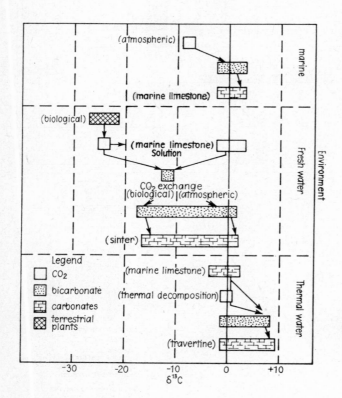

Fig. 132 The fractionation of ^{12}C and ^{13}C in carbonates of different origins and different environments (after MÜNNICH and VOGEL, 1962; from DEGENS, 1968)

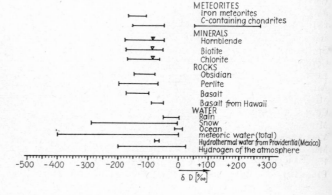

Fig. 133 Schematic representation of the variations of D related to SMOW (the values of rain and snow are quoted from FRIEDMANN, 1953, 1958; those of meteoric water from GODFREY, 1962; compiled by PILOT, 1969)

inetic process) and for isotopic exchange processes in water, e.g. with carbonates, remote
alues occur.

igs. 127 (oxygen), 126 (sulphur), and 131 (carbon) give information about the isotopic abun-
ances determined so far in various geologic units.

ig. 129 shows the development of oxygen isotope conditions with respect to the geologic
evelopment. In Figs. 132 and 133 and Table 186 specific information about the distribution
f the isotopes of carbon, hydrogen and strontium is given.

able 186. Initial $^{87}Sr/^{86}Sr$ Ratio of a Few Rocks

Locality and Type of Rock	$^{87}Sr^{86}Sr$	Authors
Granitic rock from southern British Columbia	0.705	FAIRBAIRN, HURLEY, PINSON (1964)
Palmer granite, South Australia	0.708	WHITE, COMPSTON, KLEEMANN (1967)
Granite from Bektan-Ata, Central Kazak S.S.R.	0.701	GOROKHOV, ARTEMOV (1966)
Heemskirk granite, West Tasmania		BROOKS, COMPSTON (1965)
) red granite	0.705—0.720	
) white granite A	0.734	
) white granite B	0.742	
Plutons of the Eastern Central Sierra Nevada, California (alaskite, quartz-monzonite, granodiorite)	0.707	HURLEY, BATEMAN, FAIRBAIRN, PINSON (1965)
Skaergaard intrusion, East Greenland		HAMILTON (1963)
) basic rocks	0.706	
) acid granophyre	0.710—0.730	
Magmatic province of Lebombo-Nuanetsi, South Africa		MANTON (1968)
) rhyolite and granophyre from Swaziland and Zululand	0.704,2	
) rhyolite and granites of the Nuanetsi syncline	0.708,1—0.708,5	
) basalts and gabbroes in the two regions a) and b)	0.704,2—0.712,5	

Bibliography

General Survey

FAUL, H.: Nuclear Geology. John Wiley & Sons, New York. Chapman & Hall, London 1954
JACOBSHAGEN, V.: Die Isotopenzusammensetzung natürlicher Wässer. — Geol. Rdsch. *51* (1961), 281—290
KROUSE, H. R., F. D. COOK, A. SASAKI and V. SMEJKAL: Microbiological Isotope Fractionation in Springs of Western Canada. Recent Developments in Mass Spectroscopy. Proceedings of the Int. Conf. Mass Spectroscopy, Kyoto (1970) 629—639
LEVINSON, A. A. (ed.): Proceedings of the Apollo 11 Lunar Science Conference. Chemical and Isotope Analysis. Geochim. Cosmochim. Acta, Suppl. 1, *34* (1970)
MILLER, S. L., and G. I. WASSERBURG: Isotopic and Cosmic Chemistry. North-Holland Publishing Comp., Amsterdam 1964
MÜLLER, G., and K. MAUERSBERGER: Analyse stabiler Isotope durch spezielle Methoden. Akademie-Verlag, Berlin 1964
MÜLLER, E. P., and R. WIENHOLZ: Isotopengeochemie und Erdölgenese. Z. angew. Geol. *14* (1968) 4, 176—182
PILOT, J.: Isotopengeochemie (Situation, Konzeption, Entwicklung, Möglichkeiten). Freiberger

Forschungsh., Leipzig C *255* (1970)

POLANSKI, A.: Geochimia Izotopov. Warsaw 1961 (Polish)

RANKAMA, K.: Progress in Isotope Geology. John Wiley & Sons, New York 1964

RANKAMA, K.: Isotope Geology. Pergamon Press Ltd. 1954

RÖSLER, H. J., and J. PILOT: Anwendung und Bedeutung der Isotopengeochemie. Z. angew. Geologie *15* (1969) 9, 491—500

VINOGRADOV, A. P.: Isotope and Silicate Geochemistry. Internat. Upper Mantle Project, Rep. (1965) 2, 111

VINOGRADOV, A. P.: Geochemistry of Isotopes. Vestnik Akad. Nauk SSSR (1954), 5 (Russian)

VOĬTKEVICH, G. V.: Problemy Radiogeologii. Moscow 1961 (Russian)

Sulphur

AULT, W. U.: Isotopic Fractionation of Sulfur in Geochemical Processes, in: Researches in Geochemistry by P. H. Abelson. 241—259. John Wiley & Sons, New York 1959

AULT, W. U., and J. L. KULP: Isotopic Geochemistry of Sulfur. Geochim. Cosmochim. Acta *16* (1959), 201—235

AULT, W. U., and J. L. KULP: Sulfurisotopes and Ore Desposits. Econ. Geol. *55* (1960), 73—100

CHUKHROV, F. V., V. L. VINOGRADOV, and L. P. ERMILOVA: Die genetische Deutung der Schwefel-Isotopenzusammensetzung einiger Lagerstätten Zentralkasachstans. Ber. deutsch. Ges. geol. Wiss., B, *13* (1968), 117—129

GRINENKO, L. N., and V. A. GRINENKO: Isotopic Composition of Sulphur in Connection with Problems of Ore Deposit Genesis. Report 23. Sess. Int. Geol. Congr. Section 6 (Geochem.), Prague 1968

HARZER, D., J. PILOT, and R. STARKE: Schwefel- und Sauerstoffisotopenverhältnisse von Baryten des sächsischen Erzgebirges. Bergakademie *16* (1964), 709—713

HOLSER, W. T., and I. R. KAPLAN: Isotope Geochemistry of Sedimentary Sulfates. Chem. Geol. *1* (1966), 93—135

JENSEN, M. L.: Biogeochemistry of Sulfur Isotopes. Proc. Nat. Science Foundation Symp. April 12—14, 1962. Yale University, New Haven, Conn. 1962

KAJIWARA, Y., H. R. KROUSE and A. SASAKI: Experimental Study of Sulphur Isotope Fractionation between Coexisting Sulfide Minerals. Earth and Planetary Science Letters *7* (1969), 271—277

KEMP, A. L., and H. G. THODE: The Mechanism of the Bacterial Reduction of Sulphate and of Sulphide Isotope Fractionation Studies. Geochim. Cosmochim. Acta *32* (1968) 1, 71—91

LONGINELLI, A., and G. CORTECCI: Isotope Abundance of Oxygen and Sulfur in Sulfate Ions from River Water. Earth and Planetary Science Letters *7* (1970), 376—380

MÜLLER, G., H. NIELSEN, and W. RICKE: Schwefel-Isotopen-Verhältnisse in Formationswässern und Evaporiten Nord- und Süddeutschlands. Chem. Geol. *1* (1966) 3, 211—220

NIELSEN, H.: Schwefelisotope im marinen Kreislauf und das δ^{34} S der früheren Meere. Geolog. Rundschau *55* (1965), 160—172

RÖSLER, H. J., J. PILOT, and R. GEBHARDT: Schwefel-Isotopenuntersuchungen an Magmatiten und postmagmatischen Lagerstätten des Erzgebirges und Thüringens. Bergakademie *18* (1966), 266 — 273

RÖSLER, H. J., J. PILOT, D. HARZER, and P. KRÜGER: Isotopengeochemische Untersuchungen (O, S, C) an Salinar- und Sapropelsedimenten Mitteleuropas. Report 23 Sess. Int. Geol. Congr. Section 6 (Geochem.), Prague 1968

SAKAI, H.: Isotopic Properties of Sulfur Compounds in Hydrothermal Processes. The Geochem. Soc. of Japan. Geochem. Journal *2* (1968), 29—49

SCHRAGE, I.: Schwefelisotopenuntersuchungen an einigen Lagerstättenbezirken unter besonderer Berücksichtigung der kiesigblendigen Bleierzformation der Erzlagerstätte von Freiberg. Freiberger Forschungsh. C *143* (1962)

THODE, G. H., J. MONSTER, and H. G. DUNFORD: Sulfur Isotope Geochemistry, Geochim. Cosmochim. Acta *20* (1961), 159—174

THODE, H. G.: Sulphur Isotope Geochemistry. Royal Soc. Canada, Spec. Publ. *6* (1963), 25—41

INOGRADOV, A. P., V. A. GRINENKO, and V. L. USTINOV: The Isotopic Composition of Sulphur Combi-
ations in the Black Sea. Geokhimiya (1962) *10*, 851—873 (Russian)

INOGRADOV, V. I., and L. L. SHANIN: Zur Frage der Variationen der Schwefelisotopen-Zusammen-
etzung in den alten Ozeanen. Z. angew. Geol. *15* (1969) 1, 33—36

arbon

LAYTON, R. N., and E. T. DEGENS: Use of Carbon Isotopes of Carbonates for the Differentiation of
resh-water and Marine Sediments. Bull. Am. Ass. Petrol. Geol. *43* (1959), 890-897.

LAYTON, R. N., B. F. JONES and R. A. BERNER: Isotope Studies of Dolomite Formation under Sedi-
entary Conditions. Geochim. Cosmochim. Acta *32* (1968), 415—432

OLOMBO, U., F. GAZZARINI, R. CONFIANTINI, and E. TONGIORGI: Carbon Isotope Composition of Indi-
idual Hydrocarbons from Italian Natural Gases. Nature *205* (1965) 4978

CRAIG, H.: The Geochemistry of the Stable Carbon Isotopes. Geochim. Cosmochim. Acta *3* (1953),
3—92

CRAIG, H.: Isotopic Standards for Carbon and Oxygen and Correction Factors for Mass-spectrometric
analysis of Carbon Dioxide. Geochim. Cosmochim. Acta *12* (1957), 133—149

ECKELMANN, W. R., et al.: Implications of Carbon Isotopic Composition of Total Organic Carbon of
ome Recent Sediments and Ancient Oils. Bull. Amer. Assoc. Petroleum Geologists *46* (1962) 5, 699
— 704

HAHN-WEINHEIMER, P.: Die isotopische Verteilung von Kohlenstoff und Schwefel in Marmor und
anderen Metamorphiten. Geol. Rundsch. *55* (1965) 1, 197—209

KEELING, CH. D.: A Mechanism for Cyclic Enrichment of Carbon-*12* by Terrestrial Plants. Geochim.
Cosmochim. Acta *24* (1961), 299—313

KEITH, M. L., and J. N. WEBER: Carbon and Oxygen Isotopic Composition of Selected Limestones and
Fossils. Geochim. Cosmochim. Acta *28* (1964), 1787—1816

KREJCI-GRAF, K., and F. E. WICKMANN: Ein geochemisches Profil durch den Lias alpha. Geochim.
Cosmochim. Acta, *18* (1960), 259—272

LEBEDEV, V. I.: Isotopic Composition of the Carbon in Petroleum and in Natural Gases. Geokhimiya
(1964) *11*, 1128—1137 (Russian)

MAASS, I., R. VEDDER, and H. SCHEFFLER: Untersuchungen über die Genese von Grubengasen durch
Bestimmung der isotopen Zusammensetzung des Kohlenstoffs. Abh. dtsch. Akad. Wiss. Berlin, Kl.
Chem., Geol. Biol., Special paper No. 7 (1964)

MÜLLER, E. P., I. MAASS, and H. HÜBNER: Zur Geochemie des Poreninhaltes des Hauptdolomits
unter besonderer Berücksichtigung der $^{12}C/^{13}C$-Verhältnisse. Abh. dtsch. Akad. Wiss., Kl. Chem.
Biol. Geol. (1964) 7

SILVERMAN, S. R.: Investigations of Petroleum Origin and Evolution Mechanisms by Carbon Isotope
Studies. In: H. Craig, S. L. Miller and G. I. Wasserburg, Isotopic and Cosmic Chemistry. North-Holland Publishing Comp., Amsterdam 1964

SILVERMAN, S. R., and S. EPSTEIN: Carbon-isotopic Composition of Petroleums and other Sedimentary
Organic Materials. Bull. Amer. Assoc. Petroleum Geologists *42* (1958), 998—1012

SILVERMAN, S. R.: Carbon Isotope Geochemistry of Petroleum and other Natural Organic Material.
3rd Intern. Sc. Geoch. Conf. , Vol. 2, pp. 328—339. Budapest 1963

STAHL, W., and C. H. TANG: Carbon Isotope Measurements of Methane, Higher Hydrocarbons, and of
Carbon Dioxide of Natural Gases from Northwestern Taiwan. Petrol. Geology of Taiwan *8* (1971),
77—91

VINOGRADOV, A. P., and O. I. KROPOTOVA: On the Isotopic Fractionation of Carbon in Geologic Processes. Izv. Akad. Nauk SSSR, Geol. Ser. *11* (1967), 3—13 (Russian)

WEBER, J. N., R. E. BERGENBACK, E. G. WILLIAMS and M. L. KEITH: Reconstruction of Depositional
Environments in the Pennsylvanian Vanport Basin by Carbon Isotope Ratios. Journal of Sedim.
Petrology *35* (1965), 36—48

Lead

COOPER, J. A., and I. R. RICHARDS: Lead Isotopes and Volcanic Magmas. Earth Planet. Sci. Let. *1*
(1966) 5, 259—269

DOE, B. R.: Lead Isotopes. Springer-Verlag, Berlin/Heidelberg/New York 1970

PATTERSON, C.: Characteristics of Leadotope Evolution on a Continental Scale in the Earth. In: H. Craig, S. L. Miller and G. I. Wasserburg: Isotopic and Cosmic Chemistry. North-Holland Publishing Comp., Amsterdam 1964

PILOT, J., J. LEGIERSKI and H. J. RÖSLER: Pb- und S-Isotopenuntersuchungen an Freiberger und anderen Erzlagerstätten. Geologie 19 (1970), 41—55

RICHARDS, I. R.: Major Lead Orebodies — Mantle Origin? Econ. Geology 66 (1971), 425—434

RICHARDS, I. R., J. A. COOPER, et al.: Lead Isotope Geochemistry and the Upper Mantle. Upper Mantle Project. 2nd Austral. Progr. Rep., Canberra 1967

RUSSEL, R. D., and FARQUHAR: Lead Isotopes in Geology. Elsevier 1960, 251 pp.

Strontium, Tritium, Deuterium, Nitrogen, Lithium, Potash

ABDULLAMEV, A. A., B. K. KHAITOV, A. S. ZAKHIDOV, and ANISHENKO: Die Verwendung von Tritium als Indikator unterirdischen Wassers. Z. angew. Geologie 14 (1966) 2, 86—88

COMPSTON, W., P. A. ARRIENS et al.: Strontium Isotope Geochemistry. Upper Mantle Project, 2nd Austral. Progr. Rep., Canberra 1967

FAURE, G., and P. M. HURLEY: The Isotopic Composition of Strontium in Oceanic and Continental Basalts. Application to the Origin of Igneous Rocks. J. Petrol. 4 (1963), 31—50

HEDGE, C. E., and F. G. WALTHALL: Radiogenic Strontium-87 as an Index of Geologic Processes. Science 140 (1963), 1214—1217

LÉTOLLE, R.: Recherche des Variations Isotopiques Naturelles du Potassium; Résultats, Perspectives et Interprétation. Geol. Rundsch. 55 (1965) 1, 209—216

SCHREINER, G. D. L., and H. J. F. D. WELKE: Variations in ^{39}K/^{41}K Ratio and Movement of Potassium in Heated and Stressed Xenoliths. Geochim. Cosmochim. Acta 35 (1971), 719—726

SVEC, H. I., and A. R. ANDERSON: The Absolute Abundance of the Lithium Isotopes in Natural Sources. Geochim. Cosmochim. Acta 29 (1965), 633—641

VOLYNEZH, V. F., I. K. ZADOPOZNIĬ, and K. P. FLORENSKIĬ: On the Isotopic Nitrogen Composition in the Crust of the Earth. Geokhimiya (1967) 5, 587—593 (Russian)

Oxygen and Palaeotemperature

BOWEN, R., and P. FRITZ: Oxygen Isotope Palaeotemperature Analysis. Experientia 19 (1963)

BOWEN, R.: Palaeotemperature Analysis, Vol. 2 of Methods in Geochemistry and Geophysics; Elsevier Publ. Comp., Amsterdam/London/New York 1966

CRAIG, H.: Isotopic Variations in Meteoric Waters. Science 133 (1961), 1702—1703

DEGENS, E. T: Die geochemische Verteilung von Sauerstoffisotopen in Vergangenheit und Gegenwart. Ber. Phys. Ges. zu Würzburg 69 (1958), 128—133

DONZOVA, E. I.: Sauerstoff-Isotopen und das Problem der Granitgenese. Z. Angew. Geol. 14 (1968) 12, 633—636

EPSTEIN, S.: The Variations of the ^{18}O/^{16}O-ratio in Nature and some Geologic Implications. In: Researches in Geochemistry (ed. P. H. ABELSON). John Wiley & Sons, New York; Chapman & Hall, London 1959

EPSTEIN, S., D. L. GRAF, and E. T. DEGENS: Oxygen Isotope Studies on the Origin of Dolomites. In: H. CRAIG, S. L. MILLER, and G. I. WASSERBURG, Isotopic and Cosmic Chemistry. North-Holland Publishing Comp., Amsterdam 1964

HARZER, D.: Isotopengeochemische Untersuchungen (^{18}O, ^{13}C) an hydrothermalen Ganglagerstätten. Freiberger Forschungsh. C 247 (1970)

HITCHON, B., and J. FRIEDMAN: Geochemistry and Origin of Formation Waters in the Western Canada Sedimentary Basin — I: Stable Isotopes of Hydrogen and Oxygen. Geochim. Cosmochim. Acta 33 (1969), 1321—1349

O'NEIL, J. R., and J. BARNES: ^{13}C and ^{18}O Compositions in some Fresh-Water Carbonates Associated with Ultramafic Rocks and Serpentinites: Western United States. Geochim. Cosmochim. Acta 35 (1971), 687—697

O'NEIL, J. R., and R. N. CLAYTON: Oxygen Isotope Geothermometry. In: H. CRAIG, S. L. MILLER, and G. I. WASSERBURG, Isotopic and Cosmic Chemistry. North-Holland Publishing Comp., Amsterdam 1964

SAVIN, S. M., and S. EPSTEIN: The Oxygen and Hydrogen Isotope Geochemistry of Ocean Sediments and Shales. Geochim. Cosmochim. Acta 34 (1970), 43—63

YLOR, H., and S. Epstein: Relationship between $^{18}O/^{16}O$ Ratios in Coexisting Minerals of Igneous
d Metamorphic Rocks. Bull. Geol. Soc. America 73 (1962), 461—480
YLOR, H. P. jr.: The Oxygen Isotope Geochemistry of Igneous Rocks. Contr. Mineral and Petrol.
(1968), 1—71
JRI, B., and H. P. Taylor jr.: An Oxygen and Hydrogen Isotope Study of a Granodiorite Pluton
m the Southern California Batholith. Geochim. Cosmochim. Acta 35 (1971), 383—406

).7. Physical Age Determination (Dating) of Deposits and Rocks

).7.1. Radioactive Decay Law

he possibility of a non-relative (relative within the framework of stratigraphy) dating is
ased on the fact that nuclear processes are essentially independent of geologic processes and
hysico-chemical parameters (temperature, pressure). The transformation of atomic nuclei
hich are in an energetically unstable state into the stable state is based on the radioactive decay
w which assigns macroscopically measurable quantities to time:

$$M = - \lambda M \, dt \tag{1}$$

r in the integrated form

$$I = M_0 \, e^{-\lambda t} \quad \text{or} \quad t = \frac{1}{\lambda} \ln \frac{M_0}{M}, \tag{2}$$

here:

I is the number of parent atoms present at time t (e.g. today),
I_0 is the number of parent atoms at time $t =$ zero (e.g. formation of the mineral),
is the decay constant.
I_0 is not measurable today; however, the following equation applies to a substantially closed
ystem:

$$I_0 = M + T, \tag{3}$$

here: T is the number of daughter atoms so that the decay formula can be written in the
orm suitable for dating, namely,

$$= \frac{1}{\lambda} \ln \left\{ 1 + \frac{T}{M} \right\}. \tag{4}$$

is the decay constant which is connected with the half-life $T_{1/2}$ by

$$= \frac{\ln 2}{T_{1/2}} = \frac{0.6931}{T_{1/2}}. \tag{5}$$

n the event of a dual disintegration process (e.g. in the case of ^{40}K) where two daughter
roducts T_1 and T_2 (^{40}Ar, ^{40}Ca) are created, and that at a constant branching ratio of

$$R = \frac{T_2}{T_1} = \frac{\lambda_2}{\lambda_1}, \tag{6}$$

411

then the following equation holds:

$$\frac{T}{M} = \frac{T_1 + T_2}{M} = \left(1 + \frac{T_1}{T_2}\right)\frac{T_2}{M} = \frac{1 + \dfrac{T_2}{T_1}}{T_2/T_1} \cdot \frac{T_2}{M} = \frac{1 + R}{R} \cdot \frac{T_2}{M} \tag{7}$$

$$t = \frac{1}{\lambda} \ln\left\{1 + \frac{1 + R}{R} \cdot \frac{T_2}{M}\right\} = \frac{1}{\lambda} \ln\left\{1 + \frac{\lambda_1 + \lambda_2}{\lambda_2} \cdot \frac{T_2}{M}\right\}.$$

λ is in this case the overall decay constant.

If the radioactive decay constant is known and the ratio of daughter substance to parent substance determined, the date of the last separation of daughter substance can be determined.

10.7.2. Limits of an "Absolute" Age Determination

In general, the following requirements must be complied with:

1. At time $t = 0$, a complete separation of the daughter element from the parent element should have taken place by a natural chemical or physical process (e.g. mineral formation, complete argon degassing by heating up, etc.). This process should be short as compared with the age of the mineral.

2. If, during the formation of the mineral, a chemical element of the same species as the stable end product is already present, or if it was not separated or if it will be added later, certain corrections can be made for the given quantities provided their isotopic composition is known.

3. Since time $t = 0$ no other change in the ratio of parent isotope to end product should have taken place than that which is caused by the radioactive decay (substantially closed system).

If (as for example with uranium and thorium) the end product is the result of a process of successive radioactive transformations, comprising a series of more or less short-lived intermediate elements, no gain or loss of these intermediate elements, due to chemical or physical causes, should have taken place, e.g. by diffusing away the inert gases radon, thorium, actinium. During the whole geological history of the sample, radioactive equilibrium between parent element and all of its successive products should have existed.

10.7.3. Types of Radiometric Dating

a) Age determination on the basis of storage of daughter isotopes.

Minerals with a sufficient amount of parent substance are subjected to this method by determining the ratio of parent isotope to daughter isotope.

t	0
formation of the rock, separation of daughters	today, the amounts of daughter and parent isotopes are measured

b) Age determination on the basis of the decrease of the number of parent isotopes (e.g. ^{14}C)

```
 t                                                    0
 ─────────────────────────────────────────────────────────────────────►
```

from this time there is no longer any   today
exchange with the surroundings
(dying of an organism)

The present activity of the sample is compared with that at time t (end of the exchange of parent isotopes with the surroundings).

c) Model age

Minerals which do not contain a parent isotope, but only daughter isotopes are subjected to this method.

```
 W                             P                              H
 ─────────────────────────────────────────────────────────────────────►
                                                              t
 — w                          — p
 formation of the chemical elements    crystallisation of the lead ore
 (primitive lead)                      (or Sr mineral), separation from
                                       parent substance
```

W, P and H indicate the point of time, t, w and p the date of this point of time. The results obtained by means of this method are uncertain ages and more accurate results can only be ensured on the basis of a greater number of samples.

0.7.4. The Various Dating Methods

The common lead method (model age)

Applicable to Pb free from U and Th.
On the basis of the decay series of ^{238}U, ^{235}U and ^{232}Th, the following three equations (with the denotations of the Section "model age") are obtained:

$$\left(\frac{^{206}\text{Pb}}{^{204}\text{Pb}}\right)_H = \left(\frac{^{206}\text{Pb}}{^{204}\text{Pb}}\right)_w + \left(\frac{^{238}\text{U}}{^{204}\text{Pb}}\right)_H \left\{ e^{\lambda 238 w} - e^{\lambda 238 p} \right\}, \tag{8}$$

$$\left(\frac{^{207}\text{Pb}}{^{204}\text{Pb}}\right)_H = \left(\frac{^{207}\text{Pb}}{^{204}\text{Pb}}\right)_w + \left(\frac{^{235}\text{U}}{^{204}\text{Pb}}\right)_H \left\{ e^{\lambda 235 w} - e^{\lambda 235 p} \right\}, \tag{9}$$

$$\left(\frac{^{208}\text{Pb}}{^{204}\text{Pb}}\right)_H = \left(\frac{^{208}\text{Pb}}{^{204}\text{Pb}}\right)_w + \left(\frac{^{232}\text{Th}}{^{204}\text{Pb}}\right)_H \left\{ e^{\lambda 232 w} - e^{\lambda 232 p} \right\}. \tag{10}$$

The factors in front of the brackets can be expressed by

$$\mu = \frac{^{238}\text{U}}{^{204}\text{Pb}} = \text{environmental index},$$

$$k = \frac{^{232}\text{Th}}{^{238}\text{U}}; \quad \frac{^{238}\text{U}}{^{235}\text{U}} = 139,$$

hence

$$\frac{^{235}\text{U}}{^{204}\text{Pb}} = \frac{\mu}{139}; \quad \frac{^{232}\text{Th}}{^{204}\text{Pb}} = \mu k.$$

413

If the isotopic composition of the present lead has been measured and if that of the primitive lead is known, and if further the time w is known, k and the desired model age p can be derived from the three equations. A one-stage development of the radiogenic lead is assumed.

An analogous method is that of the common strontium, though it has not been thoroughly developed up to now.

The uranium-lead and thorium-lead method

On the basis of the natural radioactive decay series

$$^{232}\text{Th} \xrightarrow{6\alpha + 4\beta} {}^{208}\text{Pb} \qquad T_{1/2} = 13.9 \cdot 10^9 \text{ a},$$

$$^{235}\text{U} \xrightarrow{7\alpha + 4\beta} {}^{207}\text{Pb} \qquad T_{1/2} = 0.71 \cdot 10^9 \text{ a},$$

$$^{238}\text{U} \xrightarrow{8\alpha + 6\beta} {}^{206}\text{Pb} \qquad T_{1/2} = 4.51 \cdot 10^9 \text{ a},$$

where the parent isotope governs the disintegration rate because it has the longest half-life, three dating methods are obtained. The amounts of parent and daughter isotopes have to be determined by means of chemical and mass-spectrometric measurements, or the amounts may be determined by mass spectrometry with the help of isotope dilution analysis (WEBSTER, 1960).

A fourth method is based on the fact that the isotopic composition of uranium is known:

$$\frac{^{238}\text{U}}{^{235}\text{U}} = 137.8$$

In this case, the amount of parent substance need not be determined, but only the isotope ratio $^{207}\text{Pb}/^{206}\text{Pb}$. The age is derived from

$$\frac{^{207}\text{Pb}}{^{206}\text{Pb}} = \Theta(t) = \frac{1}{137.8} \cdot \frac{e^{\lambda(235)t} - 1}{e^{(238)t} - 1} \tag{11}$$

The function $\Theta(t)$ is given in tables (RUSSEL and FARQUHAR, 1960). Corrections have to be made if common lead — distinguishable by the presence of ^{204}Pb — is added to the radiogenic lead. For this purpose, the isotopic composition of this common lead must be known; however, exact figures are not always available.

Variations of the isotopic ratios:

$$\frac{^{206}\text{Pb}}{^{204}\text{Pb}} = 12.60 \text{ to } 26.00$$

$$\frac{^{207}\text{Pb}}{^{204}\text{Pb}} = 14.20 \text{ to } 17.00$$

$$\frac{^{208}\text{Pb}}{^{204}\text{Pb}} = 32.8 \text{ to } 52.6$$

For the purpose of corrections, mean values of samples which have already been measured or the value of a local adjacent or pertinent lead sample are used.

The K-Ar method

Disintegration scheme

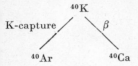

Isotopic compositions:

^{39}K 93.32	^{40}Ca 96.95	^{36}Ar 0.337
^{40}K 0.0119	^{42}Ca 0.64	^{38}Ar 0.063
^{41}K 6.66	^{43}Ca 0.138	^{40}Ar 99.6
	^{46}Ca 2.09	$\dfrac{^{40}\text{Ar}}{^{36}\text{Ar}} = 295.6$ (atmospheric argon)
	^{46}Ca 0.0032	
	^{48}Ca 0.182	

Generally used radioactive decay constants:

Anglo-America: $\lambda_K = 0.584 \cdot 10^{-10}$ a^{-1}; $\lambda = 5.30 \cdot 10^{-10}$ a^{-1}; $R = 0.124$
$\qquad\qquad\quad \lambda_\beta = 4.72 \ \cdot 10^{-10}$ a^{-1}

Canada: $\qquad \lambda_K = 0.586 \cdot 10^{-10}$ a^{-1}; $\lambda = 5.35 \cdot 10^{-10}$ a^{-1}; $R = 0.123$
$\qquad\qquad\quad \lambda_\beta = 4.76 \ \cdot 10^{-10}$ a^{-1}

USSR: $\qquad\ \lambda_K = 0.557 \cdot 10^{-10}$ a^{-1}; $\lambda = 5.28 \cdot 10^{-10}$ a^{-1}; $R = 0.118$
$\qquad\qquad\quad \lambda_\beta = 4.72 \ \cdot 10^{-10}$ a^{-1}

For the determination of the age, equation (7) is used. In many laboratories, the age is obtained from a graphical representation which is based on equation (7).

Minerals usable for K-Ar dating are mica, potash feldspar and glauconite among the silicates, and sylvine (to be used with great care), polyhalite and langbeinite among the salts.

The quality of a K-Ar dating is largely dependent on the amount of argon that has been emitted from the sample since the point of time to be dated so that the apparent age obtained from the analysis is shorter than the actual age. In general, argon is best retained by hornblende, mica, langbeinite, and polyhalite (PILOT and RÖSLER, 1967).

The ^{40}K-^{40}Ca method is rarely applicable because the radiogenic ^{40}Ca proportion is very small compared with the ^{40}Ca proportion of common calcium.

The Rb—Sr method

Isotopic compositions:

^{87}Rb 27.85 %	^{84}Sr 0.56 %	
^{85}Rb 72.15 %	^{86}Sr 9.86 %	
	^{87}Sr 7.02 %	
	^{88}Sr 82.56 %	

$T_{1/2} = 50 \cdot 10^9$ a, $\qquad \lambda = 1.39 \cdot 10^{-11}$ a^{-1}.

The following minerals are used for dating:

Lepidolite, mica, K-feldspar, frequently the whole-rock sample.

Because of uncertainties in the isotopic composition of the common strontium and for the detection of thermal perturbation (e.g. metamorphism, magmatic intrusions), the Compston-Jeffrey method shown in Fig. 134 is frequently used.

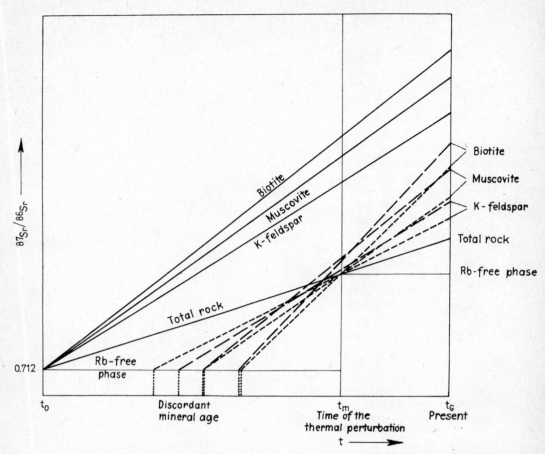

Fig. 134   Rb-Sr age of individual mineral fractions (Compston-Jeffrey Method; for explanations see text)

The straight lines plotted on the graph are the result of the following considerations: The total strontium is composed of a radiogenic proportion and a common portion:

$$^{87}Sr_{total} = {}^{87}Sr_{rad} + {}^{87}Sr_{com} .\tag{12}$$

The expansion of the exponential function within the law of radioactive disintegration results in

$$^{87}Rb_0 = {}^{87}Sr_{rad} + {}^{87}Rb = {}^{87}Rb\ e^{\lambda t} \approx {}^{87}Rb\ (1 + \lambda t) ,\tag{13}$$

$$\rightarrow {}^{87}Sr_{rad} = {}^{87}Rb\ \lambda t .\tag{14}$$

Upon substitution in equation (12) we obtain

$$\left(\frac{^{87}Sr}{^{86}Sr}\right)_{total} = \left(\frac{^{87}Rb}{^{86}Sr}\right) \lambda t + \left(\frac{^{87}Sr}{^{86}Sr}\right)_{com}\tag{15}$$

416

straight line in $\left(\frac{^{87}Sr}{^{86}Sr}\right)_{total}$ and t. The ratio $\left(\frac{^{87}Rb}{^{87}Sr}\right)$ is a function of the mineral type F and the locus (point of sampling in the rock). By measuring $\left(\frac{^{87}Sr}{^{86}Sr}\right)_{total}$ (present ratio) and $\left(\frac{^{87}Rb}{^{86}Sr}\right)$ multiplied by λ, this is the inclination of the straight line), the straight line can be determined. In the case of a common intersection, age and $(^{87}Sr/^{86}Sr)_0$ can be determined or one $(^{87}Sr/^{86}Sr)_0$ of the last thermal event (homogenisation).

The ^{14}C method

$^{12}C \sim 98.89\%$ Half-life of $^{14}C = 5570$ a.
$^{13}C \sim 1.11\%$
$^{14}C \sim 1 \cdot 10^{-10}\%$

Due to cosmic radiation, thermal (hence, slow) neutrons are formed in the upper atmospheric layers which, together with the ^{14}N contents in the air, form the radioactive isotope ^{14}C in the nuclear reaction

$$^{14}N \xrightarrow{+} {}^{14}C + p.$$

This isotope disintegrates according to

$$^{14}C \rightarrow {}^{14}N + \beta^-.$$

Production and decay of ^{14}C lead to an equilibrium to which a certain ^{14}C concentration corresponds (the produced ^{14}C immediately combines with the atmospheric oxygen into $^{14}CO_2$). The well-founded assumption that, due to *natural* processes, the abundance of ^{14}C in the atmosphere has been practically constant for at least 100,000 years, enables the determination of the age of carbonaceous materials (on the other hand, combustion of coal and oil have caused a dilution and atomic explosions an increase in the abundance of ^{14}C in the course of the last 100 years). Since, for example, living beings take up CO_2 from the atmosphere, the same ^{14}C abundance obtains within them. With the death of the living being, the exchange with the atmosphere is interrupted, the content of ^{14}C decreases according to

$$N = N_0\, e^{-\lambda t}.$$

When comparing the activity of a recent sample (as a standard, for example, wood of an age of 200 years) with that of the sample to be dated, the age can be calculated (LIBBY, 1955). Datable ages: between about 1,000 to maximum 80,000 years. Application: archaeology, sediments (glacial periods, combined with ^{18}O thermometry), age of deep waters.

Tritium method

$^1H \sim 99.9851\%$ Half-life of $^3T = 12.257$
$^2D \sim 0.0149\%$ 1 tritium unit (T.U.) $= 1 \cdot 10^{-18}\,\dfrac{T}{H}$
$^3T \sim 10^{-16}\%$

Analogous to the formation of ^{14}C, tritium ^3T is formed in the upper atmospheric layers and is practically completely combined with water (T_2O) and gets into the water cycle via rainfall. The residence time of water which has been isolated from this cycle (e.g. ground water) can be determined by determining the present and initial tritium concentration.

The short half-life and the fact that tritium is a constituent of water render this method particularly suitable for the solution of many hydrological problems (time and velocity of flow in a water course, quantity of ground water, ratio of ground water and rainwater in rivers, etc.).

Methods of minor importance are

the He—U or He—Th method (FANALE and KULP, 1962)
the Re—Os method (HIRT et al., 1963)
the ^{10}Be method (KHARKAR et al., 1963)
the luminescence method (ZELLER and RONCA, 1963)
the fossil-track method (FLEISCHER and PRICE, 1964)
the method of the pleochroic aureoles (DEUTSCH, 1957), and others

Table 187. Diffusion Constants (D or D/a^2) and Activation Energy of Various Minerals (after PILOT, 1970; see also FECHTIG and KALBITZER, 1966)

Mineral	D, D/a^2	Volume Diffusion $E \left[\dfrac{\text{kcal}}{\text{mole}}\right]$	References
Biotite	$(D/a^2)_{20\,°C} = 10^{-29}\ \text{sec}^{-1}$	69	FRECHEN, LIPPOLD (1965)
Biotite	$(D/a^2)_{20\,°C} = 10^{-34}\ \text{sec}^{-1}$	86	FRECHEN, LIPPOLD
Muscovite		72, 37, 18	GERLING et al., (1961)
Phlogopite	$D_{20\,°C} = 10^{-29}\ \text{cm}^2\ \text{sec}^{-1}$		EVERNDEN et al., (1960)
Glauconite	$D_{20\,°C} = 10^{-29}\ \text{cm}^2\ \text{sec}^{-1}$	28	EVERNDEN et al., (1960)
Margarite	$(D/a^2)_{20\,°C} = 10^{-30}\ \text{sec}^{-1}$	54	FECHTIG et al., (1960)
K-feldspar	$D_{25\,°C} = (10^{-72}),\ 10^{-27}\ \text{cm}^2\ \text{sec}^{-1}$	(106, 93) 33	AMIRKHANOV et al., (1959 b)
Microcline	$D_{20\,°C} = 10^{-26}\ \text{cm}^2\ \text{sec}^{-1}$	24	EVERNDEN et al., (1960)
Microcline	$D_{20\,°C} = 10^{-20}\ \text{cm}^2\ \text{sec}^{-1}$	18	RAADSGAARD et al., (1961)
Sanidine	$(D/a^2)_{20\,°C} = 4 \cdot 10^{-28}\ \text{sec}^{-1}$	41	FRECHEN, LIPPOLD (1965)
Sanidine	$(D/a^2)_{20\,°C} = 3 \cdot 10^{-33}\ \text{sec}^{-1}$	48	FRECHEN, LIPPOLD (1965)
Pyroxene	$D_{25\,°C} = 4 \cdot 10^{-55}\ \text{cm}^2\ \text{sec}^{-1}$	74	AMIRKHANOV et al., (1959 a)
Sylvine	$D_{40\,°C} = 10^{-19}\ \text{cm}^2\ \text{sec}^{-1}$		GENTHNER et al., (1953)
Langbeinite	$D_{40\,°C} = 10^{-34}\ \text{cm}^2\ \text{sec}^{-1}$		PILOT, RÖSLER (1967)

For a non-volume diffusion, the same authors state $(D/a^2)_{20\,°C} = 10^{-22}\ \text{sec}^{-1}$ for margarite and $(D/a^2)_{20\,°C} = 10^{-28}\ \text{sec}^{-1}$ for sanidine.

10.7.5. Geochronological Time Charts

Fig. 135 gives a survey of the geologic formations (systems) together with their geotectonic phases and their "absolute" ages as stated by several authors, enabling a comparison of the data. An expansion of these data to Precambrian formations and platforms of the earth is given in Fig. 136.

Bibliography

AFANAS'EV, G. D.: The Geological Time-scale in Terms of Absolute Dating. Report 23. Sess. Int. Geol. Congr. Section 6 (Geochem.), Prague 1968
ARSLANOV, CH. A., et al.: Absolute Age Determination According to Radiocarbon by the Scintillation Method. Geokhimiya (1968) 2, 198—206 (Russian)
BARANOV, V. I.: On the Methods of Determining the Age of the Earth. Geokhimiya (1966) 1, 15—24 (Russian)
BARANOV, V. I., and K. G. KNORRE: Age of the Earth's Crust and Dynamics of Radiogenic Gases (^{4}He and ^{40}Ar) Supply to the Atmosphere. Geokhimiya (1967) 12, 1418—1429 (Russian)

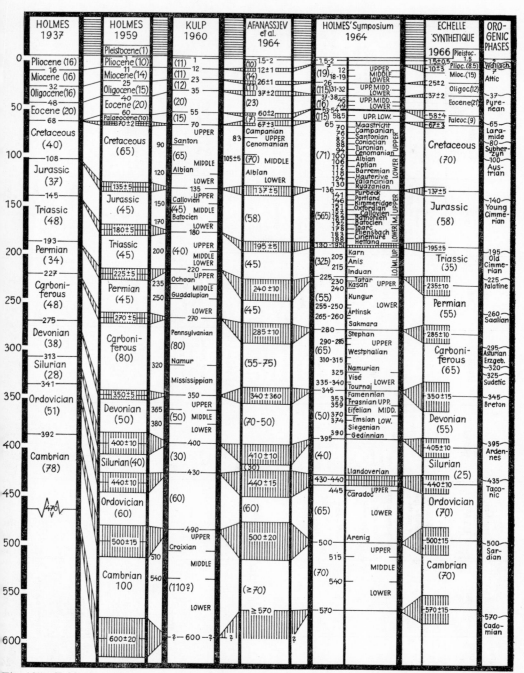

Fig. 135   Table comparing the latest published geochronological scales for the Phanerozoic time and a synthetic scale which has been recommended by the International Union for Geological Sciences, Commission for Geochronology

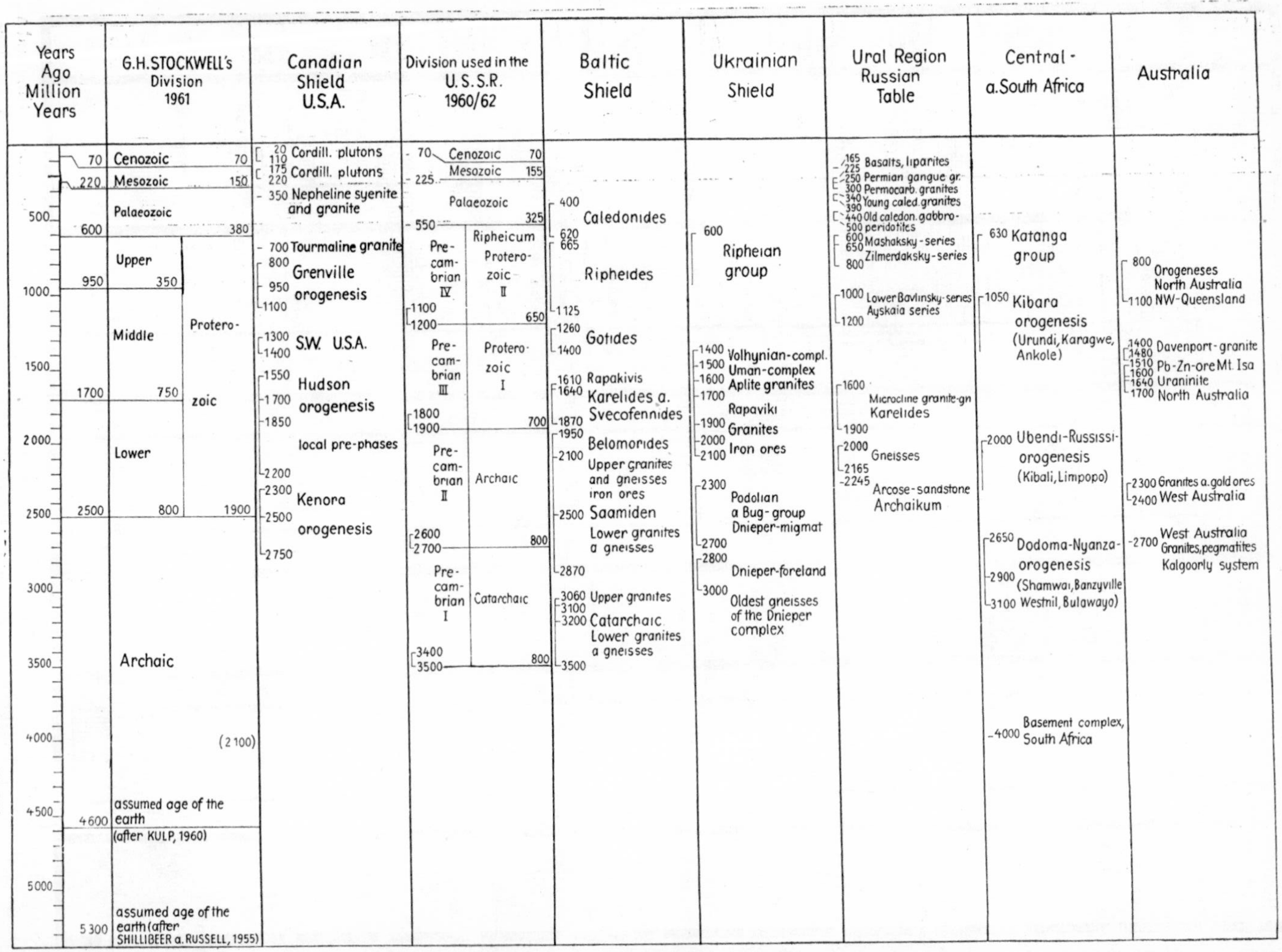

Fig. 136 Comparison of various time scales for Precambrium era (from REH, 1963)

Column: Years Ago Million Years (scale): 500, 1000, 1500, 2000, 2500, 3000, 3500, 4000, 4500, 5000

G.H. STOCKWELL's Division 1961
- 70 Cenozoic
- 220 Mesozoic 150
- Palaeozoic
- 600 — 380
- Upper 350
- Protero-zoic
- Middle 750
- Lower 800
- 1700
- 1900
- Archaic
- (2100)
- 4600 assumed age of the earth (after KULP, 1960)
- 5300 assumed age of the earth (after SHILLIBEER a. RUSSELL, 1955)

Canadian Shield U.S.A.
- 20 Cordill. plutons
- 110
- 150
- 220 Cordill. plutons
- 350 Nepheline syenite and granite
- 700 Tourmaline granite
- 800 Grenville
- 950 orogenesis
- 1100
- 1300 SW. U.S.A.
- 1400
- 1550 Hudson
- 1700 orogenesis
- 1850
- local pre-phases
- 2200
- 2300 Kenora
- 2500 orogenesis
- 2750

Division used in the U.S.S.R. 1960/62
- 70 Cenozoic 70
- 225 Mesozoic 155
- Palaeozoic 325
- 550
- Pre-cambrian IV / Ripheicum / Protero-zoic II — 650
- 1100 Pre-cambrian III / Protero-zoic I
- 1200
- 1800
- 1900 Pre-cambrian II / Archaic
- 2600
- 2700 Pre-cambrian I / Catarchaic — 800
- 3400
- 3500 — 800

Baltic Shield
- 400 Caledonides
- 620
- 665 Ripheides
- 1125
- 1260 Gotides
- 1400
- 1610 Rapakivis
- 1640 Karelides a. Svecofennides
- 1870
- 1950 Belomonides
- 2100 Upper granites and gneisses iron ores
- 2500 Saamiden Lower granites a gneisses
- 2870
- 3060 Upper granites
- 3100 Catarchaic
- 3200 Lower granites a gneisses
- 3500

Ukrainian Shield
- 600 Ripheian group
- 1400 Volhynian-compl.
- 1500 Uman-complex
- 1600 Aplite granites
- 1700
- 1900 Rapaviki Granites
- 2000
- 2100 Iron ores
- 2300 Podalian a Bug- group Dnieper-migmat
- 2700
- 2800 Dnieper-foreland
- 3000 Oldest gneisses of the Dnieper complex

Ural Region Russian Table
- 165 Basalts, liparites
- 225 Permian gangue gr.
- 250 Permocarb granites
- 300 Young caled. granites
- 340
- 440 Old caledon. gabbro-
- 500 peridotites
- 650 Moshaksky-series
- 800 Zilmerdaksky-series
- 1000 Lower-Bovinsky series Ayskaia series
- 1200
- 1600 Microcline granite-gn Karelides
- 1900 Gneisses
- 2000
- 2165 Arcose-sandstone
- 2245 Archaikum

Central- a. South Africa
- 630 Katanga group
- 1050 Kibara orogenesis (Urundi, Karagwe, Ankole)
- 2000 Ubendi-Russissi- orogenesis (Kibali, Limpopo)
- 2650 Dodoma-Nyanza- orogenesis
- 2900 (Shamwa, Banzyville)
- 3100 Westnil, Bulawayo)
- ~4400 South Africa, Basement complex

Australia
- 800 Orogeneses North Australia
- 1100 NW-Queensland
- 3400 Davenport- granite
- 1480
- 1510 Pb-Zn-ore Mt. Isa
- 1600 Uraninite
- 1640
- 1700 North Australia
- 2300 Granites a. gold ores
- 2400 West Australia
- 2700 West Australia Granites, pegmatites Kalgoorly system

420

BROECKER, W. S.: Absolute Dating and the Astronomical Dating of Glaciation. Science *151* (1966) 708, 299—304

COPPENS, R., G. DURAND, and G. JURAIN: Equilibre et Déséquilibre Radioactifs dans les Minéraux et les Roches. Conséquences Géochimiques et Géochronologiques. Sci. Terre *10* (1964/65), 107—131

DALRYMPLE, G. B., and M. A. LANPHERE: Potassium-Argon Dating. Principles, Techniques and Application to Geochemistry. Freeman and Co., San Francisco 1969

DAMON, P. E.: A Theory of "Real" K — Ar — Clocks. Eclogae geol. Helv. *63/1* (1970), 69—76

DOE, B. R.: Lead Isotopes. Springer-Verlag, Berlin/Heidelberg/New York 1970

FANALE, F. P., and J. L. KULP: The Helium Method and the Age of the Cornwall-Pennsylvania Magnetite Ore. Econ. Geol. *57* (1962), 735—746

FLEISCHER, R. L., and P. B. PRICE: Technique for Geological Dating of Minerals by Chemical Etching of Fission Fragment Tracks. Geochim. Cosmochim. Acta *28* (1964), 1705

GERLING, E. K., I. M. MOROZOVA, and V. D. SPRINTSSON: On the Nature of the Excess Argon in Some Minerals. Report 23. Sess. Int. Geol. Congr. Section 6 (Geochem.), Prague 1968

GERLING, E. K. et al.: The Determination of the Age of the Earth on the Basis of Minerals and Rocks. Khimiya Zemnoĭ Kory, Vol. I, Izd. Akad. Nauk SSSR, Moscow 1963, 366—373 (Russian)

GODWIN: Radiocarbon Dating. Nature *195* (1962), 943—945

HAMILTON, E. I.: Applied Geochronology. Academic Press, London/New York 1965

HARPER, C. T.: Graphical Solutions to the Problem of Radiogenic Argon — 40 Loss from Metamorphic Minerals. Eclogae geol. Helv. *63/1* (1970), 119—140

HINTENBERGER, H.: Die Rubidium-Strontium-Methode. Geol. Rdsch. *49* (1960), 197—223

HIRSCHMANN, G.: Eine Altersbestimmungsmethode für kristalline Gesteine mittels radioaktiver Höfe. Ber. Deutsch. Ges. Geol. Wiss., B, *12* (1967) 4, 373—387

HIRT, B., W. HERR, and W. HOFFMEISTER: Age Determinations by the Rhenium-osmium Method. In: Radioactive Dating, Proceedings of a Symposium, Athens 1962. International Atomic Energy Agency, Vienna 1963, 35—44

HOUTERMANS, F. G.: Die Bleimethoden der geologischen Altersbestimmung. Geol. Rdsch. *49* (1960), 168—196

JÄGER, E., E. NIGGLI, and H. BAETHGE: Two Standard Minerals, Biotite and Muscovite, for Rb-Sr and K-Ar Age Determinations, Sample Bern 4 B and Bern 4 M from a Gneiss from Brione, Valle Verzasca Switzerland). Schweiz. mineralog. petrogr. Mitt. *43* (1963), 465—470

KIGOSHI, K.: Ionium Dating of Igneous Rocks. Science *156* (1967) 3777, 932—934

KÖPPEL, V., and M. GRÜNENFELDER: A Study of Inherited and Newly Formed Zircons from Paragneisses and Granitised Sediments of Strona — Ceneri — Zone (Southern Alps). Schweiz. Min. Petr. Mitt. *51* (1971), 385—409

LEUTWEIN, F.: La Géochronologie des Roches. Bull. Soc. Franc. Ceram. *63* (1964), 49—64

LIBBY, E. F.: The Potential Usefulness of Natural Tritium. Proc. Natl. Acad. Sci. U.S. *39* (1953), 245

LIBBY, E. F.: Radiocarbon Dating. University of Chicago Press 1955, 175 pp.

MEIER, H.: Neuere Beiträge zur Geochronologie und Geochemie. Fortschr. chem. Forschg. *7* (1966) 2, 233—321

MÜNNICH, K. O., and J. C. VOGEL: Investigation of Meridional Transport in the Troposphere by Means of Carbon-14 Measurements. In: Radioactive Dating. I.A.E.A., Vienna 1963

PILOT, J.: Isotopengeochemie. Situation, Konzeptionen, Entwicklung, Möglichkeiten. Freiberger Forschungsh. C 255, Leipzig 1970

PILOT, J., and H. J. RÖSLER: Altersbestimmung von Kalisalzmineralen. Naturwiss. *54* (1967), 490

RÖSLER, H. J., and J. PILOT: Die zeitliche Einstufung der sächsisch-thüringischen Ganglagerstätten mit Hilfe der K-Ar-Methode. Freiberger Forsch.-H. C *209* (1967), 87—98

RUSSEL, R. D., and R. M. FABQUAR: Lead Isotopes in Geology. Interscience Publishers Inc., New York 1960

SCHAEFFER, O. A., and J. ZÄHRINGER: Potassium Argon Dating. Springer, Berlin/Heidelberg/New York 1966

TILTON, G. R., and S. R. HART: Geochronology. Science, Washington *140* (1963), 3565, 357—366

TUGARINOV, A. I., and G. V. VOĬTKEVICH: Precambrian Geochronology of the Continents. Izd. "Nedra", Moscow 1966 (Russian)

VINOGRADOV, A. P., and A. I. TUGARINOV: Geochronological Scale of the Precambrian. Report 23. Sess
Int. Geol. Congr. Section 6 (Geochem.), Prague 1968
VOĬTKEVICH, G.: Radioactive Dating. Proceedings of a Symposium, Athens 1962. Internat Atomic
Energy Agency, Vienna 1963
WEBSTER, R. K.: Mass-spectrometric Isotope Dilution Analysis. In "Methods in Geochemistry". Edit.
A. A. Smales, New York, London 1960
WETHERILL, G. W.: Radioactive Decay, Constants and Energies. Geol. Soc. Amer., Mem. (1966) 97,
513—519
ZÄHRINGER, J.: Altersbestimmungen nach der K-Ar-Methode. Geol. Rdsch. 49 (1960)
ZELLER, E. J., and L. B. RONCA: New Developments in the Thermoluminescence Method of Geologic
Age Determination. In: Radioactive Dating. Proceedings of a Symposium, Athens 1962. International
Atomic Energy Agency, Vienna 1963, 73—85
ZUBAKOV, V. A.: Geochronical Scale for the Continental Pleistocene (According to Radiometric Data).
Geokhimiya (1967) 2, 144—154 (Russian)
Group of authors: Vergleichende Tabelle der letzten veröffentlichten geochronologischen Zeitskalen
der phanerozoischen Epochen. Z. angew. Geol. 14 (1968) 8, 440—442

10.8. Geochemistry of the Anthroposphere and Technosphere

General

The natural geochemical equilibrium is more and more disturbed, directly and indirectly,
by the action of man. In many industrial states, the regional geochemical conditions are
today changed to a measurable extent. In the Soviet Union, such problems are dealt with in a
branch of geochemistry, the so-called "geochemistry of the landscape" PERELMAN, 1961);
other countries also pay particular attention, on an increasing scale, to the relation between man and geological environment within the national economic sphere (for example, in
the agricultural-medical field, WARREN, 1964, 1966; BERGMANN, 1962; VINOGRADOV, 1963).
Besides the mass handling of geological materials in the construction industry, in river regu-
lation, mining, etc., the following factors are of particular effect in this connection:

a) fertilization in agriculture and forestry,

b) injurious substances produced by the industrial society,

c) direct physiological effects on man.

Agriculture and Forestry (Agrogeochemistry)

The basis of intensive farming and forestry is the natural supply of water and mineral nutrients
and the artificial effect produced by mineral fertilisers. The influence of an insufficient or
excessive supply of nutritive matter in the soil is not terminated in the plant, but in many
cases remains active, producing also an effect, directly or via the animal, on the development
and health of man. The natural relations are revealed by VINOGRADOV (1963) in his work on
the "biogeochemical provinces". A particular influence is exerted by trace elements whose
effect has been neglected so far. An example of trace element concentrations (B, Mn, Cu, Zn)
in agricultural crops is given in Table 188.

Table 188. Contents of Trace Elements in Agricultural Crops which were Grown on Podzol under Similar Conditions (after KATALYMOV, 1955)

Plant		Yield dt/ha	% of the dry substance	Boron mg/kg	Boron g/ha	Manganese mg/kg	Manganese g/ha	Copper mg/kg	Copper g/ha	Zinc mg/kg	Zinc g/ha
Barley	grain	20.1		2.0	12.7	40.0	344	7.2	33.6	50	260
	straw	29		3.0		90.9		6.6		55	
Spring-wheat	grain	23.0		2.0	9.4	80.0	534	7.7	24.9	75	292
	straw	24.0		2.0		146.0		3.0		50	
Oats	grain	22.0		2.0	20.0	88.5	794	5.8	42.0	50	305
	straw	39.0		4.0		153.5		7.5		50	
Timothy	hay	19.0		4.0	7.6	90.9	173	5.8	11.0	40	76
Flax	grain	4.5		8.0	34.8	76.9	356	20.5	28.4	50	222
	straw	25.0		12.5		128.6		7.7		80	
Clover	hay	39.0		25.0	97.0	107.5	419	14.7	57.3	50	195
Yellow lupin	green material	70.0		20.0	140.0	434.8	3,043	18.0	126.0	110	770
Cabbage	head	350	4.7	20.0	32.9	55.8	92	6.9	11.3	35	58
Potato	tuber	270	22.9	6.0	184.1	7.0	2,230	6.0	169.4	20	1,594
	tops	500	14.7	20.0		297.6		18.0		200	
Beet	root	358	20.0	12.0	170.1	77.5	2,717	8.0	80.3	72	1,190
	tops	180	15.6	30.0		770.0		8.2		240	

423

Injurious substances

The modern industrial society pollutes the geological environment especially with the following:

a) dust and gases from metallurgical installations,
b) refuse and sewage of organic and inorganic nature from industrial works,
c) waste from sources of energy (petrol, oil, radioactive fall-out, etc.),
d) refuse, tailings from mines and dressing plants.

The effects produced on plants, animals and man are incalculable and studies are in their initial stage. The problems must be handled in close collaboration by agriculture, medicine, and state and municipal authorities. There are relations which have been discovered and which are utilised, e.g. with respect to fluorine (fluoridation of water to control caries) and iodine (control of goitre) (cf. Table 189).

Table 189. Iodine Concentration in Soils in Relation to the Incidence of the Endemic Goitre (after HERKULES, from VINOGRADOV, 1954)

Soils in New Zealand	Percentage of the Population Affected by Endemic Goitre	Iodine Concentration in Soils in %
Taranaki	4%	$1.4 \cdot 10^{-3}$
Auckland	4%	$1.2 \cdot 10^{-3}$
Dunedin	19%	$3.2 \cdot 10^{-4}$
Clutha Valley	40%	$4.0 \cdot 10^{-5}$
South Canterbury	62%	$3.0 \cdot 10^{-5}$

Bibliography

AUERMANN, E., W. BORRIS, and W. LINDELBACH: Die Trinkwasserfluoridierung. Z. gesamte Hygiene *9* (1963) 10

BERGMANN, W. (Ed.): Über die Mikronährstoffversorgung der Böden. Proc. No. 56, Deutsch. Akad. Landw. Wissensch. Berlin 1962

GERICKE, S.: Die Versorgung von Pflanze und Tier mit Mikronährstoffen. Phosphorsäure *17* (1957), 203—217

JUSATZ, H.-J.: Dreißig Jahre Geomedizin. Münch. Mediz. Wochenschrift *104* (1962) 42

KATALYMOV, V. M.: The Content of Trace Elements in Agricultural Crops. Dokl. Akad. Nauk SSSR *104* (1955) 4, 584 (Russian)

PERELMAN, A. I.: Geochemistry of the Landscape. Izd. "Nauka", Moscow 1967, 165 pp. (Russian)

PERELMAN, A. I.: Map of the Geochemical Landscapes of the USSR, 1:20,000,000. Phys. Geogr. Atlas of the Earth, Moscow 1964

POLYNOV, B. B.: Geochemical Landscapes. In: Polynov, Geographical Works, Izd. Akad. Nauk SSSR, Moscow 1952

RÖSLER, H. J., P. BEUGE, and E. MÜLLER: Einfluß des Hüttenrauches von Freiberg und Halsbrücke auf die Spurenelementgehalte der Böden. Bergakademie *21* (1969) 7, 386—397

SALMI, M.: On Relations between Geology and Multiple Sclerosis. Acta Geograph., Helsinki *17* (1963) 4

VINOGRADOV, A. P.: Biogeochemical Provinces and Their Role in the Organic Development. Geokhimiya (1963) *3*, 199—213 (Russian)

VINOGRADOV, A. P.: Geochemie seltener und nur in Spuren vorhandener Elemente im Boden. Akademie Verlag, Berlin 1954, 250 pp. (Translation from the Russian)

WARREN, H. V.: Geology, Trace Elements and Epidemiology. Geogr. J.G.B. *130* (1964) 4, 525—528

WARREN, H. V.: Symposium on Medical Geology and Geography. Science *147* (1965) 3660, 919

WEBB, J. S., I. NICHOL, and I. THORNTEN: The Broadening Scope of Regional Geochemical Reconnaissance. Report 23. Sess. Int. Geol. Congr. Section 6 (Geochem.), Prague 1968

WIETHAUPT, H.: Über die geschichtliche Seite der Luftverschmutzung. Z. Aerosol *13* (1966) 2, 166 — 175

1.1. General on Units of Measurement

Table 191. Mechanical Units in Various Systems

Quantity	Units		
	in the CGS System centimetre — gram — second	in the MKS System metre — kilogram — second	in the MKpS System metre — kilopond — second
Length	1 cm	1 m	1 m
Mass	1 g	1 kg	$\dfrac{kp \cdot s^2}{9.806,65\,m}$
Time	1 s	1 s	1 s
Velocity	1 cm/s	1 m/s	1 m/s
Acceleration	1 cm/s^2	1 m/s^2	1 m/s^2
Force	1 dyn $= 1$ g cm/s^2	1 kg m/s^2 $= 1$ N (Newton)	kp
Work	1 erg $= 1$ g cm^2/s^2	1 Joule (J) $= 1$ kg m^2/s^2	kpm
Power	1 erg/s $= 1$ g cm^2/s^3	1 Watt (W) $= 1$ kg m^2/s^3	kpm/s
Pressure	1 bar $= 10^6$ dyn/cm^2 $= 10^6$ g/cm s^2	1 kg/m s^2 $= 1$ N/m^2	kp/m^2

Table 192. Prefixes used for multiples and sub-multiples of units
(e.g. of the metric system, of mechanical units, of photometric units, etc.);
from Mitt. Bl. DAMAG No. 149, G-1-2, Berlin 1958
Example: 1 pm (picometre) $= 10^{-12}$ m

Prefix	Abbreviation	Meaning		
Tera	T	1,000,000,000,000	(10^{12})	units
Giga	G	1,000,000,000	(10^{9})	units
Mega	M	1,000,000	(10^{6})	units
Kilo	K	1,000	(10^{3})	units
Hecto	h	100	(10^{2})	units
Deca	da	10	(10^{1})	units
Deci	d	0.1	(10^{-1})	units
Centi	c	0.01	(10^{-2})	units
Milli	m	0.001	(10^{-3})	units
Micro	μ	0.000,001	(10^{-6})	units
Nano	n	0.000,000,001	(10^{-9})	units
Pico	p	0.000,000,000,001	(10^{-12})	units

11.2. Measures of Length

Table 193. Measures of Length

Ångström [Å][1])	Nanometre [nm]	Micrometre [μm]	Millimetre [mm]	Centimetre [cm]	Decimetre [dm]	Metre [m]
1	0.1	0.000,1	10^{-7}	10^{-8}	10^{-9}	10^{-10}
10	1	0.001	10^{-6}	10^{-7}	10^{-8}	10^{-9}
10,000	1,000	1	0.001	0.000,1	10^{-5}	10^{-6}
10^7	10^6	1,000	1	0.1	0.01	0.001
10^8	10^7	10,000	10	1	0.1	0.01
10^9	10^8	10^5	100	10	1	0.1
10^{10}	10^9	10^6	1,000	100	10	1

[1]) Before 1946, the lattice constants were erroneously given in so-called Å-units, whereas they are actually kX-units. From 1946, the metric Å-units have been generally used. 1 kX = 1.002,02 Å. The X-unit is the 3,029.45th part of the lattice constants of calcite at 18 °C.

Table 194. Units of Length used in Great Britain and in the U.S.A.

Name	Equivalents	Comparison with Metric Units Great Britain	U.S.A.
1 A.U. (astronomical unit)		$1.495,98 \cdot 10^8$ km	
1 international nautical mile		1.852 km	
1 mile (statute) (mi.)	1 mile = 8 furlongs = 80 chains = 320 poles (rods) = 1,760 yards	1.609,342,6 km	
1 yard (yd.)	1 yard = 3 feet	0.914,399,2 m	0.914,401,8 m
1 foot[1]) (ft.)	1 foot = 12 inches	0.304,799,7 m	0.304,800,6 m
1 inch (in.)	1 inch = 1,000 mils	2.539,998 cm	2.540,005 cm
0.621,37 mi.			1 km
1.093,61 yd.			1 m
0.393,7 in.			1 cm

[1]) e.g. 2′5″ means 2 ft. 5 in.

11.3. Measures of Area

Table 195. Important Measures of Area

Square metre [m²]	Are [a]	Hectare [ha]	Square kilometre [km²]
1	0.01	0.000,1	10^{-6}
100	1	0.01	0.000,1
10,000	100	1	0.01
10^6	10,000	100	1

S System	Metric System	US System	Metric System
in.² (sq.in.)	6.451,63 cm²	0.001,55 in.²	1 mm²
ft.² (sq.ft.)	0.092,9 m²	0.155 in.²	1 cm²
yd.² (sq.yd.)	0.836,13 m²	10.763,9 ft.²	1 m²
mi.² (sq.mi.)	2.589,98 km²	0.386,1 mi.²	1 km²

1.4. Liquid Measures and Measures of Capacity

able 197. Liquid Measures

itre¹)]	Decalitre [dal]	Hectolitre [hl]	Kilolitre [kl]
1	0.1	0.01	0.001
10	1	0.1	0.01
100	10	1	0.1
,000	100	10	1

able 198. Liquid Measures

Microlitre [μl]	Millilitre [ml]	Centilitre [cl]	Decilitre [dl]	Litre¹) [l]
,000	1	0.1	0.01	0.001
10^4	10	1	0.1	0.01
10^5	100	10	1	0.1
10^6	1,000	100	10	1

¹) 1 litre [l] = 1,000.028 cubic centimetres [cm³]

For measurements whose accuracy is not required to be more precise than 0.01%, the units below are used for

1 m³	= 1,000 l	1 cm³ = 1 ml
1 dm³ =	1 l	1 mm³ = 1 μl

Table 199. Measures of Capacity

Cubic millimetre [mm³]	Cubic centimetre [cm³]	Cubic decimetre [dm³]	Cubic metre [m³]
1	0.001	10^{-6}	10^{-9}
1,000	1	0.001	10^{-6}
10^6	1,000	1	0.001
10^9	10^6	1,000	1

Table 200. US Measures of Capacity

U.S. System	Metric System	U.S. System	Metric System
1 in.³ (cu.in.)	16.387,2 cm³	0.610,2 · 10⁻⁴ in.³	1 mm³
1 ft.³ (cu.ft.)	0.028,31 m³	0.061,02 in.³	1 cm³
1 yd.³ (cu.yd.)	0.764,56 m³	1.307,94 yd.³	1 m³

Table 201. Liquid Measures used in Great Britain and in the U.S.A.

Name	Equivalents	Comparison with Metric Units Great Britain U.S.A.	
1 barrel (bbl.)	1 barrel = 31.5 gallons		119.24 l
1 barrel petroleum	1 barrel = 42 gallons		159 l
1 gallon (gal.)	1 gallon = 4 quarts (qt.)	4.546 l	3.785,3 l
1 quart (qt.)	1 quart = 2 pints (pt.)	1.136 l	0.946 l
1 pint (pt.)		0.586 l	0.473 l

11.5. Density

$$\text{Density} = \frac{\text{mass}}{\text{volume}} \quad \varrho\,^{1)} = \frac{m}{V}\,.$$

Units of density: kilogram per cubic metre (kg/m³)

$$\text{t/m}^3 = \text{g/cm}^3 \approx \text{g/ml}^{2)} \approx \text{kg/l}^{2)}$$

[1] In mineralogy frequently denoted **D**.
[2] See note below Table of liquid measures.

Table 202. U.S. Units of Density

U.S. System	Metric System
1 lb./in.³	27.68 g/cm³
1 lb./ft.³	0.016 g/cm³
0.036,127 lb./in.³	1 g/cm³

1.6. Units of Mass

Table 203. Units of Mass

Microgram or gamma [μg] or [γ]	Milligram [mg]	Centigram [cg]	Decigram [dg]	Gram [g]	Kilogram[1] [kg]
1,000	1	0.1	0.01	0.001	10^{-6}
10,000	10	1	0.1	0.01	10^{-5}
10^5	100	10	1	0.1	0.000,1
10^6	1,000	100	10	1	0.001
10^7	10,000	1,000	100	10	0.01
10^8	10^5	10,000	1,000	100	0.1
10^9	10^6	10^5	10,000	1,000	1

1 ton [t] = 10 decitons [dt] = 1,000 kilograms [kg]
1 kilogram = 100 decagrams [dag] = 1,000 grams [g]
1 (metric) carat [c][2] = 200 mg = 0.2 g = $2 \cdot 10^{-5}$ kg
1 microgram [μg] = 0.001 kg = 10^{-6} g (also denoted [γ])
1 milligamma [γγ] = 0.001 γ = 10^{-9} g
1 microgamma [γγγ] = 0.001 γγ = 10^{-12} g

[1] The multiples and sub-multiples whose names include a prefix are not derived from the kilogram, but from its 1,000th part, the gram (Mittl. Bl. No. 149, DAMG, Berlin 1958); Table 191.
[2] According to the specifications given in Mitt. Bl. No. 149 of the DAMG (1958), the metric carat as a unit of weight is only permissible for diamonds, pearls, precious stones and precious metals. Multiples and sub-multiples of c should not be denoted by a prefix according to Table 191.

Table 204. Units of Mass of the Avoirdupois System[1]) used in Great Britain and in the U.S.A.

Name	Equivalents	Comparison with Metric Units Great Britain	U.S.A.
1 long ton (tn.lg.)	1.12 short tons = 2,240 pounds	1,016.047 kg	
1 short ton (tn.sh.)	0.893 long tons = 2,000 pounds		907.185 kg
1 pound (lb.av.)	16 ounces = 7,000 grain		453.592,4 g
1 ounce (oz.av.)	16 drams		28.349,53 g
1 dram (dr.av.)	27.343,75 grains		1.771,85 g
1 grain (gr.)			0.064,80 g
1 hundred weight (cwt.) (long)	4 quarters = 112 pounds = 0.05 tn.lg.	50.802 kg	
1 hundred weight (cwt.) (short)			45.348 kg
1 quarter (qr.)	28 pounds	12.70 kg	
1 stone	14 pounds		6.35 kg

[1] Avoirdupois or commercial system; in addition there are the troy and apothecaries' systems of weight, used for gold and silver and for medicine respectively, each use the grain which is the same as the grain avoirdupois.

1 pound troy (lb.t.) = 1 pound apothecaries (lb.ap.) = 0.823 pound avoirdupois (lb.avoir) = 373.242 g

11.7. Statements of Concentrations Usual in Geochemistry

Table 205. Statements of Concentrations Usual in Geochemistry and Their Relationship

Per cent [%]		Parts per Million[1] (ppm = g/t = mg/kg = mg/l[2])		[g/g]
100	10^2	1,000,000	10^6	10^0
10	10^1	100,000	10^5	10^{-1}
1	10^0	10,000	10^4	10^{-2}
0.1	10^{-1}	1,000	10^3	10^{-3}
0.01	10^{-2}	100	10^2	10^{-4}
0.001	10^{-3}	10	10^1	10^{-5}
0.000,1	10^{-4}	1	10^0	10^{-6}
0.000,01	10^{-5}	0.1	10^{-1}	10^{-7}
0.000,001	10^{-6}	0.01	10^{-2}	10^{-8}
0.000,000,1	10^{-7}	0.001	10^{-3}	10^{-9}
0.000,000,01	10^{-8}	0.000,1	10^{-4}	10^{-10}
0.000,000,001	10^{-9}	0.000,01	10^{-5}	10^{-11}
0.000,000,000,1	10^{-10}	0.000,001	10^{-6}	10^{-12}

[1] 10^{-3} ppm is equal to 1 ppb (part per billion)
[2] ppm = mg/l only for statements of concentrations whose error does not fall below the order of 10^{-2} % (see Table 198 Liquid Measures)

11.8. Units of Pressure and Temperature

Table 206. Units of Pressure

Microbar [μbar]	Newton per sq. metre [N/m²]	Millibar [mbar]	Bar [bar]
1	0.1	10^{-3}	10^{-6}
10	1	0.01	10^{-5}
1,000	100	1	0.001
10^6	10^5	1,000	1

Table 207. Relations between the Units of Pressure

	Torr [Torr] = Millimetre Hg [mm Hg]	Millibar [mbar]	Physical atmosphere [atm]	Technical[1] atmosphere [at]	U.S. System [lb./in.²]
1 Torr = mm Hg	1	1.333	0.001,316	0.001,359,51	0.019,34
1 mbar	0.750,06	1	0.000,987	0.001,019,72	0.014,504
1 atm	760	1,013.25	1	1.033,23	14.696
1 at	735.56	980.665	0.967,8	1	14.223
1 lb./in.²	51.714,4	68.95	0.068,046	0.070,31	1

[1] 1 technical atmosphere [at] = 10,000 kilopond/square metre

	converted into [°C]	[°F]
egree Celsius [°C]	1	$x\,°C = (\tfrac{9}{5}x + 32)\,°F$
egree Réaumur [°R]	$x\,°R = \tfrac{5}{4}x\,°C$	$x\,°R = (\tfrac{9}{4}x + 32)\,°F$
egree Fahrenheit [°F]	$x\,°F = \tfrac{5}{9}\,(x - 32)\,°C$	1
egree Kelvin [°K]	$x\,°K = (x - 273.16)\,°C$	$x\,°K = [\tfrac{9}{5}\,(x - 273.16) + 32]\,°F$
egree Rankine [°Rank]	$x\,°Rank = \tfrac{5}{9}\,(x - 491.76)\,°C$	$x\,°Rank = (x - 459.76)\,°F$

able 209. A Few Fixed Points of Temperature

	[°C]	[°R]	[°F]	[°K]	[°Rank]
oiling temperature of water at 760 Torr	100	80	212	373.2	671.8
elting temperature of ice at 760 Torr	0	0	32	273.2	491.8
ero point of the Fahrenheit scale	— 17.8	— 14.2	0	255.4	459.8
bsolute zero = zero of the Kelvin and ankine scales	—273.2	—218.6	—459.8	0	0

	°C = °K	°R	°F = °Rank
°C = 1 °K	1	0.8	1.8
°R	1.25	1	2.25
°R = 1 °Rank	0.556	0.445	1

Table 210. Comparison of Absolute Degree Intervals

The most widely used temperature scales are the international centigrade, Celsius scale and the bsolute Kelvin scale. — The "normal temperature" in volumetric analysis is $+20\,°C$.

1.9. Dimensions of the Planets, the Moon and the Sun

able 211. Earth's Dimensions (from Astronomer's Handbook, 1966, GONDOLATSCH, 1965, MAC DONALD, 1966; WEDEPOHL, 1969)

Equatorial radius a_e	6,378.388 km[1])	6,378.160 km[2])
Polar radius a_p	6,356.912 km[1])	6,356.775 km[2])
Flattening factor f	1/297[1])	1/298.25[2])
Surface	510 × 10⁶ km²	
ocean	361 × 10⁶ km² (70.8 %)	
land	149 × 10⁶ km² (29.2 %)	
Radius of sphere of equal volume a_o	6,371 km	
Volume	1.083 × 10¹² km³	
Mass	5.976 × 10¹⁸ t	
Mean density	5.517 g cm⁻³	
Gravitational constant G	6.670 × 10⁻⁸ dynes cm⁻² g⁻²	
Normal acceleration of gravity at equator g_e (based on Potsdam standard)	978.043,6 cm s⁻²	
Mean solar day d	86,400 s = 24 h	
Sidereal day S	86,164.09 s = 23 h 56 min 4.09 s	
Velocity of rotation at equator	465.12 m s⁻¹	
Mean moment of inertia C_o	8.02 × 10⁴⁴ g cm²	

[1]) I.E.R. International Ellipsoid of Reference (1924)
[2]) I.A.U. International Astronomical Union (1966)

Table 212. Dimensions of the Planets and the Moon
(from GONDOLATSCH, 1965; KUIPER, 1965; in WEDEPOHL, 1969)

Name	Symbol	a [10^6 km]	P [a]	Diameter [km]	Mass[1]) [10^{26} g]	Volume 10^{10} km^3
Mercury	☿	57.9	0.240,85	4,840	3.333	5.958
Venus	♀	108.2	0.615,21	12,228	48.70	95.765
Earth	♁	149.6	1.000,04	12,742.06	59.76	108.332
Moon	☾	0.384[2])	0.074,80[3])	3,476	0.735	2.192
Mars	♂	227.9	1.880,89	6,770	6.443	16.250
Jupiter	♃	778	11.862,23	140,720	18,993	145,923.204
Saturn	♄	1,427	29.457,7	116,820	5,684	83,469.806
Uranus	⛢	2,870	84.015,3	47,100	867.6	5,481.599
Neptune	♆	4,496	164.788,3	44,600	1,029	4,636.610
Pluto	Pl	5,881.9—5,946.5	247.7	6,000	55.3	10.833

a = semi-major axis of the orbit
P = sidereal period = true period of the planet's revolution around the Sun (with respect to the fixed star field)

[1]) Mass without moons
[2]) Mean distance from the Earth
[3]) True period of the Moon's revolution around the Earth (with respect to the fixed star field)

Name	ϱ [g/cm^3]	ϱ_{Eq} [cm/s^2]	v_e [km/s]	A	T_{max} [°K]	T_{av} [°K]	Atmosph. constituents
Mercury	5.62	380	4.29	0.056	625	—	(^{40}Ar)
Venus	5.09	869	10.3	0.76	324	229	CO_2, H_2O
Earth	5.517	978	11.2	0.39	349	246	N_2, O_2
Moon	3.35	162	2.37	0.067	387	274	—
Mars	3.97	372	5.03	0.16	306	216	N_2, CO_2, H_2O
Jupiter	1.30	2,301	57.5	0.67	131	93	H_2, CH_4, NH_3
Saturn	0.68	906	33.1	0.69	95	68	H_2, CH_4, NH_3
Uranus	1.58	972	21.6	0.93*	67*	47*	He, H_2, CH_4
Neptune	2.22	1,347	24.6	0.84*	53*	38*	He, H_2, CH_4
Pluto	—	—	—	0.14	60	43	uncertain

g_{Eq} = total acceleration, including centrifugal acceleration, at equator
v_e = velocity of escape at equator
A = Albedo = total reflectivity, wavelength = 5.500 Å
T_{max} = maximum temperature for the subsolar point of a slowly rotating planet or satellite (computed from the visual albedo)
T_{av} = average temperature of a rapidly rotating sphere
Atm. constituents = main atmospheric constituents

*) Since the albedos of Uranus and Neptune are very low in the red and infrared an effective value A = 0.7 has been adopted for calculating the temperature.

Table 213. Solar Dimensions (from WALDMEIER, 1965)

Radius	6.960×10^{10} cm
Surface area	6.087×10^{22} cm^2
Volume	1.412×10^{33} cm^3
Mass	1.989×10^{33} g
Mean density	1.409 g cm^{-3}
Density at the centre	98 g cm^{-3}
Gravitational acceleration at the solar surface	2.740×10^4 cm s^{-2}
Escape velocity at the surface	6.177×10^7 cm s^{-1}
Effective temperature	5,785 °K
Temperature at the centre	13.6×10^6 °K
Radiation	3.9×10^{33} erg s^{-1}
Specific surface emission	6.41×10^{10} erg cm^{-2} s^{-1}
Specific mean energy production	1.96 erg g^{-1} s^{-1}
Solar constant = extraterrestrial	1.39×10^6 erg cm^{-2} s^{-1}
Energy flux at the mean distance between earth and sun	2.00 cal cm^{-2} min^{-1}

11.10. Physical Constants

Table 214. Some Physical Constants (from Handbook of Chemistry and Physics, 1967/68)

Avogadro constant	N_A	$6.022,52 \times 10^{23}$ mol^{-1}
Boltzmann constant	k	$1.380,54 \times 10^{-16}$ erg °K^{-1}
Faraday constant	F	$9.648,7 \times 10^3$ cm$^{\frac{1}{2}}$ g$^{\frac{1}{2}}$ mol^{-1} (electromagn. syst.)
Planck constant	h	$6.625,6 \times 10^{-27}$ erg s
Gas constant	R	$8.314,3 \times 10^7$ erg °K^{-1} mol^{-1}
Normal volume perfect gas	V_o	$2.241,36 \times 10^4$ cm^3 mol^{-1}
Loschmidt's number	$n_o = \dfrac{N_A}{V_o}$	$2.686,99 \times 10^{19}$ cm^{-3}
Elementary charge	e	$1.602,03 \times 10^{-20}$ cm$^{\frac{1}{2}}$ g$^{\frac{1}{2}}$ (electromagn. syst.)
		$4.802,98 \times 10^{-10}$ cm$^{\frac{1}{2}}$ g$^{\frac{1}{2}}$ s^{-1} (electrostat. syst.)
Electron rest mass	m_e	$9.109,1 \times 10^{-29}$ g
Proton rest mass	m_p	$1.672,52 \times 10^{-24}$ g
Neutron rest mass	m_n	$1.674,82 \times 10^{-24}$ g
Speed of light in vacuum	c	$2.997,925 \times 10^{10}$ cm s^{-1}

Bibliography

BORCHERT, H.: Ein Vorschlag zur Verbesserung der mineralogisch-petrographisch geeigneten Körnungsstufen. Neues Jb. Mineralog., Mh. *6* (1962), 125—127

GONDOLATSCH, F.: Mechanical Data of Planets and Satellites. LANDOLDT-BÖRNSTEIN, New Series, Group VI: Astronomy, astrophysics and space research, Vol. I. Springer, Berlin/Heidelberg/New York 1965

KUIPER, G. P.: Physics of Planets and Satellites. LANDOLDT-BÖRNSTEIN, New Series, Group VI: Astronomy, astrophysics and space research, Vol. I. Springer, Berlin/Heidelberg/New York 1965

MACDONALD, G. J. F.: Geodetic Data, Handbook of Physical Constants, rev. edit. Geol. Soc. Am. Mem. *97* (1966)

11.11. Mineralogical-petrographical Grain-size Classification of Magmatites and Metamorphites

Table 215. Quantitative Mineralogical-petrographical Classification of Grain Sizes after TEUSCHER (1933), SCHNEIDERHÖHN (1961, SCHN.), BORCHERT (1962, B.). The quoted authors use the same designation for the various grain-size steps, unless otherwise stated (SCHN. or B.)

Grain-size range [mm]				Designation
TEUSCHER (1933)	Grain number/cm²	SCHNEIDERHÖHN (1961)	BORCHERT (1962)	
		>300	>300	giant-grained over 30 cm (SCHN.), giant-grained (B.)
		100—300	30—300	giant-grained up to 30 cm (SCHN.), giant-grained (B.)
		25—100		giant-grained up to 10 cm (SCHN.)
>10		10—25	10—30	large-grained (SCHN., B.)
3.3—10	(1—10)	3.3—10	3—10	coarse-grained
1—3.3	(10—100)	1—3.3	1—3	medium-grained
0.33—1	(100—1,000)	0.33—1	0.3—1	small-grained
0.1—0.33	(1,000—10,000)	0.1—0.33	0.1—0.3	fine-grained
<0.1	(>10,000)	0.033—0.1	0.03—0.1	dense (-grained) (SCHN., B.)
		0.001—0.033	0.001—0.03	microcrystalline
		<0.001	0.000,1—0.001	cryptocrystalline
			0.000,001—0.000,1	roentgenocrystalline

NATTERODT, M.: Umrechnung englisch-amerikanischer Maßeinheiten — Umrechnungswerte, Rechentafeln, Abkürzungen. Fachbuchverlag, Leipzig 1964

SCHNEIDERHÖHN, H.: Die Erzlagerstätten der Erde, Vol. 2, Die Pegmatite. Stuttgart 1961, 324 pp.

SCHUBERT, H.: Aufbereitung fester mineralischer Rohstoffe, Vol. I. VEB Deutscher Verlag für Grundstoffindustrie, Leipzig 1964

SCHÜLLER, A.: Vorschlag zur Erweiterung der Korngrößeneinteilung nach E. O. TEUSCHER (1933). Neues Jb. Mineralog. Mh. (1963), 27—30

TEUSCHER, E. O.: Methodisches zur quantitativen Strukturgliederung körniger Gesteine. Z. Kristallogr. Mineralog. Petrogr., Abt. B 44 (1933), 410—421

WALDMEIER, M.: The Quiet Sun. LANDOLDT-BÖRNSTEIN, New Series, Group VI: Astronomy, astrophysics and space research, Vol. I. Springer, Berlin/Heidelberg/New York 1965

WEDEPOHL, K. H.: Handbook of Geochemistry. Springer, Berlin/Heidelberg/New York 1969

—. Handbook of Chemistry and Physics, 48th ed. The Chemical Rubber Co., Cleveland, Ohio 1967/68

Table 216. Important Testing Screen Cloth Series (up to an aperture width of 1 mm) after SCHUBERT (1964)

International Standard ISO-TC 24 [mm]	GDR TGL 0-4188, Bl. 1 GFR DIN 4188 [mm]	Germany (before 1945) DIN 1171 [mm]	USSR GOST 3584-53 [mm]	France AFNOR X 11-501 [mm]	U.S.A. ASTM E 11-61 [mm]	(mesh No.)	Great Britain BS 410-1949 (1945 and 1955) [mm]	(mesh No.)
0.045	0.04		0.04	0.04	0.037	(400)	0.044	(350)
	0.045		0.045		0.044	(325)	0.053	(300)
	0.05		0.05	0.05	0.053	(270)		
	0.056		0.056					
0.063	0.063	0.06	0.063	0.063	0.063	(230)	0.064	(240)
	0.071	0.075	0.071		0.074	(200)	0.076	(200)
	0.08		0.08	0.08				
0.09	0.09	0.09	0.09		0.088	(170)	0.089	(170)
	0.1	0.1	0.1	0.1	0.105	(140)	0.104	(150)
			0.112					
0.125	0.125	0.12	0.125	0.125	0.125	(120)	0.124	(120)
		0.15	0.14		0.149	(100)	0.152	(100)
0.18	0.16		0.16	0.16	0.177	(80)	0.178	(85)
			0.18		0.21	(70)	0.211	(72)
	0.2	0.2	0.2	0.2				
			0.224					
0.25	0.25	0.25	0.25	0.25	0.25	(60)	0.251	(60)
		0.3	0.28		0.297	(50)	0.295	(52)
0.355	0.315		0.315	0.315	0.354	(45)	0.353	(44)
			0.355		0.42	(40)	0.422	(36)
	0.4	0.4	0.4	0.4				
		0.43	0.45					
0.5	0.5	0.5	0.5	0.5	0.5	(35)	0.5	(30)
			0.56		0.595	(30)	0.599	(25)
	0.63	0.6	0.63	0.63			0.699	(22)
0.71		0.75	0.7		0.707	(25)	0.79	1/32 *
	0.8		0.8	0.8	0.841	(20)	0.853	(18)
			0.9					
1	1	1	1	1	1	(18)	1.033	(16)

Aperture width mm, number of meshes per inch = 25.4 mm (mesh No.) in brackets; *) = dimensions in inches

Comparative Transliteration Table of Cyrillic Characters

(Proposed by the Institut für Dokumentation der Deutschen Akademie der Wissenschaften zu Berlin, 1960. Explanations see page 437)

Cyrillic Character	English			French			German			German Transcription from Russian
	from Russ.	from Bulg.	from Serb.[1]	from Russ.	from Bulg.	from Serb.	from Russ.	from Bulg.	from Serb.	
1. А а	a	a	a	a	a	a	a	a	a	a
2. Б б	b	b	b	b	b	b	b	b	b	b
3. В в	v	v	v	v	v	v	v	v	v	w
4. Г г	g	g	g	g	g	g	g	g	g	g
5. Д д	d	d	d	d	d	d	d	d	d	d
6. Ђ ђ	–	–	d	–	–	d	–	–	d	–
7. Е е	e	e	e	e	e	e	e	e	e	je
8. Ё ё	e	–	–	ё	–	–	e	–	–	jo (o after sch, sh, tsch, z)
9. Ж ж	zh	zh	ž	ž	ž	ž	ž	ž	ž	sh
10. З з	z	z	z	z	z	z	z	z	z	s
11. И и	i	i	i	i	i	–	i	i	–	i
12. Ј ј	–	–	j	–	–	j	–	–	j	–
13. Й й	ĭ	ĭ	–	j	j	–	j	j	–	i
14. К к	k	k	k	k	k	k	k	k	k	k
15. Л л	l	l	l	l	l	l	l	l	l	l
16. Љ љ	–	–	lj	–	–	lj	–	–	lj	–
17. М м	m	m	m	m	m	m	m	m	m	m
18. Н н	n	n	n	n	n	n	n	n	n	n
19. Њ њ	–	–	nj	–	–	nj	–	–	nj	–
20. О о	o	o	o	o	o	o	o	o	o	o
21. П п	p	p	p	p	p	p	p	p	p	p
22. Р р	r	r	r	r	r	r	r	r	r	r
23. С с	s	s	s	s	s	s	s	s	s	s
24. Т т	t	t	t	t	t	t	t	t	t	t
25. Ћ ћ	–	–	ć	–	–	ć	–	–	ć	–
26. У у	u	u	u	u	u	u	u	u	u	u
27. Ф ф	f	f	f	f	f	f	f	f	f	f
28. Х х	kh	kh	k	h	h	h	ch	ch	ch	ch
29. Ц ц	ts	ts	c	c	c	c	c	c	c	z
30. Ч ч	ch	ch	č	č	č	č	č	č	č	tsch
31. Џ џ	–	–	dž	–	–	dž	–	–	dž	–
32. Ш ш	sh	sh	š	š	š	š	š	š	š	sch
33. Щ щ	shch	sht	–	šč	št	–	šč	št	–	stsch
34. Ъ	"4)	ŭ	–	"4)	ă5)	–	–	ŭ	–	not represented
35. Ы	y	–	–	y	y	–	y	ÿ	–	y
36. Ь	'	'	–	'	'	–	'3)	–	–	not represented
37. Ѣ ѣ	(ê)	(ê)	–	ě	ě	–	ě	ě	–	–
38. Ю ю	yu	yu	–	ju	ju	–	ju	ju	–	ju
39. Я я	ya	ya	–	ja	ja	–	ja	ja	–	ja
40. Ѫ ѫ	–	(ū)	–	–	–	–	–	ă	–	–
41. Э э	ě	–	–	ě	–	–	ě	ě	–	é

436

Letter		Name	Transliteration
capital	small		
	α	Alpha	A, a
	β	Beta	B, b
	γ	Gamma	G, g, n
	δ	Delta	D, d
	ε	Epsilon	E, e
	ζ	Zeta	Z, z
	η	Eta	E, e
	ϑ	Theta	Th, th
	ι	Iota	I, i
	$\varkappa$	Kappa	K, k
	λ	Lambda	L, l
	μ	Mu	M, m
	ν	Nu	N, n
	ξ	Xi	X, x
	o	Omicron	O, o
	π	Pi	P, p
	ϱ	Rho	{ R, r / Rh, rh
{	σ	Sigma	S, s
{	ς		
	τ	Tau	T, t
	υ	Upsilon	Y, y, u
	φ	Phi	Ph, ph
	χ	Chi	Ch, ch
	ψ	Psi	Ps, ps
	ω	Omega	O, o

The letter gamma is transliterated n only before velars; the letter upsilon is transliterated u only as the final element in diphthongs.

Explanations to Table page 436:

From ISO/R 9: 1955, International system for the transliteration of Cyrillic characters
In the case of ascenders after, in other cases above, the preceding symbol (e.g. Ю = ń, Ь = é).
Not represented as the final word element, otherwise ''.
Not represented as the final word element, a in the middle of a word.

Authorities: DIN 1460 (draft)   B.S. 2979: 1958
    ISO/R 9: 1955   FD Zn° 46-001: 1956

Index of Authors

BOROVIK-ROMANOVA, T. F. 180
BORRIS, W. 424
BOTNEVA, T. A. 346, 347
BOTNEVA, V. V. 332
BOTOVA, M. M. 373
BOTT, M. 224
BOWEN, H. I. M. 244, 347
BOWEN, N. L. 14, 17, 18, 104
BOWEN, R. 403, 410
BOWIE, S. H. 185
BOYADIEVA, R. 197
BOYLE, R. V. 368, 379
BRÄUER, H. 254, 257, 393, 395
BRAGG, W. L. 57
BRAITSCH, O. 304, 306, 308, 313, 382, 387
BRANDT, M. P. 186
BRAUN, H. 379
BRAUNS, R. 18
BRDIČKA, R. 99
BREALEY, L. 19
BREGER, D. 344
BREGER, I. A. 341, 342, 344
BREHLER, B. 187
BREWER, L. 101
BRIDGMAN 131
BRINDLEY, G. W. 187
BRODA, E. 194
BRODE, W. H. 180
BRODSKI, A. A. 372
BROECKER, W. S. 242, 421
BROOKER, E. J. 185
BROOKS, C. K. 257
BROOKS, R. R. 182, 352
BROTZEN, O. 225, 229
BROWN, G. 187
BROWN, H. S. 213, 218, 357
BRÜGEL, W. 183
BRÜNNEE, C. 192
BRUEVICH, S. V. 316, 325
BRUNCK 170
BRUNFELT, A. O. 197
BRYAN, F. R. 182
BUBAM, W. 31
BUBELA, B. 343
BUCH, K. 313
BUCHHEIM, W. 135
BUCHSBAUM 387
BUCKENHAM, M. H. 164, 169
BUDDINGTON, A. F. 217
BUDZINSKIĬ, YU. A. 354

BUERGER, M. J. 169, 187
BUGELSKIĬ, YU. YU. 372
BULACH, A. G. 390, 396
BULLARD, E. C. 130
BULLEN, K. E. 124, 217
BUNGE, H. J. 187
BUREK, J. 368
BURKOV, V. V. 358
BURKSER, E. S. 354
BURNS, R. G. 257
BURRI, C. 383
BURTON, S. D. 354
BURYANOVA, E. Z. 353
BUTLER, J. R. 276, 277, 358

ČADEK, J. 379
CAMBEL, B. 379
CAMERON, A. G. W. 204, 207, 212
CAMERON, E. N. 203, 240, 352, 387
CAMPBELL, W. J. 185
CANNEY, F. C. 370
CAPITANT, M. 185
CARISON, M. T. 346
CARL, H. F. 185
CARMICHAEL, I. 197, 257
CARNOT, A. 344
CARPENTER, A. B. 325
CARR, M. H. 353
CARRON 276
CARTER, F. G. 185
CARTLEDGE, G. H. 69, 110
CHAIGNEAU, M. 262, 332
CHAĬNIKOV, B. I. 153
CHAKRABARTI, A. K. 379, 398
CHANDLER, W. B. 347
CHANG, R. W. H. 342
CHAO, E. C. T. 226, 227, 240
CHARLOT, G. 170
CHASE, I. W. 240
CHAYES, F. 197
CHEBOTAREV, I. I. 317, 319, 321, 325
CHELISHCHEV, N. F. 136
CHERDYNTSEV, V. V. 224
CHERNYAEV, A. M. 372
CHESTER, R. 19, 295, 387
CHEW, R. T. 368
CHILLINGAR, G. V. 291, 299, 301, 387
CHINNER, G. A. 269
CHIRVINSKIĬ, P. N. 217
CHODOS, A. A. 185, 213

441

452

Index of Subjects

459

467

Sulphur, bacteria 334
—, bibliography 358
— cycle 351
— —, microorganism 336
— isotopes in nature 402, 408
Sun, abundance of elements 204
Surface prospecting 362
Surficial water 310
Susceptibility 164, 165
—, mass 168

Tantalum, bibliography 358
Technosphere, geochemistry 422
Tectite, chemical composition 210
Tectonic processes 122
Tellurium, bibliography 358
Temperature distribution, earth 127, 128
—, lava, maximum 129
Terrestrial-limnal facies 288
Terrestrial magnetic field 125
Terrigenous mineral 287
Testing screen cloth series 435
Thalassophile elements 311
Thalassoxene elements 311
Thallium, bibliography 358
Thermal analysis 189
— water 321
Thermodynamic data 90
Thermometry, O-isotopes 403
Thixotropic property 161
Thorium, bibliography 359
— decay series 41, 44
Threshold concentration 363
Tie lines 132
Tin-bearing state 377
Tin, bibliography 358
Titanium, bibliography 359
Topazisation 377
Toxic elements 333, 367
Trace elements, definition 391
— — in rock-forming minerals 395, 399
— — in sulphide minerals 398
Transliteration table 436
Transuranic elements 23
Transvaporisation 267
Trend analysis 201
Type relation 113, 115
Typochemical elements 392

Ultrared-spectrometry 183
Unit cell 57

Unstable radiogenic isotopes 39
Upper mantle project 222
Uranium, bibliography 359
— decay series 40
—, geochemical cycle 350

Valence 53, 57, 65, 68, 118, 149, 160
—, covalence 66
—, electron 65, 68
—, ionic 65, 66
—, maximum 66, 67
—, stoichiometric 65, 66
Vanadium, bibliography 359
Van der WAAL's bond 71, 103
Variation coefficient 180
Vector analysis 201
VEGARD's rule 113
Volcanic exhalations 261, 262
— gases 329, 331

Water, balance 309, 310
—, classification 309
—, distribution 309
—, genetic types 319
—, ion-concentration 320
Weathering, chemical reactions 275
— index 274
—, relative mobility of elements 274, 275
—, resistance to 274
— solutions 311, 312
— zone 273
Well geochemistry 374
Wettability 164
Wolfram, bibliography 359

X-ray analysis 158, 186
— fluorescence analyser 184
— measurements 57
— spectral analysis 365

Yttrium, bibliography 359

Zinc, bibliography 359
Zirconium, bibliography 359
Zone of oxidation 273, 277
Zoogeochemical methods 365
Zooplankton 315, 335